Current Trends in Nonlinear Systems and Control

*In Honor of Petar Kokotović
and Turi Nicosia*

Laura Menini

Luca Zaccarian

Chaouki T. Abdallah

Editors

Birkhäuser
Boston • Basel • Berlin

Laura Menini
Dipartimento di Informatica
Sistemi e Produzione
Università di Roma "Tor Vergata"
Via del Politecnico 1
I-00133 Roma
Italia

Luca Zaccarian
Dipartimento di Informatica
Sistemi e Produzione
Università di Roma "Tor Vergata"
Via del Politecnico 1
I-00133 Roma
Italia

Chaouki T. Abdallah
Department of Electrical
  and Computer Engineering
EECE Building, MSC01 1100
University of New Mexico
Albuquerque, NM 87131-0001
USA

Mathematics Subject Classification: 34A34, 34A60, 34Cxx, 34Dxx, 34D20, 34H05, 34K35, 37-XX, 37N35, 49J24, 49K24, 68T70, 68T45, 70E60, 70Kxx, 70K20, 90B18, 90B35, 90B36, 93-XX, 93Bxx, 93Cxx, 93C05, 93C10, 93C85, 93C95, 93E12

Library of Congress Cataloging-in-Publication Data
Current trends in nonlinear systems and control : in honor of Petar Kokotović and Turi
  Nicosia / Laura Menini, Luca Zaccarian, Chaouki T. Abdallah, editors.
      p. cm. – (Systems & control)
  Includes bibliographical references and index.
  ISBN 0-8176-4383-4 (acid-free paper)
    1. Programming (Mathematics) 2. Nonlinear systems. 3. Nonlinear control theory. 4.
Control theory. I. Kokotović, Petar V. II. Nicosia, Turi. III. Menini, Laura. IV. Zaccarian,
Luca. V. Abdallah, C. T. (Chaouki T.) VI. Series.

QA402.5.C87 2005
629.8′36–dc22   2005053567

ISBN-10 0-8176-4383-4   e-ISBN 0-8176-4470–9   Printed on acid-free paper.
ISBN-13 978-0-8176-4383-6

To Petar and Turi

Foreword

This Birkhäuser series *Systems and Control: Foundations and Applications* publishes top quality state-of-the art books and research monographs at the graduate and post-graduate levels in systems, control, and related fields. Books in the series cover both foundations and applications, with the latter spanning the gamut of areas from information technology (particularly communication networks) to biotechnology (particularly mathematical biology) and economics. The series is primarily aimed at publishing authored (that is, not edited) books, but occasionally, and very selectively, high-quality volumes are published, which can be viewed as records of important scientific meetings.

One such event took place back on June 3 and 4, 2004, at Villa Mondragone in Monteporzio Catone, Rome, Italy. Several control scientists conducting cutting-edge research gathered at a workshop, "Applications of Advanced Control Theory to Robotics and Automation" (ACTRA), to present their most recent work, and more significantly, to honor two prominent control scientists, **Petar Kokotović** and **Turi Nicosia**, on the occasion of their seventieth birthdays. The meeting was very successful on all accounts, and the scientific program featured many high-quality presentations. It was therefore a foregone conclusion that this material should be made available to a broader readership—which led to the present volume.

ACTRA organizers Laura Menini, Luca Zaccarian, and Chaouki T. Abdallah undertook the task of putting together this volume by collecting individual chapters from speakers at ACTRA and some other selected authors. They were successful in producing a coherent whole with chapters organized around common themes and contributing to both theory and applications. I thank them for editing such a fine volume, which should serve as a rich source of information on the topics covered for years to come.

Tamer Başar, *Series Editor*
Urbana, IL, USA
March 1, 2005

Preface

The chapters of this book reflect the talks given during the workshop "Applications of Advanced Control Theory to Control and Automation" (ACTRA), which was held on June 3 and 4, 2004, at Villa Mondragone (Monteporzio Catone, Rome, Italy). The workshop was an opportunity to jointly honor the scientific careers of Petar Kokotović and Turi Nicosia, who coincidently reached their seventies that year, and to celebrate the significant intersection between the sets of their students and collaborators. Petar and Turi have many interests in the field of automatic control, covering many topics in control theory and several different applications. Such a variety is reflected in this book, where contributions ranging from mathematics to laboratory experiments are included. Although each chapter is self-contained, the book has been organized such that theme-related chapters are grouped together, and, in some cases, convenient reading sequences are suggested to the reader (see, e.g., the last two chapters in Part II).

The chapters in Part I deal with observer designs for nonlinear systems and linear time-delay systems, and with identification techniques for linear, nonlinear, piecewise linear, and hybrid systems. Part II is devoted to theoretical results concerned with the analysis and control of dynamic systems; its first chapter focuses on Lyapunov tools for linear differential inclusions which is followed by a chapter dealing with oscillators and synchronization. The next two chapters deal with the control of constrained systems while the last two deal with finite-time stability. Part III, devoted to robotics, is concerned with new studies concerning robot manipulators of various kinds. The first two chapters deal explicitly with parameter identification for control, the third and the fourth with advanced control techniques for robot manipulators, the fifth and sixth with mobile robots, and the last two with different classes of coordination problems. Part IV contains some modern control techniques, including interconnection and damping assignment passivity-based control, decentralized control and adaptive control, and their application to multi-machine power systems, web processing systems, a real testbed for a PVTOL aircraft, and two different marine vehicles. Part V groups together topics that have more recently been addressed by the control community: applications of the maxplus algebra to system aggregation, scheduling for machines with significant setup times and limits in the buffer capacity, and inventory control with cooperation between retailers. Finally, Part VI is devoted to the emerging control theory topic of networked control systems, i.e., systems in which the communication between different parts is affected by delays or by information losses. The chapters of Part VI deal with different analysis and design problems involving networked control systems and give a broad overview of the techniques that can be used to study such dynamic systems.

Although the scope of the book, which mirrors the interests of the two honorees, is very broad, the methodologies used by the different authors and the related tools have much in common. The book is divided into parts based on what the editors felt were major themes, keeping in mind the significant connections between the various parts. As an example, the last two chapters of Part II deal with the concept of finite-time stability, which is receiving renewed attention just in view of its recent application to networked control systems, described in the last chapter of Part VI. Another example is the problem of coordinated control of many subsystems, which is the major topic of the last chapter of Part III (focused on coordination of robot teams) but also of one of the examples in the second chapter of Part II (focused on oscillators and synchronization).

We believe that the great variety of topics covered in this book and the almost tutorial writing style that many of the authors have used will render this book pleasant reading both for experts in the field and for young researchers who seek a more intuitive understanding of these relevant topics in our research area.

We wish to thank all the speakers of the workshop ACTRA and the contributors to this volume for their constant support and encouragement during both the organization of the workshop and the editorial work for the preparation of this volume.

Rome, Italy
October 2005

Laura Menini
Luca Zaccarian
Chaouki T. Abdallah

Contents

Part I State Estimation and Identification

Part II Control and System Theory

Part III Robotics

List of Contributors

Chaouki T. Abdallah
Dept. of Electrical and Computer
Engineering
University of New Mexico
Albuquerque, NM 87131, USA
chaouki@ece.unm.edu

Pedro Albertos
Dept. of Systems Engineering
and Control
University of Valencia
C/ Vera s/n Valencia, 46021 Spain
pedro@aii.upv.es

Eleonora Alunno
Dip. di Ingegneria dell'Informazione
Università di Siena
Via Roma, 56
53100 Siena, Italia
alunno@dii.unisi.it

Francesco Amato
Corso di Laurea in Ingegneria
Informatica e Biomedica
Dip. di Medicina Sperimentale e Clinica
Università Magna Græcia
Via T. Campanella 115
88100 Catanzaro, Italia
amato@unicz.it

Murat Arcak
Dept. of Electrical, Computer, and
Systems Engineering
Rensselaer Polytechnic Institute
Troy, NY 12180, USA
arcakm@rpi.edu

Marco Ariola
Dip. di Informatica e Sistemistica
Università degli Studi di Napoli
Federico II
Via Claudio 21
80125 Napoli, Italia
ariola@unina.it

Alessandro Astolfi
Dept. of Electrical Engineering
Imperial College London
Exhibition Road
London SW7 2AZ, UK
a.astolfi@imperial.ac.uk

Dario Bauso
Dip. di Ingegneria Informatica
Università di Palermo
Viale delle Scienze
Palermo, Italia
dariobauso@libero.it

Mauro Boccadoro
Dip. di Ingegneria Elettronica e
dell'Informazione
Università di Perugia
Perugia, Italia
boccadoro@diei.unipg.it

Marc Bodson
ECE Departiment
The University of Utah
Salt Lake City, UT 84112, USA
bodson@ece.utah.edu

Basilio Bona
Dip. di Automatica e Informatica
Politecnico di Torino
Corso Duca degli Abruzzi, 24
10129 Torino, Italia
basilio.bona@polito.it

Claudio Bonivento
CASY-DEIS, Università di Bologna
Viale Risorgimento 2
40136 Bologna, Italia
cbonivento@deis.unibo.it

Marco Carbone
Dip. di Informatica, Matematica,
Elettronica e Trasporti
Università degli Studi Mediterranea di
Reggio Calabria
Via Graziella, Loc. Feo Di Vito
89100 Reggio Calabria, Italia
mcarbone@ing.unirc.it

Pedro Castillo
Heudiasyc, UMR CNRS 6599
Université de Technologie de
Compiègne, BP 20529
60205 Compiègne, France
castillo@hds.utc.fr

Nicola Ceccarelli
Dip. di Ingegneria dell'Informazione
Università di Siena
Via Roma, 56
53100 Siena, Italia
ceccarelli@dii.unisi.it

John Chiasson
ECE Department
The University of Tennessee
Knoxville, TN 37996, USA
chiasson@utk.edu

Roberto Ciferri
Dip. di Ingegneria Informatica,
Gestionale e dell'Automazione
Università Politecnica delle Marche
Via Brecce Bianche
60131 Ancona, Italia
r.ciferri@diiga.univpm.it

Guy Cohen
ENPC
6-8, avenue Blaise Pascal
Cité Descartes, Champs-sur-Marne
77455 Marne-La-Vallée
Cedex 2, France
Guy.Cohen@enpc.fr

Giuseppe Conte
Dip. di Ingegneria Informatica,
Gestionale e dell'Automazione
Università Politecnica delle Marche
Via Brecce Bianche
60131 Ancona, Italia
gconte@univpm.it

Carlo Cosentino
Dip. di Informatica e Sistemistica
Università degli Studi di Napoli
Federico II
Via Claudio 21
80125 Napoli, Italia
carcosen@unina.it

Mauro Di Marco
Dip. di Ingegneria dell'Informazione
Università di Siena
Via Roma, 56
53100 Siena, Italia
dimarco@dii.unisi.it

Peter Dorato
Dept. of Electical and Computer
Engineering
University of New Mexico
Albuquerque, NM 87131, USA
peter@ece.unm.edu

Alicia Esparza
Dept. of Systems Engineering
and Control
University of Valencia
C/ Vera s/n Valencia, 46021 Spain
alespei@isa.upv.es

Isabelle Fantoni
Heudiasyc, UMR CNRS 6599
Université de Technologie de
Compiègne, BP 20529
60205 Compiègne, France
ifantoni@hds.utc.fr

Rafael Fierro
Oklahoma State University
School of Electrical and Computer
Engineering
202 Engineering South
Stillwater, OK 74078, USA
rfierro@okstate.edu

Martha Galaz
Lab. des Signaux et Systémes
Supelec, Plateau du Moulon
91192 Gif-sur-Yvette, France
Martha.Galaz@lss.supelec.fr

Sergio Galeani
Dip. di Informatica, Sistemi e
Produzione
Università di Roma "Tor Vergata"
00133 Roma, Italia
galeani@disp.uniroma2.it

Andrea Garulli
Dip. di Ingegneria dell'Informazione
Università di Siena
Via Roma, 56
53100 Siena, Italia
garulli@dii.unisi.it

Stéphane Gaubert
INRIA-Rocquencourt
B.P. 105, 78153 Le Chesnay
Cedex, France
Stephane.Gaubert@inria.fr

Luca Gentili
CASY-DEIS, Università di Bologna
Viale Risorgimento 2
40136 Bologna, Italia
lgentili@deis.unibo.it

Antonio Giannitrapani
Dip. di Ingegneria dell'Informazione
Università di Siena
Via Roma, 56
53100 Siena, Italia
giannitrapani@dii.unisi.it

Laura Giarré
Dip. di Ingegneria dell'Automazione e
dei Sistemi
Università di Palermo
Viale delle Scienze
Palermo, Italia
giarre@unipa.it

Rafal Goebel
P.O. Box 15172
Seattle, WA 98115, USA
rafal@ece.ucsb.edu

João P. Hespanha
Center for Control Engineering and
Computation
Dept. of Electrical and Computer
Engineering
University of California
Santa Barbara, CA 93106, USA
hespanha@ece.ucsb.edu

Tingshu Hu
Center for Control Engineering and
Computation
Dept. of Electrical and Computer
Engineering
University of California
Santa Barbara, CA 93106, USA
tingshu@ece.ucsb.edu

Mikuláš Huba
University of Technology in Bratislava
Faculty of Electrical Engineering and
Information Technology
Ilkovičova 3
812 19 Bratislava,
Slovak Republic
Mikulas.Huba@stuba.sk

Marina Indri
Dip. di Automatica e Informatica
Politecnico di Torino
Corso Duca degli Abruzzi, 24
10129 Torino, Italia
marina.indri@polito.it

Gianluca Ippoliti
Dip. di Ingegneria Informatica,
Gestionale e dell'Automazione
Università Politecnica delle Marche
Via Brecce Bianche
60131 Ancona, Italia
g.ippoliti@diiga.univpm.it

Aleksandar Lj. Juloski
Dept. of Electrical Engineering
Eindhoven University of Technology
P.O. Box 513, 5600 MB Eindhoven
The Netherlands
A.Juloski@tue.nl

Alexander Leonessa
University of Central Florida
Dept. of Mechanical, Materials &
Aerospace Engineering
P.O. Box 162450
Orlando, FL 32816, USA
aleo@mail.ucf.edu

Mengwei Li
ECE Department
The University of Tennessee
Knoxville, TN 37996, USA
mwl@utk.edu

Sauro Longhi
Dip. di Ingegneria Informatica,
Gestionale e dell'Automazione
Università Politecnica delle Marche
Via Brecce Bianche
60131 Ancona, Italia
sauro.longhi@univpm.it

Rogelio Lozano
Heudiasyc, UMR CNRS 6599
Université de Technologie de
Compiègne, BP 20529
60205 Compiègne, France
rlozano@hds.utc.fr

Gian Luca Mariottini
Dip. di Ingegneria dell'Informazione
Università di Siena
Via Roma, 56
53100 Siena, Italia
gmariottini@dii.unisi.it

Silvia Mastellone
Dept. of Electrical and Computer
Engineering
University of New Mexico
Albuquerque, NM 87131, USA
silvia@ece.unm.edu

Mario Milanese
Dip. di Automatica e Informatica
Politecnico di Torino
Corso Duca degli Abruzzi, 24
10129 Torino, Italia
mario.milanese@polito.it

Ilya Miroshnik
Laboratory of Cybernetics and Control
Systems
State University of Information
Technologies, Mechanics and Optics
14, Sablinskaya
Saint Petersburg, 197101 Russia
miroshnik@yandex.ru

Yannick Morel
University of Central Florida
Dept. of Mechanical, Materials &
Aerospace Engineering
P.O. Box 162450
Orlando, FL 32816, USA
ymorel@gmail.com

Romeo Ortega
Lab. des Signaux et Systémes
Supelec, Plateau du Moulon
91192 Gif-sur-Yvette, France
Romeo.Ortega@lss.supelec.fr

Amparo Palomino
Heudiasyc, UMR CNRS 6599
Université de Technologie de
Compiègne, BP 20529
60205 Compiègne, France
apalomino@hds.utc.fr

Prabhakar R. Pagilla
School of Mechanical and Aerospace
Engineering
Oklahoma State University
Stillwater, OK 74078, USA
pagilla@ceat.okstate.edu

Simone Paoletti
Dip. di Ingegneria dell'Informazione
Università di Siena
Via Roma, 56
53100 Siena, Italia
paoletti@dii.unisi.it

Andrea Paoli
CASY-DEIS, Università di Bologna
Viale Risorgimento 2
40136 Bologna, Italia
apaoli@deis.unibo.it

Claude Pégard
CREA - EA 3299
7 rue du Moulin Neuf
80000 Amiens, France
Claude.Pegard@u-picardie.fr

Anna Maria Perdon
Dip. di Ingegneria Informatica,
Gestionale e dell'Automazione
Università Politecnica delle Marche
Via Brecce Bianche
60131 Ancona, Italia
perdon@univpm.it

Raffaele Pesenti
Dip. di Ingegneria Informatica
Università di Palermo
Viale delle Scienze
Palermo, Italia
pesenti@unipa.it

Jacopo Piazzi
Dip. di Ingegneria dell'Informazione
Università di Siena
Via Roma, 56
53100 Siena, Italia
piazzi@dii.unisi.it

Domenico Prattichizzo
Dip. di Ingegneria dell'Informazione
Università di Siena
Via Roma, 56
53100 Siena, Italia
prattichizzo@dii.unisi.it

Jean-Pierre Quadrat
INRIA-Rocquencourt
B.P. 105, 78153 Le Chesnay
Cedex, France
Jean-Pierre.Quadrat@inria.fr

Jacob Roll
Division of Automatic Control
Linköping University
SE-581 83 Linköping, Sweden
roll@isy.liu.se

Julio A. Romero
Dept. of Technology
Riu Sec Campus
University of Jaume I
12071 Castellón de la Plana, Spain
romeroj@tec.uji.es

Rodolphe Sepulchre
Electrical Engineering and Computer
Science Institute
Montefiore B28
B-4000 Liège, Belgium
r.sepulchre@ulg.ac.be

Tielong Shen
Dept. of Mechanical Engineering
Sophia University
Kioicho 7-1, Chiyoda-ku
Tokyo 102-8554, Japan
tetu-sin@sophia.ac.jp

Bruno Siciliano
Dip. di Informatica e Sistemistica
Università degli Studi di Napoli
Federico II
Via Claudio 21
80125 Napoli, Italia
siciliano@unina.it

Nilesh B. Siraskar
School of Mechanical and Aerospace
Engineering
Oklahoma State University
Stillwater, OK 74078, USA
nilesh.siraskar@okstate.edu

Nicola Smaldone
Dip. di Automatica e Informatica
Politecnico di Torino
Corso Duca degli Abruzzi, 24
10129 Torino, Italia
nicola.smaldone@polito.it

Peng Song
Rutgers, The State University of
New Jersey
Dept. of Mechanical and Aerospace
Engineering
98 Brett Road, Engineering B242
Piscataway, NJ 08854, USA
pengsong@jove.rutgers.edu

Yuanzhang Sun
Dept. of Electrical Engineering
Tsinghua University
Beijing 10084, China
yzsun@mail.tsinghua.edu.cn

Michele Taragna
Dip. di Automatica e Informatica
Politecnico di Torino
Corso Duca degli Abruzzi, 24
10129 Torino, Italia
michele.taragna@polito.it

Andrew R. Teel
Center for Control Engineering and
Computation
Dept. of Electrical and Computer
Engineering
University of California
Santa Barbara, CA 93106, USA
teel@ece.ucsb.edu

Leon M. Tolbert
ECE Department
The University of Tennessee
Knoxville, TN 37996, USA
tolbert@utk.edu

Angel Valera
Dept. of Systems Engineering
and Control
University of Valencia
C/ Vera s/n Valencia, 46021 Spain
giuprog@isa.upv.es

Paolo Valigi
Dip. di Ingegneria Elettronica e
dell'Informazione
Università di Perugia
Perugia, Italia
valigi@diei.unipg.it

Tannen VanZwieten
University of Central Florida
Dept. of Mechanical, Materials &
Aerospace Engineering
P.O. Box 162450
Orlando, FL 32816, USA
Tannen.vanzwieten@gmail.com

Antonio Vicino
Dip. di Ingegneria dell'Informazione
Università di Siena
Via Roma, 56
53100 Siena, Italia
vicino@dii.unisi.it

Luigi Villani
Dip. di Informatica e Sistemistica
Università degli Studi di Napoli
Federico II
Via Claudio 21
80125 Napoli, Italia
villani@unina.it

Yonggang Xu
Center for Control Engineering and
Computation
Dept. of Electrical and Computer
Engineering
University of California
Santa Barbara, CA 93106, USA
yonggang@ece.ucsb.edu

Kaiyu Wang
ECE Department
The University of Tennessee
Knoxville, TN 37996, USA
wkaiyu@utk.edu

Luca Zaccarian
Dip. di Informatica, Sistemi e
Produzione
Università di Roma "Tor Vergata"
00133 Roma, Italia
zack@disp.uniroma2.it

Part I

State Estimation and Identification

Circle-Criterion Observers and Their Feedback Applications: An Overview

Murat Arcak

Electrical, Computer, and Systems Engineering Department
Rensselaer Polytechnic Institute
Troy, NY 12180, USA
`arcakm@rpi.edu`

Summary. This chapter gives an overview of the "circle-criterion" design of nonlinear observers, initiated and further developed by the author in a series of papers. It summarizes these results in a concise and unified manner and illustrates them with physically motivated design examples from fuel cell power systems, ship control, and active magnetic bearing systems.

1 Introduction

The prevalent approach to observer design in the literature has been to dominate nonlinearities with high-gain linear terms or to eliminate them via geometric transformations. High-gain observers have been developed by Khalil and other authors, as surveyed in [9]. To obtain linear observer error dynamics, Krener and Isidori [12] introduced a geometric design, which has been further studied by numerous authors, including Kazantzis and Kravaris [8] who proposed a less conservative procedure. Other studies include open-loop observers, explored by Lohmiller and Slotine [13], and an $\mathcal{H}_\infty$-like design for nonlinearities with linear growth bounds, introduced by Thau [16] and extended by Raghavan and Hedrick [15].

The circle-criterion observer departed from these approaches and initiated a new line of research to exploit types of nonlinearities rather than counteract them. Its basic form, introduced in [4], is applicable to a class of systems in which the nonlinearities satisfy a monotone growth property. The observer design takes advantage of this monotonic growth by introducing a nonlinear injection term, which gives rise to a state-dependent *sector* nonlinearity in the observer error dynamics. The celebrated *circle criterion* [18, 10] states that the feedback interconnection of such a sector nonlinearity with a *strictly positive real* (SPR) linear block is asymptotically stable. Following this criterion, our design achieves convergence of the estimates to the true states with the help of a linear matrix inequality (LMI) that, if feasible, yields observer matrices that render the linear block SPR.

After briefly reviewing this basic form of the design, we proceed to survey several of its extensions. Among them is an extension to multivariable nonlinearities, in which we further relax the LMI by taking advantage of the multivariable structure. The next task addressed in this chapter is the incorporation of circle-criterion observers in output-feedback design. We show that the circle-criterion observer possesses state-dependent convergence properties and responds to the onset of plant instability with "super-exponential" convergence. This key property allows us to recover global stability of state-feedback designs with certainty-equivalence-type output-feedback controllers—a rare occurrence in nonlinear control.

2 Design Procedure

To present the basic form of the circle-criterion observer design we consider a system of the form

$$\dot{x} = Ax + G\gamma(Hx) + \beta(y, u) \tag{1}$$
$$y = Cx\,,$$

where $x \in \mathbb{R}^n$ is the state, $y \in \mathbb{R}^p$ is the measured output, $u \in \mathbb{R}^m$ is the control input, and the functions $\gamma(\cdot)$ and $\beta(\cdot, \cdot)$ are locally Lipschitz. In this section we assume that $\gamma(\cdot)$ is a scalar nonlinearity. An extension to multivariable nonlinearities is presented in Section 4.

The main restriction for observer design is that $\gamma(\cdot)$ be a nondecreasing function; that is, for all $v, w \in \mathbb{R}$, it satisfies

$$(v - w)[\gamma(v) - \gamma(w)] \geq 0\,. \tag{2}$$

When $\gamma(\cdot)$ is differentiable, (2) is equivalent to $\gamma'(v) \geq 0$ for all $v \in \mathbb{R}$. Other nonlinearities that satisfy $a \geq \gamma'(v) \geq b$, where either a or b is finite, can be brought to the nondecreasing form $a = \infty$, $b = 0$, with the help of loop transformations as discussed in [5].

The structure of our observer is

$$\dot{\hat{x}} = A\hat{x} + L(C\hat{x} - y) + \beta(y, u) + G\gamma(H\hat{x} + K(C\hat{x} - y))\,, \tag{3}$$

and the design task is to determine the matrices K and L to guarantee observer convergence. Note that we inject the output error $K(C\hat{x} - y)$ into the nonlinearity and implement $\gamma(H\hat{x} + K(C\hat{x} - y))$ instead of $\gamma(H\hat{x})$. As we shall see, this injection significantly relaxes the feasibility conditions for this observer, thus expanding its applicability.

From (1) and (3), the dynamics of the state estimation error $e = x - \hat{x}$ are given by

$$\dot{e} = (A + LC)e + G\left[\gamma(v) - \gamma(w)\right], \tag{4}$$

where

$$v := Hx \quad \text{and} \quad w := H\hat{x} + K(C\hat{x} - y). \tag{5}$$

To design K and L, we recall from (2) that $\gamma(v) - \gamma(w)$ has the same sign as $v - w$; that is, it plays the role of a sector nonlinearity with the argument

$$v - w = (H + KC)e. \tag{6}$$

The observer error dynamics (4) thus consist of the negative feedback interconnection of this sector nonlinearity with a linear block described by the transfer function

$$-(H + KC)[sI - (A + LC)]^{-1}G. \tag{7}$$

In view of the circle criterion, a design of K and L that renders this transfer function SPR guarantees exponential decay for the observer error $e(t)$

Proposition 1. ([4]) Consider the plant (1) and let $[0, t_f)$ denote the maximal interval of definition for the solution $x(t)$. If there exists a matrix $P = P^T > 0$, and a constant $\epsilon > 0$, such that

$$(A + LC)^T P + P(A + LC) + \epsilon I \leq 0 \tag{8}$$
$$PG + (H + KC)^T = 0, \tag{9}$$

then the estimation error $e = x - \hat{x}$ of the observer (3) satisfies, for all $t \in [0, t_f)$,

$$|e(t)| \leq \kappa |e(0)| \exp(-\nu t), \tag{10}$$

where $\kappa = \sqrt{\dfrac{\lambda_{max}(P)}{\lambda_{min}(P)}}$, $\nu = \dfrac{\epsilon}{2\lambda_{max}(P)}$. $\qquad\qquad\square$

Equations (8)–(9) are the Kalman-Yakubovich-Popov conditions [10, Lemma 6.3] for SPR of the transfer function (7). They constitute an LMI in the variables P, PL, ϵ, and K, which means that we can employ the numerical algorithms available for LMIs [6] to determine if (8)–(9) is feasible and, if so, to obtain a solution for K and L. The matrix $P = P^T > 0$ in this LMI yields the observer Lyapunov function

$$V_{\text{obs}} = e^T Pe, \tag{11}$$

from which the estimate (10) follows. Using this Lyapunov function it is also possible to redesign the observer for enhanced convergence properties, as presented in Section 6.

3 Feasibility and Examples

An analytical test was developed in [5] to determine feasibility of the LMI (8)–(9). This test shows that the injection term inside the nonlinearity in

(3) relaxes the feasibility conditions; that is, feasibility with $K \neq 0$ is less restrictive than with $K = 0$.

To apply this test we drop from (1) $\beta(y, u)$ and the linear terms in y because they do not affect feasibility. We then transform it to the form

$$
\begin{aligned}
y &= y_1 \\
\dot{y}_1 &= y_2 \\
\dot{y}_2 &= y_3 \\
&\cdots \\
\dot{y}_r &= \Pi\xi - \gamma(\Sigma\xi + \sigma_1 y_1 + \cdots + \sigma_r y_r) \\
\dot{\xi} &= S\xi,
\end{aligned}
\tag{12}
$$

where r is the relative degree from the output y to the nonlinearity $\gamma(\cdot)$. There is no loss of generality in assuming that $\gamma(\cdot)$ appears with a negative sign in (12) because, otherwise, we can define a new nonlinearity $\tilde{\gamma}(\cdot)$ by $\tilde{\gamma}(-v) = -\gamma(v)$, which preserves the nondecreasing property of $\gamma(v)$ and has a negative coefficient.

We decompose S, Π, and Σ as

$$
S = \begin{bmatrix} S_1 & 0 & 0 \\ 0 & S_2 & 0 \\ 0 & 0 & S_3 \end{bmatrix} \qquad
\begin{aligned}
\Pi &= [\,\Pi_1 \ \Pi_2 \ \Pi_3\,] \\
\Sigma &= [\,\Sigma_1 \ \Sigma_2 \ \Sigma_3\,],
\end{aligned}
\tag{13}
$$

where $\sigma(S_1) \subset \mathbb{C}^+$, $\sigma(S_2) \subset \mathbb{C}^0$, $\sigma(S_3) \subset \mathbb{C}^-$, where $\mathbb{C}^+$, $\mathbb{C}^0$, and $\mathbb{C}^-$ denote the open right half-plane, the imaginary axis, and the open left half-plane, respectively. We then define $U = U^T$ and $V = V^T$ by

$$
S_1^T U + U S_1 = (\Pi_1 - \Sigma_1)^T (\Pi_1 - \Sigma_1)
\tag{14}
$$
$$
S_1^T V + V S_1 = (\Pi_1 + \Sigma_1)^T (\Pi_1 + \Sigma_1).
\tag{15}
$$

Proposition 2. ([5]) Consider the system (1) in the form (12)–(13), with U and V defined as in (14)–(15). The LMI (8)–(9) is feasible with $P = P^T > 0$, $\epsilon > 0$, and $K = 0$ if and only if

$$
\sigma_r > 0, \quad \sigma_{r-1} < 0, \quad U - V > \frac{2}{\sigma_r} \Sigma_1^T \Sigma_1,
\tag{16}
$$

and

$$
w^*(\Pi_2 + \Sigma_2)^T (\Pi_2 + \Sigma_2)w < w^*(\Pi_2 - \Sigma_2)^T (\Pi_2 - \Sigma_2)w
\tag{17}
$$

for every (possibly complex) eigenvector w of S_2. If the restriction $K = 0$ is removed, then (16) is relaxed as follows: When $r = 1$, (8)–(9) is feasible if and only if $U > V$ and (17) hold for every eigenvector w of S_2. When $r = 2$, $\sigma_r > 0$ and $U - V > \frac{2}{\sigma_r} \Sigma_1^T \Sigma_1$ are also required. When $r \geq 3$, all the conditions for feasibility with $K = 0$ are required. $\qquad\square$

Example 1: Flux Observer Design for Active Magnetic Bearing Systems [17]

For the following single degree-of-freedom active magnetic bearing model

$$\dot{x}_1 = x_2$$
$$\dot{x}_2 = x_3 + x_3|x_3| \tag{18}$$
$$\dot{x}_3 = u \,,$$

the problem is to estimate the magnetic flux x_3 from measurements of the rotor position x_1 and the velocity x_2. Because $\gamma(x_3) = x_3|x_3|$ is nondecreasing we proceed with a circle-criterion observer design for the (x_2, x_3)-subsystem with $y = x_2$. To determine feasibility using Proposition 2, we omit u, define $y_1 = x_2$, $\xi = -x_3$, and represent the (x_2, x_3)-subsystem as in (12):

$$\dot{y}_1 = -\xi - \gamma(\xi)$$
$$\dot{\xi} = 0 \,. \tag{19}$$

With the restriction $K = 0$ the observer design is not feasible because $r = 1$ and $\sigma_1 = 0$ violate (16). When the restriction $K = 0$ is removed the design is feasible because (17) holds, as verified with $S_2 = 0$, $\Pi_2 = -1$, $\Sigma_2 = 1$. A solution to the LMI (8)–(9) is $L = [-1 \ -1]^T$, $K = -2$, which results in the observer

$$\dot{\hat{x}}_2 = \hat{x}_3 - (\hat{x}_2 - x_2) + (\hat{x}_3 - 2(\hat{x}_2 - x_2))|\hat{x}_3 - 2(\hat{x}_2 - x_2)|$$
$$\dot{\hat{x}}_3 = -(\hat{x}_2 - x_2) + u \,. \tag{20}$$

Using this observer and a reduced-order variant, [17] develops several output-feedback controllers, proves their stability, and tests their performance with a high-fidelity simulation model.

Example 2: Fuel Cell Hydrogen Estimation [3]

An important problem in fuel cell power systems is the estimation of the hydrogen partial pressure p_{H_2} in the anode channel, which is difficult to measure. A lumped dynamic model of this variable, obtained from the Ideal Gas Law, is

$$\dot{p}_{H_2} = \frac{RT}{V_a} \left(\frac{1}{M_{ai}} \frac{p_{H_2in}}{P_{ain}} F_{ai} - \frac{1}{M_{ao}} \frac{p_{H_2}}{P_a} F_{ao} - \frac{nI}{2F} \right), \tag{21}$$

where V_a is the anode volume; n is the number of cells; F is the Faraday constant; P_a is the total anode pressure; p_{H_2in} and P_{ain} are, respectively, the hydrogen partial pressure and the total pressure at the inlet of the anode; F_{ai} and F_{ao} are the total inlet and outlet mass flows; and M_{ai} and M_{ao} represent average molecular weights of the inlet and exit gas streams. It is physically meaningful to assume that P_{ain}, P_a, F_{ai}, F_{ao}, M_{ai}, and p_{H_2in} are known. M_{ao}, however, depends on the unmeasured p_{H_2} via

$$M_{ao} = \frac{p_{H_2}}{P_a} M_{H_2} + \frac{P_a - p_{H_2}}{P_a} \delta_a, \tag{22}$$

in which M_{H_2} is the molecular weight of hydrogen, and δ_a is an average value for the molecular weights of other gases at the anode exit.

Because hydrogen is lighter than the other molecules; that is, $M_{H_2} < \delta_a$ in (22), the ratio p_{H_2}/M_{ao} in (21) is a nondecreasing nonlinearity of p_{H_2}. Unlike the system (1) studied in the basic design procedure of Section 2, in (21) the nonlinearity has a time-varying coefficient due to the exogenous variables P_a and F_{ao}. However, because these variables are strictly positive during operation of the fuel cell, it follows from the same principles used in proving Proposition 1 that the open-loop observer

$$\dot{\hat{p}}_{H_2} = \frac{RT}{V_a} \left(\frac{1}{M_{ai}} \frac{p_{H_2in}}{P_{ain}} F_{ai} - \frac{1}{\hat{M}_{ao}} \frac{\hat{p}_{H_2}}{P_a} F_{ao} - \frac{nI}{2F} \right), \tag{23}$$

where $\hat{M}_{ao}$ is (22) with p_{H_2} replaced by $\hat{p}_{H_2}$, guarantees convergence of $\hat{p}_{H_2}$ to p_{H_2}. In [3], an additional problem is addressed in which the inlet partial pressure p_{H_2in} in (21) is an unknown parameter due to uncertainty of inlet conditions. The observer (23) is then made adaptive by using measurements of the fuel cell voltage, which is a logarithmic, thus nondecreasing, function of p_{H_2} and p_{H_2in}.

4 Extension to Multivariable Nonlinearities

The circle-criterion observer is extended in [7] to systems of the form (1) where $\gamma(\cdot) : \mathbb{R}^q \to \mathbb{R}^q$ is now a multivariable nonlinearity, the matrix H is $q \times n$, and G is $n \times q$. It is shown in [7] that Proposition 1 holds in this case if $\gamma(\cdot)$ satisfies

$$\frac{\partial \gamma(v)}{\partial v} + \left(\frac{\partial \gamma(v)}{\partial v} \right)^T \geq 0 \quad \forall v \in \mathbb{R}^q. \tag{24}$$

This condition coincides with the scalar nondecreasing property $\gamma'(v) \geq 0$ when $q = 1$ and generalizes it to multivariable nonlinearities when $q \geq 2$. An example is the system

$$\begin{aligned}
\dot{x}_1 &= x_2 \\
\dot{x}_2 &= x_2 - \tfrac{1}{3} x_2^3 - x_2 x_3^2 \\
\dot{x}_3 &= x_2 - x_3 - \tfrac{1}{3} x_3^3 - x_3 x_2^2,
\end{aligned} \tag{25}$$

in which

$$\gamma(x_2, x_3) = \begin{bmatrix} \tfrac{1}{3} x_2^3 + x_2 x_3^2 \\ x_2^2 x_3 + \tfrac{1}{3} x_3^3 \end{bmatrix} \tag{26}$$

satisfies (24), and a design as in Proposition 1 yields the observer

$$\dot{\hat{x}}_1 = \hat{x}_2 - 3\left(\hat{x}_1 - x_1\right)$$
$$\dot{\hat{x}}_2 = \hat{x}_2 - 8\left(\hat{x}_1 - x_1\right) - \tfrac{1}{3}\left(\hat{x}_2 - 2\left(\hat{x}_1 - x_1\right)\right)^3$$
$$\qquad\qquad - \left(\hat{x}_2 - 2\left(\hat{x}_1 - x_1\right)\right)\left(\hat{x}_3 - \left(\hat{x}_1 - x_1\right)\right)^2 \qquad (27)$$
$$\dot{\hat{x}}_3 = \hat{x}_2 - \hat{x}_3 - 4\left(\hat{x}_1 - x_1\right) - \tfrac{1}{3}\left(\hat{x}_3 - \left(\hat{x}_1 - x_1\right)\right)^3$$
$$\qquad\qquad - \left(\hat{x}_3 - \left(\hat{x}_1 - x_1\right)\right)\left(\hat{x}_2 - 2\left(\hat{x}_1 - x_1\right)\right)^2 .$$

A further result in [7] is to relax the LMI (8)–(9) when the multivariable nonlinearity exhibits the decoupled structure

$$\gamma(Hx) = \begin{bmatrix} \gamma_1(H_1 x) \\ \vdots \\ \gamma_k(H_k x) \end{bmatrix} ; \qquad (28)$$

that is,

$$G\gamma(Hx) := \sum_{i=1}^{k} G_i \gamma_i(H_i x), \qquad (29)$$

where $\gamma_i(\cdot) : \mathbb{R}^{q_i} \to \mathbb{R}^{q_i}$. The LMI (8)–(9) is then replaced with

$$(A + LC)^T P + P(A + LC) + \epsilon I \le 0 \qquad (30)$$
$$PG + (H + KC)^T \Lambda = 0, \qquad (31)$$

where the "multiplier" Λ is of the form

$$\Lambda = \begin{bmatrix} \lambda_1 I_{q_1 \times q_1} & 0 & \cdots & 0 \\ 0 & \lambda_2 I_{q_2 \times q_2} & \cdots & 0 \\ \vdots & \vdots & \ddots & \vdots \\ 0 & 0 & \cdots & \lambda_k I_{q_k \times q_k} \end{bmatrix}, \quad \lambda_i > 0, \;\; i = 1, 2, \cdots, k. \quad (32)$$

This multiplier takes advantage of the decoupled structure of the nonlinearity and relaxes the original LMI (8)–(9), which coincides with (30)–(31) when $\Lambda = I$. It is shown in [7] that the feasibility of (30)–(31) is indeed less restrictive than that of (8)–(9).

5 Reduced-Order Design

A reduced-order observer generates estimates only for unmeasured states. The design of such an observer starts with a preliminary change of coordinates such that the output y consists of the first p entries of the state vector $x = [y^T \; x_2^T]^T$. In the new coordinates, system (1) is

$$\dot{y} = A_1 x_2 + G_1 \gamma(H_1 y + H_2 x_2) + \beta_1(y, u)$$
$$\dot{x}_2 = A_2 x_2 + G_2 \gamma(H_1 y + H_2 x_2) + \beta_2(y, u), \qquad (33)$$

where the linear terms in y are incorporated in $\beta_1(y, u)$ and $\beta_2(y, u)$. An estimate of x_2 is then obtained via the new variable

$$\chi := x_2 + Ny, \tag{34}$$

where N is an $(n - p) \times p$ matrix to be designed. From (33), the derivative of χ is

$$\dot{\chi} = (A_2 + NA_1)\chi + (G_2 + NG_1)\gamma(H_2\chi + (H_1 - H_2N)y) + \bar{\beta}(y, u), \tag{35}$$

where $\bar{\beta}(y, u) := N\beta_1(y, u) + \beta_2(y, u) - (A_2 + NA_1)Ny$.
　　To obtain the estimate

$$\hat{x}_2 = \hat{\chi} - Ny, \tag{36}$$

we employ the observer

$$\dot{\hat{\chi}} = (A_2 + NA_1)\hat{\chi} + (G_2 + NG_1)\gamma(H_2\hat{\chi} + (H_1 - H_2N)y) + \bar{\beta}(y, u). \tag{37}$$

From (35) and (37), the dynamics of $e_2 := x_2 - \hat{x}_2 = \chi - \hat{\chi}$ are governed by

$$\dot{e}_2 = (A_2 + NA_1)e_2 + (G_2 + NG_1)[\gamma(v_2) - \gamma(w_2)], \tag{38}$$

where $v_2 := H_2\chi + (H_1 - H_2N)y$, $w_2 := H_2\hat{\chi} + (H_1 - H_2N)y$, and $[\gamma(v_2) - \gamma(w_2)]$ again plays the role of a sector nonlinearity, with argument $v_2 - w_2 = H_2e_2$. Thus, the LMI (30)–(31) is now to be replaced with

$$(A_2 + NA_1)^T P_2 + P_2(A_2 + NA_1) + \epsilon I \leq 0 \tag{39}$$

$$P_2(G_2 + NG_1) + H_2^T \Lambda = 0, \tag{40}$$

where $P_2 = P_2^T > 0$ and $\epsilon > 0$. As shown in [4], feasibility conditions for (39)–(40) and (30)–(31) are equivalent; that is, the reduced-order design is feasible whenever the full-order design is, and vice versa.

Example 3: Velocity Observer for a Class of Euler-Lagrange Systems [1]

Motivated by ship control problems, the authors of [1] studied the Euler-Lagrange model

$$\dot{q} = J(q)\nu \tag{41}$$

$$M\dot{\nu} + D\nu + d(\nu) + v(q) = \tau, \tag{42}$$

where q and ν are n-dimensional vectors of generalized positions and velocities, respectively; $M = M^T > 0$ and $D = D^T > 0$ are constant matrices; and $d(\nu)$ is a multivariable nonlinearity that consists of nonlinear hydrodynamic drag terms as well as Coriolis and centripetal forces.
　　To estimate the velocity vector ν from position measurement q, [1] exhibits a class of systems in which $d(\nu)$ satisfies (24) and proceeds with an extension

of the circle-criterion design applicable to this class. The full-order observer is

$$\dot{\hat{q}} = J(q)\hat{\nu} + L_1(q - \hat{q}) \tag{43}$$

$$M\dot{\hat{\nu}} + D\hat{\nu} + d(\hat{\nu}) + v(q) = \tau + L_2(q)(q - \hat{q}), \tag{44}$$

where $L_2(q) = J(q)^T Q$, and $Q = Q^T > 0$ and L_1 are to be designed such that $QL_1 + L_1^T Q > 0$. With this choice of the injection matrices L_1 and $L_2(q)$, the Lyapunov function

$$V = (q - \hat{q})^T Q(q - \hat{q}) + (\nu - \hat{\nu})^T M(\nu - \hat{\nu}) \tag{45}$$

proves exponential convergence of the estimates $\hat{q}$ and $\hat{\nu}$ to q and ν, respectively.

A reduced-order design is feasible in this example with $N = 0$ in (34); that is, the observer consists of a copy of (42) implemented with $\hat{\nu}$:

$$M\dot{\hat{\nu}} + D\hat{\nu} + d(\hat{\nu}) + v(q) = \tau. \tag{46}$$

Further results in [1] include an observer-based tracking controller design, its stability proof, and a case study of a spherical underwater vehicle model.

6 Output-Feedback via Certainty-Equivalence

A significant advantage of circle-criterion observers is their "super-exponential" convergence, which make them suitable for use in certainty-equivalence controllers. Indeed, numerous examples in the literature (see, e.g., [11]) have shown that exponential observer convergence may not be sufficient to preserve stability of a state-feedback design. In contrast, the convergence speed of the circle-criterion observer depends on the magnitude of the plant trajectories. The onset of instability stimulates faster observer convergence, which results in a recovery of stability achievable with the underlying state-feedback design.

While studies such as [5] and [17] incorporated circle-criterion observers in certainty-equivalence designs and gave stability proofs on a case-by-case basis, the general property was revealed recently in [14] and [2]. In particular, [14] presents analysis tools with which state-dependent convergence properties of observers can be exploited to prove stability in a certainty-equivalence design. As a special case it proves that the reduced-order circle-criterion observer achieves stability in a certainty-equivalence design, under a mild assumption on the underlying state-feedback controller (see Proposition 3 below). [2] shows that the full-order observer (3) also preserves stability when its right-hand side is augmented with the additional nonlinear term

$$-\theta P^{-1} C^T K^T [\gamma(H\hat{x} + K(C\hat{x} - y)) - \gamma(H\hat{x})], \quad \theta > 0, \tag{47}$$

where P is as in Proposition 1. This augmentation increases the negativity of the derivative of the observer Lyapunov function $V_{\mathrm{obs}} = e^T P e$ and enhances the convergence properties of the full-order observer to match those of the reduced-order design.

Proposition 3. ([2]) Consider system (1) and let $u = \alpha(x)$ be a globally asymptotically stabilizing state-feedback controller with which the closed-loop system admits a Lyapunov function with a uniformly bounded gradient. Suppose further that the function $\beta(y, u)$ in (1) is such that, for all $y, \hat{y} \in \mathbb{R}^p$, and $u \in \mathbb{R}^m$,

$$|\beta(y, u) - \beta(\hat{y}, u)| \leq \omega(|y - \hat{y}|) \tag{48}$$

for some class-$\mathcal{K}$ function $\omega(\cdot)$ that is differentiable at zero. Then the certainty-equivalence controller $u = \alpha(\hat{x})$, where $\hat{x}$ is obtained from the observer (3) augmented with (47), achieves global asymptotic stability. $\qquad\square$

As shown in [2], condition (48) can be removed if, in the certainty-equivalence design, the measured states are incorporated directly and not replaced with their observer estimates; that is, with the controller $u = \alpha(y, \hat{x}_2)$ in the coordinates of Section 5, rather than with $u = \alpha(\hat{x}) = \alpha(\hat{y}, \hat{x}_2)$.

The main assumption of Proposition 3 on the state-feedback controller is that the closed-loop system admit a Lyapunov function with a bounded gradient. Even when an available Lyapunov function does not satisfy this assumption, it may do so when scaled by a scalar nonlinear function. An example is a quadratic Lyapunov function $V = x^T Q x$ that does not have a bounded gradient. The scaled Lyapunov function $\tilde{V} = \log(1 + V)$, however, has bounded gradient

$$\frac{\partial \tilde{V}}{\partial x} = \frac{1}{1 + V}\frac{\partial V}{\partial x} = \frac{1}{1 + x^T Q x} 2 x^T Q \quad \Rightarrow \quad \left|\frac{\partial \tilde{V}}{\partial x}\right| \leq \frac{\lambda_{\max}(Q)}{\sqrt{\lambda_{\min}(Q)}}. \tag{49}$$

To illustrate Proposition 3 we now revisit the active magnetic bearing example. When augmented with the term (47), the observer (20) is

$$\dot{\hat{x}}_2 = \hat{x}_3 - (\hat{x}_2 - x_2) + (1 + 3\theta)(\hat{x}_3 - 2(\hat{x}_2 - x_2))|\hat{x}_3 - 2(\hat{x}_2 - x_2)| - 3\theta \hat{x}_3|\hat{x}_3|$$
$$\dot{\hat{x}}_3 = -(\hat{x}_2 - x_2) + u + \theta(\hat{x}_3 - 2(\hat{x}_2 - x_2))|\hat{x}_3 - 2(\hat{x}_2 - x_2)| - \theta \hat{x}_3|\hat{x}_3|, \tag{50}$$

where $\theta > 0$ is a design parameter. Furthermore, (48) is satisfied because $\beta(y, u) = u$. This means that any state-feedback controller $u = \alpha(x)$ as in Proposition 3, implemented with estimates from (50), ensures global asymptotic stability.

7 Conclusions

This chapter surveyed circle-criterion observers, which are applicable to a class of systems where the nonlinearities exhibit monotone growth properties. This class is physically relevant and arises in a number of applications. A salient feature of our design is that it exploits the nonlinearities, rather than counteract them. By doing so it achieves desirable convergence properties that make output-feedback stabilization possible via certainty-equivalence designs. It would be of interest to characterize other classes of systems in which nonlinearities can again be exploited for constructive observer schemes.

References

1. Aamo O, Arcak M, Fossen T, Kokotović P (2001) Global output tracking control of a class of Euler-Lagrange systems with monotonic nonlinearities in the velocities. International Journal of Control 74(7):649–658
2. Arcak M (2005) Certainty-equivalence output-feedback design with circle-criterion observers. IEEE Transactions on Automatic Control 50(6):905–909
3. Arcak M, Gorgun H, Pedersen L, Varigonda S (2004) A nonlinear observer design for fuel cell hydrogen estimation. IEEE Transactions on Control Systems Technology 12(1):101–110
4. Arcak M, Kokotović P (2001a) Nonlinear observers: A circle criterion design and robustness analysis. Automatica 37(12):1923–1930
5. Arcak M, Kokotović P (2001b) Observer-based control of systems with slope-restricted nonlinearities. IEEE Transactions on Automatic Control 4(7):1146–1151
6. Boyd S, El Ghaoui L, Feron E, Balakrishnan V (1994) Linear Matrix Inequalities in System and Control Theory, vol. 15 of SIAM Studies in Applied Mathematics. SIAM, Philadelphia, PA
7. Fan X, Arcak M (2003) Observer design for systems with multivariable monotone nonlinearities. Systems and Control Letters 50(4):319–330
8. Kazantzis N, Kravaris C (1998) Nonlinear observer design using Lyapunov's auxiliary theorem. Systems and Control Letters 34(5):241–247
9. Khalil H (1999) High-gain observers in nonlinear feedback control In: Nijmeijer H, Fossen T (eds), New directions in nonlinear observer design, 249–268. Springer-Verlag, New York
10. Khalil H (2002) Nonlinear systems. Prentice-Hall, Upper Saddle River, NJ
11. Kokotović P (1992) The joy of feedback: nonlinear and adaptive. IEEE Control Systems Magazine 12:7–17
12. Krener A, Isidori A (1983) Linearization by output injection and nonlinear observers. Systems and Control Letters 3:47–52
13. Lohmiller W, Slotine JJ (1998) On contraction analysis for nonlinear systems. Automatica 34:683–696
14. Praly L, Arcak M (2004) A relaxed condition for stability of nonlinear observer-based controllers. Systems and Control Letters 53(3-4):311–320
15. Raghavan S, Hedrick J (1994) Observer design for a class of nonlinear systems. International Journal of Control 59:515–528
16. Thau F (1973) Observing the state of non-linear dynamic systems. International Journal of Control 17:471–479
17. Tsiotras P, Arcak M (2005) Low-bias control of AMB subject to voltage saturation: State-feedback and observer designs. IEEE Transactions on Control Systems Technology 13(2):262–273

18. Zames G (1966) On the input-output stability of time-varying nonlinear feedback systems–Parts I and II. IEEE Transactions on Automatic Control 11:228–238, 465–476

Unknown Input Observers and Residual Generators for Linear Time Delay Systems

Giuseppe Conte and Anna Maria Perdon

Dipartimento di Ingegneria Informatica, Gestionale e dell'Automazione
Università Politecnica delle Marche
Via Brecce Bianche, 60131 Ancona, Italia
{gconte,perdon}@univpm.it

Summary. The problems of constructing observers in the presence of unknown inputs and of detecting and recognizing inputs of a special kind are considered for linear time delay systems with commensurable delays. Both problems are studied from an algebraic and geometric point of view, making use of models with coefficients in a ring, of invariant subspaces of the space module and of suitable canonical decomposition into subsystems. Feasible and constructive procedures for the analysis and solution of the problems are presented.

1 Introduction

Time delay systems of various kinds appear frequently in industrial applications, where delays are the unavoidable effects of the transportation of materials and, more generally, in control systems, where information is dispatched through slow or very long communication lines, like in teleoperated systems, systems controlled through the Internet and other large integrated communication control systems (ICCS). For this reason, a great research effort has recently been devoted to the development of analysis and synthesis techniques that, extending results from the field of linear systems, may provide tools for dealing with delay-differential systems (for an overview of the state-of-art in this field and updated references, see the Proceedings of the IFAC Workshops on Time Delay Systems for 1998, 2000, 2001, 2003).

In the general framework of time delay systems, the problem of state reconstruction, under various assumptions, has been investigated by several authors. In particular, the state reconstruction for weakly observable systems has been investigated and exact and asymptotic observers have been defined in [21]. An interesting case in a number of applications is that of the asymptotic state estimation in presence of unknown inputs, or the unknown input observation (UIO) problem. Unknown inputs, in fact, play a crucial role in decentralized control schemes, where single controllers have only partial information, as well as in general control schemes, where they model disturbances.

Moreover, the possibility of providing feasible solutions to the UIO problem allows us to deal with fault detection and isolation problems, which have been considered and approached from various points of view in connection with different classes of systems (see [20] and references therein). In particular, necessary and sufficient conditions for the existence of a dynamic gain unknown input observer for linear time delay systems are given in [11]. Here we consider the UIO problem and the related residual generation (RG) problem for linear time delay systems, using algebraic and geometric tools as in [7] and we report the results obtained in [8] and in [9].

The approach developed in [9] for studying the UIO problem relies mainly on the use of mathematical models with coefficients in a suitable ring, or systems over a ring. This choice is crucial for facilitating the introduction of a geometric point of view similar to the one described in [4] for systems without delays. In this way, in fact, one avoids the necessity of dealing with infinite dimensional vector spaces for representing the systems at issue and, in place, one can use finite dimensional modules over suitable rings. The resulting framework fits well with geometric techniques and, actually, it allows one to extend a number of results from the classical case to that of delay-differential systems (see also [6]).

The RG problem is considered from a similar point of view in [8], where it is described as the problem of constructing a sort of observer, called "residual generator," for a given delay-differential system Σ_d, whose output, called "residual output," allows one to detect the occurrence of specific inputs. This property is important when the input set of Σ_d is split into a control input subset and several so-called failure input subsets. By analyzing the residual output, failure inputs, which are unknown inputs, may be detected and recognized. Therefore, the problem of constructing a residual generator is related to the problem of state observation in presence of unknown inputs.

The chapter is organized as follows. In Section 2 we describe briefly the relationship between delay-differential systems and systems with coefficients in a ring and we recall some notions and preliminary results of the geometric approach for system with coefficients in a ring. In Section 3 we state the UIO problem and in Section 4 we develop a thorough analysis of the problem using a geometric approach. In particular, we identify an obstruction to the existence of solutions in terms of dynamic properties of a subsystem and we display the structure of possible observers. The existence of solutions is then related to the possibility of assigning the coefficients of a specific subsystem. Although restrictive conditions have to be imposed for allowing our analysis and for constructing solutions, our results provide a deep insight into the geometry of the problem and can be used for developing algorithmic procedures. The RG problem for time delay systems is described in Section 5, where a possible solution is constructed and its properties are investigated and described. The main results relate the existence of a solution to the RG problem to a geometric condition of the same kind of that found in the classical case (see [15] and [23]). Examples illustrate the construction of a possible residual generator and the relevance of the various assumptions are given. Finally, Section 6 summarizes the results.

2 Preliminaries

Let the linear, time invariant, delay-differential, dynamical system Σ_d be described by the set of equations

$$\Sigma_d = \begin{cases} \dot{x}(t) = \sum_{i=0}^{a} A_i x(t - ih) + \sum_{i=0}^{b} B_i u(t - ih) \\ y(t) = \sum_{i=0}^{c} C_i x(t - ih), \end{cases} \tag{1}$$

where, denoting by $\mathbb{R}$ the field of real numbers, x belongs to the state space $\mathbb{R}^n$; u belongs to the input space $\mathbb{R}^m$; y belongs to the output space $\mathbb{R}^p$; and A_i, $i = 0, 1, ..., a$; B_i, $i = 0, 1, ..., b$; C_i, $i = 0, 1, ..., c$, are matrices of suitable dimensions with entries in $\mathbb{R}$ and $h \in \mathbb{R}^+$ is a given delay. Introducing the delay operator δ, which is defined for any time function $f(t)$ by $\delta f(t) = f(t - h)$, we can write

$$\Sigma_d = \begin{cases} \dot{x}(t) = \sum_{i=0}^{a} A_i \delta^i x(t) + \sum_{i=0}^{b} B_i \delta^i u(t) \\ y(t) = \sum_{i=0}^{c} C_i \delta^i x(t). \end{cases} \tag{2}$$

Formally, it is possible to substitute the delay operator δ with the algebraic indeterminate Δ, so that, denoting by A, B, C the matrices with entries in $R = \mathbb{R}[\Delta]$ given, respectively, by

$$A = \sum_{i=0}^{a} A_i \Delta^i, \quad B = \sum_{i=0}^{b} B_i \Delta^i, \quad \text{and} \quad C = \sum_{i=0}^{c} C_i \Delta^i,$$

we can associate to Σ_d the system Σ over the ring $R = \mathbb{R}[\Delta]$ of real polynomials in one indeterminate, defined by the set of equations

$$\Sigma = \begin{cases} x(t + 1) = Ax(t) + Bu(t) \\ y(t) = Cx(t). \end{cases} \tag{3}$$

By abuse of notation, in the above equations we denote by x an element of the free state module $\mathcal{X} = R^n$, by u an element of the free input module $\mathcal{U} = R^m$, by y an element of to the free output module $\mathcal{Y} = R^p$.

Example 1. Consider, e.g., a delay-differential system Σ_d of the form

$$\Sigma_{\mathbf{d}} = \begin{cases} \dot{x}_1(t) = x_1(t - h) - x_2(t) \\ \dot{x}_2(t) = x_2(t) + u(t) \\ y(t) = x_1(t) + x_2(t - h). \end{cases} \tag{4}$$

We can associate to Σ_d the following system Σ over $\mathbb{R}[\Delta]$:

$$\Sigma = \begin{cases} x_1(t+1) = \Delta x_1(t) - x_2(t) \\ x_2(t+1) = x_2(t) + u(t) \\ y(t) = x_1(t) + \Delta x_2(t). \end{cases} \tag{5}$$

The transfer function of Σ_d and Σ are, respectively, given by

$$T_d = \frac{s\, e^{-sh} - e^{-2sh} - 1}{(s-1)(s-e^{-sh})} \quad \text{and} \quad T = \frac{\Delta z - \Delta^2 - 1}{(z-1)(z-\Delta)}.$$

Remark that while T_d derives from the application of the Laplace transform, T can be viewed as the result of formal computations, obtained by applying a formal z-transform to (5). Both T_d and T can be viewed as rational functions in two indeterminates, respectively, s and e^{-sh} for T_d and z and Δ for T. From this point of view, T_d and T have the same structure. In practice, Σ_d and Σ are quite different objects from a dynamical point of view, but, as their transfer functions have the same structure, they have the same signal flow graph, so that control problems concerning the input/output behavior of Σ_d can be formulated naturally in terms of the input/output behavior of Σ. Solutions found in the framework of systems over rings can often be interpreted in the original delay-differential framework, providing in this way a solution to the problem at issue. However, while s and e^{-sh} are not independent variables in the delay-differential framework, z and Δ are formally independent in the ring framework. A consequence of this fact is that the results obtained in the analysis of Σ_d by means of Σ are not influenced by the actual value of the delay h.

For delay-differential systems with noncommensurable delays, a procedure of the same kind as above gives rise to systems over the ring $R = \mathbb{R}[\Delta_1, \Delta_2, ..., \Delta_k]$ of polynomials in several indeterminates with real coefficients. Essentially, $\mathbb{R}[\Delta]$ and $R = \mathbb{R}[\Delta_1, \Delta_2, ..., \Delta_k]$ are the basic rings we will have to consider in dealing with systems associated to delay-differential systems. The main differences between the two cases are related to algebraic properties of the ring $\mathbb{R}[\Delta]$ that are not valid for $R = \mathbb{R}[\Delta_1, \Delta_2, ..., \Delta_k]$ with $k \geq 2$. Both rings are Noetherian rings, which means that increasing sequences of geometric objects (namely, submodules of a given module) converge in a finite number of steps. But only $\mathbb{R}[\Delta]$ is a principal ideal domain (PID), which implies, in particular, that the greatest common divisor of two elements is a linear combination of them (see [14]). These properties are important for assuring, respectively, the convergence of specific algorithms and the possibility of constructing suitable canonical forms.

It is important to remark that the use of systems over rings avoids the necessity of dealing with infinite dimensional vector spaces for representing time delay systems. Input, output, and especially state spaces can, in fact, be modeled as finite dimensional modules over the ring of coefficients (in our case, as already said, $\mathbb{R}[\Delta]$ or $\mathbb{R}[\Delta_1, \Delta_2, ..., \Delta_k]$). The price to pay for extending the approaches developed for systems with coefficients in the field of real numbers $\mathbb{R}$ is that ring and module algebra is more rich and complicated than linear algebra (see [1], [14]). Since nonzero elements in a ring are not necessarily invertible, a linear dependency relation like $\sum_{i=1}^{n} a_i x_i = 0$ between elements of a free module over a ring R does not imply that each x_i is a linear combination of the remaining ones. So, as opposed to what happens in the case of vector spaces, we can find sets of generators of a free module from which no basis can be extracted, as well as sets of linearly independent elements of maximal cardinality that are not sets of generators. In particular, we may have submodules of a free module that are not direct summands. From the point of view of dynamical properties, this fact has remarkable consequences, since it may prevent one from decomposing a system with respect to dynamically invariant submodules.

In order to analyze the solvability of the two problems we mentioned earlier and to investigate the construction of suitable observers, we will make use of a geometric approach that parallels the one described in [4]. To this aim, we need to recall some geometric concepts related to the classical notion of conditioned invariance.

Definition 1. ([6]) *Let Σ be a system of the form (1) over a ring R. A submodule $\mathcal{S}$ of the state module $\mathcal{X}$ is said to be*
- *(A, C) invariant or conditioned invariant if $A(\mathcal{S} \cap Ker\, C) \subseteq \mathcal{S}$;*
- *injection invariant if there exists an R-linear map $G : Y \to \mathcal{X}$ such that $(A + GC)\mathcal{S} \subseteq \mathcal{S}$.*
Any map G as above is called a friend *of $\mathcal{S}$.*

Given a submodule $\mathcal{D} \subset R^n$, the set of all conditioned invariant submodules containing $\mathcal{D}$ is closed under intersection, therefore it has a minimal element that is denoted by $\mathcal{S}^*(\mathcal{D})$.

Proposition 1. *Given a system Σ, defined over a Noetherian ring R by equations of the form (3), and a submodule $\mathcal{D}$ of its state module X, the sequence $\{\mathcal{S}_k\}_{k \geq 0}$ of submodules of X defined recursively by*

$$\begin{aligned} \mathcal{S}_0 &= \mathcal{D} \\ \mathcal{S}_{k+1} &= \mathcal{S}_k + A(\mathcal{S}_k \cap Ker\, C) \end{aligned} \tag{6}$$

converges in a finite number of steps to $\mathcal{S}^*(\mathcal{D})$.

Proof The sequence (6) is an increasing sequence of submodules, therefore the convergence in a finite number of steps is assured over any Noetherian ring.

A consequence of Proposition 1 is that (6) provides an algorithm for computing the minimum conditioned invariant submodule containing a given submodule $\mathcal{D}$.

If R is a field, as in the classical case, a conditioned invariant submodule $\mathcal{S}$ is always injection invariant, while this is not true if R is a ring. In other words, it may happen that no friend of $\mathcal{S}$ exists.

Example 2. Let Σ be a system defined over the ring $\mathcal{R}[\Delta]$ by equations of the form (3), with

$$A = \begin{pmatrix} 0 & 0 \\ 1 & 0 \end{pmatrix}, \quad C = \begin{pmatrix} \Delta & 1 \end{pmatrix}.$$

The submodule $\mathcal{S} = span\left\{\begin{pmatrix} \Delta \\ 0 \end{pmatrix}\right\}$ is an (A, C) invariant submodule for Σ,

since $Ker\ C \cap \mathcal{S} = 0$. For every R-linear map $G = \begin{pmatrix} g_1 \\ g_2 \end{pmatrix} : Y \to X$, we have

$A + GC = \begin{pmatrix} g_1\Delta & g_1 \\ g_2\Delta + 1 & g_2 \end{pmatrix}$. Then $(A + GC)\mathcal{S} = span\begin{pmatrix} g_1\Delta^2 \\ (g_2\Delta + 1)\Delta \end{pmatrix}$ is contained in $\mathcal{S}$ if and only if $g_2\Delta + 1 = 0$. Since this equation has no solution in $\mathcal{R}[\Delta]$, $\mathcal{S}$ is not an $(A + GC)$ invariant submodule.

Conditions under which a conditioned invariant submodule is feedback invariant are given in [2] and in [13].

For an (A, C) invariant submodule $\mathcal{S}$ that is not injection invariant a more general notion of "friend" can be introduced. To describe it, let us consider the quotient field of $\mathbb{R}[\Delta]$, namely, the field $Q = \mathbb{R}(\Delta)$ of rational functions in the indeterminate Δ.

Definition 2. *Given a conditioned invariant submodule $\mathcal{S}$ for a system Σ of the form (3) over the ring $R = \mathbb{R}[\Delta]$ and denoting by Q the quotient field $Q = \mathbb{R}(\Delta)$, a Q-linear map $G : Q^p \to Q^n$ is called a* generalized friend *of S if GC is an R-linear map from X to X and $(A + GC)\mathcal{S} \subseteq \mathcal{S}$.*

If we also denote by G the matrix associated (with respect to a given basis) with a generalized friend, then G has entries in Q and the requirement that GC is an R-linear map means that the entries of GC are in $\mathbb{R}[\Delta]$. In order to characterize the existence of generalized friends and to cope with situations in which a given conditioned invariant submodule is not injection invariant, we need to introduce the following notion.

Definition 3. *([5]) Let $\mathcal{S} \subseteq \mathcal{T}$ be submodules of R^n, where R is an integral domain. The* closure *of $\mathcal{S}$ in $\mathcal{T}$ is the submodule defined by*

$$\overline{\mathcal{S}}_{\mathcal{T}} = \{x \in \mathcal{T} \text{ for which there exists } a \in R,\ a \neq 0,\ \text{such that } ax \in \mathcal{S}\}.$$

The submodule $\mathcal{S}$ is said closed *in $\mathcal{T}$ (or simply closed) if $\mathcal{S} = \overline{\mathcal{S}}_{\mathcal{T}}$.*

We can easily prove that $dim\ \mathcal{S} = dim\ \overline{\mathcal{S}}$ and that $\overline{\mathcal{S}}$ is the smallest closed submodule containing $\mathcal{S}$. A key result concerning closed submodules over a PID R is the following.

Proposition 2. *([5]) Let R be a PID. Then the following are equivalent for a submodule $\mathcal{S} \subseteq \mathcal{T}$:*
i) *$\mathcal{S}$ is closed in $\mathcal{T} \subseteq R^n$;*
ii) *$\mathcal{S}$ is a direct summand of $\mathcal{T}$, i.e., there exists a submodule $\mathcal{W} \subseteq \mathcal{T}$ such that $\mathcal{T} = \mathcal{S} \oplus \mathcal{W}$;*
iii) *any basis of $\mathcal{S}$ can be completed to a basis of $\mathcal{T}$.*

A consequence of Proposition 2 and the fact that submodules of a free module over a PID are free (that is, isomorphic to R^q for some q) is that the quotient module $\mathcal{T}/\mathcal{S}$, when $\mathcal{S}$ is closed in $\mathcal{T}$, is also free.

Proposition 3. *Let $\mathcal{S}$ be a closed conditioned invariant submodule for the system Σ of the form (3) over the ring $R = \mathbb{R}[\Delta]$. Then there exists a generalized friend G for $\mathcal{S}$.*

Proof Let $\{s_1, ..., s_{k_1}, ..., s_k, ..., s_n\}$ be a basis for $\mathcal{X}$ such that $\{s_1, ..., s_k\}$ is a basis for $\mathcal{S}$ and $\{s_{k_1+1}, ..., s_k\}$ is a basis for $Ker\ C \cap \mathcal{S}$. Such a basis exists because $Ker\ C \cap \mathcal{S}$ is a direct summand of $\mathcal{S}$ and the latter is a direct summand of $\mathcal{X}$. By adding elements of the canonical basis, complete the set $\{Cs_1, ..., Cs_{k_1}\}$ to a basis $\{Cs_1, ..., Cs_{k_1}, y_{k_1+1}, ..., y_p\}$ of Q^p over the field Q. The map $G : Q^p \to Q^n$ defined by

$$G : Cs_i \longrightarrow -As_i + *, \ \forall i = 1, ..., k_1$$
$$G : y_i \longrightarrow **, \ \forall i = k_1 + 1, ..., p,$$

where $*$ indicates an arbitrary element in $\mathcal{S}$ and $**$ an arbitrary element in R^n, is a generalized friend.

It is not difficult to show that the closure $\overline{\mathcal{S}}$ of a conditioned invariant submodule $\mathcal{S}$ is conditioned invariant (see [6]). Then we may assume without loss of generality that the conditioned invariant submodules we need to deal with are closed and, as a consequence, they admit generalized friends.

3 The UIO problem

Given a delay-differential system Σ_d of the form (1), let us assume that u is an unknown input (the possible presence of additional known inputs is not relevant for our analysis and will therefore be omitted). The *UIO problem* for delay-differential systems consists essentially of finding an observer that estimates the state of the system without assuming the knowledge of u.

A preliminary analysis of the UIO problem is sufficient to understand that, due to the lack of information about u, it will usually not be possible to reconstruct the whole state module and, due to the delay structure of Σ_d, one cannot, in general, expect to avoid delays in the observed variables (think, for instance, of the case in which $y(t) = x(t - h)$ in (1)). We therefore give the following formal definition.

Definition 4. *The UIO problem for a system Σ_d of the form (1) consists of finding a linear, time invariant, delay-differential observer system Σ_o of the same general form as Σ_d, with input y (the output of Σ_d) and state z, $z \in R^n$, together with a subspace $S \subset R^n$ and, possibly, a polynomial $\phi(\delta)$ in the delay operator δ, such that dim S is minimal and the estimation error*

$$e(t) = z(t) - \phi(\delta)x(t)$$

converges to 0 *modulo* S *as* t *goes to infinity.*

The observer will provide an asymptotic estimate of the function $\phi(\delta)x(t)$ of the state $x(t)$ of Σ_d modulo S, namely of the projection of $\phi(\delta)x(t)$ on the subspace $\mathcal{X}/\mathcal{S}$. Clearly, if $\phi(\delta) = \delta^k$, we obtain an estimate of $x(t-kh)$ modulo S, namely, in a suitable basis, an estimate of a component of the state with a delay equal to kh. In the more general case in which $\phi(\delta) = \sum_j^k a_i\delta^i$, we get asymptotically $x(t - jh) = 1/a_j(z(t) - \sum_{j+1}^k a_ix(t - ih))$ modulo S.

If we consider the system Σ, associated to Σ_d, over the ring R, the UIO problem can be translated in a suitable way in the framework of this latter. It has to be remarked, however, that for a linear system with coefficients in R, the notion of asymptotic stability does not make sense, since a ring R cannot, in general, be endowed with a natural metric structure, and therefore the distance of an element from 0 cannot be defined. Convergence and stability must therefore be expressed in formal terms. Since our interest is in the framework of delay-differential systems, this can be done by selecting in $R[s]$, namely the ring of polynomials in the indeterminate s with coefficients in R, a subset $\mathcal{H}$ defined by

$$\mathcal{H} = \{p(s, \Delta) \in R[s] = \mathbb{R}[s, \Delta] \text{ such that}$$
$$p(\bar{s}, e^{-\bar{s}h}) \neq 0 \ \text{ for } \bar{s} \in \mathbf{C}, \ Re(\bar{s}) \geq 0\},$$

whose elements play the role of Hurwitz polynomials. Asymptotic stability of a system Σ_d will then be formally defined as the property of having a dynamic matrix A with $det(sI - A)$ belonging to $\mathcal{H}$. We call $\mathcal{H}$ an Hurwitz set and its elements Hurwitz polynomials.

Definition 5. *The UIO problem for a system* Σ *of the form (3) over* $R = \mathbb{R}[\Delta]$ *consists of finding an observer system* Σ_o *of the same general form as* Σ *over* R, *with input* y *(the output of* Σ*) and state* $z \in R^n$, *together with a submodule* $S \subset R^n$ *and, possibly, an element* $\phi(\Delta) \in R$, *such that* $dim\ S$ *is minimal and the estimation error*

$$e(t) = z(t) - \phi(\Delta)x(t)$$

satisfies modulo S *a dynamic equation of the form*

$$e(t + 1)_{mod S} = E(\Delta)e(t)_{mod S},$$

where $E(\Delta)$ *is a matrix with entries in* $\mathbb{R}[\Delta]$, *with* $det(sI - E(\Delta)) \in \mathcal{H}$.

Clearly, from the above discussion, the condition $det(sI - E(\Delta)) \in \mathcal{H}$ is required for assuring that solutions constructed in the ring framework give rise to solutions in the original delay-differential framework for which the error dynamic

$$\dot{e}(t)_{mod S} = E(\delta)e(t)_{mod S}$$

is asymptotically stable.

4 Problem analysis and solution

Starting from the UIO problem for a system Σ_d, let us translate it into the corresponding UIO problem for the associated system Σ, of the form (3) over the ring $R = \mathbb{R}[\Delta]$, and concentrate on the latter.

In order to construct an observer as desired, let us consider the system Σ_o defined by the equations

$$z(t+1) = (A + GC)z(t) - G\phi(\Delta)y(t), \tag{7}$$

where $\phi(\Delta)$ is a nonzero element of R, and, assuming that the output of Σ_o is equal to its state, let us define the estimation error as

$$e(t) = z(t) - \phi(\Delta)x(t).$$

By remarking that the error satisfies the dynamic equation

$$e(t+1) = (A + GC)e(t) - B\phi(\Delta)u(t), \tag{8}$$

we have that its forced component, which wholly depends on the unknown input u, remains inside the reachable submodule

$$\mathcal{S} = span\{B|\ (A+GC)B|\ldots|\ (A+GC)^{n-1}B\},$$

for any choice of $\phi(\Delta)$.

To limit the influence of the unknown input on the estimation error and to maximize the information about the state of Σ, we must therefore choose G so that $\mathcal{S}$ is as small as possible. Since $\mathcal{S}$ is $(A + GC)$ invariant, i.e., $(A + GC)\mathcal{S} \subseteq \mathcal{S}$, and therefore is conditioned invariant, the best possible choice would be to choose G as a friend of $\mathcal{S}^*(ImB)$, that is, the minimum conditioned invariant submodule for Σ containing ImB. To avoid problems related to the existence of friends and to the definition of quotient dynamics, let us assume that $\mathcal{S}^*(ImB)$ is closed (otherwise, we take its closure) and let us chose a (generalized) friend G of $\mathcal{S}^*(ImB)$ and a polynomial $\phi(\Delta)$ equal to the minimum common multiple of the denominators in G.

In this way, we can ensure that the system (7) has all its coefficients in R. In other words, this choice of $\phi(\Delta)$ avoids the use of prediction of the output $y(t)$ in the estimation of $x(t)$, at the price of delays in the observed variables. With these ingredients we can state the following result.

Proposition 4. *Given a system Σ of the form (3) over $R = \mathbb{R}[\Delta]$, let Σ_o be an observer constructed as above by means of a (generalized) friend G of $\mathcal{S}^*(ImB)$. Then, if $z(0) = \phi(\Delta)x(0)$, letting $e(t)_{mod\mathcal{S}^*(ImB)} = (z(t) - \phi(\Delta)x(t))_{mod\mathcal{S}^*(ImB)}$, we have $e(t)_{mod\mathcal{S}^*(ImB)} = 0$ for all $t \geq 0$. Moreover, $\mathcal{S}^*(ImB)$ is the smallest submodule of $\mathcal{X}$ for which this happens.*

Proof It follows from the linearity of Σ_o and the minimality of $\mathcal{S}^*(ImB)$.

The meaning of Proposition 4 is that, modulo $\mathcal{S}^*(ImB)$, the state of the observer, if this is suitably initialized, follows exactly that component of $\phi(\Delta)x(t)$ which does not depend on $u(t)$. This result represents a first step in our analysis of the UIO problem and can clearly be applied to the original framework of delay-differential systems. We need to introduce another geometric object in order to analyze the performances of the observer in case initialization is generic. Given the system Σ of the form (3) over R, we denote by $\mathcal{V}^*(\mathcal{S}^*)$ the maximum submodule $\mathcal{V}$ of $Ker\ C$ such that

$$A\mathcal{V}^* \subseteq \mathcal{V}^* + \mathcal{S}^*(ImB). \tag{9}$$

In geometric terms, $V^*(\mathcal{S}^*)$ is a controlled invariant submodule with respect to $\mathcal{S}^*(ImB)$. Notice that $\mathcal{V}^*$ can be computed, for instance, by means of the algorithm described in [3]. Let us denote $V^*(\mathcal{S}^*)$ and $\mathcal{S}^*(ImB)$ simply by $\mathcal{V}^*$ and $\mathcal{S}^*$ and remark that the following result holds.

Lemma 1. *In the above hypothesis and with the above notation, we have* $\mathcal{V}^* \cap \mathcal{S}^* = Ker\ C \cap \mathcal{S}^*$.

Proof Since, by definition, $\mathcal{V}^* \subset Ker\ C$, one inclusion is obvious. For the other one, take $x \in Ker\ C \cap \mathcal{S}^*$. Since $\mathcal{S}^*$ is (A,C) invariant, we have $Ax \in \mathcal{S}^*$ and thus $A(\mathcal{V}^* + x) \subseteq \mathcal{V}^* + \mathcal{S}^* = (\mathcal{V}^* + x) + \mathcal{S}^*$. The conclusion follows by the maximality of $\mathcal{V}^*$.

Now, to proceed in analyzing the UIO problem, we need to introduce the following technical assumption.

Assumption 1. *With the above notation, assume that the submodule* $\mathcal{V}^* + \mathcal{S}^*$ *is closed in* $\mathcal{X}$.

Since $\mathcal{V}^* \cap \mathcal{S}^*$ turns out to be closed in both $\mathcal{S}^*$ and $\mathcal{V}^*$, it is possible to choose a basis matrix $T = \begin{bmatrix} T_1 & T_2 & T_3 & T_4 \end{bmatrix}$ for $\mathcal{X}$ such that

$$\begin{aligned} Im[T_4] &= \mathcal{V}^* \cap \mathcal{S}^* & Im\begin{bmatrix} T_3 & T_4 \end{bmatrix} &= \mathcal{S}^* \\ Im\begin{bmatrix} T_2 & T_4 \end{bmatrix} &= \mathcal{V}^* & Im\begin{bmatrix} T_2 & T_3 & T_4 \end{bmatrix} &= \mathcal{S}^* + \mathcal{V}^*. \end{aligned}$$

In such basis, we write A and C as

$$A = \begin{bmatrix} A_{11} & A_{12} & A_{13} & A_{14} \\ A_{21} & A_{22} & A_{23} & A_{24} \\ A_{31} & A_{32} & A_{33} & A_{34} \\ A_{41} & A_{42} & A_{43} & A_{44} \end{bmatrix}, \quad C = \begin{bmatrix} C_1 & C_2 & C_3 & C_4 \end{bmatrix}.$$

Since $A(\mathcal{S}^* \cap Ker\ C) \subseteq \mathcal{S}^*$, we have

$$A(0,0,0,x_4)^t = (A_{14}x_4, A_{24}x_4, *, *) \in \mathcal{S}^*$$

for any choice of x_4, and thus $A_{14} = A_{24} = 0$. Since $A\mathcal{V}^* \subseteq \mathcal{V}^* + \mathcal{S}^*$, we have

$$A(0,x_2,0,x_4)^t = (A_{12}x_2 + A_{14}x_4, *, *, *) \in \mathcal{V}^* + \mathcal{S}^*$$

for any choice of x_2 and x_4, and thus $A_{12} = A_{14} = 0$. Moreover, since $\mathcal{V}^* \subseteq$ Ker C we have

$$C(0, x_2, 0, x_4)^t = C_2 x_2 + C_4 x_4 = 0$$

for any choice of x_2 and x_4, and thus $C_2 = C_4 = 0$. To better display the structure of the matrix C, let us write $\mathcal{Y} = \mathcal{W} \oplus \overline{C\mathcal{S}^*}$ for some direct summand $\mathcal{W}$. We have $dim C\mathcal{S}^* = dim\overline{C\mathcal{S}^*} = dim(Im\ T_3)$ since $Im\ T_3$ has no intersection with $Ker\ C$, and we can write, accordingly,

$$C = \begin{bmatrix} C_{11} & 0 & 0 & 0 \\ C_{21} & 0 & C_{23} & 0 \end{bmatrix},$$

with C_{23} square and nonsingular. The structure of C can be made even simpler if we make the following technical assumption (which is always satisfied over a field).

Assumption 2. *With the above notation, assume that there exist matrices K_1 and K_2 with entries in R such that $C_{21} = C_{23}K_1 + K_2 C_{11}$.*

Under Assumption 2, we can assume $C_{21} = 0$, since, otherwise, we can obtain this by applying the changes of basis in $\mathcal{X}$ and $\mathcal{Y}$ defined, respectively, by $x' = T_1 x$, and $y' = T_2 y$, with

$$T_1 = \begin{bmatrix} I & 0 & 0 & 0 \\ 0 & I & 0 & 0 \\ -K_1 & 0 & I & 0 \\ 0 & 0 & 0 & I \end{bmatrix}, \quad T_2 = \begin{bmatrix} I & 0 \\ -K_2 & I \end{bmatrix},$$

without modifying the structure of A and C and the location of their zero blocks (compare with 4.1.4 in [4]). Let the output injection G be a (generalized) friend of $\mathcal{S}^*$, where

$$G = \begin{bmatrix} g_{11} & g_{12} \\ g_{21} & g_{22} \\ g_{31} & g_{32} \\ g_{41} & g_{42} \end{bmatrix},$$

and compute the dynamic matrix $A + GC$ of the observer (7). Due to $(A + GC)\mathcal{S}^* \subseteq \mathcal{S}^*$, it turns out that $A_{13} + g_{12}C_{23} = 0$ and $A_{23} + g_{22}C_{23} = 0$, therefore one gets

$$A + GC = \begin{bmatrix} A_{11} + g_{11}C_{11} & 0 & 0 & 0 \\ A_{21} + g_{21}C_{11} & A_{22} & 0 & 0 \\ A_{31} + g_{31}C_{11} & A_{32} & A_{33} + g_{32}C_{23} & A_{34} \\ A_{41} + g_{41}C_{11} & A_{42} & A_{43} + g_{42}C_{23} & A_{44} \end{bmatrix}.$$

Taking the quotient modulo $\mathcal{S}^*$, the error equation (8) gives $e(t+1)_{mod\mathcal{S}^*} = \tilde{A}e(t)_{mod\mathcal{S}^*}$, where

$$\tilde{A} = (A + GC)|_{mod\ \mathcal{S}^*} = \begin{bmatrix} A_{11} + g_{11}C_{11} & 0 \\ A_{21} + g_{21}C_{11} & A_{22} \end{bmatrix}$$

and we see that there is a part of the error dynamics, corresponding to A_{22}, that cannot be modified by any choice of the friend G. So we can state the following proposition.

Proposition 5. *A necessary condition for the existence of solutions to the UIO problem for the system Σ of the form (3) over R is, with the above notation, that $\det(sI - A_{22})$ belongs to $\mathcal{H}$.*

Moreover, we have the following fundamental result.

Proposition 6. *If T_1 is not empty, the pair (A_{11}, C_{11}) is weakly observable, that is, the following matrix is full rank:*

$$\begin{bmatrix} C_{11} \\ C_{11}A_{11} \\ \cdots \\ C_{11}A_{11}^{n-1} \end{bmatrix}.$$

Proof Denote by $\pi : \mathcal{X} \to \mathcal{X}/\mathcal{S}^*$ the canonical projection and let $\tilde{C}$ be given by $\tilde{C} = [C_{11} \ 0]$. The submodule of $\mathcal{X}/\mathcal{S}^*$ given by $\pi(Im \ T_2)$ is contained into the nonobservable submodule

$$\mathcal{O} = \text{Ker} \begin{bmatrix} \tilde{C} \\ \tilde{C}\tilde{A} \\ \cdots \\ \tilde{C}\tilde{A}^{n-1} \end{bmatrix},$$

since it is contained in $Ker \ \tilde{C}$ and it is $\tilde{A}$ invariant. We want to show that $\pi(Im \ T_2)$ coincides with $\mathcal{O}$ and, to this aim, we consider a generic element $\pi(x) \in \mathcal{O}$. We can assume that $x = (x_1, x_2, 0, 0)^t$ and we have that

$$0 = \begin{bmatrix} \tilde{C}\tilde{A}^n \ 0 \ 0 \\ 0 \quad 0 \ 0 \end{bmatrix} x = C \begin{bmatrix} \tilde{A} \ 0 \ 0 \\ 0 \ 0 \ 0 \end{bmatrix}^n x$$

for all $n \geq 0$. It follows that the submodule V of $\mathcal{X}$ obtained by adding to V^* the span of $\left\{ \begin{bmatrix} \tilde{A} \ 0 \ 0 \\ 0 \ 0 \ 0 \end{bmatrix}^n x, \ for \ all \ n \geq 0 \right\}$ is contained in $Ker \ C$. Since, by construction, V contains V^* and it is easily seen to verify the condition $AV \subseteq V + \mathcal{S}^*(ImB)$, by the maximality of V^*, it coincides with the latter. Hence, $x_1 = 0$ and the result follows.

As a result of Proposition 6, we find that there is a part of the error dynamics, corresponding to A_{11}, that can possibly, under additional conditions, be modified by a suitable choice of the component g_{11} in G (note that the achievement of the condition $(A + GC)\mathcal{S}^* \subseteq \mathcal{S}^*$ puts constraints only on the components g_{12} and g_{22}, letting the remaining ones, and in particular g_{11}, free).

Now, the problem of satisfying the condition $det(sI - (A_{11} + g_{11}C_{11})) \in \mathcal{H}$, as required to solve the UIO problem, can be reduced to that of assigning the coefficients of $det(sI - (A_{11} + g_{11}C_{11}))$ and we can therefore invoke the results concerning the latter. By transposing the matrices involved, the problem is transformed into a Feedback Coefficient Assignment problem (see, e.g., [12]). This problem, in turn, without requiring more than the weak observability granted by Proposition 6, can be dealt with as described in [22] and [10]. Then, not all the coefficients of $det(sI - (A_{11} + g_{11}C_{11}))$ can arbitrarily be assigned and, therefore, the UIO problem is not solvable.

If we assume a stronger condition on the pair (A_{11}, C_{11}), namely if we assume that it is observable and not only weak observable, we have that, allowing a dynamical extension of the original system, all the coefficients can be arbitrarily assigned (see [12] for a discussion and a survey of available results).

Example 3. Consider a delay-differential system Σ_d of the form (1) and the associated system Σ over $\mathbb{R}[\Delta]$ of the form (3), where

$$A = \begin{bmatrix} p_1(\Delta) & 0 & 0 \\ 0 & p_2(\Delta) & 0 \\ \Delta & 1 & 1 \end{bmatrix}, \quad B = \begin{bmatrix} 0 \\ 0 \\ 1 \end{bmatrix}, \quad C = \sum_{i=0}^{c} C_i \Delta^i = \begin{bmatrix} 1 & 0 & 0 \\ 0 & 0 & 1 \end{bmatrix},$$

with $p_1(\Delta), p_2(\Delta) \in R$. We have $Ker\ C = span(0,1,0)^t$, then $Im\ B$ is conditionally invariant and it coincides with $\mathcal{S}^*$. Moreover, $A(Ker\ C) \subseteq Ker\ C + \mathcal{S}^*$, hence $Ker\ C = \mathcal{W}^*$ and $\mathcal{W}^* + \mathcal{S}^*$ is closed. Both Assumptions 1 and 2 are satisfied and we choose $T_1 = (1,0,0)^t$, $T_2 = (0,1,0)^t$, $T_3 = (0,0,1)^t$, while T_4 is empty. In $\mathcal{Y} = R^2$ we have $\overline{C\mathcal{S}^*} = C\mathcal{S}^* = span\{(0,1)^t\}$ and we choose $\mathcal{W} = span\{(1,0)^t\}$. The map $G = \begin{bmatrix} g_{11} & g_{12} \\ g_{21} & g_{22} \\ g_{31} & g_{32} \end{bmatrix}$ is a friend of $\mathcal{S}^*$ only if $g_{12} = g_{22} = 0$. Then,

$$A + GC = \begin{bmatrix} p_1(\Delta) + g_{11} & 0 & 0 \\ g_{21} & p_2(\Delta) & 0 \\ \Delta + g_{31} & 1 & 1 + g_{32} \end{bmatrix}$$

and $\phi(\Delta) = 1$, so that, in this case, no delay in the observed variables arises. Choosing, for instance, $g_{11} = -p_1(\Delta) - 1$, $g_{21} = 0$, $g_{31} = -\Delta$, $g_{32} = -2$, the observer equation in the delay-differential framework reads as

$$\begin{cases} \dot{z}_1(t) = -z_1(t) + p_1(\delta)y_1(t) + y_1(t) \\ \dot{z}_2(t) = p_2(\delta)z_2(t) \\ \dot{z}_3(t) = z_2(t) - z_3(t) + y_1(t-h) + 2y_2(t) \end{cases} \tag{10}$$

and the dynamic matrix of the estimation error modulo $\mathcal{S}^*$ is given by

$$(A + GC)|_{mod\ \mathcal{S}^*} = \begin{bmatrix} -1 & 0 \\ 0 & p_2(\delta) \end{bmatrix}.$$

Hence, the estimation error modulo S^* goes asymptotically to zero and the UIO problem is solvable (allowing us to estimate asymptotically the first two components of the state $x = (x_1, x_2, x_3)^t$ of Σ_d) if and only if $(sI - p_2(\Delta))$ is an element of the Hurwitz set $\mathcal{H}$.

5 The RG problem

Let us consider now the linear, time invariant, delay-differential, dynamical system Σ_d be described by the set of equations

$$\begin{cases} \dot{x}(t) = \sum_{i=0}^a A_i x(t - ih) + \sum_{i=0}^b B_i u(t - ih) + \\ \qquad + \sum_{i=0}^{l_1} L_i^1 m_1(t - ih) + \cdots \sum_{i=0}^{l_s} L_i^s m_s(t - ih) \\ \\ y(t) = \sum_{i=0}^c C_i x(t - ih), \end{cases} \qquad (11)$$

where u is the control input, m_j belongs to the so-called failure input space $\mathcal{M}_j = \mathbb{R}^{m_j}$, $j = 1, \ldots, s$, and L_i^j, $j = 1, \ldots, r$, $i = 1, \ldots, l_j$ are matrices of suitable dimensions with entries in $\mathbb{R}$. The occurrence of a nonzero input $m_j \in \mathcal{M}_j$ represents a failure that has to be detected and recognized. This motivates the interest in constructing, if possible, a system Σ_{RGd} that, having as input the control input of Σ_d and its output, generates an output $r(t)$ that is essentially influenced only by the failure input m_j for each specific $j = 1, \ldots, s$. Such a system is called a *residual generator* and its output is called a *residual output* (see [15]). The problem of constructing a residual generator takes the name of the RG problem. For simplicity we will consider here the case in which only two kind of failures are present, that is, $j = 1, 2$.

Definition 6. *The RG problem for a system Σ_d of the form (11), with $j = 1, 2$, consists of finding a linear, time invariant, delay-differential system Σ_{RGd} of the same general form as Σ_d with input $u \in \mathcal{U} = \mathbb{R}^m$ and $y \in \mathcal{Y} = \mathbb{R}^p$ that, in response to $(u(t), y(t))$, where $u(t)$ is the input signal to Σ_d and $y(t)$ the corresponding output signal, produces an output $r(t)$, $r \in \mathcal{R} = \mathbb{R}^r$, with the following properties:*
- *$r(t)$ is independent from m_2,*
- *$r(t)$ is asymptotically zero if $m_1(t) = 0$ for every t,*
- *there exists $k \geq 0$ such that $\dfrac{d}{dm_1}\left(\dfrac{d^k r(t)}{dt^k}\right) \neq 0$.*

Remark that, in general, the dimension of the residual generator Σ_{RGd} is smaller then n, due to the lack of information on m_1 and m_2, which can be viewed as unknown inputs.

The system Σ, associated to Σ_d, over the ring R, takes the form

$$\Sigma = \begin{cases} x(t + 1) = Ax(t) + Bu(t) + L_1 m_1(t) + \cdots + L_s m_s(t) \\ \\ y(t) = Cx(t), \end{cases} \qquad (12)$$

where, by abuse of notation, we denote by x an element of the free state module $\mathcal{X} = R^n$, by u an element of the free input module $\mathcal{U} = R^m$, by y an element of the free output module $\mathcal{Y} = R^p$, by m_i an element of the free failure module $\mathcal{M}_j = R^{m_j}$, $j = 1, \ldots, r$ and where A; B; $L_j, j = 1, \ldots, r$; C are matrices with entries in $R = \mathbb{R}[\Delta]$, given, respectively, by $A = \sum_{i=0}^{a} A_i \Delta^i$; $B = \sum_{i=0}^{b} B_i \Delta^i$; $L_j = \sum_{i=0}^{l_j} L_{1,i} \Delta^i$, $j = 1, \ldots, r$; $C = \sum_{i=0}^{c} C_i \Delta^i$. When $j = 1, 2$, the RG problem can be reformulated for Σ as follows.

Definition 7. *The RG problem for a system Σ of the form (12) over $R = \mathbb{R}[\Delta]$ consists of finding a residual generator Σ_{RG} over R of the form*

$$\begin{cases} w(t+1) = Kw(t) + \widetilde{G}y(t) + \widetilde{B}u(t) \\ \quad r(t) = Mw(t) + \widetilde{H}y(t), \end{cases} \tag{13}$$

with the following properties:
- $r(t)$ *is independent from* m_2,
- $det(sI - K) \in \mathcal{H}$,
- *there exists* $k \geq 0$ *such that* $r(t+k)$ *depends linearly on* m_1.

The linear dependency of $r(t+k)$ on m_1 allows one to detect the presence of nonzero values in the signal $m_1(t)$.

In order to construct a residual generator of the form (13) over R for a system Σ with only two failure events, let us consider an observer Σ_{obs} for Σ, defined, similarly to (7), by equations of the form

$$z(t+1) = (A + GC)z(t) + G\varphi(\Delta)y(t) + B\varphi(\Delta)u(t), \tag{14}$$

with $z \in R^n$, where G is a matrix with entries in the quotient field Q of R such that GC has entries in R and $\varphi(\Delta) \in R$ is such that $G\varphi(\Delta)$ has entries in R. The observation error $e(t) = z(t) - \varphi(\Delta)x(t)$ is easily seen to satisfy the dynamic equation

$$e(t+1) = (A + GC)e(t) - L_1\varphi(\Delta)m_1(t) - L_2\varphi(\Delta)m_2(t), \tag{15}$$

which shows that the forced component depending on the unknown input m_2 remains inside the reachable submodule $\mathcal{S} = span\{L_2|(A+GC)L_2|\ldots|(A+GC)^{n-1}L_2\}$ for any choice of $\varphi(\Delta)$.

Then, as already remarked in the construction of the solution to the UIO problem, in order to limit the influence of the unknown input m_2 on the observation error and to maximize the information about the state of Σ, in constructing Σ_{obs} we must choose G so that $\mathcal{S}$ is as small as possible. Since $\mathcal{S}$ is $(A + GC)$ invariant, i.e., $(A + GC)\mathcal{S} \subseteq \mathcal{S}$, the best one can do is to choose G as a generalized friend of $\mathcal{S}^*(Im\ L_2)$, namely the minimum conditioned invariant submodule for Σ containing $Im\ L_2$. This choice, in fact, will make $\mathcal{S}$ equal to $\mathcal{S}^*(Im\ L_2)$, which, if no confusion arises, we will denote in the following simply by $\mathcal{S}^*$.

Taking $\varphi(\Delta)$ equal to the minimum common multiple of the denominators of the entries of G, we can assure that the observer Σ_{obs} given by (14) is a system over R.

In order to proceed in the construction of a residual generator Σ_{RD}, let us consider the basis matrix $T = \begin{bmatrix} T_1 \ T_2 \ T_3 \ T_4 \end{bmatrix}$ for $\mathcal{X}$ such that $Im \begin{bmatrix} T_4 \end{bmatrix} = \mathcal{V}^* \cap \mathcal{S}^*$, $Im \begin{bmatrix} T_3 \ T_4 \end{bmatrix} = \mathcal{S}^*$, $Im \begin{bmatrix} T_2 \ T_4 \end{bmatrix} = \mathcal{V}^*$, and $Im \begin{bmatrix} T_2 \ T_3 \ T_4 \end{bmatrix} = \mathcal{V}^* + \mathcal{S}^*$. In such basis, as we have seen in the previous section, the dynamical matrix $A + GC$ of Σ_{obs} takes the form

$$A + GC = \begin{bmatrix} \widetilde{A}_{11} & 0 & 0 & 0 \\ \widetilde{A}_{21} & A_{22} & 0 & 0 \\ \widetilde{A}_{31} & A_{32} & \widetilde{A}_{33} & A_{34} \\ \widetilde{A}_{41} & A_{42} & \widetilde{A}_{43} & A_{44} \end{bmatrix},$$

with $\widetilde{A}_{i1} = A_{i1} + g_{i1}C_{11}$ for $i = 1, \dots, 4$ and $\widetilde{A}_{i3} = A_{i3} + g_{i2}C_{23}$ for $i = 3, 4$.

Now, writing $H = \begin{bmatrix} I_{k \times k} \ 0_{k \times (p-k)} \end{bmatrix}$, where $k = dim\ \mathcal{W}$, and, using the same notation as in Section 4, we have the following result.

Proposition 7. *The submodule* $(\mathcal{V}^* + \mathcal{S}^*)$ *is the submodule of unobservable states of the system* Σ' *defined by the equations*

$$\Sigma' = \begin{cases} z(t+1) = (A + GC)z(t) \\ v(t) = HCx(t). \end{cases} \tag{16}$$

Proof Remark that, by $HC = \begin{bmatrix} C_{11} \ 0 \ 0 \ 0 \end{bmatrix}$, $(\mathcal{V}^* + \mathcal{S}^*)$ is contained in $Ker\ HC$ and it is $(A + GC)$ invariant, since $(A + GC)\mathcal{N} = A\mathcal{V}^* + (A + GC)\mathcal{S}^* \subseteq \mathcal{V}^* + \mathcal{S}^* = \mathcal{N}$. Maximality follows from the properties of $\mathcal{V}^*$ and of $\mathcal{S}^*$.

Assume that $h = dim\ A_{11}$ and denote by $P : \mathcal{X} \to \mathcal{X}/\mathcal{N} = R^h$ the canonical projection on the quotient module and by M the unique map such that $MP = HC$. Then, letting $\widetilde{A}_{11} = (A_{11} + g_{11}C_{11})$, so that $P(A + GC) = \widetilde{A}_{11}P$, we have the following result.

Proposition 8. *Let* Σ_{RG} *be the system defined by the following equations*

$$\begin{cases} w(t+1) = \widetilde{A}_{11}w(t) - PG\varphi(\Delta)y(t) + PB\varphi(\Delta)u(t) \\ r(t) \quad = Mw(t) - H\varphi(\Delta)y(t), \end{cases} \tag{17}$$

where $w(t) = Pz(t) \in R^h$. *If, with the above notation, the condition*

$$(\mathcal{V}^* + \mathcal{S}^*) \bigcap (Im\ L_1) = 0$$

holds, then Σ_{RG} *initialized at* $w(0) = \varphi(\Delta)Px(0)$ *generates a residual output* $r(t)$ *for the system (12) that is independent from the failure* m_2 *and such that, for some* $k \geq 0$, $r(t + k)$ *depends linearly on* m_1.

Proof Let us define the error $\widetilde{e}(t)$ as $\widetilde{e}(t) = Pe(t) = w(t) - \varphi(\Delta)Px(t)$ and remark that, from (15) and from the fact that $Im\ L_2 \subseteq \mathcal{N}$ implies $PL_2 = 0$, we have

$$\widetilde{e}(t+1) = Pe(t+1) = P(A+GC)e(t) - \varphi(\Delta)PL_1 m_1(t) =$$
$$= \widetilde{A}_{11} Pe(t) - \varphi(\Delta)PL_1 m_1(t) = \widetilde{A}_{11}\widetilde{e}(t) - \varphi(\Delta)PL_1 m_1(t).$$

Hence, the residual output $r(t)$ verifies the relation

$$r(t) = MPz(t) - \varphi(\Delta)HCx(t)(t) = MPe(t) = M\widetilde{e}(t)$$

and the conclusion follows.

This result can be straightforwardly translated into the framework of delay-differential systems, providing, in the case of correct initialization, a residual generator Σ_{RGd} of the form

$$\Sigma_{RGd} = \begin{cases} \dot{w}(t) = Kw(t) + \widetilde{G}y(t) + \widetilde{B}u(t) \\ r(t) = Mw(t) + \widetilde{H}y(t), \end{cases}$$

for the system Σ_d given by (11). In case Σ_{RGd} cannot be correctly initialized, the behavior of the residual output $r(t)$ in the delay-differential framework depends on the stability of Σ_{RGd}, that is, on the fact that $det(sI - \widetilde{A}_{11}) = det(sI - (A_{11} - g_{11}C_{11}))$ belongs to the Hurwitz set $\mathcal{H}$. The stability issue is related to the possibility of assigning the coefficients of the characteristic polynomial of $A_{11}+g_{11}C_{11}$ by means of a suitable choice of g_{11}, and the same results discussed in the previous section apply here.

Example 4. Consider a system Σ of the form (12) over R with, in particular, $s = 2$ and

$$A = \begin{bmatrix} p_1(\Delta) & 0 & 0 & 0 \\ 0 & p_2(\Delta) & 0 & 0 \\ 1 & 1 & p_3(\Delta) & 0 \\ 1 & 1 & p_4(\Delta) & 1 \end{bmatrix}, \ L_2 = \begin{bmatrix} 0 \\ 0 \\ 0 \\ 1 \end{bmatrix} \text{ and } C = \begin{bmatrix} 1 & 0 & 0 & 0 \\ 0 & 0 & 0 & 1 \\ 0 & 1 & 0 & 0 \end{bmatrix}.$$

We have $Ker\ C = span\{(0,0,1,0)^t\}$, hence $Ker\ C \cap ImL_2 = 0$. As a consequence, ImL_2 is conditioned invariant and it coincides with $\mathcal{S}^*$. Moreover, $A(Ker\ C) = span(0,0,p(\Delta),\Delta)^t) \subseteq Ker\ C + \mathcal{S}^*$, thus $Ker\ C = \mathcal{V}^*$ and $(\mathcal{V}^* + \mathcal{S}^*)$ is closed in $\mathcal{X}$. Now, we choose $T_1 = \{(1,0,0,0)^t; (0,1,0,0)^t\}$, $T_2 = (0,0,1,0)^t$, $T_3 = (0,0,0,1)^t$, while T_4 is empty (actually $T = I_{4\times4}$). In $\mathcal{Y} = (\mathbb{R}[\Delta])^3$ we have $\overline{C\mathcal{S}^*} = C\mathcal{S}^* = span\{(0,1,0)^t\}$, so that Assumption 2 is satisfied, and, choosing $\mathcal{W} = span\{(1,0,0)^t, (0,0,1)^t\}$, we can rewrite the matrix C (with a slight abuse of notation) as

$$C = \begin{bmatrix} 1 & 0 & 0 & 0 \\ 0 & 1 & 0 & 0 \\ 0 & 0 & 0 & 1 \end{bmatrix}.$$

If, for instance, L_1 is such that $(\mathcal{V}^*+\mathcal{S}^*)\cap(Im\ L_1) = 0$ holds, we can construct an observer system of the form (14) choosing G as

$$G = \begin{bmatrix} \gamma_{11} & \gamma_{12} & 0 \\ \gamma_{21} & \gamma_{22} & 0 \\ \gamma_{31} & \gamma_{32} & 0 \\ \gamma_{41} & \gamma_{42} & \gamma_{43} \end{bmatrix}.$$

Remark that, since the elements of GC must belong to $R[\Delta]$, $\varphi(\Delta) = 1$. Letting $H = \begin{bmatrix} 1 & 0 & 0 \\ 0 & 1 & 0 \end{bmatrix}$, we get the residual generator

$$\Sigma_{RG} = \begin{cases} \dot{w}(t) = Kw(t) + \tilde{G}y(t) \\ r(t) = w(t) + \tilde{H}y(t), \end{cases}$$

where

$$K := A_{11} = \begin{bmatrix} p_1(\Delta) + \gamma_{11} & \gamma_{12} \\ \gamma_{21} & p_2(\Delta) + \gamma_{22} \end{bmatrix},$$

$\tilde{G} := -PG$, $\tilde{H} = -H$. If we take, for instance, $\gamma_{12} = \gamma_{21} = 0$, $\gamma_{11} = -1 - p_1(\Delta)$ and $\gamma_{22} = -1 - p_2(\Delta)$, Σ_{RG} solves the FPRG for Σ. Remark that in this case, there exists a stable observer for Σ only if $p_3(\Delta)$ is in $\mathcal{H}$.

6 Conclusion

It has been shown that, by exploiting the natural correspondence between delay-differential systems and systems over rings, the UIO problem and the related RG problem for time delay systems can be dealt with by means of algebraic and geometric techniques. This choice provides feasible procedures for tackling both problems, analyzing possible obstructions to the existence of solutions and, if possible, constructing solutions.

References

1. Assan J (1999) Analyse et synthèse de l'approche géométrique pour les systèmes linéaires su un anneau, Thèse de Doctorat, Université de Nantes
2. Assan J, Lafay JF, Perdon AM (2002) Feedback invariance and injection invariance for systems over rings, Proc. 15th IFAC World Congress, Barcelona, Spain
3. Assan J, Lafay JF, Perdon AM, Loiseau JJ (1999) Effective computation of maximal controlled invariant submodules over a Principal Ring, Proc. 38th IEEE CDC, Phoenix, AZ, 4216–4221
4. G. Basile, G. Marro (1992) Controlled and conditioned invariants in linear system theory, Prentice-Hall, Upper Saddle River, NJ
5. Conte G, Perdon AM (1982) Systems over a principal ideal domain. A polynomial model approach, SIAM Journal on Control and Optimization 20(1):112–124

6. Conte G, Perdon AM (1997) Noninteracting control problems for delay-differential systems via systems over rings, European Journal of Automation 31(6):1059–1076
7. Conte G, Perdon AM (2000) Systems over rings: geometric theory and applications Annual Reviews in Control 24(1):113–124
8. Conte G, Perdon AM (2003) The fundamental problem of residual generation linear time delay systems, IFAC workshop on Time Delay Systems, Rocquencourt, France
9. Conte G, Perdon AM, Guidone-Peroli G (2003) Unknown input observers for linear delay systems: a geometric approach, Proc. 42nd IEEE CDC, Maui, HI
10. Conte G, Perdon AM, Neri F (2001) Remarks and results about the finite spectrum assignment problem, Proc. 40th IEEE CDC, Orlando, FL
11. Sename O, Fattouh A, Dion JM (2001) Further results on unknown input observers design for time-delay systems, Proc. 40th IEEE CDC, Orlando, FL
12. Kamen EW (1991) Linear system over rings: from R. E. Kalman to the present, Mathematical System Theory–The Influence of R. E. Kalman, Springer-Verlag, Berlin
13. Ito N, Schmale W, Wimmer HK (2000) (C,A)-invariance of modules over principal ideal domains, SIAM Journal on Control and Optimization 38(6):1859–1873
14. Lang S (1984) Algebra, Addison-Wesley, Reading, MA
15. Massoumnia MA (1986) A geometric approach to the synthesis of failure detection filters, IEEE Trans. Automatic Control, 34:316–321
16. Proc. of the 1st IFAC Workshop on Linear Time Delay Systems (1998) Grenoble, France
17. Proc. of the 2nd IFAC Workshop on Linear Time Delay Systems (2000) Ancona, Italy
18. Proc. of the 3rd IFAC Workshop on Linear Time Delay Systems (2001) Albuquerque, NM
19. Proc. of the 4th IFAC Workshop on Linear Time Delay Systems (2003) Rocquencourt, France
20. Patton RJ, Frank PM, Clark RN (2000) Issues of fault diagnosis for dynamical systems, Springer-Verlag, New York
21. Picard P (1996) Sur l'observabilité et la commande des systèmes linéaires à retards modélisés sur un anneau, Thèse de Doctorat, Université de Nantes
22. Sename O, Lafay JF (1996) A new result on coefficient assignment for linear multivariable systems with delays, Proc. 35th IEEE CDC, New Orleans, LA
23. Szabó Z, Bokor J, Balasy G (2002) Detection filter design for LPV systems–A geometric approach, 15th IFAC Triennial World Congress, Barcelona, Spain

Set Membership Identification: The H_∞ Case*

Mario Milanese and Michele Taragna

Dipartimento di Automatica e Informatica
Politecnico di Torino
Corso Duca degli Abruzzi, 24
10129 Torino, Italia
{mario.milanese,michele.taragna}@polito.it

Summary. Robustness had become a central issue in system and control theory, focusing the researchers' attention from the study of a single model to the investigation of a set of models, described by a set of perturbations of a "nominal" model. This set, often indicated as the uncertainty model set, has to be suitably constructed to describe the inherent uncertainty about the system under consideration and to be used for analysis and design purposes. H_∞ identification methods deliver uncertainty model sets in a form suitable to be used by well-established robust design techniques, based on H_∞ or μ optimization methods. The literature on H_∞ identification is now very extensive. Some of the most relevant contributions related to assumption validation, evaluation of bounds on unmodeled dynamics, convergence analysis, and optimality properties of different algorithms are here surveyed from a deterministic point of view.

1 Introduction

A general problem appearing in many scientific and technical fields is to make some kind of inference on a dynamical system S^o, starting from some general information on it and from a finite number of noisy measurements. Typical examples of inference are smoothing, filtering, prediction, control design, decision making, fault detection, and diagnosis. The usual approach is to estimate a model M of S^o and to make the inference on M. Due to the unavoidable discrepancies between the identified model M and the actual system S^o, it is of paramount importance to evaluate the inference error, i.e., the error in making inference on M instead of S^o.

Consider in particular the control design problem. Typically, the system to be controlled is not completely known and a control law has to be designed, able to drive the plant to reach, if possible, given performance specifications.

---

*This research was supported in part by funds of Ministero dell'Università e della Ricerca Scientifica e Tecnologica under the Project "Robustness and optimization techniques for control of uncertain systems".

The classical approach consists of building a mathematical model of the plant, on the basis of the available information (priors and measurements), and then designing a control that meets the desired performance specifications for the identified model. However, this method does not take into account that any identified model is only an approximation of the actual system. Indeed, the performances actually achievable on the plant may be very poor, according to the size of the modeling error, and even the closed-loop stability may be missed. In order to face these problems, robustness had become in recent years a central issue in system and control theory, focusing the researchers' attention from the study of a single model M to the investigation of a set of models $\mathcal{M}$. Such a set, often indicated as *uncertainty model set* or model set for short, has to be suitably constructed to describe the inherent uncertainty about the system under consideration and to be used for analysis and design purposes. Typically, model sets are described by a set of perturbations of a "nominal" model. Thus, in the control problem, the inference error corresponds to the difference between the performances predicted on the nominal model and those actually achievable on the real system. Since the real system is not known, the actually achievable performances are unknown and then those guaranteed for all systems belonging to the model set are used instead.

Since data are corrupted by noise and provide only limited information, no finite bound on the inference error can be derived if no prior assumption on S^o and on noise is made. In particular, some information on S^o is required, e.g., by assuming that S^o belongs to some subset K of dynamical systems. Different theories have been developed, according to different assumptions on the set K and the noise.

Classical statistical identification theory gives deep and extensive results for the case that K is a set of parametric models $M(p)$, $p \in \Re^q$, and that noise is stochastic with known probability density function, possibly filtered by a parametric noise model. As a matter of fact, in most practical applications there exists no p^o such that $S^o = M(p^o)$ and the problem arises of considering that only approximated models can be estimated and that the effects of un-modeled dynamics have to be accounted for. This appears to be a formidable problem. Some results are available in statistical identification and learning literature, essentially related to asymptotic analysis, see, e.g., [22, 45, 21]. Indeed, evaluation of the identification accuracy with a finite number of samples is of great importance. Set membership (SM) identification methods have been developed in the last 20 years in order to deal with unmodeled dynamics and finite samples, see, e.g., [31, 41, 23, 33, 25, 37, 11, 10, 6] and the references therein.

H_∞ identification, where the modeling error is measured by the H_∞ norm, is among the most investigated SM methods in the literature on identification for robust control. Indeed, H_∞ identification methods deliver uncertainty model sets in a form suitable to be used by well-established robust design techniques, based on H_∞ or μ optimization methods. The literature on H_∞ identification is now very extensive, considering time and/or frequency domain measurements and under different prior assumptions. In this chapter, some of the most relevant contributions related to assumption validation, evaluation of bounds on unmodeled dynamics, convergence analysis, and optimality properties are surveyed from a deterministic point of view.

Needless to say, any overview reflects the authors' outlook on the topic, and this is not an exception. Moreover, space limitations and uniformity of exposition do not allow a complete coverage of the literature. In particular, for model set validation results under both deterministic and stochastic frameworks, closely related to the assumption validation problem investigated in this chapter, the interested reader can refer, e.g., to [42, 47, 39, 5, 46, 48] and the references cited there. Other important issues are not reviewed as well, such as the probabilistic and mixed probabilistic-deterministic approaches to model set identification, the derivation of reduced order model sets, the structure of unfalsified models, the sample complexity, the experiment design, the input selection, the presence of outliers, and mixed time-frequency measurements. Moreover, only pointwise bounds on errors are considered, while other classes of bounded noise, e.g., ℓ_2 or ℓ_1 bounded noise, have been investigated. Finally, it must be remarked that some of the results reported in the chapter are specific H_∞ instances of other more general referenced results.

2 Prior and experimental information

In this section, the main concepts of SM H_∞ identification are introduced. Causal, discrete-time, single-input single-output, linear time-invariant, ℓ_2 BIBO stable, possibly distributed parameter, dynamical systems are considered. Any such system S is uniquely determined by its impulse response $h^S \doteq \left\{ h_k^S \right\}_{k=0}^\infty$, whose power-series representation of the transfer function $S(z) \doteq \sum_{k=0}^\infty h_k^S z^k$ is an element of the Hardy space $\mathcal{H}_\infty(\mathbf{D})$ defined as $\mathcal{H}_\infty(\mathbf{D}) \doteq \left\{ f : \mathbf{D} \to \mathbf{C} \mid f \text{ analytic in } \mathbf{D} \text{ and } \|f\|_\infty \doteq \sup_{z \in \mathbf{D}} |f(z)| < \infty \right\}$, where $\mathbf{D} \doteq \{ z \in \mathbf{C} : |z| < 1 \}$ is the open unit disk. Note that $S(z)$ as defined here merely denotes the standard z-transform of h^S evaluated at z^{-1}.

Let $S^o \in \mathcal{H}_\infty(\mathbf{D})$ be the actual plant to be identified using both experimental information and prior information (or assumptions).

Experimental information
The noisy measurements are represented as

$$y^N = F_N(S^o) + e^N,$$

where $y^N = [y_0 \ \dots \ y_{N-1}]^T \in \Re^N$ is a known vector depending on the actual data, F_N is a known operator called "information operator" indicating how the measurements depend on S^o, and $e^N \in \Re^N$ is an unknown vector representing the measurement noise. The following experimental settings are considered.

- Time domain measurements of N samples of the output y of the system S^o, initially at rest, fed by a known one-sided input u ($u_\ell = 0$ for $\ell < 0$, $u_0 \neq 0$):

$$y_\ell = \sum_{k=0}^{\ell} h_k^{S^o} u_{\ell-k} + e_\ell, \quad \ell = 0, \ldots, N-1. \tag{1}$$

In this case, the information operator is given by $F_N(S) = F_N h^S$, where

$$F_N = \begin{bmatrix} T_u & 0_{N \times \infty} \end{bmatrix} \in \Re^{N \times \infty} \tag{2}$$

and $T_u \in \Re^{N \times N}$ is the lower triangular Toeplitz matrix formed by the input vector $u^N = [u_0 \ \ldots \ u_{N-1}]^T$.

- Measurements of real and imaginary part of the complex-valued samples of the system frequency response $S^o(\omega_k)$, $k = 1, \ldots, N/2$:

$$\begin{aligned} y_{2k-2} &= \Re e\left(S^o(\omega_k)\right) + e_{2k-2} \\ y_{2k-1} &= \Im m\left(S^o(\omega_k)\right) + e_{2k-1}. \end{aligned} \tag{3}$$

In this case, the information operator is given by $F_N(S) = F_N h^S$, where

$$F_N = \begin{bmatrix} \Omega^T(\omega_1) & \cdots & \Omega^T(\omega_{N/2}) \end{bmatrix}^T \in \Re^{N \times \infty} \tag{4}$$

$$\Omega(\omega) = \begin{bmatrix} \Omega_1(\omega) \\ \Omega_2(\omega) \end{bmatrix} = \begin{bmatrix} \Re e\left(\Psi(\omega)\right) \\ \Im m\left(\Psi(\omega)\right) \end{bmatrix} \in \Re^{2 \times \infty} \tag{5}$$

$$\Psi(\omega) = \begin{bmatrix} 1 & e^{j\omega} & e^{j2\omega} & \cdots \end{bmatrix} \in \mathbf{C}^{1 \times \infty}. \tag{6}$$

- Mixed time and frequency domain measurements, consisting of both (1) and (3), can be considered as well. In such a case, the information operator can be obtained by stacking together the information operators (2) and (4). ∎

Prior information on plant S^o

Assumption on S^o is typically given as $S^o \in K \subset \mathcal{H}_\infty(\mathbf{D})$, where K is a nonfinitely parametrized subset of dynamical systems. The following subsets K investigated in the literature are considered in the chapter:

1. $K_{\rho,L}^{(1)} \doteq \left\{ S \in \mathcal{H}_\infty(\mathbf{D}) : \sup_{z \in \overline{\mathbf{D}}_\rho} |S(z)| \leq L \right\}$
   with $L > 0$, $\rho > 1$, and $\overline{\mathbf{D}}_\rho \doteq \{z \in \mathbf{C} : |z| \leq \rho\}$ the closed disk of radius ρ.
2. $K_{\rho,L}^{(2)} \doteq \left\{ S \in \mathcal{H}_\infty(\mathbf{D}) : \left|h_k^S\right| \leq L\rho^{-k}, \ \forall k \geq 0 \right\}$
   with $L > 0$ and $\rho > 1$.
3. $K_\gamma^{(3)} \doteq \left\{ S \in \mathcal{H}_\infty(\mathbf{D}) : \sup_{z \in \mathbf{D}} \left|\dfrac{dS(z)}{dz}\right| \leq \gamma \right\}$
4. $K_\gamma^{(4)} \doteq \left\{ S \in \mathcal{H}_\infty(\mathbf{D}) : \sup_{\omega \in [0,\pi]} \left|\dfrac{dS_{R/I}(\omega)}{d\omega}\right| \leq \gamma \right\}$
   where the notation $S_R(\omega) \doteq \Re e\left[S(e^{-j\omega})\right]$ and $S_I(\omega) \doteq \Im m\left[S(e^{-j\omega})\right]$ is used. The notation $(\cdot)_{R/I}$ indicates both $(\cdot)_R$ and $(\cdot)_I$.

The following result, derived by standard properties of analytic functions, may prove useful to understand the relations among these sets.

Result 1.

$$K_{\rho,L}^{(1)} \subset K_{\rho,L}^{(2)} \subset K_\gamma^{(3)} \subset K_\gamma^{(4)}, \quad \text{for } \gamma = \frac{L\rho}{(\rho-1)^2}. \qquad \blacksquare$$

It has to be pointed out that the computational complexity of the assumption validation and derivation of almost optimal algorithms is highly dependent on the assumed subset K. In fact, as shown in Sections 3 and 7 and discussed in

Section 8, the choice of K has a direct effect on the number of data that can be reasonably processed and on the tightness of the identified model sets.

Prior information on noise e^N

$$e^N \in \mathcal{B}_e = \left\{ \tilde{e}^N = [\tilde{e}_0 \ \ldots \ \tilde{e}_{N-1}]^T \in \Re^N : \|A\tilde{e}^N\|_\infty^{W_e} \leq \varepsilon \right\}, \qquad (7)$$

where $\|A\tilde{e}^N\|_\infty^{W_e} = \|W_e^{-1} A\tilde{e}^N\|_\infty = \max_{0 \leq k \leq l-1} w_{e,k}^{-1} |(A\tilde{e}^N)_k|$, $A \in \Re^{l \times N}$ is a given matrix of rank N and $W_e = \text{diag}(w_{e,0}, \ldots, w_{e,l-1}) \in \Re^{l \times l}$ is a given weighting matrix with $w_{e,k} > 0 \ \forall k$.

By taking $A = W_e = I_{N \times N}$, such an assumption accommodates for constant magnitude bound on noise: $|\tilde{e}_k| \leq \varepsilon \ \forall k$. By suitably choosing W_e, it is possible to consider noise bounds dependent on k: $|\tilde{e}_k| \leq w_{e,k} \cdot \varepsilon \ \forall k$, to account, e.g., for relative measurement errors. By suitably choosing A, it is possible to account for information on deterministic uncorrelation properties of the noise, see, e.g., [17, 34, 44]. ∎

3 Validation of prior assumptions

As is typical in any identification theory, the problem of checking the validity of prior assumptions arises. The only thing that can be actually done is to check if prior assumptions are invalidated by the data, evaluating if no system exists consistent with data and assumptions. To this end, it is useful to consider the feasible systems set (FSS), often called *unfalsified systems set*, i.e., the set of all systems consistent with prior information and measured data.

Definition 1. *Feasible Systems Set*

$$FSS\left(K, \mathcal{B}_e, F_N, y^N\right) \doteq \left\{ S \in K : y^N = F_N(S) + \tilde{e}^N, \ \ \tilde{e}^N \in \mathcal{B}_e \right\}. \qquad ∎$$

Thus, prior assumptions are invalidated if FSS is empty. However, it is usual to introduce the concept of prior assumption validation to denote the consistency with the measured data, i.e., that FSS is not empty.

Definition 2. *Validation of prior assumptions*
Prior assumptions are considered validated *if $FSS \neq \emptyset$.* ∎
Note that the fact that the prior assumptions are consistent with the present data does not exclude that they may be invalidated by future data.

The following results show how to validate different prior assumptions.

Result 2 [8, 9]. Given $N/2$ frequency-domain measurements, the prior assumptions $S^o \in K_{\rho,L}^{(1)}$ and $e^N \in \mathcal{B}_e$ with $A = W_e = I_{N \times N}$ are validated if and only if there is a vector $\eta^N \in \mathcal{B}_e$ such that

$$\begin{bmatrix} -Q^{-1} & -L^{-1}(D_y - D_\eta)^H \\ -L^{-1}(D_y - D_\eta) & -Q \end{bmatrix} \leq 0, \qquad (8)$$

where $D_y \doteq \text{diag}(\{y_{2k-2} + j \cdot y_{2k-1}\}_{k=1}^{N/2})$, $D_\eta \doteq \text{diag}(\{\eta_{2k-2} + j \cdot \eta_{2k-1}\}_{k=1}^{N/2})$, and $Q \doteq \left[\frac{1}{1 - \rho^{-2} z_i \bar{z}_l}\right]$, with $z_k = e^{j\omega_k}$, for $i, l, k = 1, \ldots, N/2$. ∎

Result 3 [7]. Given N time-domain measurements, the prior assumptions $S^o \in K_{\rho,L}^{(1)}$ and $e^N \in \mathcal{B}_e$ with $A = W_e = I_{N \times N}$ are validated if and only if there is a vector $\eta^N \in \mathcal{B}_e$ such that

$$\begin{bmatrix} -T_u^T D_\rho^2 T_u & -(T_y - T_\eta)^T \\ -(T_y - T_\eta) & -L^2 D_\rho^{-2} \end{bmatrix} \leq 0, \tag{9}$$

with $D_\rho \doteq \mathrm{diag}\left(1, \rho, \cdots, \rho^{N-1}\right)$ and T_u, T_y, and T_η given by the lower triangular Toeplitz matrices associated with u^N, y^N and η^N, respectively. ∎

These two validation results for $K = K_{\rho,L}^{(1)}$ are based on two different interpolation techniques, Nevanlinna-Pick (see, e.g., [4]) and Carathéodory-Fejér (see, e.g., [40]), respectively. The interpolation conditions are converted into the linear matrix inequality (LMI) problems (8) and (9), which appear to be computationally more efficient.

Let S^n be the FIR_n system with impulse response

$$h^{S^n} \doteq \left\{ h_0^S, h_1^S, \ldots, h_{n-1}^S, 0, \ldots \right\}.$$

Result 4 [27]. Given time or frequency domain measurements, let n be a given positive integer. Conditions for validating prior assumptions $S^o \in K_{\rho,L}^{(2)}$ and $e^N \in \mathcal{B}_e$ are

i) $\varepsilon^* \leq \varepsilon$ is a sufficient condition, being ε^* solution of the problem

$$\varepsilon^* = \min_{\tilde{\varepsilon}, S^n : \left\{ \begin{array}{l} S^n \in K_{\rho,L}^{(2)} \\ \|A \cdot (y^N - F_N h^{S^n})\|_\infty^{W_e} \leq \tilde{\varepsilon} \end{array} \right.} \tilde{\varepsilon}. \tag{10}$$

ii) $\varepsilon^* \leq \varepsilon + \varepsilon_n \|W_e^{-1} A\|_{\infty,\infty}$ is a necessary condition, with $\varepsilon_n = \dfrac{L\rho^{1-n}}{\rho - 1}$ and

$\|W_e^{-1} A\|_{\infty,\infty} = \max_{0 \leq i \leq l-1} \sum_{k=0}^{N-1} |(W_e^{-1} A)_{ik}|$.

iii) $\varepsilon^* \leq \varepsilon$ is a necessary and sufficient condition for the case of time domain data, if $n \geq N$ is chosen. ∎

Note that the problem (10) can be solved by linear programming techniques. By choosing n sufficiently large, the "gap" between sufficient and necessary conditions can be made arbitrarily small.

Result 5 [26]. Given $N/2$ frequency domain measurements, conditions for validating prior assumptions $S^o \in K_\gamma^{(4)}$ and $e^N \in \mathcal{B}_e$ with $A = I_{N \times N}$ are

i) $\xi_{R/I,k} \geq \underline{h}_{R/I,k}$, for $k = 1, \ldots, N/2$, is a necessary condition, where

$$\xi_{R/I,k} \doteq \min_{l=1,\ldots,N/2} \left(\overline{h}_{R/I,l} + \gamma \left| \omega_l - \omega_k \right| \right)$$

$$\underline{h}_{R,k} \doteq y_{2k-2} - w_{e,2k-2} \cdot \varepsilon, \quad \underline{h}_{I,k} \doteq y_{2k-1} - w_{e,2k-1} \cdot \varepsilon$$

$$\overline{h}_{R,k} \doteq y_{2k-2} + w_{e,2k-2} \cdot \varepsilon, \quad \overline{h}_{I,k} \doteq y_{2k-1} + w_{e,2k-1} \cdot \varepsilon.$$

ii) $\xi_{R/I,k} > \underline{h}_{R/I,k}$, for $k = 1, \ldots, N/2$, is a sufficient condition. ∎

These latter validation conditions can be easily checked by straightforward computation.

4 Model sets, identification algorithms, and errors

The $FSS\left(K, \mathcal{B}_e, F_N, y^N\right)$ summarizes the overall information on the system to be identified, i.e., prior assumptions on system and noise $(K, \mathcal{B}_e)$ and information coming from experimental data $\left(F_N, y^N\right)$, thus describing the uncertainty about the system to be identified. If prior assumptions are "true," FSS includes S^o and, in the line with the robustness paradigm, control should be designed to be robust versus such an uncertainty model set. However, FSS is in general not represented in a suitable form to be used by robust control design techniques, and model sets with such a property have to be looked for, e.g., described by linear fractional transformations. In order to be consistent with robust control design philosophy, *model sets including the set of unfalsified systems* have to be looked for. This is formalized by the following definition.

Definition 3. *Model set*
   A set of models $\mathcal{M} \subseteq \mathcal{H}_\infty\left(\mathbf{D}\right)$ *is called a* model set *for S^o if*
$$\mathcal{M} \supseteq FSS. \qquad \blacksquare$$
In this chapter, additive model sets are considered, of the form
$$\mathcal{M}(M, W) = \{M + \Delta : |\Delta\left(\omega\right)| \leq W\left(\omega\right), \ \forall \omega \in [0, 2\pi]\}, \qquad (11)$$
where M is called nominal model. In this case, the following result is an immediate consequence of Definition 3.

Result 6. For given nominal model M:
-   $\mathcal{M}(M, W)$ is a model set for S^o if and only if:
$$W\left(\omega\right) \geq W^*\left(\omega, M\right) \doteq \sup_{S \in FSS} |S\left(\omega\right) - M\left(\omega\right)|, \quad \forall \omega \in [0, 2\pi].$$
-   $\mathcal{M}(M, W^*)$ is the smallest model set for S^o of the form (11), i.e., for any model set $\mathcal{M}(M, W)$ it results that
$$\mathcal{M}(M, W^*) \subseteq \mathcal{M}(M, W). \qquad \blacksquare$$
   The nominal model can be obtained by some identification algorithm, i.e., an operator ϕ mapping the available information, represented by the quadruple $\left(K, \mathcal{B}_e, F_N, y^N\right)$, into a model $M \in \mathcal{H}_\infty\left(\mathbf{D}\right)$:
$$\phi\left(K, \mathcal{B}_e, F_N, y^N\right) = M.$$
For notational simplicity, the dependence on y^N only will be usually indicated.
   Some of the main features of an identification algorithm can be summarized as follows.

Definition 4. *Linear/nonlinear, untuned/tuned, interpolatory algorithm*
   - An algorithm ϕ is said to be linear *if it is a linear function of the data y^N; otherwise, it is said to be* nonlinear.
   - An algorithm ϕ is said to be untuned *if it does not depend on plant and noise information, i.e., if $\phi\left(K, \mathcal{B}_e, F_N, y^N\right)$ is actually not dependent on the constants involved in K and $\mathcal{B}_e$ definitions; otherwise, it is said to be* tuned.
   - An algorithm ϕ^I is said to be interpolatory *if it always gives models consistent with prior information and measured data, i.e., if $M^I = \phi^I\left(y^N\right) \in FSS$.*

■

Given an algorithm ϕ, the error $||S^o - \phi(y^N)||_\infty$ cannot be exactly known. The tightest upper bound on this error *for given data record* is $\sup_{S \in FSS} ||S - \phi(y^N)||_\infty$, while *for any possible system and noise* it is given by $\sup_{S \in K} \sup_{\tilde{e}^N \in B_e} ||S - \phi(F_N(S) + \tilde{e}^N)||_\infty$. This motivates the definition of the following two identification errors.

Definition 5. *Local and global identification errors*
- *The* local identification error *of the algorithm ϕ and of the identified model $M = \phi(y^N)$ is*
$$E_l(\phi) = E(M) = \sup_{S \in FSS} ||S - M||_\infty .$$
- *The* global identification error *of the algorithm ϕ is*
$$E_g(\phi) = \sup_{S \in K} \sup_{\tilde{e}^N \in B_e} ||S - \phi(F_N(S) + \tilde{e}^N)||_\infty .$$

■

Note that $E_g(\phi) \geq E_l(\phi)$, since the following result holds.

Result 7 [30].
$$E_g(\phi) = \sup_{y^N} E_l(\phi(y^N)) .$$

■

The local error $E_l(\phi)$, contrary to the global error $E_g(\phi)$, is not worst-case with respect to the noise. This fact has important implications in optimality and convergence properties, as shown in the next two sections.

In Section 7, bounds on $E_g(\phi)$ for different prior assumptions and algorithms are reported.

For given identification algorithm ϕ, providing the model $M = \phi(y^N)$, the evaluation of local error $E_l(\phi)$ is important because it represents the tightest bound on model error $||S^o - M||_\infty$. Since $E_l(\phi) = \sup_\omega \sup_{S \in FSS} |S(\omega) - \hat{M}(\omega)| = \sup_\omega W^*(\omega, M)$, evaluation of $E_l(\phi)$ can be made by computing $W^*(\omega, M)$ for a sufficiently coarse set of frequencies. The following result shows how to compute $W^*(\omega, M)$ in the case that $K = K_{\rho,L}^{(2)}$.

Result 8 [29]. Assume time or frequency domain measurements and $K = K_{\rho,L}^{(2)}$. Let $m \geq 3$ and n be such that there exists a FIR_n system $S^n \in FSS$. Then, for given model M:
$$\underline{W}_m^n(\omega, M) \leq W^*(\omega, M) \leq \overline{W}_m^n(\omega, M)$$
$$\lim_{n,m \to \infty} \overline{W}_m^n(\omega, M) = \lim_{n,m \to \infty} \underline{W}_m^n(\omega, M) = W^*(\omega, M) ,$$
where
$$\overline{W}_m^n(\omega, M) = \max_{k=1,\ldots,m} ||M(\omega) - \overline{v_k}(\omega)||_2 + \frac{L\rho^{1-n}}{\rho - 1}$$
$$\underline{W}_m^n(\omega, M) = \max_{k=1,\ldots,m} ||M(\omega) - \underline{t_k}(\omega)||_2$$
$$\overline{v_k}(\omega) = \begin{bmatrix} \sin(s_k) & \cos(s_k) \\ \sin(s_{k+1}) & \cos(s_{k+1}) \end{bmatrix}^{-1} \cdot \begin{bmatrix} [\sin(s_k) & \cos(s_k)] \cdot \overline{t_k}(\omega) \\ [\sin(s_{k+1}) & \cos(s_{k+1})] \cdot \overline{t_{k+1}}(\omega) \end{bmatrix} \quad (12)$$
$$s_k = 2\pi k/m$$

$$\overline{t_k}(\omega) = \Omega(\omega)\cdot \operatorname*{argmin}_{S^n\in\overline{FSS}^n} \left[-\Omega_1(\omega)\cdot\sin(s_k)+\Omega_2(\omega)\cdot\cos(s_k)\right]\cdot h^{S^n} \in \Re^2 \qquad (13)$$

$$\underline{t_k}(\omega) = \Omega(\omega)\cdot \operatorname*{argmin}_{S^n\in\underline{FSS}^n} \left[-\Omega_1(\omega)\cdot\sin(s_k)+\Omega_2(\omega)\cdot\cos(s_k)\right]\cdot h^{S^n} \in \Re^2 \qquad (14)$$

$$\overline{FSS}^n = \left\{ S^n\in K_{\rho,L}^{(2)} : \|(W_e + \widetilde{W})^{-1}\cdot A\cdot(y^N - F_N h^{S^n})\|_\infty\leq \varepsilon\right\}$$

$$\underline{FSS}^n = \left\{ S^n\in K_{\rho,L}^{(2)} : \|W_e^{-1}A\cdot(y^N - F_N h^{S^n})\|_\infty\leq \varepsilon\right\}$$

$$\widetilde{W} = \frac{L\rho^{1-n}}{\rho-1}\cdot \operatorname{diag}\left(\|a_1\|_1,\ldots,\|a_l\|_1\right)\in\Re^{l\times l}, \text{ with } a_\ell = \ell\text{th row of } A. \quad\blacksquare$$

Note that the optimization problems (13) and (14) are linear programs. Moreover, the value of n as required by this result can always be found, provided that prior assumptions are validated, i.e., $FSS \neq \emptyset$.

From the previous result it follows that, in the case $K = K_{\rho,L}^{(2)}$,

$$\underline{E}_m^n(M) \doteq \sup_{0\leq\omega\leq 2\pi} \underline{W}_m^n(\omega,M)\leq E(M)\leq \sup_{0\leq\omega\leq 2\pi} \overline{W}_m^n(\omega,M)\doteq\overline{E}_m^n(M)$$

$$\lim_{n,m\to\infty}\overline{E}_m^n(M) = \lim_{n,m\to\infty}\underline{E}_m^n(M) = E(M).$$

The computation of $W^*(\omega,M)$ allows also us to derive, for given nominal model M, the smallest model set guaranteeing to contain the feasible systems set, according to Result 6.

5 Optimality properties

Algorithms minimizing the global and local identification errors lead to the following optimality concepts.

Definition 6. *Algorithm local and global optimality*
 - *An algorithm ϕ^* is called ℓ-optimal if for all $K,\mathcal{B}_e,F_N,y^N$:*
$$E_l(\phi^*) = \inf_\phi E_l(\phi\left(K,\mathcal{B}_e,F_N,y^N\right)) \doteq r\left(K,\mathcal{B}_e,F_N,y^N\right).$$

$r\left(K,\mathcal{B}_e,F_N,y^N\right)$ *is called* local radius of information.
 - *An algorithm ϕ^g is called g-optimal if for all $K,\mathcal{B}_e,F_N,y^N$:*
$$E_g(\phi^g) = \inf_\phi E_g(\phi\left(K,\mathcal{B}_e,F_N,y^N\right)) \doteq R\left(K,\mathcal{B}_e,F_N\right).$$

$R\left(K,\mathcal{B}_e,F_N\right)$ *is called* global radius of information. $\blacksquare$

Local optimality is a stronger optimality concept than global optimality. In fact, if an algorithm is ℓ-optimal, then it is g-optimal, but the converse implication is not true.

It is also useful to define the optimality of the identified model $M = \phi\left(y^N\right)$ as follows.

Definition 7. *Optimal model*
 A model $M = \phi\left(y^N\right)$ is called optimal *if for the given $K,\mathcal{B}_e,F_N,y^N$:*
$$E(M) = \inf_{\widetilde{M}\in\mathcal{H}_\infty(\mathbf{D})} E(\widetilde{M}) = r\left(K,\mathcal{B}_e,F_N,y^N\right).$$

A basic result in information-based complexity relates ℓ-optimal algorithms, optimal model, and the H_∞ Chebicheff center M^c of FSS, defined as

$$M^c = \arg\inf_{\widetilde{M} \in \mathcal{H}_\infty(\mathbf{D})} \sup_{S \in FSS} \| S - \widetilde{M} \|_\infty.$$

Result 9 [43]. If M^c exists, then it is an optimal model and the algorithm $\phi^c\left(y^N\right) = M^c$, called *central*, is an ℓ-optimal algorithm.   ∎

Note that the central algorithm ϕ^c is g-optimal, but there exist other g-optimal algorithms. In particular, while ϕ^c is nonlinear, linear g-optimal algorithms exist, as given by the following result.

Result 10 [24, 30]. There exist linear g-optimal algorithms.   ∎

Computing central or linear g-optimal algorithms is not known in the present H_∞ setting. This motivates the interest in deriving algorithms having lower complexity, at the expense of some degradation in the accuracy of the identified model set. The following definition is introduced to give a measure of such a degradation of the local error of a given algorithm with respect to the minimal error obtained by a central algorithm.

Definition 8. *Algorithm deviation*
  The deviation $dev(\phi)$ *of the algorithm* ϕ *is*

$$dev(\phi) = \sup_{y^N} \frac{E_l\left(\phi(y^N)\right)}{r\left(y^N\right)}.$$

Note that $dev(\phi) \geq 1 \ \forall\phi$ and $dev(\phi^c) = 1$. Higher values of $dev(\phi)$ mean worse identified models when any possible set of measurements is processed.

The following result shows that, in the present H_∞ setting, linear algorithms, though possibly g-optimal, may give large degradation of the local error with respect to the minimal error obtained by a central algorithm.

Result 11 [43, 20]. No linear algorithm with finite deviation exists.   ∎

The question arises if it is possible to derive computable algorithms with finite and possibly "small" deviation. This question is answered by the following result.

Result 12 [43]. For any interpolatory algorithm ϕ^I it holds:

$$dev(\phi^I) \leq 2.$$ ∎

For this reason, interpolatory algorithms are often called *almost optimal*. Methods for computing interpolatory algorithms for different prior assumptions are presented in Section 7.

A given algorithm ϕ, by processing *any possible information* $K, \mathcal{B}_e, F_N, y^N$, gives an identified model $M = \phi(y^N)$ for which the ratio $E(M)/r$ is bounded as

$$E(M)/r \leq dev(\phi).$$

However, *for given information* $K, \mathcal{B}_e, F_N, y^N$, the *actual ratio* $E(M)/r$ may be significantly lower than $dev(\phi)$. Then, for given identified model, it is of interest to evaluate the actual value of this ratio, called *model optimality level*.

Definition 9. *Model optimality level*

The optimality level $\alpha(M)$ *of a model* $M = \phi\left(y^N\right)$ *is*

$$\alpha(M) = E(M)/r. \qquad \blacksquare$$

Note that $\alpha(M) \geq 1\ \forall M \in \mathcal{H}_\infty\,(\mathbf{D})$ and $\alpha(M^c) = 1$. The model optimality level is actually a measure of the degradation of the identification accuracy: the higher $\alpha(M)$ is, the worse the model M is meant to be, on the basis of the available information.

A result on evaluation of model optimality level is available for $K = K^{(2)}_{\rho,L}$.

Let $V \subset \Re^2$ be a polytope. Its radius $r_2\,[V]$ in the Euclidean norm is defined as $r_2\,[V] = \inf_{s \in \Re^2} \sup_{v \in V} \|s - v\|_2$ and can be easily computed via standard algorithms available in computational geometry literature.

Result 13 [29]. Assume time or frequency domain measurements and $K = K^{(2)}_{\rho,L}$. Let $m \geq 3$ and n be such that there exists a FIR_n system $S^n \in FSS$. Then, for given model M:

$$\underline{\alpha}^n_m(M) \leq \alpha(M) \leq \overline{\alpha}^n_m(M),$$

where

$$\underline{\alpha}^n_m(M) = \max\left\{1,\ \underline{E}^n_m(M)\,\Big/\,\Big[\sup_{0 \leq \omega \leq 2\pi} r_2\,[VO^n_m(\omega)] + \tfrac{L\rho^{1-n}}{\rho-1}\Big]\right\}$$
$$\overline{\alpha}^n_m(M) = \overline{E}^n_m(M)\,\Big/\sup_{0 \leq \omega \leq 2\pi} r_2\,[VI^n_m(\omega)]\,,$$

with $VO^n_m(\omega)$ and $VI^n_m(\omega)$ the convex hulls of points $\overline{v_k}(\omega)$ and $\underline{t_k}(\omega)$, $k = 1,\ldots,m$, defined in (12) and (14), respectively. $\qquad \blacksquare$

Polytopes $VO^n_m(\omega)$ and $VI^n_m(\omega)$ are outer and inner convergent approximations of the value set $V(\omega)$, the set in the complex plane of $S(e^{-j\omega})$ for all $S \in FSS$.

6 Convergence properties

In order to investigate algorithm convergence when applied to any $S \in K$ and any $e^N \in \mathcal{B}_e$, conditions for convergence of the global error to zero as $N \to \infty$ are sought. In general, this convergence cannot hold, unless $\varepsilon \to 0$, as shown by the following result.

Result 14 [43].

$$E_g(\phi) \geq \sup_{S \in FSS(K,\mathcal{B}_e,F_N,0)} \|S\|_\infty > 0,\ \forall N. \qquad \blacksquare$$

Different kinds of algorithm convergence can be defined.

Definition 10. *Algorithm convergence and robust convergence*

- *An algorithm ϕ is said to be* convergent *if*

$$\lim_{\varepsilon \to 0} \lim_{N \to \infty} E_g(\phi) = 0.$$

- *An algorithm ϕ is said to be* robustly convergent *if it converges regardless of a priori information.*

By Definition 4, tuned algorithms are not robustly convergent.

The following strong negative result holds for linear algorithms.

Result 15 [36]. No robustly convergent linear algorithm exists.   ■

This implies that any untuned linear algorithm is not convergent. In particular, least squares algorithms are not convergent, since they are linear and untuned, i.e., not dependent on prior assumptions on system and noise.

Convergent algorithms can be obtained by interpolation, as shown by the next result.

Result 16 [6]. Any interpolatory algorithm is convergent.   ■

In Section 7, interpolatory algorithms are presented for the different prior assumptions. These algorithms are tuned and then are not robustly convergent.

Two questions may be of interest:
- Do there exist convergent linear algorithms?
- Do there exist robustly convergent algorithms?

The answer is affirmative for both questions, as shown in Section 7 where convergent tuned linear and robustly convergent nonlinear algorithms are presented.

Contrary to the global error $E_g(\phi)$, the local error $E_l(\phi)$ may converge to zero for finite values of ε, under suitable deterministic uncorrelation assumptions on noise and for suitable inputs. This may happen because the local error $E_l(\phi)$, contrary to the global error $E_g(\phi)$, is not worst-case with respect to the noise, and then can account for information on its uncorrelation properties. In particular, by suitably choosing A and W_e, the noise set $\mathcal{B}_e$ can arbitrarily approximate the set

$$\widetilde{\mathcal{B}}_e = \left\{ e^N \in \Re^N : \left| \sup_\omega \sum_{k=0}^{N-1} e_k \exp^{j\omega k} \right| N^\alpha \leq \varepsilon, \ \alpha > 1/2 \right\},$$

which is composed of deterministic counterparts of uncorrelated noise. For example, sequences of independently identically distributed bounded random variables asymptotically belong to $\widetilde{\mathcal{B}}_e$ with probability 1, see, e.g., [17, 44]. If noise in time domain experiments belongs to this set, the FSS asymptotically shrinks to a singleton for any ε, as shown by the following result.

Result 17 [44]. Let measurements be in time domain and $e^N \in \widetilde{\mathcal{B}}_e$. Then, an input sequence u can be found such that

$$\lim_{N \to \infty} r\left(y^N\right) = 0.$$ ■

Looking for convergence to zero of local error for given ε leads to the following convergence concept.

Definition 11. *Strong convergence*

An algorithm ϕ is said to be strongly convergent *if*

$$\lim_{N \to \infty} E_l(\phi) = 0.$$ ■

Results 17 and 12 imply that, in the case of deterministic uncorrelated noise belonging to $\widetilde{\mathcal{B}}_e$, an input sequence u can be found such that any central and interpolatory algorithm using such an input is strongly convergent.

7 Identification algorithm properties

In this section, the main algorithms available in the literature are reviewed, starting from the simplest ones (linear algorithms) and finishing with the highest performing ones (nonlinear interpolatory algorithms). The features analyzed are convergence, tightness in error evaluation, order of the identified model, and computational complexity.

Linear algorithms

Linear algorithms operate linearly on the experimental data. *Untuned* linear algorithms, based on least squares or polynomial approximation techniques (see, e.g., [35]), are independent of the prior information available on system and noise and, as consequence of Result 15, they cannot be convergent. Indeed, their global identification error may be divergent for finite ε [1, 38, 2].

Convergent *tuned* linear algorithms have been obtained based on least squares optimization, with constraints or penalty terms depending on plant and noise prior information [15, 19, 13]. The following result is obtained by minimizing least squares with a quadratic penalty term.

Result 18 [19, 13]. Assume $S^o \in K_{\rho,L}^{(1)}$, $e^N \in \mathcal{B}_e$ with $A = W_e = I_{N \times N}$ and $N/2$ equispaced frequency domain measurements. Then the linear algorithm

$$\hat{M}(z) = \sum_{k=0}^{n(N)-1} q_k^* \, z^k, \quad q_k^* = \frac{c_k(y^N)}{1 + \left(\varepsilon/L + \rho^{-n(N)}\right)^2 \rho^{2k}},$$

with $c_k(y^N) = \frac{2}{N} \sum_{l=1}^{N/2} (y_{2l-2} + j \cdot y_{2l-1}) \left(e^{j4\pi/N}\right)^{(l-1)k}$ the inverse discrete Fourier transform (DFT) coefficients of y^N, has a global error which, for $N/2 \geq n(N) > 0$, is bounded as

$$E_g(\phi) \leq L\rho^{-n(N)} + \left(1 + \sqrt{2}\right) L \sqrt{\tfrac{\rho+1}{\rho-1}} \left(\varepsilon/L + \rho^{-n(N)}\right)^{1/2}.$$

If $\lim_{N \to \infty} n(N) = \infty$, the algorithm is convergent. ∎

This result can be extended to nonuniformly spaced frequency domain measurements.

Linear algorithms are simple and easily computed, but have some important drawbacks. In particular, they cannot be robustly convergent (Result 15) and do not have finite deviation (Result 11), i.e., give identified models whose optimality properties can be arbitrarily bad. In order to achieve the robust convergence or finite deviation, it is necessary to resort to more sophisticated nonlinear algorithms.

Nonlinear "two-stage" algorithms

To overcome the robust convergence limitations of linear algorithms, nonlinear untuned algorithms have been derived for frequency domain measurements, performing the following "two step" procedure:

- stage 1: a noncausal preliminary model $\hat{M}^{(0)} \in \mathcal{L}_\infty$ is derived through a "untuned" linear algorithm performing a bilateral interpolation in $\mathcal{L}_\infty$ by means of trigonometric polynomials

$$\hat{M}^{(0)}(z) = \sum_{k=-n+1}^{n-1} w_{k,n} \, c_k(y^N) \, z^k,$$

where $\{w_{k,n}\}_{k=0}^{n-1}$ is a weighting (or *window*) sequence independent of prior information;

- stage 2: the identified model is chosen as the best (causal) approximation of $\hat{M}^{(0)}$ in $\mathcal{H}_\infty\left(\mathbf{D}\right)$, by solving the nonlinear Nehari approximation problem

$$\hat{M}\left(z\right) = \arg\min_{M\in\mathcal{H}_\infty\left(\mathbf{D}\right)}\|\hat{M}^{(0)} - M\|_\infty.$$

The solution is given by the Nehari's theorem [32]:

$$\hat{M}\left(z\right) = \hat{M}^{(0)}\left(z\right) - \bar{\sigma}\cdot\frac{\sum_{k=1}^{n-1}\psi_{n-k}\cdot z^k}{z^{n-1}\sum_{k=1}^{n-1}\zeta_k\cdot z^k},$$

with $\zeta = [\zeta_1,\ldots,\zeta_{n-1}]^T$ and $\psi = [\psi_1,\ldots,\psi_{n-1}]^T$ the right and left singular vectors of the Hankel matrix associated to the coefficients $w_{k,n}c_k(y^N)$, $k = -1,-2,\ldots,-n+1$, and $\bar{\sigma}$ the corresponding maximum singular value.

The two-stage algorithms proposed in the literature differ from one another in the first step, since the approximation in $\mathcal{L}_\infty$ can be performed using different weighting sequences, even symmetric with respect to k (i.e., sinc-square, triangular, cosine, trapezoidal windows) and truncated for $k\geq n$.

Result 19 [18, 14, 36, 15]. Assume $S^o \in K_{\rho,L}^{(1)}$, $e^N \in \mathcal{B}_e$ with $A = W_e = I_{N\times N}$ and $N/2$ equispaced frequency domain measurements. Then, the global identification error of a two-stage algorithm is bounded as follows

i) if $w_{k,n} = \sin\left(2k\pi/N\right)^2/\left(2k\pi/N\right)^2$ for $|k| < n$ and $w_{k,n} = 0$ for $|k| \geq n$ (sinc-square window), then

$$E_g\left(\phi\right) \leq 2\left[\min\left(\frac{8L\pi}{N\left(\rho-1\right)},\frac{4L\pi^2\left(\rho+1\right)}{N^2\left(\rho-1\right)^2}\right) + \frac{N^2\left(L+\varepsilon\right)}{2n\pi^2} + \varepsilon\right].$$

ii) if $w_{k,n} = 1 - |k|/n$ for $|k| < n$ and $w_{k,n} = 0$ for $|k| \geq n$ (triangular window), then, for any $m \leq \left(N+2\right)/4$

$$E_g\left(\phi\right) \leq 2\left[\varepsilon + 2L\rho^{-m} + L\rho\left(1 - \rho^{-m}\right)\left(\rho-1\right)^{-2}/n\right].$$

iii) if $w_{k,n} = \cos\left(k\pi/\left(2n+1\right)\right)$ for $|k| < n$ and $w_{k,n} = 0$ for $|k| \geq n$ (cosine window), then, for any $m \leq \left(N+2\right)/4$

$$E_g\left(\phi\right) \leq 2\left[\frac{L}{\rho^m} + \left(\pi-1\right)\left(\varepsilon + \frac{L}{\rho^m}\right) + \frac{\pi^2 L\rho\left(1+\rho\right)}{8n^2\left(\rho-1\right)^3}\right].$$

iv) if $w_{k,n}$ is a trapezoidal window defined by

$$w_{k,n} = \begin{cases} 1 + k/n & -n < k < -1 \\ 1 & 0 \leq k \leq m-1 \\ 1 - \left(k-m+1\right)/n & m \leq k \leq n+m-1 \\ 0 & \text{elsewhere} \end{cases},$$

where $m + n \leq N/2 + 1$, then

$$E_g\left(\phi\right) \leq \varepsilon\sqrt{2N/n} + \left(1 + \sqrt{2N/n}\right)L\rho^{-m}.\qquad\blacksquare$$

From the above result it turns out that, to achieve robust convergence, the number n of inverse DFT samples to be computed (and successively smoothed) varies significantly according to the chosen window sequence. With the sinc-square window, n must be such that $\lim_{N\to\infty} N^2/n\left(N\right) = 0$: this condition is met, for example, for any choice of n such that $n\left(N\right) = \mathcal{O}\left(N^3\right)$. In the other cases, the identification error approaches 0 as $1/n$, $1/n^2$, and $\sqrt{N/n}$, respectively; then, in the first two cases it is only requested that $n \to \infty$, while in the latter case n can be chosen such that $n = n(N) = \mathcal{O}(N^2)$. Since the order of the identified model $\hat{M}(z)$ is $2n - 3$, the benefit obtained by suitably smoothing the inverse DFT coefficients of data can be well understood. At the same time, the trade-off between the rate of convergence of the identification error bound as a function of the number of data and the magnitude of the bound on the worst-case error of the approximation algorithm can be observed. It turns out that, similar to what happens in classical statistical spectral analysis, while sophisticated windows providing fast convergence on the algorithm can be found, they necessarily imply worse errors than simpler windows for a small number of data.

To summarize the positive features of "two-stage" nonlinear algorithms, they are robustly convergent if the weighting sequence $\{w_{k,n}\}$ is even symmetric with respect to k, truncated for $k \geq n$ and thus independent of the prior information. Their identification error bound and model order are highly dependent on the chosen window sequence. The computational complexity of these algorithms is relatively small. As main drawbacks, their deviation is unknown and, even more relevant, the identified model may not belong to the set FSS of systems consistent with the overall priors available on the system to be identified.

Interpolatory algorithms

These nonlinear tuned algorithms identify models belonging to the FSS:
$$\phi^I\left(K, \mathcal{B}_e, F_N, y^N\right) = \hat{M}^I \in FSS,$$
and they are able to interpolate the experimental data in an approximated way, taking explicitly into account the available prior information. From Result 12, their deviation is not greater than 2 and for this reason they are often called "almost optimal" or "2-optimal." Moreover, these interpolatory algorithms are convergent but not robustly, since they are tuned.

In general, a two-step procedure is carried out:
- step 1: validation of prior information,
- step 2: identification of a model $\hat{M}^I \in FSS$ by means of nonlinear interpolation techniques.

Result 20 [16, 9]. Assume $S^o \in K_{\rho,L}^{(1)}$, $e^N \in \mathcal{B}_e$ with $A = W_e = I_{N \times N}$ and $N/2$ equispaced frequency domain measurements. Then, an interpolatory algorithm is given by the following procedure
- step 1: find a solution $\eta^N \in \mathcal{B}_e$ of the consistency problem represented by the LMI (8) in Result 2;
- step 2: by means of the standard Nevanlinna-Pick's algorithm, build a function $\hat{M}(z) \in K_{\rho,L}^{(1)}$ interpolating $\tilde{y}^N \doteq y^N + \eta^N$ and use it as the identified model.

The global identification error is bounded by
$$E_g(\phi) \leq 2\varepsilon\sqrt{\tfrac{l+m}{l-m}} + 2L\rho^{-(2m+1)}\left(1 + \sqrt{\tfrac{l+m}{l-m}}\right),$$

where l, m are arbitrary integers satisfying $0 < m < l \leq N/2 - m$. ∎

Result 21 [7]. Assume $S^o \in K_{\rho,L}^{(1)}$, $e^N \in \mathcal{B}_e$ with $A = W_e = I_{N \times N}$ and N time domain measurements. Then, an interpolatory algorithm is given by the following procedure:

- step 1: find a solution $\eta^N \in \mathcal{B}_e$ of the consistency problem represented by the LMI (9) in Result 3;
- step 2: by means of the standard Carathéodory-Fejér procedure, build a function $\hat{M}(z) \in K_{\rho,L}^{(1)}$ interpolating $\tilde{y}^N \doteq y^N + \eta^N$ and use it as the identified model.

The global identification error is bounded by

$$E_g(\phi) \leq 2 \left[\sum_{k=0}^{N-1} \min \left\{ \varepsilon \sum_{i=0}^{k} |\tau_i|, \frac{L}{\rho^k} \right\} + \frac{L}{\rho^{N-1}(\rho-1)} \right],$$

where $\{\tau_k\}_{k=0}^{N-1}$ are the elements of the first column of T_u^{-1}. ∎

The order of the identified model is equal to the number of data N, except in singular cases where the order may be lower. Under the computational point of view, these two interpolatory algorithms are very much burdensome when the number of data is high. Moreover, the Pick's matrix may easily result to be ill-conditioned, while similar comments hold for the time domain algorithm.

Result 22 [28]. Given time or frequency domain data and $K = K_{\rho,L}^{(2)}$, let n be a positive integer such that there exists a FIR_n system $S^n \in FSS$. Then, for given positive integers m and q, an interpolatory algorithm is $\phi^{no}(y^N) = M_n^{no}$, where M_n^{no} is the FIR_n model whose impulse response is obtained as solution of the problem

$$h^{M_n^{no}} = \arg \min_{S^n \in FSS} \|s^* - \Omega^* \cdot h^{S^n}\|_\infty, \tag{15}$$

where

$$s^* = \begin{bmatrix} s^*(\tilde{\omega}_1) \\ \vdots \\ s^*(\tilde{\omega}_q) \end{bmatrix}, \quad s^*(\omega) = \begin{bmatrix} \max_{s=\begin{bmatrix} s_1 \\ s_2 \end{bmatrix} \in VO_m^n(\omega)} s_1 \\ \max_{s=\begin{bmatrix} s_1 \\ s_2 \end{bmatrix} \in VO_m^n(\omega)} s_2 \\ \min_{s=\begin{bmatrix} s_1 \\ s_2 \end{bmatrix} \in VO_m^n(\omega)} s_1 \\ \min_{s=\begin{bmatrix} s_1 \\ s_2 \end{bmatrix} \in VO_m^n(\omega)} s_2 \end{bmatrix}, \quad \Omega^* = \begin{bmatrix} \Omega^*(\tilde{\omega}_1) \\ \vdots \\ \Omega^*(\tilde{\omega}_q) \end{bmatrix}, \quad \Omega^*(\omega) = \begin{bmatrix} \Omega^n(\omega) \\ \Omega^n(\omega) \end{bmatrix}$$

with $\tilde{\omega}_k \in [0, \pi]$ for $k = 1, \ldots, q$; VO_m^n the convex hull of points $\overline{v}_k(\omega)$, $k = 1, \ldots, m$, defined in (12) in Result 8; $\Omega^n(\omega)$ given by the first n columns of the matrix $\Omega(\omega)$ defined in (5). ∎

Solution of problem (15) can be performed by linear programming. Since ϕ^{no} is interpolatory, it follows from Result 12 that $\alpha(M_n^{no}) \leq 2$. Indeed, the actual value of $\alpha(M_n^{no})$ can be evaluated from Result 13 and it can be expected to be near to 1, since model M_n^{no} is derived as an approximation of the optimal model M^c.

Result 23 [12]. Given $N/2$ equispaced frequency domain measurements, $K = K_\gamma^{(3)}$ and $A = W_e = I_{N \times N}$, an interpolatory algorithm is $\phi^I(y^N) = M_n$, where M_n is the FIR_n model whose impulse response is obtained as solution of the problem

$$h^{M_n} = \arg \min_{S^n \in FSS} \max_{\theta \in [0,\pi)} \left| \sum_{k=0}^{n-1} k \cdot h_k^{S^n} \cdot e^{jk\theta} \right|$$

s.t. $\left| \sum_{k=0}^{n-1} h_k^{S^n} e^{j2k\pi\ell/N} - (y_{2\ell-2} + j \cdot y_{2\ell-1}) \right| \leq \varepsilon + \gamma \cdot \left(\frac{1}{n} + \sqrt{\frac{\pi}{2nN}} \right), \ \ell \in \left[0, \frac{N}{2} - 1\right].$

The global identification error is bounded by

$$E_g(\phi) \leq 2\varepsilon + \gamma \cdot \left[1/n + \pi/N + \sqrt{\pi/(nN)}\right]. \qquad \blacksquare$$

The above polynomial minimization problem can be solved by standard convex optimization methods.

Result 24 [26]. Given $N/2$ frequency domain data, $K = K_\gamma^{(4)}$ and $A = I_{N \times N}$, let n be a positive integer such that there exists a FIR_n system $S^n \in FSS$. Then, for given positive integer q, an interpolatory algorithm is $\phi^*(y^N) = M_n^*$, where M_n^* is the FIR_n model whose impulse response is obtained as solution of the problem

$$h^{M_n^*} = \arg \min_{S^n \in FSS} \left\| s^* - \Omega^* \cdot h^{S^n} \right\|_\infty, \qquad (16)$$

where

$$s^* = \left[\Re e\left[S^{no}(\widetilde{\omega}_1)\right], \Im m\left[S^{no}(\widetilde{\omega}_1)\right], \ldots, \Re e\left[S^{no}(\widetilde{\omega}_q)\right], \Im m\left[S^{no}(\widetilde{\omega}_q)\right] \right]^T$$

$$S^{no}(\omega) = \frac{1}{2} \left\{ \overline{S}_R^*(\omega) + \underline{S}_R^*(\omega) + j\left[\overline{S}_I^*(\omega) + \underline{S}_I^*(\omega)\right] \right\}$$

$$\underline{S}_{R/I}^*(\omega) = \max_{k=1,\ldots,N/2} \left(\underline{h}_{R/I,k} - \gamma |\omega - \omega_k| \right)$$

$$\overline{S}_{R/I}^*(\omega) = \min_{k=1,\ldots,N/2} \left(\overline{h}_{R/I,k} + \gamma |\omega - \omega_k| \right)$$

$$\Omega^* = \left[\Omega^n(\widetilde{\omega}_1)^T, \ldots, \Omega^n(\widetilde{\omega}_q)^T \right]^T$$

with $\widetilde{\omega}_k \in [0, \pi]$ for $k = 1, \ldots, q$; $\underline{h}_{R/I,k}$ and $\overline{h}_{R/I,k}$ defined as in Result 5; $\Omega^n(\omega)$ defined as in Result 22.

The local identification error is bounded by

$$E_l(\phi) \leq \frac{1}{2} \sup_{0 \leq \omega \leq 2\pi} \sqrt{\left[\overline{S}_R^*(\omega) - \underline{S}_R^*(\omega)\right]^2 + \left[\overline{S}_I^*(\omega) - \underline{S}_I^*(\omega)\right]^2} \leq \sqrt{2} E_l(\phi). \qquad \blacksquare$$

Solution of problem (16) can be obtained by linear programming.

8 Discussion

Some main features of the results presented are now discussed from a user point of view.

The main ingredients on which the different results are built are

- *type of experiment*: time domain ((1)–(2)), frequency domain ((3)–(6));

- *type of algorithm*: linear or nonlinear, tuned or untuned, interpolatory (Definition 4);
- *type of prior information on plant*, i.e., type of subset K such that $S^o \in K$;
- *type of prior information on noise*. In this chapter, for the reasons discussed in the Introduction, methods assuming only pointwise bounded noise as described by (7) are presented. Thus, the main distinction between the methods is their ability in dealing with $A \neq I_{N \times N}$, allowing to account for information on deterministic uncorrelation noise properties.

The main features on which the "goodness" of the methods is evaluated are

- *convergence*: simple, robust (Definition 10) or strong (Definition 11).
- *a priori optimality*: measured by the algorithm deviation $dev(\phi)$ (Definition 8), which gives the maximal degradation for any possible set of measurements of the local identification error guaranteed by the algorithm ϕ with respect to the minimal error achievable by an optimal algorithm.
- *tightness in error evaluation*. The identification algorithm ϕ, after processing the available data y^N, gives a model $M = \phi(y^N)$ and the local error $E_l(\phi)$ represents the tightest bound on the model error $||S^o - M||_\infty$. If this information is used e.g. for robust control design, the tighter the evaluation of $E_l(\phi)$ is, the less conservative is the design.
- *frequency shaping of uncertainty*. Many methods provide only upper bounds on the identification error $E_g(\phi) \leq \overline{E}$. This way, model sets of the form (11) with $W(\omega) = \overline{E}$, $\forall \omega \in [0, 2\pi]$ are obtained. According to Result 6, methods able to obtain tight evaluation of $W^*(\omega, M)$ can deliver smaller model sets, which in turn, if used for robust control, give rise to less conservative design.
- *a posteriori optimality*: measured by the model optimality level $\alpha(M)$ (Definition 9), which gives the actual value of the identification error degradation of the model M identified using the available data y^N with respect to the optimal model.
- *computational complexity*.

Let us now summarize the main properties of the different types of algorithms.

Linear algorithms require low computational effort, allowing them to work with very large number of data (up to several thousands), but they are not robustly convergent (Result 15), i.e., in order to guarantee convergence they have to be tuned to the prior assumptions on the system to be identified and on noise (Definition 4). A significant drawback of linear algorithms is that they have no finite deviation (Result 11), i.e., the local identification error of the identified models may be arbitrarily larger than the minimal possible one.

Nonlinear two-stage algorithms have been devised that are robustly convergent, i.e., the convergence is guaranteed for any value of the constants appearing in the prior assumptions on the system to be identified and on noise (Result 19). Their computational effort is still relatively low, since they require, in addition to the computation of a linear untuned algorithm, the solution of a Nehari problem. Thus, two-stage algorithms can process quite large amounts of data (up to some thousands). No optimality property of two-stage algorithms is known. In particular, no bound on their deviation is known, so that it is unknown how far the identified models are from being optimal.

For both types of algorithms, linear and two-stage, bounds $E_g(\phi) \leq \overline{E}$ on their global error are provided assuming that $S^o \in K^{(1)}_{\rho,L}$ and $A = I_{N \times N}$ in noise assumption (7). These bounds are useful to prove their convergence properties. However, their tightness is unknown and, in view of Result 7, they cannot be tight bounds on the local error $E_l(\phi)$. Thus, model sets $\mathcal{M}(M, \overline{E}) = \left\{ M + \Delta : |\Delta(\omega)| \leq \overline{E}, \ \forall \omega \in [0, 2\pi] \right\}$ derived from these bounds may be largely conservative.

Nonlinear interpolatory algorithms are convergent (Result 16), but not robustly, since they are tuned to the prior assumptions on the system to be identified and on noise. However, in case of deterministic uncorrelated noise, they are strongly convergent, since from Results 17 and 12 it follows that the local error $E(M)$ of identified models converges to zero for finite values of ε. Another important property of interpolatory algorithms is that their deviation is bounded by 2 (Result 12), thus guaranteeing that also the optimality level of identified models is not greater than 2. For this reason, interpolatory algorithms are often indicated as almost-optimal, since deviation 1 is guaranteed by optimal algorithms.

In conclusion, interpolatory algorithms have excellent convergence and optimality features. Their properties in relation to the other features (computational complexity, tightness in error evaluation, frequency shaping, "a posteriori" optimality) are highly dependent on the assumed set K:

- *computational complexity*
  - If $K = K^{(1)}_{\rho,L}$, algorithms based on Nevanlinna-Pick and Carathéodory-Fejér interpolation are used for both steps, validation (Results 2 and 3) and algorithm computation (Results 20 and 21). Computational problems may arise in processing more than moderate number of data (some decades).
  - If $K = K^{(2)}_{\rho,L}$, linear programming optimization has to be performed both in validation (Result 4) and algorithm computation (Result 22). The required computational effort allows to process up to several hundreds of data.
  - If $K = K^{(3)}_\gamma$, convex optimization methods are used (Result 23). Also in this case several hundreds of data may be processed with a reasonable computational effort.
  - If $K = K^{(4)}_\gamma$, the validation is computationally trivial (Result 5) and the algorithm computation requires the solution of one linear programming problem (Result 24). Then, a very large amount of data (up to several thousands) can be processed.

- *tightness in error evaluation*
  - If $K = K^{(1)}_{\rho,L}$, only bounds $E_g(\phi) \leq \overline{E}$ on their global error are provided (Results 20 and 21). Their tightness is unknown and, in view of Result 7, they cannot be tight bounds on the local error $E_l(\phi)$. Moreover,

these bounds do not account for possible deterministic uncorrelation information on noise.

- If $K = K_{\rho,L}^{(2)}$, the error $E(M)$ of the model M identified by the interpolatory algorithm of Result 22 can be evaluated as tightly as desired by means of Result 8, possibly taking into account noise uncorrelation properties.
- If $K = K_\gamma^{(3)}$, only a bound $E_g(\phi) \leq \overline{E}$ on the global error is provided (Result 23). Possible deterministic uncorrelation information on noise is not accounted for.
- If $K = K_\gamma^{(4)}$, an upper bound on the local error is provided, whose overbounding is not greater than $\sqrt{2}$ (Result 24). Also in this case, possible uncorrelation information on noise is not accounted for.

- *frequency shaping of uncertainty*

  Methods for tight evaluation of the frequency shaping of uncertainty of the identified model are available only for the case $K = K_{\rho,L}^{(2)}$ (Result 8).
- *a posteriori optimality*

  Methods for evaluating the optimality level of the identified model are available only for the case $K = K_{\rho,L}^{(2)}$ (Result 13).

Table 1 summarizes the main results presented and discussed here.

Algorithm	Convergence	"A priori" optimality	Error tightness	Frequency shaping	"A posteriori" optimality	Computational complexity
Linear "untuned"	No	No	No	No	No	Low
Linear "tuned"	Simple	No	No	No	No	Low
Nonlinear "two-stage"	Robust	No	No	No	No	Low
Nonlinear "interpolatory"						
with $K = K_{\rho,L}^{(1)}$	Simple/strong*	Almost	No	No	No	High
with $K = K_{\rho,L}^{(2)}$	Simple/strong*	Almost	Yes	Yes	Yes	Medium
with $K = K_\gamma^{(3)}$	Simple/strong*	Almost	No	No	No	Medium
with $K = K_\gamma^{(4)}$	Simple/strong*	Almost	$\sqrt{2}$	No	No	Low

* If deterministically uncorrelated noise is assumed.

Table 1. Comparison among different algorithms.

References

1. Akçay H, Hjalmarsson H (1994) The least-squares identification of FIR systems subject to worst-case noise. Systems Control Lett., 23:329–338
2. Akçay H, Ninness B (1998) On the worst-case divergence of the least-squares algorithm. Systems Control Lett., 33:19–24

3. Andersson L, Rantzer A, Beck C (1999) Model comparison and simplification. International Journal of Robust and Nonlinear Control, 9:157–181

4. Ball JA, Gohberg I, Rodman L (1990) Interpolation of rational matrix functions. Birkhäuser, Cambridge, MA

5. Chen J (1997) Frequency-domain tests for validation of linear fractional uncertain models. IEEE Trans. Automat. Control, AC-42(6):748–760

6. Chen J, Gu G (2000) Control-oriented system identification: an H_∞ approach. John Wiley & Sons, Inc., New York

7. Chen J, Nett CN (1995) The Carathéodory-Fejér problem and H_∞/ℓ_1 identification: a time domain approach. IEEE Trans. Automat. Control, AC-40(4):729–735

8. Chen J, Nett CN, Fan MKH (1992) Worst-case system identification in H_∞: validation of *a priori* information, essentially optimal algorithms, and error bounds, 251–257. In: Proc. of the American Control Conference, Chicago, IL

9. Chen J, Nett CN, Fan MKH (1995) Worst case system identification in H_∞: validation of *a priori* information, essentially optimal algorithms, and error bounds. IEEE Trans. Automat. Control, AC-40(7):1260–1265

10. Garulli A, Tesi A, Vicino A (eds), (1999) Robustness in identification and control, vol. 245 of Lecture Notes in Control and Inform. Sci. Springer-Verlag, Godalming, UK

11. Giarré L, Milanese M, Taragna M (1997) H_∞ identification and model quality evaluation. IEEE Trans. Automat. Control, AC-42(2):188–199

12. Glaum M, Lin L, Zames G (1996) Optimal H_∞ approximation by systems of prescribed order using frequency response data, 2318–2321. In: Proc. of the 35th IEEE Conf. on Decision and Control, Kobe, Japan

13. Gu G, Chu CC, Kim G (1994) Linear algorithms for worst case identification in h_∞ with applications to flexible structures, 112–116. In: Proc. of the American Control Conference, Baltimore, MD

14. Gu G, Khargonekar PP (1992a) A class of algorithms for identification in H_∞. Automatica, 28(2):299–312

15. Gu G, Khargonekar PP (1992b) Linear and nonlinear algorithms for identification in H_∞ with error bounds. IEEE Trans. Automat. Control, AC-37(7):953–963

16. Gu G, Xiong D, Zhou K (1993) Identification in H_∞ using Pick's interpolation. Systems Control Lett., 20:263–272

17. Hakvoort RG, van den Hof PMJ (1995) Consistent parameter bounding identification for linearly parametrized model sets. Automatica, 31(7):957–969

18. Helmicki AJ, Jacobson CA, Nett CN (1991) Control oriented system identification: a worst-case/deterministic approach in H_∞. IEEE Trans. Automat. Control, AC-36(10):1163–1176

19. Helmicki AJ, Jacobson CA, Nett CN (1993) Least squares methods for H_∞ control-oriented system identification. IEEE Trans. Automat. Control, AC-38(5):819–826

20. Kon MA, Tempo R (1989) On linearity of spline algorithms. Journal of Complexity, 5(2):251–259

21. Ljung L, Guo L (1997) The role of model validation for assessing the size of the unmodeled dynamics. IEEE Trans. Automat. Control, AC-42(9):1230–1239

22. Ljung L, Yuan ZD (1985) Asymptotic properties of black-box identification of transfer functions. IEEE Trans. Automat. Control, AC-30(6):514–530

23. Mäkilä PM, Partington JR, Gustafsson TK (1995) Worst-case control-relevant identification. Automatica, 31(12):1799–1819

24. Marchuk AG, Oshipenko KY (1975) Best approximation of functions specified with an error at a finite number of points. Mat. Zametki, 17:359–368 (in Russian; English Transl., *Math. Notes*, vol. 17, 207-212, 1975)

25. Milanese M, Norton J, Piet-Lahanier H, Walter É (eds) (1996) Bounding approaches to system identification. Plenum Press, New York

26. Milanese M, Novara C, Taragna M (2001) "Fast" set membership H_∞ identification from frequency-domain data, 1698–1703. In: Proc. of European Control Conf. 2001, Porto, Portugal

27. Milanese M, Taragna M (2000) Set membership identification for H_∞ robust control design. In: Proc. of 12th IFAC Symposium on System Identification SYSID 2000, Santa Barbara, CA

28. Milanese M, Taragna M (2001) Nearly optimal model sets in H_∞ identification, 1704–1709. In: Proc. of the European Control Conf. 2001, Porto, Portugal

29. Milanese M, Taragna M (2002) Optimality, approximation, and complexity in set membership H_∞ identification. IEEE Trans. Automat. Control, AC-47(10):1682–1690

30. Milanese M, Tempo R (1985) Optimal algorithms theory for estimation and prediction. IEEE Trans. Automat. Control, AC-30(8):730–738

31. Milanese M, Vicino A (1991) Optimal estimation theory for dynamic systems with set membership uncertainty: an overview. Automatica, 27(6):997–1009

32. Nehari Z (1957) On bounded bilinear forms. Ann. Math., 65:153–162

33. Ninness B, Goodwin GC (1995) Estimation of model quality. Automatica, 31(12):1771–1797

34. Paganini F (1996) A set-based approach for white noise modeling. IEEE Trans. Automat. Control, AC-41(10):1453–1465

35. Parker PJ, Bitmead RR (1987) Adaptive frequency response identification, 348–353. In: Proc. of the 26th IEEE Conf. on Decision and Control, Los Angeles, CA

36. Partington JR (1992) Robust identification in H_∞. J. Math. Anal. Appl., 166:428–441

37. Partington JR (1997) Interpolation, Identification, and Sampling, vol. 17 of London Math. Soc. Monographs New Series. Clarendon Press - Oxford, New York

38. Partington JR, Mäkilä PM (1995) Worst-case analysis of the least-squares method and related identification methods. Systems Control Lett., 24:193–200

39. Poolla K, Khargonekar P, Tikku A, Krause J, Nagpal K (1994) A time-domain approach to model validation. IEEE Trans. Automat. Control, AC-39(5):951–959

40. Rosenblum M, Rovnyak J (1985) Hardy classes and operator theory. Oxford Univ. Press, New York

41. Smith RS, Dahleh M (eds) (1994) The modeling of uncertainty in control systems, vol. 192 of Lecture Notes in Control and Inform. Sci. Springer-Verlag, London

42. Smith RS, Doyle JC (1992) Model validation: a connection between robust control and identification. IEEE Trans. Automat. Control, AC-37(7):942–952

43. Traub JF, Wasilkowski GW, Woźniakowski H (1988) Information-based complexity. Academic Press, New York

44. Venkatesh SR, Dahleh MA (1997) Identification in the presence of classes of unmodeled dynamics and noise. IEEE Trans. Automat. Control, AC-42(12):1620–1635

45. Vidyasagar M (1996) A theory of learning and generalization with application to neural networks and control systems. Springer-Verlag, New York

46. Zhou T (2001) On the consistency between an LFT described model set and frequency domain data. IEEE Trans. Automat. Control, AC-46(12):2001–2007
47. Zhou T, Kimura H (1993) Time domain identification for robust control. Systems Control Lett., 20(3):167–178
48. Zhou T, Wang L, Sun Z (2002) Closed-loop model set validation under a stochastic framework. Automatica, 38(9):1449–1461

Algebraic Methods for Nonlinear Systems: Parameter Identification and State Estimation

John Chiasson,[1] Kaiyu Wang,[1] Mengwei Li,[1] Marc Bodson,[2] and Leon M. Tolbert[1,3]

[1]ECE Department, The University of Tennessee, Knoxville, TN 37996, USA
  {chiasson,wkaiyu,mwl,tolbert}@utk.edu
[2]ECE Department The University of Utah, Salt Lake City, UT 84112, USA
  bodson@ece.utah.edu
[3]Oak Ridge National Laboratory, NTRC, 2360 Cherahala Boulevard, Knoxville,
  TN 37932, USA
  tolbertlm@ornl.gov

Summary. Algebraic methods are presented for solving nonlinear least-squares type problems that arise in the parameter identification of nonlinear systems. The tracking of the induction motor rotor time constant is solved in detail. Also, an approach to estimating state variables using algebraic relationships (in contrast to dynamic observers) is discussed in the context of speed estimation for induction motors.

1 Introduction

Algebraic methods have long been used as tools for problems in linear systems theory. In this chapter, some algebraic tools are explored as methods for nonlinear parameter identification and for constructing observers for nonlinear systems.

Parameter identification and state estimation continue to be an important areas of research precisely because they are used in many practical engineering problems. For example, the parameters characterizing the internal working of a physiological system are almost never available for direct measurement and therefore must be approached indirectly as a parameter estimation problem. The recent Bode Lecture by Ljung outlined many of the challenges of nonlinear system identification as well as its particular importance to biological systems [18]. In these types of problems, the model developed for analysis is typically a nonlinear state space model with unknown parameter values. For example, in [24] a 15th order nonlinear differential equation model based on first principles is developed to model blood glucose uptake in humans, but only a few of the parameter values are directly measurable. In addition to the parameters not being directly measured, very few of the state variables are measurable, so the model must be reformulated as an input-output model, which invariably results in overparameterization. Though a standard least squares (regressor) approach can "theoretically" be applied, the overparameterized model typically results in an ill-conditioned problem and unreliable results. In this chapter a method is proposed that can be used for parameter identification of a significant class of such systems of the type just described.

This approach is related to the differential algebra tools for analysis of nonlinear systems developed by Fliess [9][26], which has led to a clearer understanding of the nonlinear identification problem. Ollivier [22] as well as Ljung and Glad [19] have developed the use of the characteristic set of an ideal as a tool for identification problems. In particular, Ollivier developed the notion of the exhaustive summary of the model [22]. The use of these differential algebraic methods for system identification has also been considered in [20][25]. The focus of that research has been the determination of a priori identifiability of a given system model. However, as stated in [25], the development of an efficient algorithm using these differential algebraic techniques is still unknown. Though related, the approach proposed in this chapter is different from that in [20][25]. The approach presented in [25] typically leads to an overparameterized system that is well known to be ill-conditioned. In [20] the emphasis is on a priori global identifiability conditions.

Here it is shown, using the nonlinear techniques of elimination theory, that a significant class of nonlinear identification problems can be formulated as a nonlinear least-squares problem whose minimum value can be guaranteed to be found in a finite number of steps. The proposed methodology starts with an input-output linearly overparameterized model whose parameters are *rationally* related. (This is not atypical in many engineering examples as will be shown below.) After making appropriate substitutions, the problem is transformed into a *nonlinear* least-squares problem that is not overparameterized. It is then shown how the nonlinear least-squares problem can be solved using elimination theory. The computational issues are the *symbolic* computation of the Sylvester matrices to compute resultant polynomials based on elimination theory and the numerical computation of the roots of polynomials of high degree in a *single* variable. These issues are addressed below. Another important distinction of this work is our concern about actually computing the parameter values needed to characterize the input-output response rather than determining whether a system is a priori identifiable (cf. [20]). Even if a nonlinear state space model is not a priori identifiable in the sense of [20][25] (i.e., all of the parameters of the *state space* model being determined from input-output data), the model can be still be entirely adequate for the application at hand. For example, the induction motor is not a priori identifiable, yet the parameters that can be identified are complete enough to characterize the system for control purposes [27][36].

The technique described in this chapter has potential to be an efficient and practical method for use in a variety of real world systems, especially in biological systems (see [2][3][4][15]). Unlike man-made systems such as electric machines, airplanes, and space vehicles, where the use of sensors is often a cost issue, in biological systems the technology is often not available to fully sense (measure) all the variables in a living organism. Consequently, one can only characterize such systems using input-output data.

2 Examples of Nonlinear System Models

In this section, a set of mathematical models is given that characterize various biological systems. This is done to show that the class of problems to be considered here is significant.

Example 1. The model for a glucose uptake system is (from [18])

$$\frac{dx_1}{dt} = -\theta_1 \frac{x_1/\theta_2 - x_2/\theta_3}{1 + x_1/\theta_2 + x_2/\theta_3} + \theta_4 \frac{u - x_1}{1 + u/\theta_5 + x_1/\theta_5 + x_1 u/\theta_5}$$

$$\frac{dx_2}{dt} = \theta_1 \frac{x_1/\theta_2 - x_2/\theta_3}{1 + x_1/\theta_2 + x_2/\theta_3} - \theta_6 \frac{x_2/\theta_7 - \theta_8}{1 + x_2/\theta_2 + \theta_8}$$

$$y = x_2.$$

The state space model is nonlinear in the unknown parameters θ_i.

Example 2. A two compartment model describing the kinetics of a drug in the human body is given by (from [25])

$$\frac{dx_1}{dt} = -\frac{k_{21} + V_M}{K_m + x_1} x_1 + k_{12}x_2 + b_1 u$$

$$\frac{dx_2}{dt} = k_{21}x_1 - (k_{02} + k_{12})x_2 \tag{1}$$

$$y = c_1 x_1.$$

If K_m were known a priori, then this model could be written using a linear regressor formulation if *both* state variables were measurable. On the other hand, as shown in [25], this can be rewritten as the input-output model

$$0 = \frac{d^2y}{dt^2}y^2 + k_{21}k_{02}y^3 - c_1b_1(k_{21} + k_{02})y^2u + (k_{21} + k_{12} + k_{02})\frac{dy}{dt}y^2$$

$$- c_1b_1y^2\frac{du}{dt} - K_mc_1^3b_1\frac{du}{dt} + (2k_{21}K_mk_{02} + k_{12}V_M + k_{02}V_M)c_1y^2$$

$$- 2(k_{12} + k_{02})c_1^2b_1K_myu + 2K_mc_1(k_{21} + k_{12} + k_{02})y\frac{dy}{dt} + 2K_mc_1y\frac{d^2y}{dt^2}$$

$$- 2K_mc_1^2b_1y\frac{du}{dt} + K_mc_1^2(k_{21}k_{02} + k_{12}V_M + k_{02}V_M)y - K_m^2b_1c_1^3(k_{12} + k_{02})u$$

$$+ K_m^2c_1^2\frac{d^2y}{dt^2} + K_m^2c_1^2(k_{21} + k_{12} + k_{02} + V_M)\frac{dy}{dt},$$

where it is now nonlinear in the parameters even if K_m is known.

Example 3. A model describing tumor targeting by antibodies is given by (from [30] where α, β, δ are known)

$$\frac{dx_1}{dt} = -k_4 x_2 - (k_3 + k_7) x_1 + u$$

$$\frac{dx_2}{dt} = k_3 x_1 - k_4 x_2 + k_5 x_2 (\alpha - x_3) + k_6 x_3 - k_5 x_2 \beta (\delta - x_4) + k_6 x_4$$

$$\frac{dx_3}{dt} = k_5 x_2 (\alpha - x_3) - k_6 x_4$$

$$\frac{dx_4}{dt} = k_5 x_2 \beta (\delta - x_4) - k_6 x_4$$

$$\frac{dx_5}{dt} = k_7 x_1$$

$$y = c_1 x_1.$$

This state space model is linear in the unknown parameters so that *if* all the state variables were measurable, linear least-squares techniques could be used to estimate the parameters. However, this is typically not the case and, as in the previous example, an input-output model results in a system that is nonlinear in the unknown parameters.

Example 4. A model to describe microbial growth in a batch reactor is (from [10])

$$\frac{dx_1}{dt} = \frac{p_1 x_2}{p_2 + x_1} x_1 - p_3 x_1$$

$$\frac{dx_2}{dt} = -p_4 \frac{p_1 x_2}{p_2 + x_1} x_1$$

$$x_1(0) = a$$

$$x_2(0) = b.$$

Though these systems are all rational functions of the state variables and the unknown parameter values, this is not an a priori condition to be able to apply the methodology proposed below. The core requirements are that the coefficients of the *input-output model* are rational functions of the unknown parameters with no requirement that the model itself be a rational function of the inputs and outputs. The methodology provides a way to compute the identifiable parameters, that is, those that characterize the input-output model.

3 Nonlinear Least-Squares Parameter Identification

To introduce the methodology, an explicit example is considered, namely the induction motor. Though biological systems represent a rich new area for the methodology, the electric machine model of an induction motor will serve to illustrate the power of the proposed approach. Experimental results are also presented.

Standard models of induction machines are available in the literature. For the development here, a two-phase equivalent state space model in the rotor coordinate system is the most convenient as given by [27]:

$$\frac{di_{Sx}}{dt} = \frac{1}{\sigma L_S}u_{Sx} - \gamma i_{Sx} + \frac{\beta}{T_R}\psi_{Rx} + n_p\beta\omega\psi_{Ry} + n_p\omega i_{Sy} \tag{2}$$

$$\frac{di_{Sy}}{dt} = \frac{1}{\sigma L_S}u_{Sy} - \gamma i_{Sy} + \frac{\beta}{T_R}\psi_{Ry} - n_p\beta\omega\psi_{Rx} - n_p\omega i_{Sx} \tag{3}$$

$$\frac{d\psi_{Rx}}{dt} = \frac{M}{T_R}i_{Sx} - \frac{1}{T_R}\psi_{Rx} \tag{4}$$

$$\frac{d\psi_{Ry}}{dt} = \frac{M}{T_R}i_{Sy} - \frac{1}{T_R}\psi_{Ry} \tag{5}$$

$$\frac{d\omega}{dt} = \frac{Mn_p}{JL_R}(i_{Sy}\psi_{Rx} - i_{Sx}\psi_{Ry}) - \frac{\tau_L}{J}, \tag{6}$$

where $\omega = d\theta/dt$ with θ the position of the rotor, n_p is the number of pole pairs, and i_{Sx}, i_{Sy} are the (two-phase equivalent) stator currents in the rotor coordinate system, and ψ_{Rx}, ψ_{Ry} are the (two-phase equivalent) rotor flux linkages also in the rotor coordinate system. The parameters of the model are the five electrical parameters, R_S and R_R (the stator and rotor resistances), M (the mutual inductance), L_S and L_R (the stator and rotor inductances), and the two mechanical parameters, J (the inertia of the rotor) and τ_L (the load torque). The symbols $T_R = L_R/R_R$, $\beta = M/(\sigma L_S L_R)$, $\sigma = 1 - M^2/(L_S L_R)$ and $\gamma = R_S/(\sigma L_S) + M^2 R_R/(\sigma L_S L_R^2)$ have been used to simplify the expressions. T_R is referred to as the *rotor time constant* while σ is called the *total leakage factor*.

To simplify the presentation, the (important) special case in which only the two parameters R_S, T_R need be identified is considered. These two parameters will vary due to Ohmic heating, and a method for online tracking of the values of T_R and R_S as they change due to Ohmic heating is presented. The electrical parameters M, L_S, σ are now assumed to be known and not varying. Measurements of the stator currents i_{Sa}, i_{Sb} and voltages u_{Sa}, u_{Sb} as well as the position θ of the rotor are assumed to be available with velocity reconstructed from the position measurements. However, the rotor flux linkages ψ_{Rx}, ψ_{Ry} are *not* assumed to be measured.

3.1 Linear Overparameterized Model

Standard *linear* least-squares methods for parameter estimation are based on equalities where known signals depend *linearly* on unknown parameters. The induction motor model does not fit in this category unless the rotor flux linkages are measured. However, as the rotor flux linkages are not usually measured, the first step is to develop an input-output model in which the

64   J. Chiasson, K. Wang, M. Li, M. Bodson, and L. M. Tolbert

fluxes ψ_{Rx}, ψ_{Ry} and their derivatives $d\psi_{Rx}/dt, d\psi_{Ry}/dt$ are eliminated. The four equations (2)–(5) can be used to solve for $\psi_{Rx}, \psi_{Ry}, d\psi_{Rx}/dt, d\psi_{Ry}/dt$, but one is left without another independent equation to set up a regressor system for the identification algorithm. A new set of independent equations is found by differentiating (2) and (3) to obtain

$$\frac{1}{\sigma L_s}\frac{du_{Sx}}{dt} = \frac{d^2 i_{Sx}}{dt^2} + \gamma\frac{di_{Sx}}{dt} - \frac{\beta}{T_R}\frac{d\psi_{Rx}}{dt} - n_p\beta\omega\frac{d\psi_{Ry}}{dt} - n_p\beta\psi_{Ry}\frac{d\omega}{dt}$$
$$- n_p\omega\frac{di_{Sy}}{dt} - n_p i_{Sy}\frac{d\omega}{dt} \tag{7}$$

and

$$\frac{1}{\sigma L_s}\frac{du_{Sy}}{dt} = \frac{d^2 i_{Sy}}{dt^2} + \gamma\frac{di_{Sy}}{dt} - \frac{\beta}{T_R}\frac{d\psi_{Ry}}{dt} + n_p\beta\omega\frac{d\psi_{Rx}}{dt} + n_p\beta\psi_{Rx}\frac{d\omega}{dt}$$
$$+ n_p\omega\frac{di_{Sx}}{dt} + n_p i_{Sx}\frac{d\omega}{dt}. \tag{8}$$

Next, (2)–(5) are solved for $\psi_{Rx}, \psi_{Ry}, d\psi_{Rx}/dt, d\psi_{Ry}/dt$ and substituted into (7) and (8) to obtain

$$0 = -\frac{d^2 i_{Sx}}{dt^2} + \frac{di_{Sy}}{dt}n_p\omega + \frac{1}{\sigma L_S}\frac{du_{Sx}}{dt} - \left(\gamma + \frac{1}{T_R}\right)\frac{di_{Sx}}{dt} - i_{Sx}\left(-\frac{\beta M}{T_R^2} + \frac{\gamma}{T_R}\right)$$
$$+ i_{Sy}n_p\omega\left(\frac{1}{T_R} + \frac{\beta M}{T_R}\right) + \frac{u_{Sx}}{\sigma L_S T_R} + n_p\frac{d\omega}{dt}i_{Sy} - n_p\frac{d\omega}{dt}\frac{1}{\sigma L_S(1 + n_p^2\omega^2 T_R^2)} \times$$
$$\left(-\sigma L_S T_R\frac{di_{Sy}}{dt} - \gamma i_{Sy}\sigma L_S T_R - i_{Sx}n_p\omega\sigma L_S T_R - \frac{di_{Sx}}{dt}n_p\omega\sigma L_S T_R^2\right.$$
$$\left. - \gamma i_{Sx}n_p\omega\sigma L_S T_R^2 + i_{Sy}n_p^2\omega^2\sigma L_S T_R^2 + n_p\omega T_R^2 u_{Sx} + T_R u_{Sy}\right) \tag{9}$$

$$0 = -\frac{d^2 i_{Sy}}{dt^2} - \frac{di_{Sx}}{dt}n_p\omega + \frac{1}{\sigma L_S}\frac{du_{Sy}}{dt} - \left(\frac{di_{Sy}}{dt} - i_{Sy}\left(-\frac{\beta M}{T_R^2} + \frac{\gamma}{T_R}\right)\right.$$
$$- i_{Sx}n_p\omega\left(\frac{1}{T_R} + \frac{\beta M}{T_R}\right) + \frac{u_{Sy}}{\sigma L_S T_R} - n_p\frac{d\omega}{dt}i_{Sx} + n_p\frac{d\omega}{dt}\frac{1}{\sigma L_S(1 + n_p^2\omega^2 T_R^2)} \times$$
$$\left(-\sigma L_S T_R\frac{di_{Sx}}{dt} - \gamma i_{Sx}\sigma L_S T_R + i_{Sy}n_p\omega\sigma L_S T_R + \frac{di_{Sy}}{dt}n_p\omega\sigma L_S T_R^2\right.$$
$$\left. + \gamma i_{Sy}n_p\omega\sigma L_S T_R^2 + i_{Sx}n_p^2\omega^2\sigma L_S T_R^2 - n_p\omega T_R^2 u_{Sy} + T_R u_{Sx}\right). \tag{10}$$

This set of equations may be rewritten in regressor form as

$$y(t) = W(t)K, \tag{11}$$

where $W(t) \in \Re^{2\times 8}, K \in \Re^8$, and $y(t) \in \Re^2$ are given by

$$W(t) = \begin{bmatrix} -\dfrac{di_{Sx}}{dt} & -\dfrac{di_{Sx}}{dt} + n_p\omega i_{Sy} + n_p\omega M\beta i_{Sy} + \dfrac{u_{Sx}}{\sigma Ls} & M\beta i_{Sx} & -i_{Sx} \\[3mm] -\dfrac{di_{Sy}}{dt} & -\dfrac{di_{Sy}}{dt} - n_p\omega i_{Sx} - n_p\omega M\beta i_{Sx} + \dfrac{u_{Sy}}{\sigma Ls} & M\beta i_{Sy} & -i_{Sy} \end{bmatrix}$$

$$n_p\frac{di_{Sy}}{dt}\frac{d\omega}{dt} + n_p^2\left(\omega i_{Sx}\frac{d\omega}{dt} - \omega^2\frac{di_{Sx}}{dt}\right) \quad +n_p^3\omega^3 i_{Sy}(1 + M\beta)+$$

$$-n_p\frac{di_{Sx}}{dt}\frac{d\omega}{dt} + n_p^2\left(\omega i_{Sy}\frac{d\omega}{dt} - \omega^2\frac{di_{Sy}}{dt}\right) \quad -n_p^3\omega^3 i_{Sx}(1 + M\beta)+$$

$$\frac{1}{\sigma Ls}\left(n_p^2\omega^2 u_{Sx} - n_p u_{Sy}\frac{d\omega}{dt}\right) \quad n_p i_{Sy}\frac{d\omega}{dt} - n_p^2\omega^2 i_{Sx} \quad n_p^2\left(i_{sx}\omega\frac{d\omega}{dt} - \omega^2\frac{di_{Sx}}{dt}\right)$$

$$\frac{1}{\sigma Ls}\left(n_p^2\omega^2 u_{Sy} + n_p u_{Sx}\frac{d\omega}{dt}\right) \quad -n_p i_{Sx}\frac{d\omega}{dt} - n_p^2\omega^2 i_{Sy} \quad n_p^2\left(i_{sy}\omega\frac{d\omega}{dt} - \omega^2\frac{di_{Sy}}{dt}\right)$$

$$n_p^2\left(\omega\frac{di_{Sx}}{dt}\frac{d\omega}{dt} - \omega^2\frac{d^2 i_{Sx}}{dt^2}\right) + \frac{di_{Sy}}{dt}n_p^3\omega^3 - \frac{n_p^2}{\sigma Ls}\left(\omega u_{Sx}\frac{d\omega}{dt} - \omega^2\frac{du_{Sx}}{dt}\right)$$

$$n_p^2\left(\omega\frac{di_{Sy}}{dt}\frac{d\omega}{dt} - \omega^2\frac{d^2 i_{Sy}}{dt^2}\right) - \frac{di_{Sx}}{dt}n_p^3\omega^3 - \frac{n_p^2}{\sigma Ls}\left(\omega u_{Sy}\frac{d\omega}{dt} - \omega^2\frac{du_{Sy}}{dt}\right)$$

$$\tag{12}$$

$$K \triangleq \left[\gamma \ \frac{1}{T_R} \ \frac{1}{T_R^2} \ \frac{\gamma}{T_R} \ T_R \ \gamma T_R \ \gamma T_R^2 \ T_R^2\right]^T \tag{13}$$

and

$$y \triangleq \begin{bmatrix} \dfrac{d^2 i_{Sx}}{dt^2} - n_p i_{Sy}\dfrac{d\omega}{dt} - n_p\omega\dfrac{di_{Sy}}{dt} - n_p^2\omega^2 M\beta i_{Sx} & -\dfrac{du_{Sx}/dt}{\sigma Ls} \\[3mm] \dfrac{d^2 i_{Sy}}{dt^2} + n_p i_{Sx}\dfrac{d\omega}{dt} + n_p\omega\dfrac{di_{Sx}}{dt} - n_p^2\omega^2 M\beta i_{Sy} & -\dfrac{du_{Sy}/dt}{\sigma Ls} \end{bmatrix}.$$

Recall that $\beta = M/(\sigma L_S L_R)$ and $\gamma = R_S/(\sigma L_S) + M^2 R_R/(\sigma L_S L_R^2)$. Because $M^2/L_R = (1 - \sigma)L_S$ and $M\beta = (1 - \sigma)/\sigma$, it is seen that y and W depend only on *known* quantities while the unknowns R_S, T_R are contained only within K.

Though the system regressor (11) is linear in the parameters, a standard linear least-squares technique will give unreliable results as the system is overparameterized. The overparameterization is seen by the relationships

$$K_3 = K_2^2, K_4 = K_1 K_2, K_5 = 1/K_2, K_6 = K_1/K_2, K_7 = K_1/K_2^2, K_8 = 1/K_2^2,$$
$$\tag{14}$$

which shows that only the two parameters K_1, K_2 are independent. These two parameters determine R_S and T_R by

$$T_R = 1/K_2, R_S = \sigma L_S K_1 - (1 - \sigma)L_S K_2. \tag{15}$$

3.2 Nonlinear Least-Squares Identification

Using the model (11), the nonlinear least-squares method involves minimizing

$$E^2(K) = \sum_{n=1}^{N} \left| y(n) - W(n)K \right|^2 = R_y - 2R_{Wy}^T K + K^T R_W K \qquad (16)$$

subject to the constraints (14). On physical grounds, the parameters K_1, K_2 are constrained to

$$0 < K_1 < \infty, 0 < K_2 < \infty. \qquad (17)$$

Also, based on physical grounds, the squared error $E^2(K)$ will be minimized in the interior of this region. Define

$$E^2(K_1, K_2) \triangleq \sum_{n=1}^{N} \left| y(n) - W(n)K \right|^2 \Big|_{\substack{K_3=K_2^2 \\ K_4=K_1 K_2}}$$

$$\vdots$$

$$= R_y - 2R_{Wy}^T K \Big|_{\substack{K_3=K_2^2 \\ K_4=K_1 K_2}} + \left(K^T R_W K \right) \Big|_{\substack{K_3=K_2^2 \\ K_4=K_1 K_2}} \qquad (18)$$

$$\vdots \qquad \qquad \vdots$$

As just explained, the minimum of (18) must occur in the interior of the region and therefore at an extremum point. This then entails solving the two equations

$$\frac{\partial E^2(K_1, K_2)}{\partial K_1} = 0 \quad \text{and} \quad \frac{\partial E^2(K_1, K_2)}{\partial K_2} = 0. \qquad (19)$$

The partial derivatives in (19) are *rational* functions in the parameters K_1, K_2. Defining

$$p_1(K_1, K_2) \triangleq K_2^4 \frac{\partial E^2(K_1, K_2)}{\partial K_1} \quad \text{and} \quad p_2(K_p) \triangleq K_2^5 \frac{\partial E^2(K_1, K_2)}{\partial K_2} \qquad (20)$$

results in the $p_i(K_1, K_2)$ being *polynomials* in the parameters K_1, K_2 and having the same positive zero set (i.e., the same roots satisfying $K_i > 0$) as the system (19). The degrees of the polynomials p_i are given in Table 1.

	deg K_1	deg K_2
$p_1(K_1, K_2)$	1	7
$p_2(K_1, K_2)$	2	8

Table 1. Degrees of the polynomials in (20).

The point of this previous development was to show that the online identification of the parameters T_R and R_S can be reduced to solving two polynomials simultaneously. This is an important development because all possible solutions to this set may be found using elimination theory as is now summarized.

Digression—Solving Systems of Polynomial Equations [5][13]

To explain, let $a(K_1, K_2)$ and $b(K_1, K_2)$ be polynomials in K_1 whose coefficients are polynomials in K_2. Then, for example, if $a(K_1, K_2)$ and $b(K_1, K_2)$ have degrees 3 and 2, respectively, in K_1, they may be written in the form

$$a(K_1, K_2) = a_3(K_2)K_1^3 + a_2(K_2)K_1^2 + a_1(K_2)K_1 + a_0(K_2)$$
$$b(K_1, K_2) = b_2(K_2)K_1^2 + b_1(K_2)K_1 + b_0(K_2).$$

The $n \times n$ *Sylvester* matrix, where $n = \deg_{K_2}\{a(K_1, K_2)\} + \deg_{K_2}\{b(K_1, K_2)\} = 3 + 2 = 5$, is defined by

$$S_{a,b}(K_2) = \begin{bmatrix} a_0(K_2) & 0 & b_0(K_2) & 0 & 0 \\ a_1(K_2) & a_0(K_2) & b_1(K_2) & b_0(K_2) & 0 \\ a_2(K_2) & a_1(K_2) & b_2(K_2) & b_1(K_2) & b_0(K_2) \\ a_3(K_2) & a_2(K_2) & 0 & b_2(K_2) & b_1(K_2) \\ 0 & a_3(K_2) & 0 & 0 & b_2(K_2) \end{bmatrix}. \tag{21}$$

The *resultant* polynomial is then defined by

$$r(K_2) = \text{Res}\left(a(K_1, K_2), b(K_1, K_2), K_1\right) \triangleq \det S_{a,b}(K_2) \tag{22}$$

and is the result of *eliminating* the variable K_1 from $a(K_1, K_2)$ and $b(K_1, K_2)$. In fact, the following is always true.

Theorem 1. *Any solution* (K_{10}, K_{20}) *of* $a(K_1, K_2) = 0$ *and* $b(K_1, K_2) = 0$ *must have* $r(K_{20}) = 0$.

Using the polynomials (20), the variable K_1 is eliminated to obtain

$$r(K_2) \triangleq \text{Res}\left(p_1(K_1, K_2), p_2(K_1, K_2), K_1\right)$$

where it turns out that $\deg_{K_2}\{r(K_2)\} = 20$. The parameter K_2 was chosen as the variable *not* eliminated because its degree is much higher than K_1, meaning it would have a larger (in dimension) Sylvester matrix. The positive roots of $r(K_2) = 0$ are found, which are then substituted back into $p_1(K_1, K_2) = 0$ (or $p_2(K_1, K_2) = 0$) to find the positive roots in K_1, etc. By this method of back solving, all possible (finite number) candidate solutions are found and one simply chooses the one that gives the smallest squared error.

   Remark Rather than use resultants, one could also consider computing a Gröbner basis to find the zero set [5][6][13]. It has been our experience that the standard resultant computation works more efficiently than the standard computations of a Gröbner basis. However, this needs to be more fully researched. From a numerical point of view, there are results showing that homotopy methods can guarantee that all solutions can be found to systems of *polynomial* equations [21].

68   J. Chiasson, K. Wang, M. Li, M. Bodson, and L. M. Tolbert

After finding the solution that gives the minimal value for $E^2(K_1, K_2)$, one needs to know if the solution makes sense. For example, in the *linear* least-squares problem, there is a unique well-defined solution provided that the regressor matrix R_W is nonsingular (or, in practical terms, its condition number is not too large). In the nonlinear case here, a Taylor series expansion in $K_p \triangleq (K_1, K_2)$ about the computed minimum point $K_p^* \triangleq (K_1^*, K_2^*)$ gives $(i, j = 1, 2)$

$$E^2(K_p) = E^2(K_p^*) + \frac{1}{2} \left[K_p - K_p^* \right]^T \frac{\partial^2 E^2(K_p^*)}{\partial K_i \partial K_j} \left[K_p - K_p^* \right] + \cdots . \quad (23)$$

One then checks that the Hessian matrix $\partial^2 E^2(K_p^*)/\partial K_i \partial K_j$ is positive definite as well as its condition number to ensure that the data are sufficiently rich to identify the parameters.

Experimental Results

A three-phase, 230 V, 0.5 Hp, 1735 rpm ($n_p = 2$ pole-pair) induction machine was used to carry out the experiments. A 4096 pulse/rev optical encoder was attached to the motor for position measurements. The motor was connected to a three-phase 60 Hz source through a switch. When the switch was closed, the stator voltages (see Figure 1), the stator currents (see Figure 2), and the rotor speed (see Figure 3) were sampled at 4 kHz. The collected data were processed to obtain $y(t), W(t)$ for (11). Using (18), the polynomials (20) were formed and their solutions computed using resultants. There were three

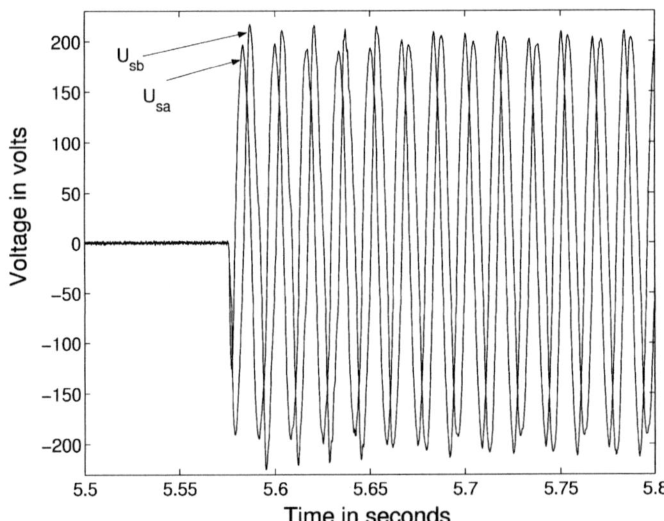

Fig. 1. Sampled two-phase equivalent voltages u_{Sa} and u_{Sb}.

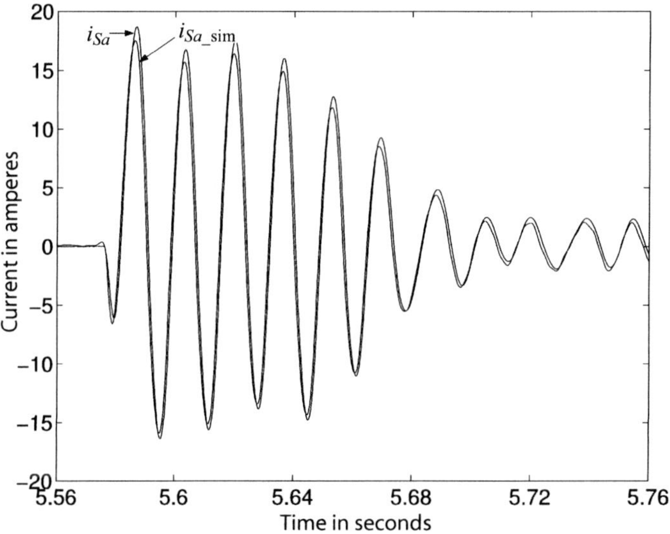

Fig. 2. Phase a current i_{Sa} and its simulated response i_{Sa_sim}.

different sets of solutions with $K_1 > 0, K_2 > 0$ and the solution pair that minimized $E^2(K_1, K_2)$ was then chosen resulting in $K_1 = 241.1$ and $K_2 = 7.6$. The estimated machine parameters using (15) are

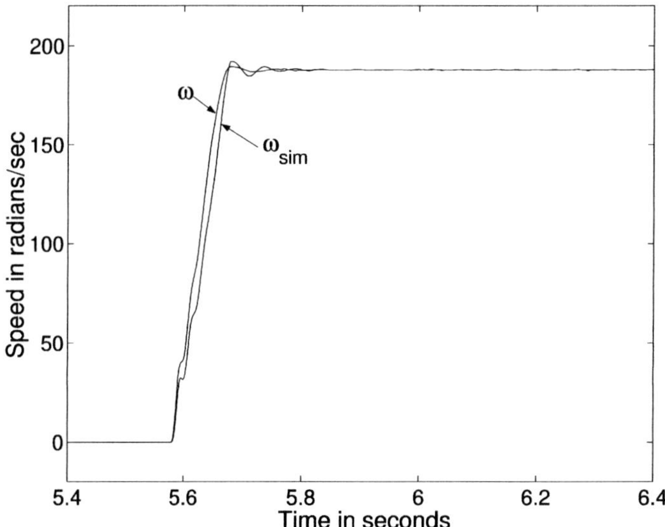

Fig. 3. Calculated speed ω and simulated speed ω_{sim}.

$$T_R = 0.132 \text{ s and } R_S = 5.1 \ \Omega. \tag{24}$$

For comparison, the stator resistance was measured using an ohmmeter giving a value of 4.9 ohms. The sampled two-phase equivalent current i_{Sa} and its simulated response i_{Sa_sim} are shown in Figure 2. The current i_{Sa_sim} is from a simulation using the measured input voltages from the experiment and the identified parameters from (24)—the other parameters are known. Note that both T_R and R_S are in the stator current equations (2) and (3). Further, the Hessian matrix was calculated at the minimum point according to (23) and was positive definite with a condition number of 295.

It turns out that for this system to be sufficiently excited (i.e., the Hessian is positive definite so that T_R and R_S are identifiable), it is enough that the motor operates at constant speed under load (nonzero rotor currents).

4 Computational Issues

As seen from the development of the proposed estimation method, the main computational issues are the *symbolic* computation of the Sylvester matrices to compute resultant polynomials based on elimination theory and the numerical computation of the roots of polynomials of high degree in a single variable. To increase the capability of the proposed method, these issues must be addressed.

The results in [11][12] for the symbolic computation of the determinant of a matrix show the potential for speeding up this computation by orders of magnitude over existing methods. The idea of the algorithm in [11][12] is based on polynomial methods in control and the discrete Fourier transform. To summarize the approach, rewrite the resultant polynomial (22) in the form

$$r(K_1) = \sum_{i=0}^{N} p_i K_1^i, \tag{25}$$

where the unknowns p_i and N are to be found. Any upper bound of the actual degree of $r(K_1)$ can be used for N. Such an upper bound is easily computed by finding the minimum of the sum of either the row or the column degrees of the Sylvester matrix [14]. Let $K_{1k} = e^{-j\frac{2\pi k}{N+1}}, n = 0, 1, ..., N$ be $N+1$ different values of K_1. Then the discrete Fourier transform (DFT) of the set of numbers $\{p_0, p_1, ..., p_N\}$ is

$$y_k = \sum_{i=0}^{N} p_i e^{-j\frac{2\pi k}{N+1}i} = \sum_{i=0}^{N} p_i \left(e^{-j\frac{2\pi}{N+1}}\right)^k,$$

with inverse DFT given by

$$p_i \triangleq \frac{1}{N+1} \sum_{k=0}^{N} y_k e^{j\frac{2\pi i}{N+1}k}.$$

Here y_k is just (25) evaluated at $K_{1k} = e^{-j\frac{2\pi k}{N+1}}$. That is, one computes the *numerical* determinant of Sylvester matrix of the two polynomials (see (21) and (22)) at the $N+1$ points K_{1k} (this is fast) and obtains the DFT of the coefficients of (25). Then the p_i are computed using the inverse DFT. That is, the symbolic calculation of the determinant is reduced to a finite number of fast *numerical* calculations. Such an approach has been shown to be as much as 500 times faster than existing methods [11].

The computation of the roots of a polynomial in one variable has been the object of research by numerical analysts for many years and is well documented in the literature (e.g., see [16]). In the research [36] on identifying the four (identifiable) parameters of the induction motor, it was necessary to use rational arithmetic to obtain accurate roots of a polynomial of degree 104. Care must be taken to ensure that the computed roots are indeed roots to the polynomial. Typically in an identification problem, one is looking for real roots that are restricted to a finite range simplifying the numerical search for the roots. (In particular, Sturm's theorem [see [29], p. 5] can be used to find the precise number of *real* roots in a given interval.)

5 Algebraic State Estimators

In this section, the idea of using purely algebraic methods to estimate the state variables of a system is discussed. The method involves computing derivatives of the output variables and solving for the unknown state variable. However, as the equations are nonlinear, the goal is not a dynamic observer whose stability must be ascertained, but rather an algebraic observer with no stability issue. The issue of numerically differentiating the output signals is of course an important practical concern, and methods to address this problem are considered in [7][8].

As in the previous section, the induction motor is used as an example. If the speed of an induction motor is measured (along with stator currents), then the field-oriented control methodology provides the capability to precisely control the torque of the machine. The speed (along with the rotor time constant and the stator currents) is required for constructing a stable observer to estimate the rotor flux linkages in an induction motor. However, the speed/position sensor adds significant cost to the system and is one of the least reliable parts of the system. As a consequence, it is of interest to develop a speed sensorless field-oriented controller.

Many different techniques have been proposed to estimate speed of an induction motor without a shaft sensor, but none has emerged as being completely satisfactory. Much literature exists in this area and the reader is referred to [1][17][23][28][31]–[35] for an exposition of many of the existing approaches. Currently, induction motors are used throughout industry in pumps, fans, manufacturing machinery, conveyor belt drives, etc., where they are run in open loop. However, in many of these situations a speed sensorless field-oriented controller would be a distinct advantage. That is, it is desirable to have a field-oriented controller for its performance without the cost and reliability issues of having a speed/position sensor. For example, a conveyor belt in a mine that brings out coal in buckets is typically powered by induction motors running in open loop from a 60 Hz voltage source. When power is lost, the coal buckets must all be emptied to reduce the load so that the motors can bring the conveyor belt system back up to speed. After the system is up to speed, the buckets may be refilled so that the motor is again fully loaded. The availability of a sensorless field-oriented controller capable of performing such a start up under full load would greatly reduce the down-time. Such reliable speed sensorless control algorithms would find application in a huge number of different industrial applications.

5.1 Space Vector Model of the Induction Motor

In the space vector model of the induction motor [17], one lets $\underline{i}_S = i_{Sa} + j i_{Sb}$, $\underline{\psi}_R = \psi_{Ra} + j\psi_{Rb}$, and $\underline{u}_S = u_{Sa} + j u_{Sb}$ and the induction motor model may be written as

$$\frac{d}{dt}\underline{i}_S = \frac{\beta}{T_R}\left(1 - jn_P\omega T_R\right)\underline{\psi}_R - \gamma\underline{i}_S + \frac{1}{\sigma L_S}\underline{u}_S \tag{26}$$

$$\frac{d}{dt}\underline{\psi}_R = -\frac{1}{T_R}\left(1 - jn_P\omega T_R\right)\underline{\psi}_R + \frac{M}{T_R}\underline{i}_S \tag{27}$$

$$\frac{d\omega}{dt} = \frac{n_p M}{J L_R}\left(i_{Sb}\psi_{Ra} - i_{Sa}\psi_{Rb}\right) - \frac{\tau_L}{J}. \tag{28}$$

To replace the speed sensor, the estimation of rotor speed is based on measured stator voltages and currents at the motor terminals. The first step is to eliminate the flux linkage $\underline{\psi}_R$ and its derivative $d\underline{\psi}_R/dt$ as they are not available measurements. The two equations (26) and (27) can be used to solve for $\underline{\psi}_R, d\underline{\psi}_R/dt$, but one is left without another independent equation to set up a speed estimator. A new independent equation is found by differentiating (26), which gives

$$\frac{d^2}{dt^2}\underline{i}_S = \frac{\beta}{T_R}\left(1 - jn_P\omega T_R\right)\frac{d}{dt}\underline{\psi}_R - jn_P\beta\underline{\psi}_R\frac{d\omega}{dt} - \gamma\frac{d}{dt}\underline{i}_S + \frac{1}{\sigma L_S}\frac{d}{dt}\underline{u}_S. \tag{29}$$

Using these three (complex valued) equations, one can eliminate $\underline{\psi}_R$ and $\frac{d}{dt}\underline{\psi}_R$ to obtain

$$\frac{d^2}{dt^2}\underline{i}_S = -\frac{1}{T_R}\left(1 - jn_P\omega T_R\right)\left(\frac{d}{dt}\underline{i}_S + \gamma\underline{i}_S - \frac{1}{\sigma L_S}\underline{u}_S\right) + \frac{\beta M}{T_R^2}\left(1 - jn_P\omega T_R\right)\underline{i}_S$$
$$- \gamma\frac{d}{dt}\underline{i}_S + \frac{1}{\sigma L_S}\frac{d}{dt}\underline{u}_S - \frac{jn_P T_R}{1 - jn_P\omega T_R}\left(\frac{d}{dt}\underline{i}_S + \gamma\underline{i}_S - \frac{1}{\sigma L_S}\underline{u}_S\right)\frac{d\omega}{dt}.$$

The equation for $d\omega/dt$ can be written as

$$\frac{d\omega}{dt} = -\frac{(1-jn_P\omega T_R)^2}{jn_P T_R^2} + \frac{1-jn_P\omega T_R}{jn_P T_R} \times$$

$$\frac{\dfrac{\beta M}{T_R^2}(1-jn_P\omega T_R)\underline{i}_S - \gamma\dfrac{d}{dt}\underline{i}_S + \dfrac{1}{\sigma L_S}\dfrac{d}{dt}\underline{u}_S - \dfrac{d^2}{dt^2}\underline{i}_S}{\dfrac{d}{dt}\underline{i}_S + \gamma\underline{i}_S - \dfrac{1}{\sigma L_S}\underline{u}_S}. \tag{30}$$

(These calculations are equivalent to those carried out for the parameter identification work in the previous section.)

Equation (30) actually leads to two possibilities for a speed observer. If the signals are measured exactly and the motor satisfies its dynamic model, the right-hand side must be real. Breaking down the right side of (30) into its real and imaginary parts, the *real* part has the form

$$\frac{d\omega}{dt} = a_2\left(u_{Sa}, u_{Sb}, i_{Sa}, i_{Sb}\right)\omega^2 + a_1\left(u_{Sa}, u_{Sb}, i_{Sa}, i_{Sb}\right)\omega + a_0\left(u_{Sa}, u_{Sb}, i_{Sa}, i_{Sb}\right). \tag{31}$$

The expressions for a_2, a_1, a_0 are lengthy and therefore not explicitly presented here. (The notational dependence is a little misleading as they depend on the derivatives of the currents and voltages as well.) Note that (31) is singular, i.e., the denominator of (31) is zero, if and only if $\left|\underline{\psi}_R\right| \equiv 0$. The equation (31) could be used as a dynamic speed observer. Specifically, one computes coefficients a_2, a_1, a_0 at each time step and then integrates the *nonlinear time-varying* differential equation (31) in real-time for an estimate $\hat{\omega}$. However, its stability must be ascertained, that is, if $\hat{\omega}(t_0) \neq \omega(t_0)$, will $\hat{\omega}(t) \longrightarrow \omega(t)$? For example, if $u_{Sa} = \text{constant}$ and $u_{Sb} = 0$ (during initial flux build up in the machine), it turns out that $a_2(u_{Sa}, u_{Sb}, i_{Sa}, i_{Sb}) = a_1(u_{Sa}, u_{Sb}, i_{Sa}, i_{Sb}) = a_0(u_{Sa}, u_{Sb}, i_{Sa}, i_{Sb}) \equiv 0$ and therefore $d\omega/dt \equiv 0$, which is an unstable observer under these conditions.[1]

On the other hand, the *imaginary* part of (30) must be zero leading to an algebraic equation satisfied by ω of the form

$$q(t, \omega) \triangleq q_2(u_{Sa}, u_{Sb}, i_{Sa}, i_{Sb})\omega^2 + q_1(u_{Sa}, u_{Sb}, i_{Sa}, i_{Sb})\omega +$$
$$q_0(u_{Sa}, u_{Sb}, i_{Sa}, i_{Sb})$$
$$\equiv 0. \tag{32}$$

If $q_2 \neq 0$, (32) can be solved to obtain

$$\omega_1 \triangleq \frac{-q_1 + \sqrt{q_1^2 - 4q_2 q_0}}{2q_2} \tag{33}$$

$$\omega_2 \triangleq \frac{-q_1 - \sqrt{q_1^2 - 4q_2 q_0}}{2q_2}. \tag{34}$$

---

[1] However, under normal operating conditions of an induction motor, the a_i are not identically zero.

At least one of these two solutions must track the motor speed. This speed estimator does not have any stability issue, but a procedure to determine whether ω_1 or ω_2 is the actual motor speed is required. Further, there are circumstances for which (32) is singular, that is, $q_2(u_{Sa}, u_{Sb}, i_{Sa}, i_{Sb}) = q_1(u_{Sa}, u_{Sb}, i_{Sa}, i_{Sb}) = q_0(u_{Sa}, u_{Sb}, i_{Sa}, i_{Sb}) = 0$. For example, during flux build up where $u_{Sa} = $ constant and $u_{Sb} = 0$, the q_i are all identically zero and the value of ω is not determined by (32).

In order to determine the correct solution ω_1 or ω_2, one can proceed as follows. Noting that $q(t, \omega) \equiv 0$, its derivative is also identically zero, that is,

$$(2q_2\omega + q_1)\frac{d\omega}{dt} + \dot{q}_2\omega^2 + \dot{q}_1\omega + \dot{q}_0 \equiv 0. \tag{35}$$

Next, substitute the right-hand side of (31) for $d\omega/dt$ to obtain

$$r(t, \omega) \triangleq 2q_2 a_2 \omega^3 + (2q_2 a_1 + q_1 a_2 + \dot{q}_2)\,\omega^2 + (2q_2 a_0 + q_1 a_1 + \dot{q}_1)\,\omega +$$
$$q_1 a_0 + \dot{q}_0 \equiv 0. \tag{36}$$

$r(t, \omega)$ in (36) is a third degree polynomial in ω. Dividing it by $q(t, \omega)$, which is a second degree polynomial in ω, the division algorithm gives

$$r(t, \omega) = (2a_2\omega + 2a_1 - q_1 a_2/q_2 + \dot{q}_2/q_2)\, q(t, \omega)$$
$$+ \left(2q_2 a_0 - q_1 a_1 + \dot{q}_1 - 2q_0 a_2 + q_1^2 a_2/q_2 - q_1\dot{q}_2/q_2\right)\omega$$
$$+ (q_1 a_0 + \dot{q}_0 - 2q_0 a_1 + q_0 q_1 a_2/q_2 - q_0 \dot{q}_2/q_2) \equiv 0. \tag{37}$$

As $q(t, \omega) \equiv 0$, it follows that

$$\left(2q_2 a_0 - q_1 a_1 + \dot{q}_1 - 2q_0 a_2 + q_1^2 a_2/q_2 - q_1\dot{q}_2/q_2\right)\omega$$
$$+ (q_1 a_0 + \dot{q}_0 - 2q_0 a_1 + q_0 q_1 a_2/q_2 - q_0 \dot{q}_2/q_2) \equiv 0. \tag{38}$$

Equation (38) is now a *first* order polynomial in ω with a unique solution provided the coefficient of ω in (38) is nonzero. This last equation involves 3rd derivatives of the stator currents and 2nd derivatives of the stator voltages and therefore will be noisier due to the differentiations compared to the solutions of (32) whose coefficients have 2nd derivatives of the stator currents and only one derivative of the stator voltages. To avoid this additional noise, one could just use the solution of (38) to determine which of ω_1, ω_2 is the correct motor speed. Specifically, choose either ω_1 or ω_2 as the speed estimate $\hat{\omega}$ depending on which is closest to the solution of (38). As the simulations below show, one of the two solutions ω_1, ω_2 is typically far from the actual motor speed.

5.2 Simulation of the Algebraic Estimator

As a first look at the viability of an algebraic observer, simulations are carried out. Here, a three-phase (two-phase equivalent) induction motor model was simulated with machine parameter values given by $n_p = 2, R_S = 5.12$

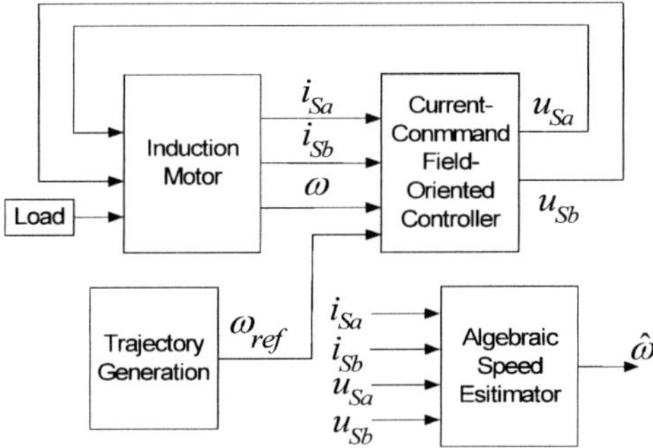

Fig. 4. Algebraic speed estimator.

ohms, $L_S = 0.2919$ H, $J = 0.0021$ kgm^2, $\tau_{L\_rated} = 2.0337$ Nm, $I_{max} = 2.77$ A, $V_{max} = 230$ V. Figure 4 shows the block diagram of the system. In this system, a current command field-oriented controller is used [17]. The induction motor model is based on (26)–(28). Figure 5 shows the simulation results of the motor speed and the two estimated speeds ω_1, ω_2 for a low speed trajectory with full rated load on the motor. From $t = 0$ to $t = 0.4$ seconds,

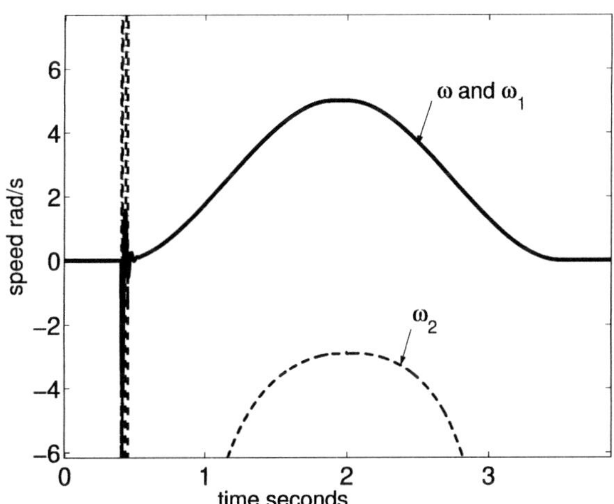

Fig. 5. ω, ω_1 and ω_2 for a low speed trajectory ($\omega_{max} = 5$) with full rated load.

a constant u_{Sa} is applied to the motor to build up the flux with $\omega \equiv 0$. At $t = 0.4$ seconds, the machine starts on the low speed trajectory ($\omega_{max} = 5$ rad/s) with *full load* at the start. Figure 6 is an enlarged view of Figure 5 between 0.3 seconds and 0.6 seconds. Note that the correct solution for the speed does alternate between ω_1 and ω_2.

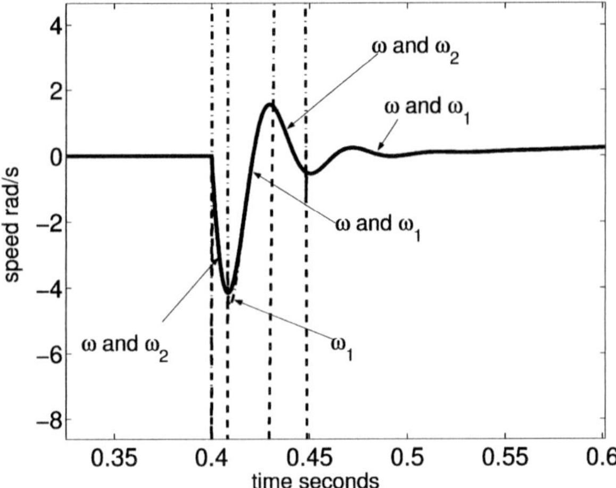

Fig. 6. An enlarged view of Figure 5.

6 Conclusions

In this work, the problem of parameter identification and state estimation for
nonlinear systems has been considered. It was shown that for a class of systems
whose input-output models are nonlinear in the parameters, but rationally related, the parameters could be identified by solving a nonlinear least-squares
problem using elimination theory. The problem of state estimation was considered from the point of view of finding an algebraic relationship between
the unknown state variable and the known outputs. Such an approach entails
computing higher order derivatives of the measured outputs and eliminating
the derivatives of the state variable to be estimated.

References

1. Bodson M, Chiasson J (2002) A comparison of sensorless speed estimation
   methods for induction motor control. In: Proceedings of the 2002 American
   Control Conference, Anchorage, AK
2. Chauvet GA (1995a) Theoretical systems in biology: hierarchical and functional
   integration: Volume I–Molecules and cells. Pergamon, Oxford, UK
3. Chauvet GA (1995b) Theoretical systems in biology: hierarchical and functional
   integration: Volume II–Tissues and organs. Pergamon, Oxford, UK
4. Chauvet GA (1995c) Theoretical systems in biology: hierarchical and functional
   integration: Volume III–Organisation and regulation. Pergamon, Oxford, UK
5. Cox D, Little J, O'Shea D (1996) Ideals, varieties, and algorithms: an introduction to computational algebraic geometry and commutative algebra, 2nd
   edition. Springer-Verlag, New York

6. Cox D, Little J, O'Shea D (1998) Using algebraic geometry. Springer-Verlag, New York

7. Diop S, Grizzle JW, Chaplais F (2000) On numerical differentiation algorithms for nonlinear estimation. In: Proceedings of the IEEE Conference on Decision and Control, Sydney, Australia

8. Diop S, Grizzle JW, Moraal PE, Stefanopoulou A (1994) Interpolation and numerical differentiation algorithms for observer design. 1329–1333. In: American Control Conference preprint

9. Fliess M, Glad ST (1993) An algebraic approach to linear and nonlinear control, In: Essays on control: perspectives in the theory and applications 223–267 Birkhäuser, Cambridge, MA

10. Holmberg A (1982) On the practical identifiability of microbial growth models incorporating Michaelis-Menten type nonlinearities. Mathematica Bioscience, 62:23–43

11. Hromcik M, Šebek M (1999a) New algorithm for polynomial matrix determinant based on FFT. In: Proceedings of the European Conference on Control ECC'99, Karlsruhe, Germany

12. Hromcik M, Šebek M (1999b) Numerical and symbolic computation of polynomial matrix determinant. In: Proceedings of the 1999 Conference on Decision and Control, Tampa, FL

13. Von zur Gathen J, Gerhard J (1999) Modern computer algebra. Cambridge University Press, Cambridge, UK

14. Kailath T (1980) Linear systems. Prentice-Hall, Englewood Cliffs, NJ

15. Keener J, Sneyd J (1998) Mathematical physiology. Springer-Verlag, New York

16. Kincaid D, Cheney W (2002) Numerical analysis: mathematics of scientific computing. Brooks/Cole, Pacific Grove, CA

17. Leonhard W (2001) Control of electrical drives, 3rd edition. Springer-Verlag, New York

18. Ljung L (2003) Challenges of nonlinear identification, Bode Lecture, IEEE Conference on Decision and Control, Maui, HI

19. Ljung L, Glad ST (1994) On global identifiability for arbitrary model parameterisations. Automatica 30(2):265–276

20. Margaria G, Riccomagno E, Chappell MJ, Wynn HP (2004) Differential algebra methods for the study of the structural identifiability of biological rational polynomial models Preprint

21. Morgan AP, Sommese AJ, Watson LT (1989) Finding all isolated solutions to polynomial systems using HOMPACK. ACM Transactions on Mathematical Software 15(2):93–122

22. Ollivier F (1990) Le problème de l'identifiabilité structurelle globale: Ètude thèorique, mèthodes effective et bornes de complexitè PhD thesis, École Polytéchnique, Paris, France

23. Rajashekara K, Kawamura A, Matsuse K (eds) (1996) Sensorless control of AC motor drives–speed and position sensorless operation, IEEE Press, Piscataway, NJ

24. Ruiz-Velázquez E, Femat R, Campos-Delgado DU (2004) Blood glucose control for type I diabetes mellitus: a robust tracking H_∞ problem. Control Engineering Practice, 12(9):1179–1195

25. Saccomani M (2004) Some results on parameter identification of nonlinear systems. Cardiovascular Engineering: An International Journal, 4(1):95–102

26. Sira-Ramirez H, Agrawal SK (2004) Differentially flat systems. Marcel-Dekker, New York
27. Stephan J, Bodson M, Chiasson J (1994) Real-time estimation of induction motor parameters. IEEE Transactions on Industry Applications 30(3):746–759
28. Strangas EG, Khalil HK, Oliwi BA, Laubnger L, Miller JM (1999) A robust torque controller for induction motors without rotor position sensor: analysis and experimental results. IEEE Transactions on Energy Conversion 14(4):1448–1458
29. Sturmfels B (2002) Solving systems of polynomial equations. CBMS Regional Conference Series in Mathematics, American Mathematical Society, Providence, RI
30. Thomas G, Chappell M, Dykes P, Ellis J, Ramsden D, Godfrey K, Bradwell A (1989) Effect of dose, molecular size, affinity and protein binding on tumor uptake of antibody or ligand: A biomathematical model. Cancer Research 49:3290–3296
31. Vas P (1998) Sensorless vector control and direct torque control. Oxford University Press, Oxford, UK
32. Vélez-Reyes M (1992) Decomposed algorithms for parameter estimation PhD thesis, Massachusetts Institute of Technology
33. Vélez-Reyes M, Fung WL, Ramos-Torres JE (2001) Developing robust algorithms for speed and parameter estimation in induction machines. In: Proceedings of the IEEE Conference on Decision and Control, Orlando, FL
34. Vélez-Reyes M, Minami K, Verghese G (1989) Recursive speed and parameter estimation for induction machines. In: Proceedings of the IEEE Industry Applications Conference, San Diego, CA
35. Vélez-Reyes M, Verghese G (1992) Decomposed algorithms for speed and parameter estimation in induction machines. 156–161. In: Proceedings of the IFAC Nonlinear Control Systems Design Symposium, Bordeaux, France
36. Wang K, Chiasson J, Bodson M, Tolbert LM (2004) A nonlinear least-squares approach for estimation of the induction motor parameters. In: Proceedings of the IEEE Conference on Decision and Control, to appear

Recent Techniques for the Identification of Piecewise Affine and Hybrid Systems

Aleksandar Lj. Juloski,[1] Simone Paoletti,[2] and Jacob Roll[3]

[1]Department of Electrical Engineering, Eindhoven University of Technology,
P.O. Box 513, 5600 MB Eindhoven, The Netherlands
A.Juloski@tue.nl
[2]Dipartimento di Ingegneria dell'Informazione, Università di Siena,
Via Roma 56, 53100 Siena, Italia
paoletti@dii.unisi.it
[3]Division of Automatic Control, Linköping University,
SE-581 83 Linköping, Sweden
roll@isy.liu.se

Summary. The problem of piecewise affine identification is addressed by studying four recently proposed techniques for the identification of PWARX/HHARX models, namely a Bayesian procedure, a bounded-error procedure, a clustering-based procedure and a mixed-integer programming procedure. The four techniques are compared on suitably defined one-dimensional examples, which help to highlight the features of the different approaches with respect to classification, noise and tuning parameters. The procedures are also tested on the experimental identification of the electronic component placement process in pick-and-place machines.

1 Introduction

This chapter focuses on the problem of identifying piecewise affine (PWA) models of discrete-time nonlinear and hybrid systems from input-output data. PWA systems are obtained by partitioning the state and input space into a finite number of nonoverlapping convex polyhedral regions, and by considering linear/affine subsystems sharing the same continuous state variables in each region. The interest in PWA identification techniques is motivated by several reasons. Since PWA maps have universal approximation properties [18, 10], PWA models represent an attractive black-box model structure for nonlinear system identification. In addition, given the equivalence between PWA systems and several classes of hybrid systems [4, 14], the many different analysis, synthesis, and verification tools for hybrid systems (see, e.g., [2, 21, 28] and references therein) can be applied to the identified PWA models. PWA systems have indeed many applications in different contexts such as neural networks, electrical networks, time-series analysis and function approximation.

In the extensive literature on nonlinear black-box identification (see, e.g., [27] and references therein), a few techniques can be found that lead to PWA models of nonlinear dynamical systems. An overview and classification of them is presented in [25]. Recently, novel contributions to this topic have been also proposed in the hybrid systems community [5, 6, 13, 16, 22, 26, 29]. Identification of PWA models is a challenging problem that involves the estimation of both the parameters of the affine submodels and the coefficients of the hyperplanes defining the partition of the state and input space (or the regressor space, for models in regression form). The main difficulty is that the identification problem includes a classification problem, in which each data point must be associated to one region and to the corresponding submodel. The problem is even harder when the number of submodels must also be estimated. In this chapter, four recently proposed techniques for the identification of (possibly) discontinuous PWA models are considered, namely the Bayesian procedure [16], the bounded-error procedure [5, 6], the clustering-based procedure [13], and the mixed-integer programming (MIP) procedure [26]. While the MIP procedure formulates the identification problem as a mixed-integer linear or quadratic program (MILP/MIQP) that can be solved for the global optimum, the other three procedures can only guarantee suboptimal solutions. On the other hand, the very high worst-case computational complexity of MILP/MIQP problems makes the approach in [26] affordable only when few data are available, or when data are clustered together. The four procedures are studied here for the classification accuracy and the effects of noise, overestimated model orders, and varying the tuning parameters on the identification results. The study of specific cases can indeed shed some light on the properties of the different techniques and guide the user in their application to practical situations.

This chapter is organized as follows. The PWA identification problem is formulated and discussed in Section 2. Section 3 describes the four compared procedures and introduces several quantitative measures for assessing the quality of the identified models. The different approaches of the four procedures to data classification are addressed in Section 4. The effects of the overestimation of model orders on the identification accuracy are investigated in Section 5, while Section 6 studies the effects of noise. In Section 7 the sensitivity of the identification results to tuning parameters is analyzed for the Bayesian, bounded-error, and clustering-based procedures. In Section 8 the four procedures are tested on experimental data from the electronic component placement process in pick-and-place machines [15]. Finally, conclusions are drawn in Section 9.

2 Problem formulation

Piecewise affine autoregressive exogenous (PWARX) models can be seen as collections of affine ARX models equipped with the switching rule determined by a polyhedral partition of the regressor set. Letting $k \in \mathbb{Z}$ be the time index, and $u(k) \in \mathbb{R}$ and $y(k) \in \mathbb{R}$ the system input and output, respectively, a PWARX model establishes a relationship between past observations and future outputs in the form

$$y(k) = f(x(k)) + e(k), \tag{1}$$

where $e(k) \in \mathbb{R}$ is the prediction error, $f(\cdot)$ is the PWA map

$$f(x) = \begin{cases} [x' \ 1]\theta_1 \ \text{if } x \in \mathcal{X}_1 \\ \quad \vdots \\ [x' \ 1]\theta_s \ \text{if } x \in \mathcal{X}_s, \end{cases} \tag{2}$$

defined over the regressor set $\mathcal{X} = \bigcup_{i=1}^s \mathcal{X}_i \subseteq \mathbb{R}^n$ on which the PWARX model is valid, and $x(k) \in \mathbb{R}^n$ is the regression vector with fixed structure depending only on past n_a outputs and n_b inputs:

$$x(k) = [y(k-1) \ \ldots \ y(k-n_a) \ u(k-1) \ \ldots \ u(k-n_b)]' \tag{3}$$

(hence, $n = n_a + n_b$). In (2) s is the number of submodels and $\theta_i \in \mathbb{R}^{n+1}$ are the parameter vectors (PVs) of the affine ARX submodels. The regions $\mathcal{X}_i$ are convex polyhedra that do not overlap, i.e., $\mathcal{X}_i \cap \mathcal{X}_j = \emptyset$ for all $i \neq j$. Hence, for a given data point $(y(k), x(k))$, the corresponding active *mode* $\mu(k)$ can be uniquely defined as

$$\mu(k) = i \ \ \text{iff} \ \ x(k) \in \mathcal{X}_i. \tag{4}$$

Given the data set $\mathcal{D} = \{(y(k), x(k))\}_{k=1}^N$, the considered identification problem consists of finding the PWARX model that best matches the data according to some specified criterion of fit (e.g., the minimization of the sum of absolute or squared prediction errors, see [27]). For fixed model orders n_a and n_b, this problem involves the estimation of the number of submodels s, the PVs $\{\theta_i\}_{i=1}^s$, and the polyhedral partition $\{\mathcal{X}_i\}_{i=1}^s$. It also includes a classification problem in which each data point is associated to one region and the corresponding submodel. In general, the simultaneous optimal estimation of all the quantities above leads to very complex, nonconvex optimization problems with potentially many local minima, which complicate the use of local search minimization algorithms. One of the main difficulties concerns the selection of the number of submodels s. Constraints on s must be introduced in order to keep the number of submodels low and to avoid overfit. Heuristic and suboptimal approaches to the identification of PWARX models have been proposed in the literature (see [25] for an overview). Most of these approaches either assume a fixed s, or adjust s iteratively (e.g., by adding one submodel at a time) in order to improve the fit. When s is fixed, the identification of a PWARX model amounts to a PWA regression problem, namely the problem of reconstructing the PWA map $f(\cdot)$ from the finite data set $\mathcal{D}$.

Note that, if the partition of the regressor set is either known or fixed a priori, the problem complexity reduces to that of a linear identification problem, since the data points can be classified to corresponding data clusters $\{\mathcal{D}_i\}_{i=1}^s$, and standard linear identification techniques can be applied to estimate the PVs for each submodel [19].

3 The compared procedures

In this section, four recently proposed procedures for the identification of PWA models are briefly introduced and described: the Bayesian procedure [16], the bounded-error procedure [5, 6], the clustering-based procedure [13], and the MIP procedure [26].

In its basic formulation, the MIP procedure considers hinging-hyperplane ARX models, which form a subclass of PWARX models with continuous PWA map $f(\cdot)$ [10]. For this class of models, the identification problem is formulated as a mixed-integer linear or quadratic program that can be solved for the global optimum.

The Bayesian, bounded-error, and clustering-based procedures identify models in PWARX form. The basic steps that these procedures perform are data classification and parameter estimation, followed by the reconstruction of the regions. The bounded-error procedure also estimates the number of submodels. The first two steps are performed in a different way by each procedure, as described in the following sections, while the estimation of the regions can be carried out in the same way for all procedures. Basically, given the clusters $\{\mathcal{D}_i\}_{i=1}^s$ of data points provided by the data classification phase, the corresponding clusters of regression vectors $\mathcal{R}_i = \{x(k) \mid (y(k), x(k)) \in \mathcal{D}_i\}$ are constructed. Then, for all $i \neq j$ a separating hyperplane of the clusters $\mathcal{R}_i$ and $\mathcal{R}_j$ is sought, i.e., a hyperplane

$$M'_{ij}x = m_{ij}, \tag{5}$$

with $M_{ij} \in \mathbb{R}^n$ and $m_{ij} \in \mathbb{R}$, such that $M'_{ij}x(k) < m_{ij}$ for all $x(k) \in \mathcal{R}_i$ and $M'_{ij}x(k) > m_{ij}$ for all $x(k) \in \mathcal{R}_j$. If such a hyperplane cannot be found (i.e., the data set is not linearly separable), one is interested in finding a generalized separating hyperplane that minimizes the number of misclassified data points or some misclassification cost. Robust linear programming (RLP) [7] and support vector machines (SVM) [11] methods (and their extensions to the multiclass case [8, 9]) can be employed. The minimization of the number of misclassifications is equivalent to solving a maximum feasible subsystem (MAX FS) problem for a system of linear inequalities (see [24] and references therein). The interested reader is referred to [5, 13, 23] for a detailed overview.

3.1 MIP procedure

The procedure proposed in [25, 26] is, in its basic formulation, an algorithm for optimal identification of hinging-hyperplane ARX (HHARX) models [10], which are described by

$$y(k) = f(x(k); \theta) + e(k)$$

$$f(x(k); \theta) = \varphi(k)'\theta_0 + \sum_{i=1}^{M} \sigma_i \max\{\varphi(k)'\theta_i, 0\}, \tag{6}$$

where $\varphi(k)' = [x(k)'\ 1]$, $\theta' = [\theta_0'\ \theta_1' \ldots \theta_M']$, and $\sigma_i \in \{-1, 1\}$ are fixed a priori. It is easy to see that HHARX models form a subclass of PWARX models for which the PWA map $f(\cdot)$ is continuous. The number of submodels s is bounded by the quantity $\sum_{j=0}^{n} \binom{M}{j}$, which only depends on the dimension n of the regressor space and the number M of hinge functions.

The identification problem considered in [25, 26] selects the optimal parameter vector θ^* by solving

$$\theta^* = \arg\min_{\theta} \sum_{k=1}^{N} |y(k) - f(x(k); \theta)|^p, \tag{7}$$

where $p = 1$ or 2. Assuming a priori known bounds on θ (which may be taken arbitrarily large), problem (7) can be reformulated as a MILP/MIQP problem by introducing binary variables

$$\delta_i(k) = \begin{cases} 0 & \text{if } \varphi(k)'\theta_i \leq 0 \\ 1 & \text{otherwise,} \end{cases} \tag{8}$$

and auxiliary continuous variables $z_i(k) = \max\{\varphi(k)'\theta_i, 0\}$. The MILP/MIQP problems can then be solved for the global optimum.

The optimality of the described algorithm comes at the cost of a theoretically very high worst-case computational complexity, which means that it is mainly suitable for small-scale problems. To be able to handle somewhat larger problems, different suboptimal approximations were proposed in [25]. Various extensions are also possible so as to handle nonfixed σ_i, discontinuities, general PWARX models, etc., again at the cost of increased computational complexity. For more details, see [25, 26].

3.2 Bayesian procedure

The Bayesian procedure [16, 17] is based on the idea of exploiting the available prior knowledge about the modes and the parameters of the hybrid system. The PVs θ_i are treated as random variables and described through their probability density functions (*pdf*s) $p_{\theta_i}(\cdot)$. A priori knowledge on the parameters can be supplied to the procedure by choosing appropriate a priori parameter *pdf*s $p_{\theta_i}(\cdot; 0)$. Various parameter estimates, such as expectation or maximum a posteriori probability estimate, can be easily obtained from the parameter *pdf*s. The data classification problem is posed as the problem of finding the data classification with the highest probability. Since this problem is combinatorial, an iterative suboptimal algorithm is derived in [16, 17].

Data classification and parameter estimation are carried out through sequential processing of the collected data points. At iteration k the data point $(y(k), x(k))$ is considered and attributed to the mode $\hat{\mu}(k)$ using maximum likelihood. Then, the a posteriori *pdf* of $\theta_{\hat{\mu}(k)}$ is computed using as a fact that $(y(k), x(k))$ was generated by mode $\hat{\mu}(k)$. To numerically implement the described procedure, particle filtering algorithms are used (see, e.g., [3]). After the parameter estimation phase, each data point is attributed to the mode that most likely generated it.

To estimate the regions, a modification of the standard multicategory RLP (MRLP) method [8] is proposed in [16, 17]. For each data point $(y(k), x(k))$ attributed to mode i, the price for misclassification into mode j is defined as

$$\nu_{ij}(x(k)) = \log \frac{p\Big((y(k), x(k)) \mid \mu(k) = i\Big)}{p\Big((y(k), x(k)) \mid \mu(k) = j\Big)}, \tag{9}$$

where $p\Big((y(k), x(k)) \mid \mu(k) = \ell\Big)$ is the likelihood that $(y(k), x(k))$ was generated by mode ℓ. Prices for misclassification are plugged into MRLP.

The Bayesian procedure requires that the model orders n_a and n_b and the number of submodels s are fixed. The most important tuning parameters are the a priori parameter *pdfs* $p_{\theta_i}(\cdot; 0)$, and the *pdf* $p_e(\cdot)$ of the error term.

3.3 Bounded-error procedure

The main feature of the bounded-error procedure [5, 6, 23] is to impose that the error $e(k)$ in (1) is bounded by a given quantity $\delta > 0$ for all the samples in the estimation data set $\mathcal{D}$.

At *initialization*, the estimation of the number of submodels s, data classification, and parameter estimation are performed simultaneously by partitioning the (typically infeasible) set of N linear complementary inequalities

$$|y(k) - \varphi(k)'\theta| \leq \delta, \quad k = 1, \ldots, N, \tag{10}$$

where $\varphi(k)' = [x(k)' \ 1]$, into a minimum number of feasible subsystems (MIN PFS problem). Then, an iterative *refinement* procedure is applied in order to deal with data points $(y(k), x(k))$ satisfying $|y(k) - \varphi(k)'\theta_i| \leq \delta$ for more than one θ_i. These data are termed *undecidable*. The refinement procedure alternates between data reassignment and parameter update. If desirable, it enables the reduction of the number of submodels. For given positive thresholds α and β, submodels i and j are merged if $\alpha_{i,j} < \alpha$, where

$$\alpha_{i,j} = \|\theta_i - \theta_j\|_2 / \min\{\|\theta_i\|_2, \|\theta_j\|_2\}. \tag{11}$$

Submodel i is discarded if the cardinality of the corresponding data cluster $\mathcal{D}_i$ is less than βN. Data points that do not satisfy $|y(k) - \varphi(k)'\theta_i| \leq \delta$ for any θ_i are discarded as *infeasible* during the classification process, making it possible to detect outliers. In [5, 6] parameter estimates are computed through the ℓ_∞ projection estimator, but any other projection estimate, such as least squares, can be used [20].

The bounded-error procedure requires that the model orders n_a and n_b are fixed. The main tuning parameter is the bound δ: the larger δ, the smaller the required number of submodels at the price of a worse fit of the data. The optional parameters α and β, if used, also implicitly determine the final number of submodels returned by the procedure. Another tuning parameter is the number of nearest neighbors c used to attribute undecidable data points to submodels in the refinement step.

3.4 Clustering-based procedure

The clustering-based procedure [13] is based on the rationale that small subsets of regression vectors that lie close to each other could be very likely attributed to the same region and the same submodel. The main steps of the procedure are hence the following:

- *Local regression.* For $k = 1, \ldots, N$, a local data set $\mathcal{C}_k$ is built by collecting $(y(k), x(k))$ and the data points $(y(j), x(j))$ corresponding to the $c-1$ regression vectors $x(j)$ that are nearest to $x(k)$. Local parameter vectors θ_k^{LS} are then computed for each local data set $\mathcal{C}_k$ through least squares.
- *Construction of feature vectors.* The centers $m_k = \frac{1}{c} \sum_{(y,x)\in\mathcal{C}_k} x$ are computed, and the feature vectors $\xi_k = [(\theta_k^{LS})' \; m_k']'$ are formed.
- *Clustering.* Feature vectors are partitioned into s groups $\{\mathcal{F}_i\}_{i=1}^s$ through clustering. To this aim, a "K-means"-like algorithm exploiting suitably defined confidence measures on the feature vectors is used. The confidence measures reduce the influence of outliers and poor initializations.
- *Parameter estimation.* Since the mapping of the data points onto the feature space is bijective, the data clusters $\{\mathcal{D}_i\}_{i=1}^s$ can be easily built according to the rule: $(y(k), x(k)) \in \mathcal{D}_i \leftrightarrow \xi_k \in \mathcal{F}_i$. The PVs $\{\theta_i\}_{i=1}^s$ are estimated from data clusters through weighted least squares.

The clustering-based procedure requires that the model orders n_a and n_b and the number of submodels s are fixed. The parameter c, defining the cardinality of the local data sets, is its main tuning knob. A modification to the clustering-based procedure is proposed in [12] to allow for the simultaneous estimation of the number of submodels.

3.5 Quality measures

In the following sections the four procedures described above will be compared on suitably defined test examples. To this aim, some quantitative measures for assessing the quality of the identification results are introduced here.

When the true system generating the data is known and belongs to the considered model class, the accuracy of the estimated PVs can be evaluated by computing the quantity:

$$\Delta_\theta = \max_{i=1,\ldots,s} \frac{\|\theta_i - \bar{\theta}_i\|_2}{\|\bar{\theta}_i\|_2}, \tag{12}$$

where $\bar{\theta}_i$ and θ_i are the true and identified PVs for mode i, respectively. Δ_θ is zero for the perfect estimates and increases as estimates get worse. In general, a sensible quality measure for the estimated regions is harder to define. For the one-dimensional bimodal case ($n = 1$ and $s = 2$) the following measure

$$\Delta_\chi = \left| \frac{m_{12}}{M_{12}} - \frac{\bar{m}_{12}}{\bar{M}_{12}} \right| \tag{13}$$

is used, where $\bar{M}_{12}$, $\bar{m}_{12}$, M_{12}, m_{12} are the coefficients of the true and estimated separating hyperplane (5), respectively.

A general quality measure, which is also applicable when the true system is not known, is provided by the averaged sum of the squared residuals:

$$\hat{\sigma}_e^2 = \frac{1}{s} \sum_{i=1}^{s} \frac{\mathrm{SSR}_i}{|\mathcal{D}_i|}, \tag{14}$$

where the set $\mathcal{D}_i$ contains the data points classified to mode i, $|\cdot|$ here denotes the cardinality of a set, and the sum of squared residuals (SSR) for mode i is defined as follows:

$$\mathrm{SSR}_i = \sum_{(y(k),x(k))\in\mathcal{D}_i} (y(k) - [x(k)'\ 1]\theta_i)^2. \tag{15}$$

The quality of the identified model is considered acceptable if $\hat{\sigma}_e^2$ is small and/or close to the expected noise variance of the true system.

Models with good one-step ahead prediction properties may perform poorly in simulation. To evaluate the model performance in simulation, a suitable measure of fit is

$$FIT = 100 \cdot \left(1 - \frac{\|\hat{\mathbf{y}} - \mathbf{y}\|_2}{\|\mathbf{y} - \bar{y}\|_2} \right), \tag{16}$$

where $\mathbf{y} = [y(1) \ldots y(N)]'$ is the vector of system outputs, $\bar{y}$ is the mean value of $\mathbf{y}$, and $\hat{\mathbf{y}} = [\hat{y}(1) \ldots \hat{y}(N)]'$ is the vector of simulated outputs, obtained by building regression vectors $x(k)$ from real inputs and previously simulated outputs. Equation (16) can be interpreted as the percentage of the output variation that is explained by the model.

In experimental identification, (14) and (16) are useful for selecting good models from a set obtained by applying each identification procedure with different values of the tuning parameters and/or of the model orders.

4 Intersecting hyperplanes

From the descriptions in the previous section, it is evident that each identification procedure implements a different approach to parameter estimation and data classification. The aim of this section is to evaluate how the different procedures are able to deal with data points that are consistent with more than one submodel, namely data points lying in the proximity of the intersection of two or more submodels. Wrong attribution of these data points may indeed lead to misclassifications when estimating the polyhedral regions.

In order to illustrate this problem, an example where the submodels of the true system intersect over the regressor set $\mathcal{X}$ is designed. Consider the one-dimensional PWARX system $y(k) = \bar{f}(x(k)) + \eta(k)$, where the additive noise $\eta(k)$ is normally distributed with zero mean and variance $\sigma_\eta^2 = 0.005$, and the PWA map $\bar{f}(\cdot)$ is defined as

$$\bar{f}(x) = \begin{cases} 0.5x + 0.5 & \text{if } x \in [-2.5,\, 0] \\ -x + 2 & \text{if } x \in (0,\, 2.5]. \end{cases} \tag{17}$$

$N = 100$ regressors $x(k)$ are generated. The 80% is uniformly distributed over $[-2.5,\, 2.5]$, and the remaining 20% over $[0.85,\, 1.15]$, so that the intersection of the two submodels is excited thoroughly. Results for the bounded-error, clustering-based, and MIP procedures are shown in Figure 1 (left). Note that an extension of the MIP procedure described in Section 3.1 was applied to handle discontinuities in the model.

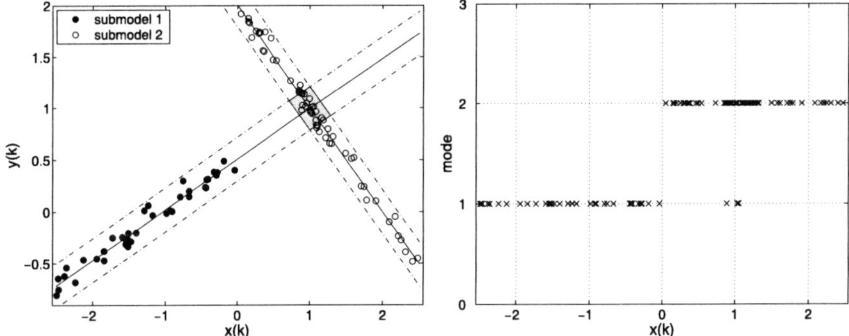

Fig. 1. Left: Results of data classification for the bounded-error, clustering-based, and MIP procedures. All the three procedures yield $\Delta_\theta = 0.0186$ (using least squares) and $\Delta_\mathcal{X} = 0.0055$ (using RLP). **Right**: Data classification by attributing each data point to the submodel that generates the smallest prediction error.

In this example the three procedures classify correctly all the data points, and both the PVs and the switching threshold are estimated accurately. However, this might not be the case in general. For the clustering-based procedure, the quality of the results depends on the choice of the cardinality c of the local data sets (see Section 7). In addition, the clustering may fail when there is a large variance on the centers m_k corresponding to similar θ_k^{LS}.

The bounded-error procedure is applied with $\delta = 3\,\sigma_\eta$ and $c = 10$. The gray area in Figure 1 (left) represents the region of all possible *undecidable* data points for the fixed δ. In [5] undecidable data points were discarded during the classification process to avoid errors in the region estimation phase. However, in this way a nonnegligible amount of information is lost when a large number of undecidable data points shows up. In [6, 23] undecidable data points are attributed to submodels by considering the assignments of the c nearest neighbors. Also for this method, classification results depend on the choice of c (see again Section 7). It is interesting to point out that, if parameter estimates are computed through the ℓ_∞ projection estimator, one gets $\Delta_\theta = 0.0671$ in this example. As expected, parameter estimates are worse than using least squares, since the noise of the true system is normally distributed.

Classification results of the three procedures are compared to those obtained by attributing each data point to the submodel that generates the smallest prediction error [29]. Results using this approach are shown in Figure 1, right. Three data points around the intersection of the two submodels are misclassified. This leads to nonlinearly separable classes that determine a larger error in the estimation of the switching threshold ($\Delta_\chi = 0.0693$ in this example using RLP).

The Bayesian procedure is applied on a different data set, shown in Figure 2 (right). The procedure is initialized with a priori parameter *pdf*s $p_{\theta_1}(\cdot;0) = p_{\theta_2}(\cdot;0) \sim \mathcal{U}([-2.5, 2.5] \times [-2.5, 2.5])$, where $\mathcal{U}(I)$ denotes the uniform distribution over the set I. Note that a priori parameter *pdf*s overlap.

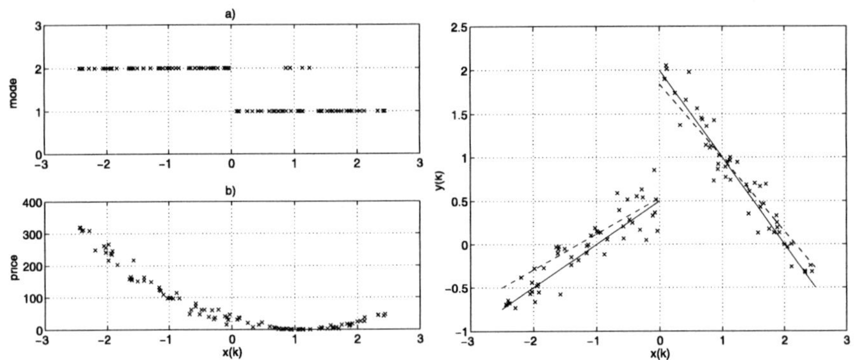

Fig. 2. Left: a) Data points attributed to modes. b) Pricing function for misclassifications. **Right:** Estimation data set (crosses), true system (solid lines) and estimated model (dashed lines). The Bayesian procedure yields $\Delta_\theta = 0.1366$ and $\Delta_\chi = 0.0228$.

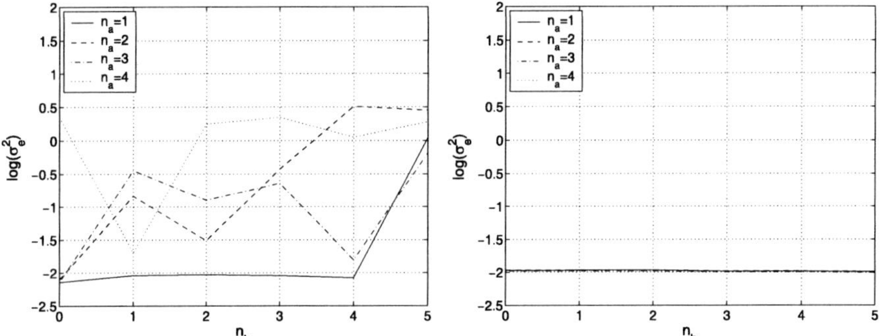

Fig. 3. Left: $\hat{\sigma}_e^2$ for the clustering-based procedure ($s = 2$, $c = 20$). **Right:** $\hat{\sigma}_e^2$ for the MIP procedure (2-norm criterion, $M = 1$).

Results of data classification are shown in Figure 2 (left(a)). There are five wrongly classified data points. These points are close to the intersection of the two submodels. The misclassification pricing function (9) is plotted in Figure 2 (left(b)). The weight for misclassification of the wrongly attributed data points is small compared to the weight for misclassification of those correctly attributed.

5 Overestimation of model orders

The four identification procedures require that the model orders n_a and n_b are fixed. In order to investigate the effects of overestimated model orders, the one-dimensional system $y(k) = \bar{f}(y(k-1)) + \eta(k)$ is considered, where the additive noise $\eta(k)$ is normally distributed with zero mean and variance $\sigma_\eta^2 = 0.01$, and the PWA map $\bar{f}(\cdot)$ is defined as

$$\bar{f}(x) = \begin{cases} 2x + 10 & \text{if } x \in [-10, 0) \\ -1.5x + 10 & \text{if } x \in [0, 10]. \end{cases} \tag{18}$$

The sequence $y(k)$ is generated with initial condition $y(0) = -10$. A fictitious input sequence is also generated as $u(k) \sim \mathcal{U}([-10, 10])$. The four identification procedures are applied for all combinations of $n_a = 1, \ldots, 4$ and $n_b = 0, 1, \ldots, 5$, and $\hat{\sigma}_e^2$ is computed for each identified model. Note that the true system orders are $n_a = 1$ and $n_b = 0$.

Figure 3 (left) shows the log-values of $\hat{\sigma}_e^2$ for models with different model orders identified by the clustering-based procedure. For true system orders the procedure identifies a model with $\hat{\sigma}_e^2$ close to the noise variance, but the performance significantly deteriorates when the model orders are overestimated. This is due to the adopted rationale that data points close to each other in the regressor space could be very likely attributed to the same submodel. When overestimating the model orders, the regression vector is extended with elements that do not contain relevant information for identification, but alter the distances in the feature space. This may determine misclassifications during clustering and consequent bad estimates of the PVs and of the regions.

Since the MIP procedure solves the optimization problem (7) at the optimum ($p = 2$ is considered since the noise is normally distributed), from Figure 3 (right), it is apparent that the procedure has no difficulties in estimating the overparameterized models.

Results for the bounded-error procedure are shown in Figure 4 (left). For the case $n_a = 1$, $n_b = 0$, a value of δ allowing to obtain $s = 2$ submodels is sought. The procedure is then applied to the estimation of the overparameterized models using the same δ. When extending the regression vector, the minimum number of feasible subsystems of (10) does not increase and remains equal in this example. Hence, the minimum partition obtained for $n_a = 1$, $n_b = 0$ is also a solution in the overparameterized case. This explains the very good results shown in Figure 4 (left), which are comparable to those obtained by the MIP procedure. The enhanced version [6, 23] of the greedy algorithm [1] is applied here for solving the MIN PFS problem. For completeness, it is reported that the corresponding values of Δ_θ obtained[1] range between 0.005 and 0.012.

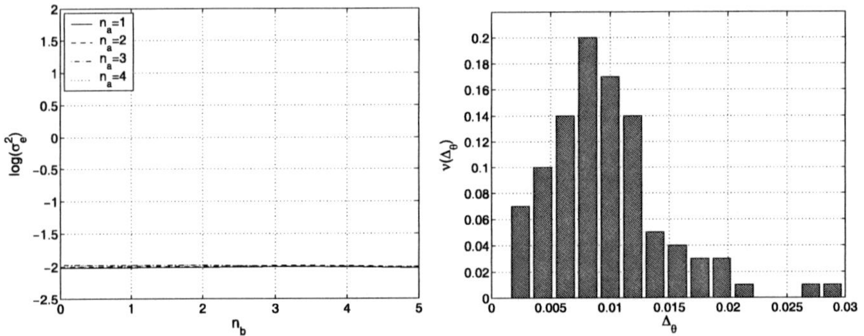

Fig. 4. Left: $\hat{\sigma}_e^2$ for the bounded-error procedure ($\delta = 3\,\sigma_\eta$, $c = 10$, α and β not used). **Right:** Approximate distribution of Δ_θ over $Q = 100$ runs using the bounded-error procedure and different realizations of Gaussian noise with $\sigma_\eta^2 = 0.075$.

Values of $\hat{\sigma}_e^2$ for the Bayesian procedure with two different initializations are shown in Figure 5. In the left plot, the a priori parameter pdfs for the case $n_a = 1$, $n_b = 0$ are chosen as $p_{\theta_1}(\cdot\,; 0) = p_{\theta_2}(\cdot\,; 0) \sim \mathcal{U}([-5, 5] \times [-20, 20])$.

---

[1]To compute Δ_θ, the entries of the true parameter vectors corresponding to superfluous elements in the regression vector are set to 0.

For increased orders, added elements in the parameter vectors are taken to be uniformly distributed in the interval $[-5, 5]$ (while their "true" value should be 0). In the right plot, the a priori parameter pdfs for the case $n_a = 1$, $n_b = 0$ are chosen as $p_{\theta_1}(\cdot; 0) \sim \mathcal{U}([0, 4] \times [8, 12])$ and $p_{\theta_2}(\cdot; 0) \sim \mathcal{U}([-4, 0] \times [8, 12])$, and all added elements are taken to be uniformly distributed in the interval $[-0.5, 0.5]$. This example shows the importance of proper choices for the initial parameter pdfs in the Bayesian procedure. With precise initial pdfs the algorithm estimates relatively accurate overparameterized models. When the a priori information is not adequate, the performance rapidly deteriorates.

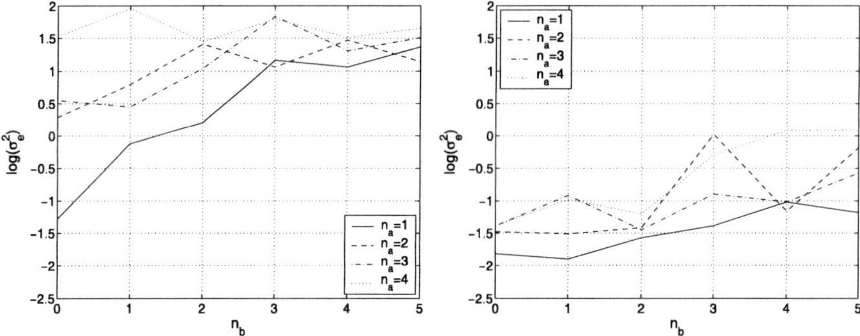

Fig. 5. Left: $\hat{\sigma}_e^2$ for the Bayesian procedure with $s = 2$ and unprecise initial parameter pdfs. **Right:** With precise initial parameter pdfs.

6 Effects of noise

This section addresses the effects of noise on the identification accuracy. The first issue of interest is the effect that different noise realizations with the same statistical properties have on the identification results. The second issue is how different statistical properties of the noise affect the identification results.

To shed some light on these issues, an experiment is designed with the one-dimensional PWARX system $y(k) = \bar{f}(x(k)) + \eta(k)$ and the PWA map $\bar{f}(\cdot)$ defined as in (18). The additive noise $\eta(k)$ is normally distributed with zero mean and variance σ_η^2. A noiseless data set of $N = 100$ data points is generated with $x(k) \sim \mathcal{U}([-10, 10])$. For fixed σ_η^2, $Q = 100$ noise realizations are drawn and added to the noiseless data set. For each noise realization, a model is identified using the different procedures, and the value of Δ_θ is computed for the identified models. In this way an approximate distribution of Δ_θ for each σ_η^2 and each procedure can be constructed, and its mean and variance can be estimated. Figure 4 (right) shows one such distribution obtained using the bounded-error procedure.

Figure 6 shows the estimated means and variances of the Δ_θ distributions as functions of σ_η^2 for the four procedures. From the analysis of the plots, it is apparent that the MIP procedure achieves the best performance with respect to noise. The bounded-error procedure performs well when δ is chosen close to $3\sigma_\eta$ and the true PVs are quite different, as in this example. However, in practical situations such a value is unlikely to be available, and several trials are needed to find a suitable value for δ. As the noise level increases, the clustering-based procedure requires an increase in the cardinality c of the local data sets in order to reduce the variance of the estimates (see Section 7 and the discussion in [13, Section 3.1]). With precise initialization as in Section 5, the Bayesian procedure achieves comparable performance to the other procedures, while with imprecise initialization the quality measures are the worst of all procedures (not shown in Figure 6).

7 Effects of varying the tuning parameters

The four identification procedures described in Section 3 require that some parameters that directly determine the structure of the identified models are fixed a priori. These are the model orders n_a and n_b for all procedures, the number of modes s for the Bayesian and the clustering-based procedure, and the number of hinge functions M for the MIP procedure. The Bayesian, bounded-error, and clustering-based procedures also have several tuning parameters whose influence on identification results is not immediately obvious. In this section, the effects of varying the tuning parameters will be illustrated for these procedures by means of examples.

To investigate the role of the parameter c of the clustering-based procedure, an experiment is designed where the approximate distribution of Δ_θ is computed as described in Section 6 for different values of c and $\sigma_\eta^2 = 0.075$.

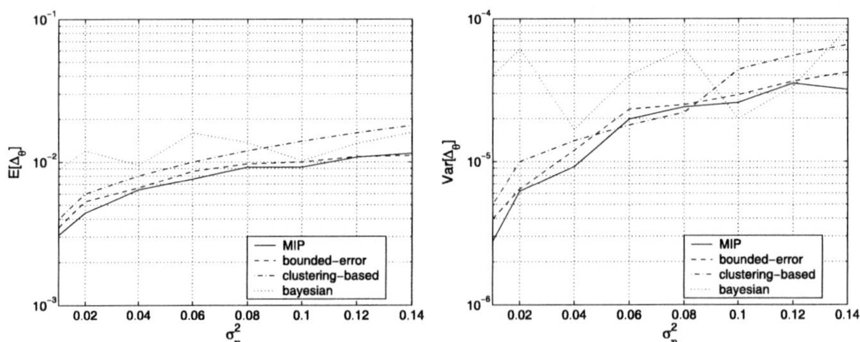

Fig. 6. Left: Estimated means of the Δ_θ distributions for several values of the noise variance σ_η^2. **Right:** Estimated variances.

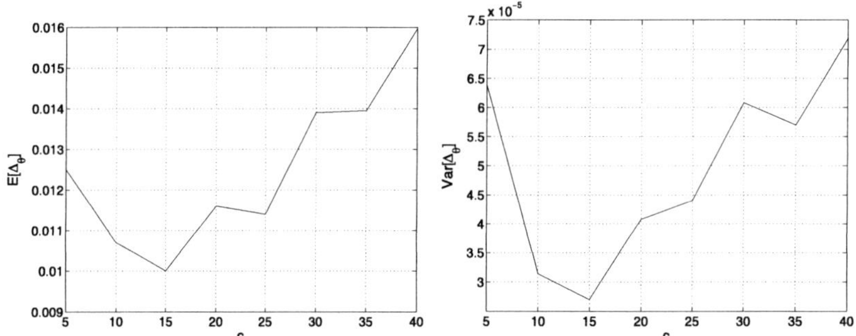

Fig. 7. Left: Estimated means of the Δ_θ distributions for several values of c using the clustering-based procedure and $\sigma_\eta^2 = 0.075$. **Right:** Estimated variances.

The estimated means and variances of such distributions are shown in Figure 7. It is apparent that there exists an optimal value of c for which the identified model is most accurate. In the considered example the procedure gives the best results for $c = 15$, corresponding to the minimum of both curves in Figure 7. For the selection of c in practical cases, see [13].

The bounded-error procedure has several tuning parameters, and finding their right combination is not always straightforward. The role of the parameters δ, α, and c will be investigated here by considering the one-dimensional PWARX system $y(k) = \bar{f}(x(k)) + \eta(k)$, where the additive noise $\eta(k)$ is normally distributed with zero mean and variance $\sigma_\eta^2 = 0.01$, and the PWA map $\bar{f}(\cdot)$ is defined as

$$\bar{f}(x) = \begin{cases} 0.3x + 10 & \text{if } x \in [-2.5, 0] \\ x + 10 & \text{if } x \in (0, 2.5]. \end{cases} \tag{19}$$

$N = 100$ regressors $x(k)$ are generated uniformly distributed over $[-2.5, 2.5]$. Figure 8 (upper left) shows a plot of the estimated minimum number of feasible subsystems of (10) as a function of δ. In general, such a plot can be used to select an appropriate value for δ close to knee of the curve. In this example, choosing δ in the suggested way allows one to estimate the correct number of modes $s = 2$. Using $\delta = 0.3$, the initialization provides parameter estimates with $\alpha_{1,2} = 5.9\%$. If α is selected greater than $\alpha_{1,2}$, the two submodels are merged into one during the refinement phase. In this case, a high number of infeasible data points shows up (Figure 8 (upper right)), which indicates poor fit of the data. Hence, the refinement procedure can be repeated using a smaller value of α. The role of c with respect to classification is illustrated in Figure 8 (bottom). On the left, perfect classification is obtained using $c = 5$, whereas on the right three data points are misclassified using $c = 40$, which causes a larger error when estimating the switching threshold.

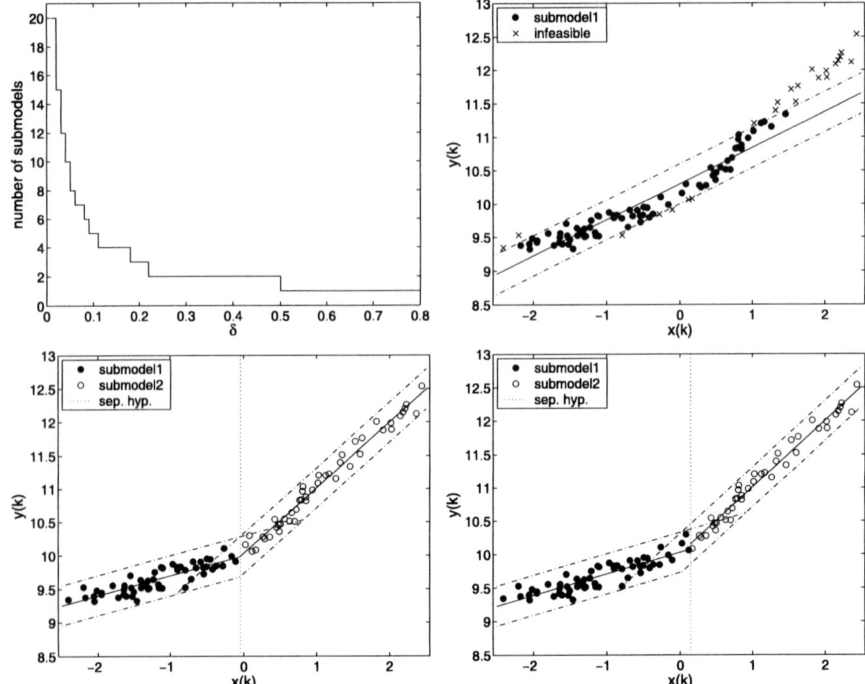

Fig. 8. Upper left: Plot of the estimated minimum number of submodels as a function of δ for the bounded-error procedure. **Upper right:** Results of data classification with $\delta = 0.3$ and $\alpha = 10\%$. **Lower left:** With $\delta = 0.3$, $\alpha = 2\%$, and $c = 5$, yielding $\Delta_\theta = 0.0035$ and $\Delta_\chi = 0.0411$. **Lower right:** With $\delta = 0.3$, $\alpha = 2\%$, and $c = 40$, yielding $\Delta_\theta = 0.0035$ and $\Delta_\chi = 0.1422$.

For the Bayesian procedure the most important tuning knobs are the a priori *pdf*s of the parameters. Effects of improper choices are clearly illustrated in Sections 5 and 6. The more precise the a priori knowledge on the system, the better the procedure is expected to perform (see also Section 8).

8 Experimental example

In this section the four procedures are applied to the identification of the electronic component placement process in pick-and-place machines.[2] Pick-and-place machines are used to automatically place electronic components on printed circuit boards. The process consists of a mounting head carrying the electronic component, which is pushed down until it comes in contact with

---

[2] The authors would like to thank Hans Niessen for some of the results presented in this section.

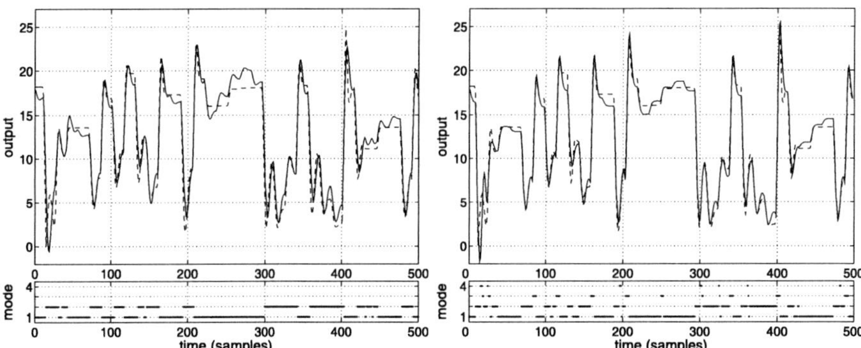

Fig. 9. Left: Simulation results on validation data using the PWARX model identified by the clustering-based procedure with $n_a = 2$, $n_b = 2$, $s = 2$ and $c = 90$ ($FIT = 74.7127\%$). **Right:** Simulation results on validation data using the HHARX model identified by the MIP procedure with $n_a = 2$, $n_b = 1$ and $M = 2$ ($FIT = 81.5531\%$). Solid—simulated output, dashed—system output.

the circuit board, and then is released. The input to the system is the voltage applied to the motor driving the mounting head. The output of the system is the position of the mounting head. A detailed description of the process and of the experimental setup can be found in [15, 17].

A data record over an interval of $T = 15$ s is available. The data sets used for identification with the Bayesian, clustering-based, and MIP procedures are sampled at 50 Hz, while the data set used for identification with the bounded-error procedure is sampled at 150 Hz. Two modes of operation can be distinguished through physical insight into the process. In the *free mode*, the mounting head moves unconstrained, whereas in the *impact mode* the carried component is in contact with the circuit board. Hard nonlinear phenomena due to dry friction are also present.

Results obtained by applying the four identification procedures are shown in Figures 9, 10, and 11. Note that the aim in this section is only to show that all the four procedures are able to estimate sensible models of the experimental process. A fair comparison of models identified by different procedures is not always possible here, mainly because they were obtained using different model structures and/or different data sets.

For identification using the Bayesian and clustering-based procedures, a data set consisting of 750 samples is partitioned into two overlapping sets of 500 points each. The first set is used for estimation and the second for validation. Figure 9 (left) shows the simulation results on validation data for the best model identified by the clustering-based procedure. The best model is obtained for a high value of c. A possible explanation of this, given in [15, 17], is that by using large local data sets the effects of dry friction can be "averaged out" as a process noise. Differences between the measured and simulated responses due to unmodeled dry friction are clearly visible on the time interval $(225, 300)$.

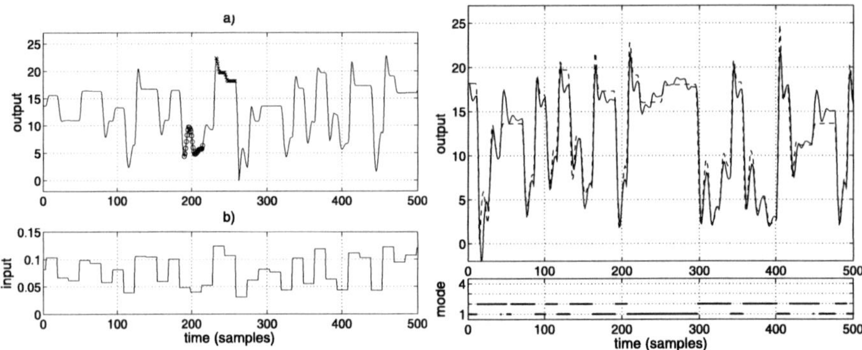

Fig. 10. Left: Data set used for identification with the Bayesian procedure. a) Output signal. Data marked with $\circ$ and $\times$ are those used for initializing the free mode and the impact mode, respectively. b) Input signal. **Right:** Simulation results on validation data using the identified PWARX model with $n_a = 2$, $n_b = 2$ and $s = 2$ ($FIT = 77.1661\%$). Solid — simulated output, dashed — system output.

Physical insight into the process helps the initialization of the Bayesian procedure. Although the mode switch does not occur at a fixed height of the mounting head, data points below a certain height can be most likely attributed to the free mode. A similar consideration holds for the impact mode, see Figure 10 (left(a)). The a priori information is exploited to obtain rough estimates θ_i^{LS} of the PVs of the two modes through least squares. As in [13], the empirical covariance matrix V_i of θ_i^{LS} is computed. The a priori parameter *pdf*s are then taken to be normal distributions with means θ_i^{LS} and covariance matrices V_i. Simulation results on validation data are shown in Figure 10 (right). It is apparent that the identified model benefits from the prior knowledge and yields a higher value of FIT than the model obtained through the clustering-based procedure.

The MIP procedure identifies a model with $n_a = 2$, $n_b = 1$, and $M = 2$ using $N = 150$ estimation data points. The 2-norm criterion is considered, and $s = 4$ submodels are obtained. Simulation results on the same validation data set as for the Bayesian and clustering-based procedures are shown in Figure 9 (right). Although a smaller number of data points is used to estimate the model, it is apparent that the use of more than two submodels is favorable to improve the fit.

The bounded-error procedure identifies models with orders $n_a = 2$ and $n_b = 2$ using $N = 1000$ estimation data. Two models with $s = 2$ and $s = 4$ modes are identified for $\delta = 0.06$ and $\delta = 0.04$, respectively. For $s = 4$, multi-category RLP [8] is used for region estimation. Simulation results on validation data are shown in Figure 11. Again, it is apparent that the fit improves as the number of submodels increases, i.e., as δ decreases. The active mode evolution at the bottom of Figure 11 (left) clearly shows that one of the two submodels is active in situations of high incoming velocity of the mounting head (i.e., rapid transitions from low to high values of the mounting head position). One submodel modeling the same situation is also present in the identified model with $s = 4$ modes.

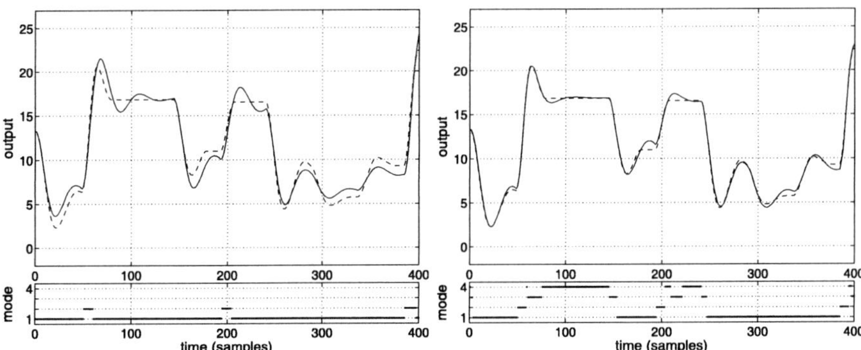

Fig. 11. Left: Simulation results on validation data using the model with $s = 2$ modes identified by the bounded-error procedure ($FIT = 81.3273\%$). **Right:** Using the model with $s = 4$ modes ($FIT = 93.4758\%$). Solid — simulated output, dashed — system output.

9 Conclusions

In this chapter the problem of PWA identification has been addressed by studying four recently proposed techniques for the identification of PWARX or HHARX models. The four techniques have been compared on suitably defined test examples and tested on the experimental identification of the electronic component placement process in pick-and-place machines.

Identification using the clustering-based procedure is straightforward, as only one parameter has to be tuned. However, poor results are obtained when the model orders are overestimated, since distances in the feature space become corrupted by irrelevant information. The bounded-error procedure gives good results when the right combination of the tuning parameters is found, but several attempts are often needed for finding such a combination. The Bayesian procedure is designed to take advantage of prior knowledge on the system and has been shown to be very effective in the pick-and-place machine identification. The MIP procedure provides globally optimal results and needs no parameter tuning, but requires the solution of MILP/MIQP problems whose theoretical worst-case computational complexity is very high.

References

1. Amaldi E, Mattavelli M (2002) The MIN PFS problem and piecewise linear model estimation. Discrete Applied Mathematics 118:115–143
2. Antsaklis PJ, Nerode A (eds) (1998) Special issue on hybrid systems. IEEE Transactions on Automatic Control 43(4):457–579
3. Arulampalam MS, Maskell S, Gordon N, Clapp T (2002) A tutorial on particle filters for online nonlinear/nongaussian Bayesian tracking. IEEE Transactions on Signal Processing 50(2):174–188
4. Bemporad A, Ferrari-Trecate G, Morari M (2000) Observability and controllability of piecewise affine and hybrid systems. IEEE Transactions on Automatic Control 45(10):1864–1876
5. Bemporad A, Garulli A, Paoletti S, Vicino A (2003) A greedy approach to identification of piecewise affine models. In: Maler O, Pnueli A (eds), Hybrid systems: computation and control. Lecture Notes on Computer Science, 97–112. Springer Verlag, New York
6. Bemporad A, Garulli A, Paoletti S, Vicino A (2004) Data classification and parameter estimation for the identification of piecewise affine models. In: Proceedings of the 43rd IEEE Conference on Decision and Control, Atlantis, Bahamas
7. Bennett KP, Mangasarian OL (1992) Robust linear programming discrimination of two linearly inseparable sets. Optimization Methods and Software 1:23–34
8. Bennett KP, Mangasarian OL (1994) Multicategory discrimination via linear programming. Optimization Methods and Software 3:27–39
9. Bredensteiner EJ, Bennett KP (1999) Multicategory classification by support vector machines. Computational Optimization and Applications 12:53–79
10. Breiman L (1993) Hinging hyperplanes for regression, classification, and function approximation. IEEE Transactions on Information Theory 39(3):999–1013
11. Cortes C, Vapnik V (1995) Support-vector networks. Machine Learning 20:273–297
12. Ferrari-Trecate G, Muselli M (2003) Single-linkage clustering for optimal classification in piecewise affine regression. In: Engell S, Gueguen H, Zaytoon J (eds), Proceedings IFAC Conference on Analysis and Design of Hybrid Systems, Saint–Malo Brittany, France
13. Ferrari-Trecate G, Muselli M, Liberati D, Morari M (2003) A clustering technique for the identification of piecewise affine systems. Automatica 39(2):205–217
14. Heemels WPMH, De Schutter B, Bemporad A (2001) Equivalence of hybrid dynamical models. Automatica 37(7):1085–1091
15. Juloski A, Heemels WPMH, Ferrari-Trecate G (2004) Data-based hybrid modelling of the component placement process in pick-and-place machines. Control Engineering Practice 12(10):1241–1252
16. Juloski A, Wieland S, Heemels WPMH (2004) A Bayesian approach to identification of hybrid systems. In: Proceedings 43rd IEEE Conference on Decision and Control, Atlantis, Bahamas

17. Juloski A (2004) Observer design and identification methods for hybrid systems. PhD thesis, Department of Electrical Engineering, Eindhoven University of Technology, Eindhoven, The Netherlands
18. Lin JN, Unbehauen R (1992) Canonical piecewise-linear approximations. IEEE Transactions on Circuits and Systems–I: Fundamental Theory and Applications 39(8):697–699
19. Ljung L (1999) System identification: theory for the user. 2nd ed., Prentice Hall, Englewood Cliffs, NJ
20. Milanese M, Vicino A (1991) Optimal estimation theory for dynamic systems with set membership uncertainty: an overview. Automatica 27(6):997–1009
21. Morse AS, Pantelides CC, Sastry SS, Schumacher JM (eds) (1999) Special issue on hybrid systems. Automatica 35(3):347–535
22. Münz E, Krebs V (2002) Identification of hybrid systems using apriori knowledge. In: Proceedings 15th IFAC World Congress, Barcelona, Spain
23. Paoletti S (2004) Identification of piecewise affine models. PhD thesis, Department of Information Engineering, University of Siena, Siena, Italy
24. Pfetsch ME (2002) The maximum feasible subsystem problem and vertex-facet incidences of polyhedra. PhD thesis, Technischen Universität Berlin, Berlin, Germany
25. Roll J (2003) Local and piecewise affine approaches to system identification. PhD thesis, Department of Electrical Engineering, Linköping University, Linköping, Sweden
26. Roll J, Bemporad A, Ljung L (2004) Identification of piecewise affine systems via mixed-integer programming. Automatica 40(1):37–50
27. Sjöberg J, Zhang Q, Ljung L, Benveniste A, Delyon B, Glorennec P, Hjalmarsson H, Juditsky A (1995) Nonlinear black-box modeling in system identification: a unified overview. Automatica 31(12):1691–1724
28. Van der Schaft AJ, Schumacher JM (2000) An introduction to hybrid dynamical systems. Vol. 251 of Lecture Notes in Control and Information Sciences. Springer Verlag, New York
29. Vidal R, Soatto S, Ma Y, Sastry S (2003) An algebraic geometric approach to the identification of a class of linear hybrid systems. In: Proceedings 42nd IEEE Conference on Decision and Control, Maui, HI

Part II

Control and System Theory

Dual Matrix Inequalities in Stability and Performance Analysis of Linear Differential/Difference Inclusions[*]

Rafal Goebel,[1] Tingshu Hu,[2] and Andrew R. Teel[2]

[1]P.O. Box 15172
Seattle, WA 98115, USA
`rafal@ece.ucsb.edu`
[2]Center for Control Engineering and Computation
Electrical and Computer Engineering
University of California, Santa Barbara, CA 93106, USA
`{rafal,tingshu,teel}@ece.ucsb.edu`

Summary. This chapter provides numerical examples to illustrate the recent results by the authors relating asymptotic stability and dissipativity of a linear differential or difference inclusion to these properties for the corresponding dual linear differential or difference inclusion. It is shown how this duality theory broadens the applicability of numerical algorithms for stability and performance analysis that have appeared previously in the literature.

1 Introduction

Perhaps the simplest pair of dual linear matrix inequalities for control systems is

$$A^T P + PA < 0 \tag{1}$$

for a given matrix A and a symmetric and positive definite P, and

$$AQ + A^T Q < 0, \tag{2}$$

for a symmetric and positive definite Q. The two inequalities are equivalent through $Q = P^{-1}$, and then each of them characterizes the stability of both the linear system $\dot{x}(t) = Ax(t)$ and its dual system $\dot{x}(t) = A^T x(t)$. Another pair of dual matrix inequalities is

$$\begin{bmatrix} A^T P + PA + C^T C & PB \\ B^T P & -\gamma^2 I \end{bmatrix} < 0, \tag{3}$$

---

[*]Research by T. Hu and A.R. Teel was supported in part by the ARO under Grant no. DAAD19-03-1-0144, the NSF under Grant no. ECS-0324679, and by the AFOSR under Grant no. F49620-03-1-0203.

and

$$\begin{bmatrix} AQ + QA^T + BB^T & QC^T \\ CQ & -\gamma^2 I \end{bmatrix} < 0. \tag{4}$$

They are equivalent through $Q = \gamma^2 P^{-1}$, and then each characterizes the finite L^2-gain, bounded by γ, for the pair of dual systems (A, B, C) and (A^T, C^T, B^T). Further examples of pairs of dual matrix inequalities come from characterizing other input-output performances of linear systems, such as passivity or the Hankel norm.

The stated matrix inequalities arise from stability and performance analysis of linear systems through quadratic Lyapunov or storage functions. The use of such functions in the analysis of differential inclusions, leading to matrix inequalities like these in (1) to (4) but holding for all A (or all A, B, C) in a certain set, is possible; see [3].

While quadratic Lyapunov functions lead to easily tractable linear matrix inequalities and simplify computational issues a great deal, it has been realized that they can yield conservative evaluation of stability for linear differential inclusions (see, e.g., [3, 5, 6, 14, 16, 23]). It is now well established that convex homogeneous — but not necessarily quadratic — Lyapunov functions are sufficient to characterize the stability of linear differential/difference inclusions (LDIs); see [15, 6]. Recent years have witnessed an extensive search for homogeneous Lyapunov functions. Particular types of functions studied are piecewise quadratic Lyapunov functions ([15, 22]), polyhedral Lyapunov functions ([2, 4]), and homogeneous polynomial Lyapunov functions ([5, 14, 23]).

Recently, more attention was given in [7, 13, 11, 10] to two particular classes of convex homogeneous Lyapunov functions: the functions given as the pointwise maximum of a family of quadratic functions, and those given as the convex hull of a family of quadratic functions. For simplicity, we refer to functions in these classes as max functions and convex hull functions. It is shown with an example in [3] (p. 73) that a max function may validate the stability of an LDI even when all quadratics fail. The convex hull function was first used for stability purposes in [11, 10] to estimate the domain of attraction for saturated linear systems and systems with a generalized sector condition (where it was called a "composite quadratic" function). Finally, [7, 13] noted and explored the convex duality between the two classes and used it to enhance stability analysis of LDIs and saturated linear systems.

Furthermore, [7] established a set of important symmetric relationships between dual linear differential inclusions. Stability of the LDI[1]

$$\dot{x}(t) \in \mathrm{co}\{A_i\}_{i=1}^m x(t) \tag{5}$$

was shown to be equivalent to stability of the dual LDI

$$\dot{\xi}(t) \in \mathrm{co}\{A_i^T\}_{i=1}^m \xi(t). \tag{6}$$

---

[1] The LDI means that $\dot{x}(t)$ is an element of the convex hull of points $A_i x(t)$, $i = 1, 2, \ldots, m$, for almost all t.

Based on the max function and the convex hull function, the following matrix inequalities are suggested:

$$A_i^T P_j + P_j A_i \leq \sum_{k=1}^{l} \lambda_{ijk}(P_k - P_j) - \gamma P_j \tag{7}$$

for all $i = 1, 2, ..., m$, $j = 1, 2, ..., l$ and

$$Q_j A_i^T + A_i Q_j \leq \sum_{k=1}^{l} \lambda_{ijk}(Q_k - Q_j) - \gamma Q_j \tag{8}$$

for all $i = 1, 2, \ldots, m$, $j = 1, 2, \ldots, l$. If there exist positive definite and symmetric matrices $P_1, P_2, \ldots, P_l$ and nonnegative numbers λ_{ijk} solving (7) or matrices $Q_1, Q_2, \ldots, Q_l$ and numbers $\lambda_{ijk} \geq 0$ solving (8), then both (5) and (6) are stable, similarly to what occurs for linear systems.

However, the matrix inequalities (7) and (8) are no longer equivalent. This might be unexpected but can be explained since either (7) or (8) is only a sufficient condition for stability of the LDIs. For linear systems, existence of a solution to (1) or (2) is sufficient, but also necessary for stability.

Even though the dual matrix inequalities (7) and (8) may still be conservative for stability analysis, numerical examples have shown that they may significantly improve on what can be achieved by quadratics. Furthermore, by duality, we can combine (7) and (8) to obtain a better estimate: if either inequality is satisfied, then the stability of both LDIs is confirmed.

Similar matrix inequalities can be stated to evaluate the L^2-gain, dissipativity, and other input-output performance measures for LDIs. We give several examples in this chapter. As can be seen from (7) and (8), the matrix inequalities based on max functions or convex hull functions are bound to be more complicated than their counterparts derived from quadratic functions, as a price for reducing conservatism. In fact, (7) and (8) are bilinear matrix inequalities (BMIs) instead of LMIs. Despite the well-known fact that BMIs are NP-hard, attempts have been made to make them more tractable; see, e.g., [1, 8, 9]. By using the path-following method presented in [9], we have developed a handful of algorithms to solve (7), (8), and other dual matrix inequalities arising from performance analysis. Our numerical experience shows that the path-following method is very effective. (We used straightforward iterative schemes in our earlier computation without much success.)

The purpose of this chapter is to present several matrix inequalities, some of which were previously stated in [7], for stability and performance analysis of LDIs and illustrate their applicability and effectiveness by way of examples. Through this, we also justify the application of convex duality theory and motivate the interest in the class of convex functions given by a maximum of quadratics or the functions given by a convex hull of quadratics.

The material is organized as follows. Section 2 is theoretical and outlines how the use of convex conjugate functions leads to a duality theory for LDIs;

the details are in [7]. Our numerical tools are based on the two classes of Lyapunov and storage functions described in Section 3. Various examples related to stability and relying on (7) and (8) are also given there. For further details, see [7, 13]. A general reference for the convex analysis materials used is [17]; see also [18]. In Section 4, we discuss examples related to dissipativity properties of LDIs, more specifically the L^2-gain and passivity. General results for such properties are the topic of a forthcoming work by the authors; for dissipativity concepts for linear systems, consult [20], [21]. Section 5 focuses on discrete-time systems and includes an example of an application of the matrix inequalities technology to estimation of domains of attraction of nonlinear systems (for details on the continuous-time case, see [13]).

2 Duality of Lyapunov functions

In the analysis of linear differential and difference inclusions, in contrast to equations, relying on quadratic Lyapunov functions is not sufficient. For example, the linear differential inclusion (5) can be asymptotically (and then in fact exponentially) stable when no quadratic Lyapunov function exists, i.e., when for no symmetric and positive definite P we have $A_i^T P + P A_i < 0$ for $i = 1, 2, \ldots, m$.[2] Existence of such a matrix leads to stronger stability of (5), called quadratic stability. See Example 2. Consequently, in order to establish a duality theory for linear differential and difference inclusions, for example, results stating that (5) is asymptotically stable if and only if (6) is asymptotically stable, one may need to look at Lyapunov functions that are not quadratic, and moreover, to find a relationship for such functions corresponding to that between P and P^{-1} for quadratic ones.

If (5) is asymptotically stable, then there exists $\gamma > 0$ and a differentiable, strictly convex, positive definite, and homogeneous of degree 2 function V such that

$$\nabla V(x)^T A x \leq -\gamma V(x) \quad \text{for all } x, \tag{9}$$

for all $A \in \mathrm{co}\{A_i\}_{i=1}^m$.[3] Homogeneity of degree 2 of V means that for all $\lambda \in \mathbb{R}$, $V(\lambda x) = \lambda^2 V(x)$. In what follows, we will denote by $\mathcal{L}$ the class of all differentiable, strictly convex, positive definite, and homogeneous of degree 2 functions.[4] Of course, every quadratic function $\frac{1}{2} x^T P x$ with a symmetric and positive definite P is in $\mathcal{L}$.

---

[2] The Lyapunov inequality holding at each of A_is is sufficient for it to hold at each element of $\mathrm{co}\{A_i\}_{i=1}^n$.

[3] The inequality (9) corresponds to the bound $\|x(t)\| \leq c\|x(0)\|e^{-\frac{1}{2}\gamma t}$ on solutions of the LDI. Slightly abusing the notation, we will refer to γ as the convergence rate, or the constant of exponential stability.

[4] Asymptotic stability of (5) is equivalent to exponential. If (5) is exponentially stable with a coefficient $\bar{\gamma}/2 > 0$, then γ in (9) can be chosen arbitrarily close, but smaller than, $\bar{\gamma}$. If one does not insist on differentiability of V, a Lyapunov inequality with $\bar{\gamma}$ can be written, using the convex subdifferential of V in place of ∇V.

It turns out that the convex functions of class $\mathcal{L}$ are very well suited to support a duality theory for linear differential and difference inclusions. Moreover, the key construction, leading from one such convex function to another and reflecting the relationship between P and P^{-1} for quadratic Lyapunov functions, exists and is well appreciated in convex analysis and optimization. It is the construction of a convex conjugate function. For any $V \in \mathcal{L}$, its convex conjugate is defined as

$$V^*(\xi) = \sup_x \left\{ \xi^T x - V(x) \right\}. \tag{10}$$

The function V^* is also convex, and in fact $V^* \in \mathcal{L}$. For example, verifying that it is homogeneous can be done directly: for any $\lambda \neq 0$, we have

$$V^*(\lambda \xi) = \sup_x \left\{ (\lambda \xi)^T x - V(x) \right\} = \lambda^2 \sup_x \left\{ \frac{1}{\lambda} \xi^T x - \frac{1}{\lambda^2} V(x) \right\}$$
$$= \lambda^2 \sup_x \left\{ \xi^T \left(\frac{x}{\lambda} \right) - V \left(\frac{x}{\lambda} \right) \right\} = \lambda^2 \sup_x \left\{ \xi^T x - V(x) \right\}.$$

The last expression is exactly $\lambda^2 V^*(\xi)$. Showing the other properties of V^* requires more elaborate techniques.[5] The relationship between V and V^* is one-to-one, and in fact it is symmetric: the conjugate of V^* is V:

$$(V^*)^*(x) = \sup_\xi \left\{ x^T \xi - V^*(\xi) \right\} = V(x). \tag{11}$$

In particular, $V \in \mathcal{L}$ if and only if $V^* \in \mathcal{L}$. This, in a sense, generalizes the fact that P is symmetric and positive definite if and only if P^{-1} is.

Even more importantly, the convex conjugate turns out to be exactly the object one needs when passing from a Lyapunov inequality verifying stability of a linear differential or difference inclusion to a Lyapunov function verifying the stability of a dual inclusion. This is particularly striking for difference inclusions; nothing more than the very definition of V^* is then needed. Indeed, suppose that for some $V \in \mathcal{L}$ and some $\gamma \in (0, 1)$ we have

$$V(Ax) \leq \gamma V(x) \quad \text{for all } x. \tag{12}$$

We will now show that this condition is equivalent to

$$V^*(A^T \xi) \leq \gamma V^*(\xi) \quad \text{for all } \xi. \tag{13}$$

The argument is a direct computation. Suppose that (12) is true. Then

$$V^*(A^T \xi) = \sup_x \left\{ (A^T \xi)^T x - V(x) \right\} \leq \sup_x \left\{ \xi^T A x - \frac{1}{\gamma} V(Ax) \right\},$$

---

[5]In general, for finite convex functions, conjugacy gives an equivalence between strict convexity of a function and differentiability of its conjugate, and vice versa. Thus, the conjugate of a strictly convex and differentiable function is differentiable and strictly convex.

where the inequality comes from replacing $V(x)$ by the bound coming from (12). The expression on the right above can be rewritten as a supremum taken over the range of A, which is not greater than the supremum taken over the whole space. That is,

$$V^*(A^T\xi) \le \frac{1}{\gamma} \sup_{z \in \mathrm{rge}\,A} \left\{ (\gamma\xi)^T z - V(z) \right\} \le \frac{1}{\gamma} \sup_{x} \left\{ (\gamma\xi)^T x - V(x) \right\}.$$

The supremum on the right is exactly $V^*(\gamma\xi)$. As we already know that V^* is positively homogeneous of degree 2, we obtain $V^*(A^T\xi) \le \gamma V^*(\xi)$, which is the inequality in (13).

An immediate and important consequence of the fact that a Lyapunov inequality (12) translates to the dual inequality (13) is that stability of a linear difference inclusion is equivalent to stability of its dual. More specifically,

(i) the linear difference inclusion[6]

$$x^+ \in \mathrm{co}\{A_i\}_{i=1}^m x \tag{14}$$

is exponentially stable with constant γ

if and only if

(ii) the dual linear difference inclusion

$$\xi^+ \in \mathrm{co}\{A_i^T\}_{i=1}^m \xi \tag{15}$$

is exponentially stable with constant γ.

A corresponding result for continuous time relies on the relationship between ∇V and ∇V^*. These mappings are inverses of one another. Further refinement of that fact for positively homogeneous functions leads to the desired result: a function $V \in \mathcal{L}$ is such that the Lyapunov inequality (9) holds if and only if $V^* \in \mathcal{L}$ is such that the dual Lyapunov inequality,

$$\nabla V^*(\xi)^T A^T \xi \le -\gamma V^*(\xi) \quad \text{for all } \xi, \tag{16}$$

is satisfied. Consequently, as in discrete time, we have

(i) the linear differential inclusion (5) is exponentially stable with constant γ

if and only if

(ii) the dual linear differential inclusion (6) is exponentially stable with constant γ.

---

[6]The linear difference inclusion (14) means that $x(k+1)$ is an element of the convex hull of points $A_i x(k)$, $i = 1, 2, \ldots, m$, for all k.

An immediate benefit of the equivalence of stability, for both differential and difference inclusions, is that it doubles the number of numerical tools one can use to establish stability. Given a particular numerical technique to test whether a differential (or difference) inclusion given by A_1, A_2, ..., A_m is asymptotically stable, one can apply it as well to the transposes. If either test—for the original matrices or for the transposes—shows stability, the theory we described concludes that in fact both inclusions—the original one and the dual one—are stable. This simple trick can lead to very surprising results.

Example 1. Consider

$$A_1 = \begin{bmatrix} 0 & 1 & 0 \\ 0 & 0 & 1 \\ -1 & -2 & -4 \end{bmatrix}, \; M = \begin{bmatrix} -2 & 0 & -1 \\ 1 & -10 & 3 \\ 3 & -4 & 2 \end{bmatrix}, \; A_2 = A_1 + aM, \qquad (17)$$

with $a > 0$. Then consider the LDI with the state matrix belonging to the set $\mathrm{co}\{A_1, A_2(a)\}$. One of the previously used numerical tests for stability of LDIs, proposed in [5], searches for the existence of a homogeneous polynomial Lyapunov function (HPLF). This test, when applied using fourth-order HPLFs to A_1, $A_2(a)$, shows that the LDI is stable for positive a up to 75.1071. That same test, applied to A_1^T, $A_2^T(a)$, and thus testing the stability of an LDI given by $\mathrm{co}\{A_1^T, A_2(a)^T\}$, shows it is stable for all positive a! By duality, the original inclusion is also stable for all such as.

Another benefit is that duality helps in designing numerical tests and identifying favorable classes of potential Lyapunov functions. We illustrate this in the next section.

3 Classes of potential Lyapunov functions

A linear differential inclusion (5) is asymptotically, and then in fact exponentially stable, if and only if there exists a convex, positive definite, and homogeneous of degree 2 function V such that the Lyapunov inequality (9) holds for all $A \in \mathrm{co}\{A_i\}_{i=1}^m$.[7] To construct numerical tools that search for functions verifying stability of a given LDI, one needs to restrict attention to particular classes of potential Lyapunov functions. Here, we discuss two such classes, general enough to approximate any convex, positive definite, and homogeneous of degree 2 function, while also amenable to numerical methods.

First we note that, as it could be expected from the discussion at the beginning of Section 2, for quadratic functions we have

---

[7]From now on, we do not insist that V be differentiable. Consequently, inequality (9) should be understood to hold at all points where V is differentiable, which is almost everywhere. This is in fact equivalent to (9) being valid with ∇V replaced by the subdifferential of V in the sense of convex analysis. The subdifferential may be a set, and then (9) is understood to hold for every element of the subdifferential.

$$V(x) = \frac{1}{2}x^T P x, \quad V^*(\xi) = \frac{1}{2}\xi^T P^{-1}\xi,$$

when P is symmetric and positive definite. It can be verified by a direct computation. Thus, quadratic functions given by symmetric and positive definite matrices form a class that is conjugate to itself.

The two classes of convex functions we want to use in stability analysis are conjugate to one another. They are the functions given by a pointwise maximum of a family of quadratic functions and the functions given as the convex hull of a family of quadratic functions. More specifically, given symmetric and positive definite $P_1, P_2, \ldots, P_l$, by the max function we mean the pointwise maximum of the quadratic functions given by P_js, that is,

$$V_{\max}(x) = \max_{j=1,2,\ldots,l} \frac{1}{2}x^T P_j x. \tag{18}$$

This function is strictly convex, positive definite, and homogeneous of degree 2.

With the help of the max function, and also the convex hull function we discuss later, convenient BMI conditions for stability can be obtained. For example, suppose we want to verify the stability of an LDI given by two matrices A_1, A_2, using the maximum of two quadratic functions. That is, we want to find P_1, P_2 such that $V_{\max}$ given by (18) is a Lyapunov function for the LDI under discussion. Pick a point x where $V_{\max}$ is differentiable, and suppose that $\frac{1}{2}x^T P_1 x \geq \frac{1}{2}x^T P_2 x$. Then $\nabla V_{\max}(x) = P_1 x$, and for $V_{\max}$ to satisfy (9) at x, we need

$$x^T \left(P_1 A_i^T + A_i P_1\right) x \leq -\gamma x^T P_1 x$$

for $i = 1, 2$. On the other hand, we do not need $x^T \left(P_2 A_i^T + A_i P_2\right) x$ to be negative. Symmetrically, at points where $\frac{1}{2}x^T P_2 x \geq \frac{1}{2}x^T P_1 x$, we need

$$x^T \left(P_2 A_i^T + A_i P_2\right) x \leq -\gamma x^T P_2 x$$

for $i = 1, 2$. This suggests the following sufficient condition for $V_{\max}$ given by P_1, P_2 to be a Lyapunov function for $\dot{x}(t) \in \mathrm{co}\{A_1, A_2\}x(t)$: for some nonnegative $\lambda_1, \lambda_2, \lambda_3, \lambda_4$,

$$\begin{aligned}
P_1 A_1^T + A_1 P_1 &\leq \lambda_1 (P_2 - P_1) - \gamma P_1, \\
P_1 A_2^T + A_2 P_1 &\leq \lambda_2 (P_2 - P_1) - \gamma P_1, \\
P_2 A_1^T + A_1 P_2 &\leq \lambda_3 (P_1 - P_2) - \gamma P_2, \\
P_2 A_2^T + A_2 P_2 &\leq \lambda_4 (P_1 - P_2) - \gamma P_2.
\end{aligned}$$

In other words, if P_js and λ_ps solving the system above exist, $\dot{x}(t) \in \mathrm{co}\{A_1, A_2\}x(t)$ is stable. This can be easily generalized: if (7) has a solution, then the LDI (5) is exponentially stable (with constant γ), and a Lyapunov function verifying it is the max function (18). We stress that in general, existence of a solution to (7) is not necessary for stability of (5). In the case of $l = 2$, by the S-procedure, the existence of a solution is necessary for (18) to be a Lyapunov function, but this is still only sufficient for stability. For a stable LDI, while there always exists a Lyapunov function given by the point-wise maximum of quadratic functions, there may not exist one given by the maximum of two such functions.

Example 2. Consider the LDI $\dot{x}(t) \in \mathrm{co}\{A_1, A_2(a)\}$, where

$$A_1 = \begin{bmatrix} -1 & -1 \\ 1 & -1 \end{bmatrix}, \quad A_2(a) = \begin{bmatrix} -1 & -a \\ 1/a & -1 \end{bmatrix} \tag{19}$$

for $a > 1$. This LDI was used in [6] to show that the existence of a common quadratic Lyapunov function is not necessary to guarantee the exponential stability of the LDI. The maximal a that ensures the existence of a common quadratic Lyapunov function was found to be $3 + \sqrt{8} = 5.8284$. With the phase plane method, it was confirmed that the LDI is still stable for $a = 10$. However, as pointed out in [6], the analytical method is highly unlikely to be feasible for general systems.

Here we illustrate how increasing the number of matrices defining the max-imum function (18) improves the estimates of parameter a for which the LDI remains stable. In particular, our computation—that is, solving the system of BMIs (7)—carried out with 7 matrices P_j verifies stability for $a = 10.108$. Table 1 shows the maximal a (denoted $a_{\max}$) verified by $V_{\max,l}$ given by $l = 1, 2, \ldots, 7$ matrices, which guarantees the stability of the LDI.

l	1	2	3	4	5	6	7
$a_{\max}$	5.8284	8.109	8.955	9.428	9.731	9.948	10.108

Table 1. Maximal as for the LDI in Example 2.

The seven matrices defining $V_{\max,7}$ that verify the stability of the LDI at $a = 10.108$, listed for verification, are

$$\begin{bmatrix} 0.2854 & -0.7282 \\ -0.7282 & 7.6744 \end{bmatrix}, \begin{bmatrix} 0.5899 & -0.0010 \\ -0.0010 & 5.9677 \end{bmatrix}, \begin{bmatrix} 0.4925 & -0.3298 \\ -0.3298 & 6.7759 \end{bmatrix},$$

$$\begin{bmatrix} 0.6699 & 0.3275 \\ 0.3275 & 4.9847 \end{bmatrix}, \begin{bmatrix} 0.7257 & 0.5792 \\ 0.5792 & 3.9458 \end{bmatrix}, \begin{bmatrix} 0.3900 & -0.5799 \\ -0.5799 & 7.3360 \end{bmatrix}, \begin{bmatrix} 0.7592 & 0.7279 \\ 0.7279 & 2.8877 \end{bmatrix}.$$

As a visual verification of stability for $a = 10.108$, we sketch in Figure 1 the vectors $A_1 x$ and $A_2 x$ at points x on the boundary of the 1-level set of $V_{\max,7}$; that is, points where $V_{\max,7}(x) = 1$. By linearity and homogeneity, verifying one level set is sufficient.

While Lyapunov functions, computed for example by solving (7) as de-scribed above, verify stability, they can be also used to confirm instability. We illustrate this below, where a Lyapunov function for an LDI with a cer-tain parameter value is used to show instability when the parameter is varied.

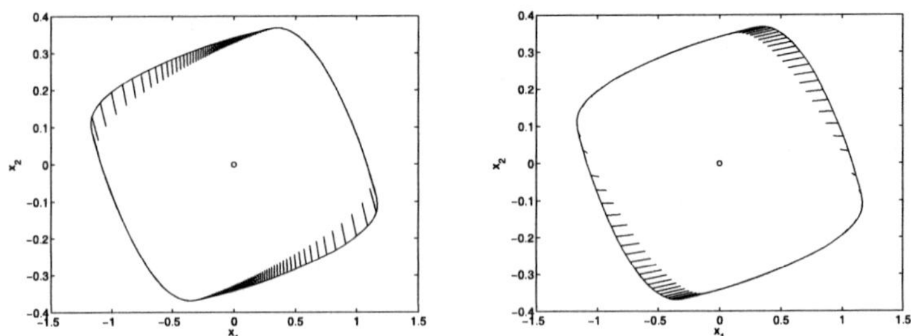

Fig. 1. $\dot{x} = A_1 x$ (left) and $\dot{x} = A_2 x$ (right) on the boundary of a level set.

Example 3. Recall the LDI from Example 2. In [6], it is pointed out that it may be stable for $a > 10$ (while it is only verified for a up to 10). Here we would like to estimate a lower bound on as that destabilize the LDI.

Suppose that $V_{\max}$ verifies stability of the LDI for a up to $\bar{a}$. At each x, we find the index i that maximizes $\nabla V_{\max}^T(x) A_i x$, use this to choose the "worst" switching among the vertices of the LDI, and try to produce potentially diverging trajectories for values of a larger than $\bar{a}$.

Below, $V_{\max,l}$ denotes the max function given by l matrices, as obtained in Example 2. By using $V_{\max,2}$, the lower bound on as (denoted as $\underline{a}_{\min}$) that guarantees instability of the LDI is detected as $\underline{a}_{\min} = 12.175$. A closed limit trajectory resulting from the worst switching law at $a = 12.175$ is plotted as the outer curve in the right box in Figure 2. The inner closed curve in the same box is the boundary of the 1-level set of $V_{\max,2}$. With $V_{\max,7}$, the lower bound for a that guarantees instability is $\underline{a}_{\min} = 11.292$. The corresponding "worst" switching leads to a closed limit trajectory plotted in the middle box in Figure 2, along with the 1-level set of $V_{\max,7}$.

Plotted in the right box in Figure 2 is a diverging trajectory corresponding to $l = 7$ and $a = 11.5$ (initial state marked with "*"). As we can see from the left box and the middle box, the difference between the limit trajectory and the boundary of the level set for $l = 7$ is smaller than that for $l = 2$. It is expected that as l is increased, the boundary of the level set can be made even closer to a limit trajectory, indicating that the Lyapunov function would

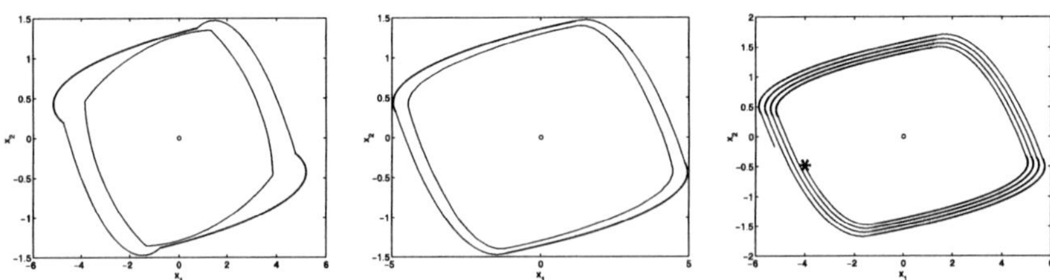

Fig. 2. Trajectories. Left: $l = 2, a = 12.175$; middle: $l = 7, a = 11.292$; right: $l = 7, a = 11.5$.

give a better estimation of stability. As expected, with an increased l, the difference between $a_{\max}$ and $\underline{a}_{\min}$ will get smaller.

We now describe the conjugate of the max function (18). First we need an additional construction. Given symmetric and positive definite matrices Q_1, $Q_2, \ldots, Q_l$, the minimum of the quadratic functions given by them needs not be convex. The convex hull function (determined by Q_js),

$$V_{\mathrm{co}}(\xi) = \mathrm{co} \min_{j=1,2,\ldots,l} \frac{1}{2}\xi^T Q_j \xi, \tag{20}$$

is the greatest convex function bounded above by the aforementioned minimum (equivalently, by each of the quadratic functions). This function is convex, differentiable, positive definite, and homogeneous of degree 2. The conjugacy relationship between the max function and the convex hull function is as follows: if $Q_j = P_j^{-1}$, then

$$V_{\max}^*(\xi) = V_{\mathrm{co}}(\xi), \quad V_{\mathrm{co}}^*(x) = V_{\max}(x).$$

In other words, the conjugate of the maximum of quadratics is the convex hull of quadratics given by the inverses of the original matrices.[8] If either $V_{\max}$ or V_{co} is quadratic, then both of them are, and their conjugacy reduces to the conjugacy for quadratic functions as stated at the beginning of this section.

With the help of the convex hull function V_{co} and the duality theory of Section 2, a condition for stability of an LDI, "dual" to matrix inequality (7), can be derived. Note that as (7) leads to stability of the linear differential inclusion (5), verified through $V_{\max}$ given by P_js, the inequality (8) leads to stability of the dual linear differential inclusion (6), verified through $V_{\max}$ given by Q_js. But by duality, this is equivalent to stability of (5), verified through V_{co} given by Q_j^{-1}s. In short, existence of a solution to (8) is a sufficient condition for stability of (5). We stress again that in general, this condition is not equivalent to (7).

Just as using the same numerical test for a given LDI and its dual inclusion may result in different stability estimates (recall Example 1), solving the systems of inequalities (7) and (8) may lead to different conclusions.

Example 4. Recall the linear differential inclusion from Example 1. We used both (7) and (8), with two unknown matrices, to estimate the range of the parameter a for which the LDI is stable. Using (7), we verified stability for all $a > 0$. Using (8), stability is verified for a up to 441. In other words, the matrix inequality based on the max function performs better than the inequality based on the convex hull function for this particular LDI. Conversely, the convex hull function performs better than the max function for the dual LDI. See also Example 7.

---

[8]A more general relationship is valid. The conjugate of a pointwise maximum of a family of convex functions is the convex hull of their conjugates.

Example 5. Consider the differential inclusion $\dot{x}(t) \in \mathrm{co}\{A_1, A_2, A_3\}x(t)$ where

$$A_1 = \begin{bmatrix} -2 & -2 \\ 4 & -2 \end{bmatrix}, \quad A_2 = \begin{bmatrix} -1 & 0 \\ 6 & -1 \end{bmatrix}, \quad A_3 = \begin{bmatrix} -5 & -3 \\ 2 & -1 \end{bmatrix}.$$

The system is not quadratically stable, but is asymptotically (and thus exponentially) stable. It can be confirmed with the convex hull function $V_{\mathrm{co},2}$ given by (20) with two matrices Q_j. The convergence rate turns out to be $\gamma = 0.0902$, that is,

$$\frac{d}{dt}V_{\mathrm{co},2}(x(t)) \leq -0.0902 V_{\mathrm{co},2}(x(t))$$

for all solutions to the LDI. However, stability cannot be confirmed by any of the max functions $V_{\max,2}$ given by (18) with two matrices. The maximal γ satisfying

$$\frac{d}{dt}V_{\max,2}(x(t)) \leq -\gamma V_{\max,2}(x(t))$$

for some $V_{\max,2}$ is -0.1575. Note that this in particular verifies that the LDI is not quadratically stable.

We carried out a similar experiment after doubling the dimension of the state space. Solving the BMIs of Example 5 took approximately 5 to 7 seconds; solving the corresponding ones for Example 6 took approximately twice that much time.

Example 6. Consider $\dot{x}(t) \in \mathrm{co}\{A_1, A_2, A_3\}x(t)$ where

$$A_1 = \begin{bmatrix} -2 & -2 & 1 & 0 \\ 4 & -2 & 0 & 1 \\ 0 & 0 & -2 & -2 \\ -1 & -1 & 4 & -2 \end{bmatrix}, \quad A_2 = \begin{bmatrix} -6 & -3 & 0 & 0 \\ 2 & -1 & 2 & 0 \\ 0 & -2 & -6 & -3 \\ 0 & -1 & 2 & -1 \end{bmatrix}, \quad A_3 = \begin{bmatrix} -1 & 0 & 1 & 0 \\ 5 & -1 & 4 & 1 \\ -1 & 0 & -1 & 0 \\ -1 & -1 & 5 & -1 \end{bmatrix}.$$

It is not quadratically stable. Stability can be confirmed with $V_{\mathrm{co},2}$, and the convergence rate turns out to be $\gamma = 0.1948$. However, stability cannot be confirmed with any $V_{\max,2}$. The maximal γ satisfying

$$\frac{d}{dt}V_{\max,2}(x(t)) \leq -\gamma V_{\max,2}(x(t))$$

is -0.0428.

4 Dissipativity properties

The duality theory outlined in Section 2 and the tools for verifying stability as discussed in Section 3 can be extended and applied to treat dissipativity properties of linear differential inclusions with disturbances.

Consider the following LDI with external disturbance:

$$\begin{bmatrix} \dot{x}(t) \\ y(t) \end{bmatrix} = \mathrm{co} \left\{ \begin{bmatrix} A & B \\ C & D \end{bmatrix}_i \right\}_{i=1}^{m} \begin{bmatrix} x(t) \\ d(t) \end{bmatrix}. \tag{21}$$

It is called dissipative with (positive semidefinite) storage function V and supply rate h if for all x, d,

$$\nabla V(x)^T (Ax + Bd) \leq h(Cx + Dd, d), \tag{22}$$

for all

$$\begin{bmatrix} A & B \\ C & D \end{bmatrix} \in \mathrm{co} \left\{ \begin{bmatrix} A & B \\ C & D \end{bmatrix}_i \right\}_{i=1}^{m}.$$

For example, consider $h(c,d) = -\frac{1}{2}\|c\|^2 + \frac{1}{2}\beta^2\|d\|^2$. Dissipativity with this supply rate means exactly that the LDI (21) has finite L^2-gain, bounded above by β. Later, we also discuss passivity and passivity with extra feedforward.

Following the ideas of Section 3, one can state sufficient conditions for dissipativity, based on considering storage functions given by a maximum of quadratic functions (18) or by a convex hull function (20). For example, if there exist symmetric and positive definite $P_1, P_2, \ldots, P_l$ and numbers $\lambda_{ijk} \geq 0$ for $i = 1, 2, \ldots, m$, $j, k = 1, 2, \ldots, l$ such that

$$\begin{bmatrix} A_i^T P_j + P_j A_i + \sum_{k=1}^{l} \lambda_{ijk}(P_j - P_k) + C_i^T C_i & P_j B_i + C_i^T D_i \\ B_i^T P_j + D_i^T C_i & -\beta^2 I + D_i^T D_i \end{bmatrix} < 0, \tag{23}$$

then the LDI (21) has finite L^2-gain of at most β. Furthermore, the max function (18) is a storage function verifying this property.

Example 7. Consider the LDI $\dot{x}(t) \in \mathrm{co}\{A_1, A_2(a)\}$ given by matrices (17). The LDI is stable for all $a > 0$, see Example 1 or 4. Here we illustrate how considering the max function $V_{\mathrm{max},l}$ with different l yields better convergence rate estimates, for the case of $a = 10000$. Furthermore, we introduce one-dimensional disturbance and observation to the system, by considering $B_1 = B_2 = [1\ 1\ 1]^T$, $C_1 = C_2 = [1\ 1\ 1]$, and $D_1 = D_2 = 0$ in (21). For this LDI, we rely on (23) to estimate the L^2-gain. The table below shows how broadening the class of Lyapunov and storage functions improves the convergence rate and L^2-gain estimates.

l	1	2	3	4	5
γ	-1.4911	0.0531	0.1416	0.1642	0.1772
L^2-gain	N/A	57.6956	21.6832	18.7191	17.3278

Table 2. Convergence rate and L^2-gain estimates for the LDI in Example 7.

Duality of dissipativity properties can also be established, similarly to what we outlined for stability. Consider the dual of (21):

$$\begin{bmatrix} \dot{\xi}(t) \\ z(t) \end{bmatrix} = \mathrm{co} \left\{ \begin{bmatrix} A^T & C^T \\ B^T & D^T \end{bmatrix}_i \right\}_{i=1}^{m} \begin{bmatrix} \xi(t) \\ w(t) \end{bmatrix}. \tag{24}$$

Suppose that the supply rate is given by

$$h(c,d) = \frac{1}{2} \begin{bmatrix} c \\ d \end{bmatrix}^T M \begin{bmatrix} c \\ d \end{bmatrix} \quad \text{for} \quad M = \begin{bmatrix} -R & Z \\ Z^T & S \end{bmatrix},$$

with R and S symmetric, positive semidefinite and that M is invertible. Note that this is the case, for example, for the supply rate corresponding to finite L^2-gain. Let the dual supply rate $h^\sharp$ be given by[9]

$$h^\sharp(v,w) = -\frac{1}{2} \begin{bmatrix} w \\ -v \end{bmatrix}^T M^{-1} \begin{bmatrix} w \\ -v \end{bmatrix}.$$

If V is a convex, positive definite, and positively homogeneous of degree 2 function, then the LDI (21) is dissipative with storage function V and supply rate h if and only if the dual LDI (24) is dissipative with the storage function V^* and supply rate $h^\sharp$, that is, for all ξ, w,

$$\nabla V^*(\xi)^T (A^T \xi + C^T w) \le h^\sharp(B^T \xi + D^T w, w). \tag{25}$$

A more general equivalence can be shown, where the terms $-\gamma V(x)$ and $-\gamma V^*(\xi)$ are added to the right-hand sides of inequalities (22) and (25). Thus, the stated equivalence generalizes the one between the stability of the LDI (5) and of its dual (6), as verified by inequalities (9), (16).

In particular, consider $h(c,d) = -\frac{1}{2}\|c\|^2 + \frac{1}{2}\beta^2\|d\|^2$ corresponding to finite L^2-gain. Then $M = \begin{bmatrix} -I & 0 \\ 0 & \beta^2 I \end{bmatrix}$, $M^{-1} = \begin{bmatrix} -I & 0 \\ 0 & \beta^{-2} I \end{bmatrix}$, and $h^\sharp(v,w) = \frac{1}{2}\|w\|^2 - \frac{1}{2}\beta^{-2}\|v\|^2$. The general facts stated above imply that, for a convex, positive definite, and homogeneous of degree 2 function V,

$$\nabla V(x)^T (Ax + Bd) \le -\frac{1}{2}\|Cx + Dd\|^2 + \frac{1}{2}\beta^2\|d\|^2$$

for all x, d if and only if for all ξ, w

$$\nabla V^*(\xi)^T (A^T \xi + C^T w) \le -\frac{1}{2}\beta^{-2}\|B^T \xi + D^T w\|^2 + \frac{1}{2}\|w\|^2.$$

This equivalence suggests that a dual to the sufficient condition (23) for the LDI (21) to have finite L^2-gain can be stated. The dual condition relies on

---

[9]Note that h is a function concave in one variable, convex in the other. A conjugacy theory for such functions, extending that for convex functions, does exist. In particular, a concave/convex function conjugate to h can be defined. The dual supply rate $h^\sharp$ is closely related to it.

the convex hull function (20) serving as the storage function and corresponds to (23) just as the dual stability condition (8) corresponds to (7). It is

$$\begin{bmatrix} Q_j A_i^T + A_i Q_j + \sum_{k=1}^{l} \lambda_{ijk}(Q_j - Q_k) + B_i B_i^T & Q_j C_i^T + B_i D_i^T \\ C_i Q_j + D_i B_i^T & -\beta^2 I + D_i D_i^T \end{bmatrix} < 0. \quad (26)$$

We use these dual conditions in the following example.

Example 8. Consider the differential inclusion

$$\dot{x}(t) \in co\{A_1, A_2\}x(t) + Bd(t), \quad y(t) = Cx(t),$$

where

$$A_1 = \begin{bmatrix} 0 & 1 & 0 \\ 0 & 0 & 1 \\ -1 & -2 & -3 \end{bmatrix}, \quad A_2 = \begin{bmatrix} 0 & 1 & 0 \\ 0 & 0 & 1 \\ -2 & -3 & -1 \end{bmatrix}, \quad B = \begin{bmatrix} 0 \\ -1 \\ 1 \end{bmatrix}, \quad C = \begin{bmatrix} 1 & 0 & 1 \end{bmatrix}.$$

The matrices A_1 and A_2 are taken from an example in [19].

With zero disturbance d, the system is stable, but not quadratically. Stability is confirmed with both the max function (18) and the convex hull function (20) with $l = 2$. The convergence rate verified by $V_{max,2}$ and that with $V_{co,2}$ are both equal to 0.0339, that is, (9) holds with $\gamma = 0.0339$ for some max function and some convex hull function given by two quadratics.

However, the estimates of the L^2-gain of the full system, obtained by the max function and the convex hull function with two quadratics, differ. They are, respectively, 16.7337 and 30.6556.

Now consider $h(c, d) = c^T d - \frac{1}{2}\eta\|d\|^2$. Then (22) represents passivity of the inclusion with disturbance (21) when $\eta = 0$, and passivity with extra feedforward when $\eta > 0$. Arguments as those outlined for the L^2-gain show that for a convex, positive definite, and homogeneous of degree 2 function V,

$$\nabla V(x)^T (Ax + Bd) \le (Cx + Dd)^T d - \frac{1}{2}\eta\|d\|^2 \quad (27)$$

for all x, d if and only if for all ξ, w

$$\nabla V^*(\xi)^T (A^T\xi + C^T w) \le (B^T\xi + D^T w)^T w - \frac{1}{2}\eta\|w\|^2.$$

Thus passivity (with feedforward) of (21) is equivalent to passivity (with feedforward) of (24). Similarly one obtains that the existence of symmetric and positive definite $P_1, P_2, \dots, P_l$ and numbers $\lambda_{ijk} \ge 0$ for $i = 1, 2, \dots, m$, $j, k = 1, 2, \dots, l$ such that

$$\begin{bmatrix} A_i^T P_j + P_j A_i + \sum_{k=1}^{N} \lambda_{ijk}(P_j - P_k) & P_j B_i - C_i^T \\ B_i^T P_j - C_i & \eta - D_i^T - D_i \end{bmatrix} < 0 \quad (28)$$

is a sufficient condition for passivity of (21). Of course, a dual condition can be given. We do not state it explicitly, but apply it, and (28), in examples.

118   R. Goebel, T. Hu, and A.R. Teel

Example 9. Consider (21) given by

$$A_1 = \begin{bmatrix} 0 & 1 & 0 \\ 0 & 0 & 1 \\ -1 & -2 & -3 \end{bmatrix}, \quad B_1 = \begin{bmatrix} 0.6 \\ 0 \\ 0.6 \end{bmatrix}, \quad C_1 = \begin{bmatrix} 0 & -1 & 1 \end{bmatrix}, \quad D_1 = 2,$$

$$A_2 = \begin{bmatrix} 0 & 1 & 0 \\ 0 & 0 & 1 \\ -1 & -1 & -3 \end{bmatrix}, \quad B_2 = B_1, \quad C_2 = \begin{bmatrix} 0 & -1 & -1 \end{bmatrix}, \quad D_2 = D_1.$$

The matrices A_1 and A_2 are taken from [19], where it was shown that the system (without disturbance) is quadratically stable. Passivity of the inclusion, however, cannot be confirmed with a quadratic function: the largest η for which the passivity inequality (27) holds is -0.601. When the maximum of two quadratic functions is used as a storage function, that is, the matrix inequality (28) is solved, the maximal η is -0.0797. When the convex hull function is used, the maximal η is 0.0891. This confirms passivity of the LDI, in fact with feedforward.

Example 10. Consider the LDI (21) with $m = 2$, and with B_2 and C_2 depending on a parameter a. The matrices are

$$A_1 = A_2 = \begin{bmatrix} 0 & 1 & 0 \\ 0 & 0 & 1 \\ -1 & -2 & -3 \end{bmatrix}, \quad B_1 = \begin{bmatrix} 0 \\ 0 \\ 1 \end{bmatrix}, \quad B_2 = B_1 + a \begin{bmatrix} 0 \\ -1 \\ 1 \end{bmatrix},$$

$$C_1 = \begin{bmatrix} 1 & 0 & 0 \end{bmatrix}, \quad C_2 = C_1 + a \begin{bmatrix} 0 & 1 & 1 \end{bmatrix}, \quad D_1 = D_2 = 2.$$

For $a = 0$, the LDI reduces to a linear system, which is passive (and this can be verified with a quadratic storage function). In fact, quadratic functions verify passivity of the LDI for all $a \in [0, 2.097]$. Using the maximum of two quadratic functions increases the upper limit to 2.3706, the convex hull of two quadratics yields 2.3497.

5 Discrete-time systems

As we showed in Section 2, convex conjugate functions also allow for establishing a duality theory for linear difference inclusions, that is, for discrete-time counterparts of the systems discussed so far. The functions proposed in Section 3 can of course also be used as Lyapunov or storage functions in discrete time, and they suggest matrix inequalities for stability and dissipativity properties.

First consider the LDI (14). If there exist symmetric and positive definite $P_1, P_2, \ldots, P_l$ and numbers $\lambda_{ijk} \geq 0$ with $\sum_{k=1}^{l} \lambda_{ijk} = 1$ for $i = 1, 2, \ldots, m$, $j, k = 1, 2, \ldots, l$ such that

$$A_i^T P_j A_i < \gamma \sum_{k=1}^{l} \lambda_{ijk} P_k$$

for all $i = 1, 2, \ldots, m$, $j = 1, 2, \ldots, l$, then the discrete-time Lyapunov inequality (12) holds for the max function (18), for all $A \in \mathrm{co}\{A_i\}_{i=1}^m$. Dually, if matrices P_j and numbers λ_{ijk} satisfy

$$A_i P_j A_i^T < \gamma \sum_{k=1}^{l} \lambda_{ijk} P_k,$$

the Lyapunov inequality (12) holds for the convex hull function (20) given by $Q_j = P_j^{-1}$. Solution of either system of matrix inequalities verifies stability of both (14) and (15).

Example 11. Consider the difference inclusion (14) given by

$$A_1 = \begin{bmatrix} 0.3 & 1 & 0 \\ 0 & 0.6 & 1 \\ 0 & 0 & 0.7 \end{bmatrix}, \quad A_2 = \begin{bmatrix} 0.3 & 0 & 0 \\ -0.5 & 0.7 & 0 \\ -0.2 & -0.5 & 0.7 \end{bmatrix}.$$

The smallest γ for which the Lyapunov inequality (12) holds with a quadratic function is 1.2353, so a quadratic function does not verify stability of the inclusion. (For stability, we need $\gamma \in (0, 1)$.) The smallest such γ (i.e., the convergence rate) estimated with the max function with two quadratics is 0.9570; a convex hull of two quadratics yields 1.0231. In this case, only the max function verifies stability of the inclusion.

For LDIs with disturbance

$$\begin{bmatrix} x^+ \\ y \end{bmatrix} \in \mathrm{co}\left\{ \begin{bmatrix} A & B \\ C & D \end{bmatrix} \right\}_{i=1}^{m} \begin{bmatrix} x \\ d \end{bmatrix}, \tag{29}$$

dissipativity properties can be verified through the following matrix inequalities. Let $\gamma > 0$. If there exist symmetric, positive definite $P_1, P_2, \ldots, P_l$ and numbers $\lambda_{ijk} \geq 0$ with $\sum_{k=1}^{l} \lambda_{ijk} = 1$ for $i = 1, 2, \ldots, m$, $j, k = 1, 2, \ldots, l$ such that

$$\begin{bmatrix} A_i^T P_j A_i - \sum_{k=1}^{l} \lambda_{ijk} P_k + C_i^T C_i & A_i^T P_j B_i + C_i^T D_i \\ B_i^T P_j A_i + D_i^T C_i^T & -\gamma^2 I + B_i^T P_j B_i + D_i^T D_i \end{bmatrix} < 0 \tag{30}$$

for all $i = 1, 2, \ldots, m$, $j = 1, 2, \ldots, l$, then γ is an upper bound for the l^2-gain of (29). This is verified by the max function $V_{\max}$ given by (18), that is, for $V = V_{\max}$ and all x, d, x^+, y satisfying (29), the following inequality holds:

$$V(x^+) - V(x) < -\frac{1}{2}\|y\|^2 + \frac{1}{2}\gamma^2\|d\|^2.$$

Similarly, if there exist P_js and λ_{ijk}s such that

$$\begin{bmatrix} A_i^T P_j A_i - \sum_{k=1}^{l} \lambda_{ijk} P_k & A_i^T P_j B_i - C_i^T \\ B_i^T P_j A_i - C_i & \eta + B_i^T P_j B_i - D_i^T - D_i \end{bmatrix} < 0, \tag{31}$$

then, for $V = V_{\max}$,

$$V(x^+) - V(x) \le y^T d - \frac{1}{2}\eta\|d\|^2, \tag{32}$$

and thus (29) is passive (if $\eta = 0$) and passive with feedforward (if $\eta > 0$). Duality for discrete-time systems with disturbance can also be established, and inequalities dual to (30) and (31) can be used.

Example 12. Consider the inclusion (29) given by

$$A_1 = \begin{bmatrix} 0 & 1 & 0 \\ 0 & 0 & 1 \\ 0.16 & 0 & -0.32 \end{bmatrix}, \quad B_1 = \begin{bmatrix} 2.5 \\ 0 \\ 0.4 \end{bmatrix}, \quad C_1 = \begin{bmatrix} 1 & 0 & 0 \end{bmatrix},$$

$$A_2 = A_1 + \begin{bmatrix} 1 & -1 & -1 \\ 1 & -3 & 0 \\ 0 & -1.5 & 2 \end{bmatrix} \times a, \quad B_2 = \begin{bmatrix} a \\ 0 \\ 2.5 \end{bmatrix}, \quad C_2 = \begin{bmatrix} a & -a & 1 \end{bmatrix},$$

where the parameter $a \ge 0$. The maximal a ensuring stability verifiable by quadratics is 0.3153. The maximal a ensuring stability verifiable by $V_{\mathrm{co},2}$ or $V_{\max,2}$ is 0.4475. At $a = 0.4475$, the convergence rate, i.e., the minimal β such that $V(x^+) \le \beta V(x)$, by quadratics is 1.3558, while that by $V_{\mathrm{co},2}$ or $V_{\max,2}$ is 1. At $a = 0.3153$, the convergence rate by quadratics is 1, while the convergence rate by $V_{\mathrm{co},2}$ or $V_{\max,2}$ is 0.6873. With $V_{\mathrm{co},2}$, the l^2-gain is bounded by 7.038, with $V_{\max,2}$, the l^2-gain is bounded by 8.3615.

Example 13. Consider the inclusion (29) given by

$$A_1 = \begin{bmatrix} 0 & 1 & 0 \\ 0 & 0 & 1 \\ 0.16 & 0 & -0.32 \end{bmatrix}, \quad B_1 = \begin{bmatrix} 2.5 \\ 0 \\ 0.4 \end{bmatrix}, \quad C_1 = \begin{bmatrix} 1 & 0 & 0 \end{bmatrix}, \quad D_1 = 5,$$

$$A_2 = \begin{bmatrix} 0.4 & 0.6 & -0.4 \\ 0.4 & -1.2 & 1 \\ 0.16 & -0.6 & 0.48 \end{bmatrix}, \quad B_2 = \begin{bmatrix} 0.4 \\ 0 \\ 2.5 \end{bmatrix}, \quad C_2 = \begin{bmatrix} 0.4 & -0.4 & 1 \end{bmatrix}, \quad D_2 = 6.$$

Stability is not confirmed by quadratics while $V_{\mathrm{co},2}$ and $V_{\max,2}$ both ensure a convergence rate of 0.7530. With $V_{\max,2}$, the maximal η such that (32) is -0.0148. With $V_{\mathrm{co},2}$, the maximal η is 0.5078.

In the concluding example, we outline an application of the dual BMI technology to estimation of the domain of attraction of a nonlinear system.

Example 14. Consider a second-order saturated linear system

$$x^+ = Ax + B\mathrm{sat}(Fx),$$

where

$$A = \begin{bmatrix} 0.8 & 0 \\ 0 & 1.2 \end{bmatrix}, \quad B = \begin{bmatrix} 0.6 \\ -1 \end{bmatrix}, \quad F = \begin{bmatrix} 0.5 & 1 \end{bmatrix},$$

and sat is the standard saturation function, that is $\mathrm{sat}(s) = s$ when $s \in [-1, 1]$ and $\mathrm{sat}(s) = 1$ (respectively, -1) when $s > 1$ (respectively, $s < -1$). We want to estimate the domain of attraction of this system with contractively invariant 1-level set of certain type of Lyapunov functions. For a given x_0, the Lyapunov function is optimized so that its invariant level set contains αx_0 with a maximal α.

We first consider quadratic functions. Using the algorithm from [12], the maximal α for $x_{01} = \begin{bmatrix} 1 & 0 \end{bmatrix}^T$ is $\alpha_1 = 5.0581$ and the maximal α for $x_{02} = \begin{bmatrix} 0 & 1 \end{bmatrix}^T$ is $\alpha_2 = 3.6402$.

Next we consider $V_{\mathrm{co},2}$. An algorithm to maximize α so that αx_0 is inside an invariant level set of $V_{\mathrm{co},2}$ can be developed similarly to that in Section 3 of [13]. From this algorithm, the maximal α for x_{01} is $\tilde{\alpha}_1 = 7.1457$ and the maximal α for x_{02} is $\tilde{\alpha}_2 = 5$. In the left box of Figure 3, we plot the level set (thick solid curve) as the convex hull of two ellipsoids (dash-dotted line). The point on the thick curve marked with " $*$ " is $\tilde{\alpha}_2 x_{02}$. The inner ellipsoid in solid line is the maximal invariant ellipsoid where the point marked with " $*$ " is $\alpha_2 x_{02}$. To demonstrate that the level set is actually invariant, we plot the image of the boundary under the next step map $x \mapsto Ax + B\mathrm{sat}(Fx)$ (see the thin solid curve in the right box of Figure 3). For comparison, we also plot the image under the linear map $x \mapsto (A + BF)x$ in dashed curve. Parts of the thin solid curve overlap with the thick solid curve. This means that some trajectories overlap parts of the boundary of the level set. As a matter of fact, for any $\alpha > \tilde{\alpha}_2$, a trajectory starting from $\alpha \begin{bmatrix} 0 & 1 \end{bmatrix}^T$ will diverge.

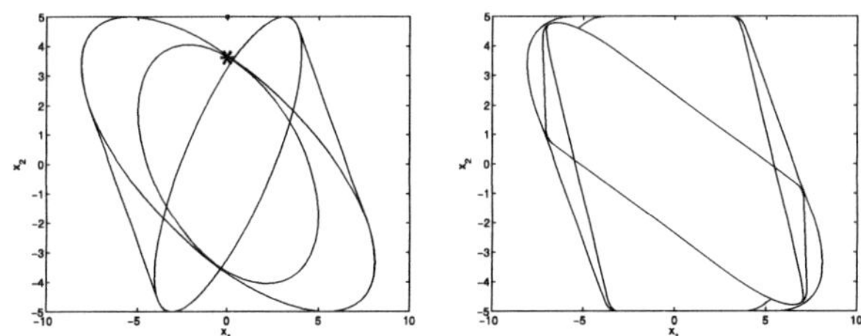

Fig. 3. Left: the invariant level set; right: the next step map.

References

1. Beran E, Vandenberghe L, Boyd S (1997) A global BMI algorithm based on the generalized Benders decomposition. In: Proc. of the European Control Conference, Brussels, Belgium, paper no. 934
2. Blanchini F (1995) Nonquadratic Lyapunov functions for robust control. Automatica 31:451–461

3. Boyd S, El Ghaoui L, Feron E, Balakrishnan V (1994) Linear matrix inequalities in system and control theory. SIAM, Philadelphia, PA
4. Brayton R, Tong C (1979) Stability of dynamical systems: a constructive approach. IEEE Trans. on Circuits and Systems 26:224–234
5. Chesi G, Garulli A, Tesi A, Vicino A (2003) Homogeneous Lyapunov functions for systems with structured uncertainties. Automatica 39:1027–1035
6. Dayawansa W, Martin C (1999) A converse Lyapunov theorem for a class of dynamical systems which undergo switching. IEEE Trans. Automat. Control 44(4):751–760
7. Goebel R, Teel A, Hu T, Lin Z (2004) Dissipativity for dual linear differential inclusions through conjugate storage functions. In: Proc. of the 43rd IEEE Conference on Decision and Control, Bahamas
8. Goh K, Safonov M, Papavassilopoulos GP (1994) A global optimization approach for the BMI problem. 2009–2014. In: Proc. of the 33rd IEEE Conference on Decision and Control, Lake Buena Vista, FL
9. Hassibi A, How J, Boyd S (1999) A path-following method for solving BMI problems in control. 1385–1389. In: Proc. of American Control Conference, San Diego, CA
10. Hu T, Huang B, Lin Z (2004) Absolute stability with a generalized sector condition. IEEE Trans. Automatic Control 49(4):535–548
11. Hu T, Lin Z (2003) Composite quadratic Lyapunov functions for constrained control systems. IEEE Trans. Automatic Control 48(3):440–450
12. Hu T, Lin Z, Chen B (2002) Analysis and design for linear discrete-time systems subject to actuator saturation. Systems & Control Lett. 45(2):97–112
13. Hu T, Lin Z, Goebel R, Teel A (2004) Stability regions for saturated linear systems via conjugate Lyapunov functions. In: Proceedings of the 43rd IEEE Conference on Decision and Control, Bahamas
14. Jarvis-Wloszek Z, Packard A (2002) An LMI method to demonstrate simultaneous stability using nonquadratic polynomial Lyapunov functions. 287–292. In: Proc. of the 41st IEEE Conference on Decision and Control, Las Vegas, NV
15. Molchanov A, Pyatnitskiy Y (1989) Criteria of asymptotic stability of differential and difference inclusions endountered in control theory. Systems & Control Lett. 13:59–64
16. Power H, Tsoi A (1973) Improving the predictions of the circle criterion by combining quadratic forms. IEEE Trans. Automat. Control 18(1):65–67
17. Rockafellar R (1970) Convex analysis. Princeton University Press
18. Rockafellar R, Wets RJB (1998) Variational analysis. Springer, New York
19. Shorten RN, Narendra KS (2003) On common quadratic Lyapunov functions for pairs of stable LTI systems whose system matrices are in companion form. IEEE Trans. Automat. Control 48(4):618–621
20. Willems J (1972a) Dissipative dynamical systems, part I: General theory. Arch. Rational Mech. Anal. 45:321–351
21. Willems J (1972b) Dissipative dynamical systems, part II: Linear systems with quadratic supply rates. Arch. Rational Mech. Anal. 45:352–393
22. Xie L, Shishkin S, Fu M (1997) Piecewise Lyapunov functions for robust stability of linear time-varying systems. Systems & Control Lett. 31:165–171
23. Zelentsovsky A (1994) Nonquadratic Lyapunov functions for robust stability analysis of linear uncertain systems. IEEE Trans. Automat. Control 39(1):135–138

Oscillators as Systems and Synchrony as a Design Principle*

Rodolphe Sepulchre

Electrical Engineering and Computer Science Institute
Montefiore B28, B-4000 Liège, Belgium
r.sepulchre@ulg.ac.be

Summary. The chapter presents an expository survey of ongoing research by the author on a system theory for oscillators. Oscillators are regarded as open systems that can be interconnected to robustly stabilize ensemble phenomena characterized by a certain level of synchrony. The first part of the chapter provides examples of design (stabilization) problems in which synchrony plays an important role. The second part of the chapter shows that dissipativity theory provides an interconnection theory for oscillators.

1 Introduction

Oscillators are dynamical systems that exhibit stable limit cycle oscillations. The emphasis in this chapter is on oscillators as *open* systems, that is, as systems that can be interconnected to other systems. Synchrony refers to the tendency of interconnected oscillators to produce ensemble phenomena, that is, to phase lock as if an invisible conductor was orchestrating them. The emphasis in this chapter is on synchrony as a design principle, that is, on the use of synchrony to achieve stable oscillations in interconnected systems.

The manifestations of synchrony are numerous both in nature and in engineered devices. The interested reader will find several compelling illustrations in the stimulating recent essay by Strogatz [30]. As narrated in this essay and elsewhere, the accidental discovery by Huygens that two clocks in the same room tend to synchronize was soon regarded as the discovery of an undesirable phenomenon, revealing the sensitivity of clocks to external small perturbations at a time where the challenge was to engineer robust devices that could travel the ocean and provide a precise measure of longitude. Today, the growing interest for synchrony in engineering applications is precisely due

---

*This chapter presents research results of the Belgian Program on Interuniversity Attraction Poles, initiated by the Belgian Federal Science Policy Office. The scientific responsibility rests with its author.

to the robustness of collective phenomena, making an ensemble phenomenon insensitive to individual failures. In this sense, synchrony is a system concept.

In order to become a design principle, synchrony requires an interconnection theory for oscillators. Detailed models of oscillators abound in the literature, most frequently in the form of a set of nonlinear differential equations whose solutions robustly converge to a limit cycle oscillation. Local stability analysis is possible by means of Floquet theory, but global convergence analysis is usually restricted to second-order models and uses phase plane techniques. When analyzing collective phenomena in possibly large ensembles of interconnected oscillators, the dynamical model for each oscillator is usually further simplified, such as in phase models [35] where the state variable of each oscillator is a single phase variable on the circle.

The objective of this chapter is twofold: first, to motivate the use of synchrony as a design principle and the need for an interconnection theory of oscillators; second, to propose an external characterization of oscillators based on dissipativity theory and to examine its implications for the stability and synchrony analysis of interconnected oscillators.

In the first part of the chapter, we describe two stabilization problems in which synchrony plays an important role. Section 2 studies the stabilization of a bounce juggler, illustrative of rhythmic control tasks encountered in multileg robotics. We show how the stabilization of period two orbits (which mimics the shower pattern of a juggler) is best understood as achieving a phase-locking property for two impact oscillators. A distinctive feature of the proposed control is that it uses no feedback (sensorless control), even though the orbit is exponentially unstable in the unactuated system. The phase-locked property of the impact oscillators is induced by suitable oscillatory forcing of their input.

Section 3 describes a collective stabilization problem for N particles that move at unit speed in the plane with steering control. (Relative) equilibria of the model correspond to parallel or circular motions of the group. The orientation of each particle is a phase variable on the circle. Treating the orientation variables of the particles as phase variables of oscillators, parallel motion corresponds to synchronization whereas circular motion can be understood as a form of desynchronization. The synchrony measure is here the velocity of the center of mass of the group. It is maximal in parallel motions and minimal in circular motions. It coincides with a usual measure of synchrony in phase models of oscillators [31].

The two examples illustrate the role of synchrony as a design principle. Their ad hoc treatment also underlines the lack of interconnection theory for oscillators. In Section 4, we further illustrate with models from neurodynamics the persistent gap between physical models of oscillators and abstract models used to study their interconnections. This prompts us to introduce in Section 5 an external characterization of oscillators that fits their description by physical state space models but at the same time has implications for the stability and synchrony analysis of their interconnections. Following the dissipativity approach introduced by Willems [33], the external characterization is in the form of a dissipation inequality, with a new supply rate enabling a limit cycle behavior for the solutions of the isolated oscillator. We examine the implications of this dissipativity characterization for (i) (global) stability analysis of an isolated oscillator, (ii) (global) analysis of interconnections of N identical oscillators, and (iii) (global) synchrony analysis of interconnections of N identical oscillators. The theory covers two basic oscillation mechanisms, illustrated in the simplest way by van der Pol model and by Fitzhugh–Nagumo model, respectively.

This chapter should be regarded as an expository survey of ongoing research. The simplest examples are employed to illustrate the concepts and we refer the reader to more technical papers for the general treatment and for complete statements of the results.

2 Sensorless stabilization of rhythmic tasks

Synchrony plays an essential role in the robust coordination of rhythmic tasks. Neuroscientists have long identified the role of central pattern generators in living organisms as autonomous neural clocks that provide the rhythmic signals necessary to coordinate multileg locomotion such as walking, hopping, or swimming. We describe here a manifestation of synchrony as a design principle in a very contrived but illuminating example: the stabilization of periodic orbits in the bounce juggler model illustrated in Figure 1.

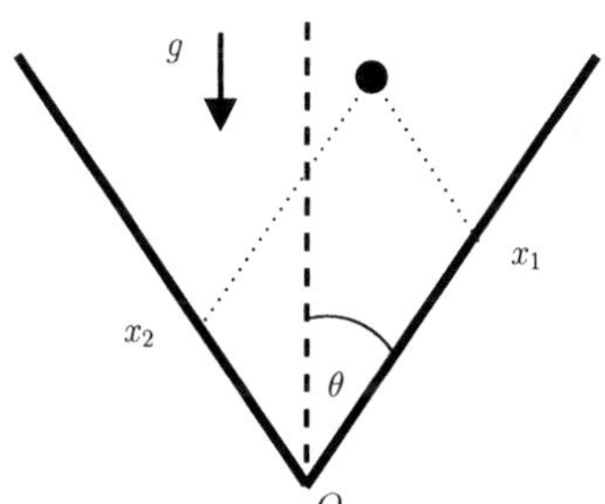

Fig. 1. A bounce juggler model.

This toy stabilization problem captures several important issues of impact control problems and is the subject of ongoing research [23, 3, 20, 21]. Here we only describe the problem in its simplest configuration and underline the phase-locking properties of the stabilized system.

The bounce juggler model describes the dynamics of a point mass (ball) in the plane under the action of a constant gravitational field. The ball undergoes elastic collisions with two intersecting edges, an idealization of the juggler's two arms. The two edges form a fixed angle θ with the direction of gravity. In a coordinate system aligned with the edges, the (nondimensional) equations of motion write

$$\ddot{x}_1 = -1, \; x_1 \geq 0$$
$$\ddot{x}_2 = -1, \; x_2 \geq 0. \tag{1}$$

A collision occurs when either x_1 or x_2 becomes zero, in which case Newton's rule is applied, that is, the normal velocity is reversed and the tangential velocity is conserved. The solution of (1) is then continued (restarted) with this new initial condition until a new collision occurs. The system is conservative (energy is conserved both during the frictionless flight of the ball and through the elastic collisions), and, except for the collision times, the two degrees of freedom of the system are decoupled. In spite of this apparent simplicity, this Hamiltonian system with collisions exhibits very rich dynamics [13, 36]. In particular, for $\theta > 45\,\mathrm{deg}$, the system possesses an infinite family of periodic orbits, each of which is exponentially unstable. In recent work [20, 21], we have shown that period one and period two orbits can be stabilized by proper oscillatory actuation of the wedge. This sensorless stabilization phenomenon is rather surprising because an exponentially unstable periodic orbit of the unactuated wedge becomes exponentially stable in the actuated wedge in spite of any feedback measurement. The analysis in [20, 21] shows that the phenomenon persists over a broad range of angles θ and, when the collisions are nonelastic, over a broad range of coefficients of restitution. Recent experimental validation of this sensorless stabilization suggests that the phenomenon is also quite robust.

The essence of the just-described sensorless stabilization phenomenon is best understood by considering the model in the special configuration illustrated in Figure 2, that is, for the particular angle $\theta = 45\,\mathrm{deg}$ (orthogonal wedge), and when the actuation of each edge is restricted to the direction orthogonal to the edge.

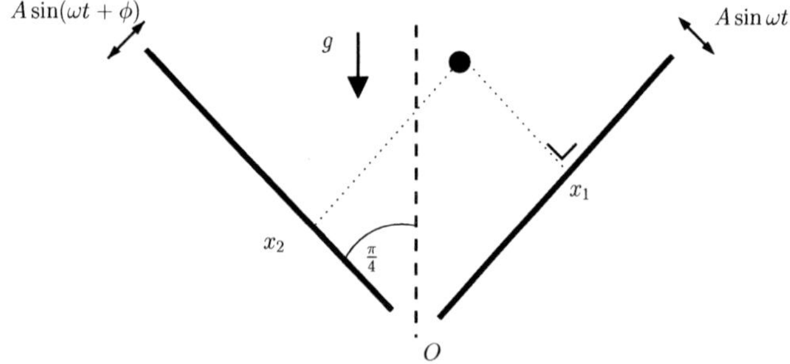

Fig. 2. A cartoon of the orthogonal wedge with orthogonal actuation of the two edges, leading to decoupled closed-loop dynamics.

In this special configuration, the dynamics of the vibrating wedge write

$$\ddot{x}_1 = -1, \quad x_1 \geq A\sin(\omega t)$$
$$\ddot{x}_2 = -1, \quad x_2 \geq A\sin(\omega t + \phi), \tag{2}$$

where A and ω are the amplitude and the vibrating frequency of each edge and where ϕ is the phase shift between the vibration of the two edges.

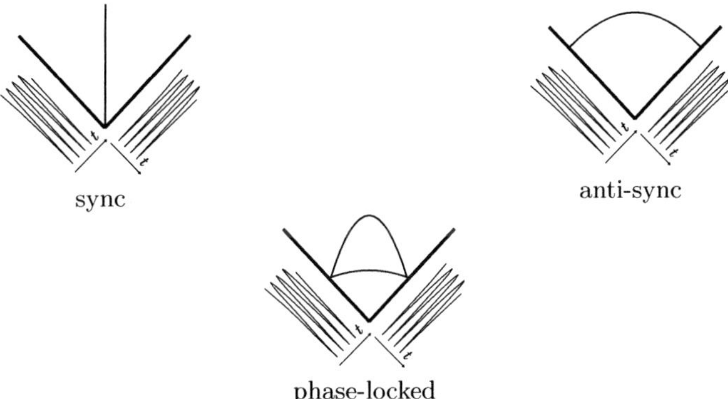

Fig. 3. The equivalence between period two orbits of the orthogonal wedge and two phase-locked impact oscillators (bouncing balls).

Because the two edges vibrate orthogonally to each other, the collision rules reduce to the very simple expression:

$$t: \quad x_1(t) = A\sin\omega t \Rightarrow \dot{x}_1(t^+) = -e\,\dot{x}_1(t^-)$$
$$t: \quad x_2(t) = A\sin(\omega t + \phi) \Rightarrow \dot{x}_2(t^+) = -e\,\dot{x}_2(t^-),\ 0 < e < 1, \tag{3}$$

which means that the two degree-of-freedom dynamics of the bounce juggler decouple into two one degree-of-freedom dynamics. The dynamics of each one degree-of-freedom subsystem are the bouncing ball dynamics first studied by Holmes [6]. It is well known that the bouncing ball dynamics exhibit a stable period one orbit in a suitable parameter range of A (or ω), within which the period between two successive collisions locks with a multiple of the forcing period $T = \frac{2\pi}{\omega}$. In the same parameter range, the orthogonally vibrating bounce juggler exhibits a stable period two orbit. The additional parameter ϕ determines the phase shift between the collisions with each edge, as illustrated in Figure 3.

The period two orbit of the orthogonally vibrating bounce juggler is thus equivalent to the period one motion of two phase-locked bouncing balls or impact oscillators. Because of the coupling, this transparent description of the dynamics is lost when the edges are not orthogonal to each other and when the actuation of the wedge is an oscillatory motion around the fixed vertex instead of an axial vibration of each edge separately. The analysis in [20, 21] nevertheless shows that the exponential stabilization of period two orbits persists over a broad range of parameters, even in this generalized situation.

Stable and unstable periodic orbits have been recently described in models of insect locomotion [7]. These models provide practical and relevant examples of Hamiltonian systems in which the different degrees of freedom are coupled only through collisions. Our current work investigates whether periodic forcing of some parameters can act as a (sensorless) stabilization mechanism in these models similarly to the bounce juggler example described in this chapter. To the best of our knowledge, the stability analysis in all reported examples in the literature is based on (tedious) calculations of Poincare maps that can be determined analytically only in overly simplified situations. A general interconnection theory for such rhythmic oscillators is lacking at the present time.

3 Collective stabilization

Another illustration of synchrony as a design principle is the task of stabilizing a large collection of identical control systems (agents) around a collective motion. This problem has received considerable attention over the last years (see, e.g., [11] and the references therein) and includes numerous engineering applications in unmanned sensor platforms. For example, autonomous underwater vehicles (AUVs) are used to collect oceanographic measurements in network formations that maximize the information intake, see, e.g., [15]. In ongoing work [25, 17, 26], we study a continuous-time kinematic model of N identical, self-propelled particles subject to planar steering controls, first considered in [9, 10]. In complex notation, this model is given by

$$\dot{r}_k = e^{i\theta_k} \tag{4}$$
$$\dot{\theta}_k = u_k, \tag{5}$$

where $r_k \in \mathbb{R}^2$ and $\theta_k \in S^1$ are the position and heading of the kth particle. Unless otherwise indicated, $k = 1, \ldots, N$. The steering control law is denoted by u_k. If we define the relative position and orientation variables, $r_{jk} = r_j - r_k$ and $\theta_{jk} = \theta_j - \theta_k$, then the control, u_k, can be decomposed into relative spacing and alignment terms, i.e.,

$$u_k = u_k^{spac}(r_{jk}, \theta_{jk}) + u_k^{align}(\theta_{jk}). \tag{6}$$

The alignment control is a function of the relative orientation, θ_{jk}, whereas the spacing control is function of both the relative position r_{jk}, and orientation θ_{jk}.

The particle model is a specialization of the Frenet-Serret equations in $SE(2)$, the group of planar rigid motions, restricted to a constant velocity, see, e.g., [22]. The Lie group structure of the state space has important implications. If the control law (6) depends only on the relative orientations and positions of the particles, then the system is invariant under the action of the group $SE(2)$ (i.e., there is a planar rotation and translation symmetry). Under this assumption, the configuration space of the particles can be described on a reduced shape space. Justh and Krishnaprasad [9] show that fixed points in the shape space, which correspond to relative equilibria, occur for

$$u_1 = u_2 = \ldots = u_N. \tag{7}$$

In particular, the relative equilibrium with $u_1 = u_2 = \ldots = u_N = 0$ results in parallel trajectories of the group; the relative equilibrium with $u_1 = u_2 = \ldots = u_N \neq 0$ results in all the vehicles orbiting the same point at the same (constant) radius. The control problem is to design a feedback (6) that stabilizes a particular relative equilibrium of the model, that necessarily correspond either to a parallel motion or a circular motion for the group.

A key parameter for the stabilization of the group is the velocity of the center of mass

$$v = |\dot{R}| = \left| \frac{1}{N} \sum_k \dot{r}_k \right| = \left| \frac{1}{N} \sum_{k=1}^{N} e^{i\theta_k} \right|. \tag{8}$$

The velocity v is maximal ($v = 1$) for parallel motion whereas it is minimal ($v = 0$) for circular motion around the (fixed) center of mass. This suggests to control the potential

$$U = \frac{N}{2} v^2 = \frac{1}{2N} \sum_{k=1}^{N} \sum_{j=1}^{N} \cos \theta_{kj}.$$

Gradient dynamics with respect to U yield

$$\dot{\theta}_k = K \frac{\partial U}{\partial \theta_k} = -\frac{K}{N} \sum_{j=1}^{N} \sin(\theta_k - \theta_j). \tag{9}$$

The only critical points of U are its minima, corresponding to $v = 0$, that is, motions around a fixed center of mass, and its maxima, corresponding to $v = 1$, that is, parallel motion.

The parameter v thus provides a good measure of synchrony for the group. Its interpretation in connection with the literature of phase models of coupled oscillators will be discussed in the next section. It prompts us to choose the orientation control

$$u_k^{align}(\theta_{jk}) = K \frac{\partial U}{\partial \theta_k}. \tag{10}$$

When K is positive, the orientation control (10) stabilizes parallel motions of the group. The spacing control acts as an additive correction to control the shape of the group formation, e.g., the relative distance between particles.

When K is negative, the orientation control (10) stabilizes group motions around a fixed center of mass. The spacing control acts as an additive correction to control the shape of the group formation, e.g., the distance to the center of mass in order to stabilize a circular motion.

The reader is referred to [25, 17] for further details on how to stabilize particular relative equilibria. We provide one illustration below to show that the design of the spacing control becomes somewhat decentralized once the group parameter v has been stabilized. In [10], (a variant of) the following control law is proposed to stabilize circular motion of particle k around a fixed beacon R_0:

$$u_k = -f(\rho_k) < \frac{\tilde{r}_k}{\rho_k}, ie^{i\theta_k} > - < \frac{\tilde{r}_k}{\rho_k}, e^{i\theta_k} >, \tag{11}$$

with $\tilde{r}_k = r_k - R_0$ and $\rho_k = \| \tilde{r}_k \|$. The second term of the control law (11) stabilizes circular motions: it vanishes when the velocity vector is orthogonal to the relative position vector. The function $f(\cdot)$ in (11) provides attraction to the beacon when the distance ρ_k exceeds the equilibrium distance d_0 and repulsion otherwise. (The choice $f(\rho_k) = 1 - (d_0/\rho_k)^2$ is proposed in [10].)

The control law (11) is a single particle control law: u_k only depends on the state (r_k, θ_k) of particle k. But stabilization of the center of mass by means of the orientation control (10) suggests that the beacon control law (11) may serve as a spacing control law if the beacon R_0 is replaced by the center of mass $R = \frac{1}{N} \sum_{k=1}^{N} r_k$ in the definition of $\tilde{r}_k$. One then obtains a composite control law

$$u_k = \frac{1}{N} \sum_{j=1}^{N} \left(-K \sin(\theta_k - \theta_j) - f(\rho_k) < \frac{r_{kj}}{\rho_k}, ie^{i\theta_k} > - < \frac{r_{kj}}{\rho_k}, e^{i\theta_k} > \right) \tag{12}$$

to stabilize circular motions of the group on a unique circle centered at the (fixed) center of mass. Convergence analysis is provided in [25] for large negative values of the parameter K, based on a time-scale separation between the fast convergence of the group center of mass and the slow(er) convergence of each particle relative to the center of mass.

In ongoing work, we have obtained a Lyapunov analysis showing almost global convergence for any value of $K < 0$ for a variant of the control law (12); we are also studying extensions of this basic design to more specific collective motions such as the splay state formation, which is a circular relative equilibrium characterized by N particles equally spaced on the circle [26]. Central to all these designs is the decoupling between a prescribed level of synchrony for the group (achieved by the orientation control) and the relative spacing of individual particles relative to the center of mass (achieved by a decentralized spacing control).

4 Oscillators, accumulators (storages), and phasors

Models of oscillators abound in physics and biology. In this section we briefly review common ways of simplifying these models when it comes to studying their stability properties or the stability properties of their interconnections. We illustrate our point with models from neurodynamics, which have been studied extensively in the literature and appear with many variants depending on the context. For the sake of illustration, we briefly review a biophysical model of the action potential, its two-dimensional simplification used for stability analysis, and two distinct one-dimensional simplifications used for interconnection analysis. The situation is exemplative of most dynamical models of oscillators in physics and biology (see, for instance, [4] for several models of biochemical oscillators).

Nerve cells (neurons) *fire*, that is, a current stimulus above threshold at their input triggers a succession of short electric pulses (action potentials) at their output. Even though action potentials do not persist forever, they are conveniently modeled as a sustained limit cycle oscillation of the electric potential across the cell membrane. The physical basis for this oscillatory mechanism is provided by the celebrated model of Hodgkin and Huxley [5]. The membrane is modeled as a capacitive circuit and the membrane potential depends on several ionic currents (mainly sodium and potassium) flowing through the membrane. Ion channels regulate the flow of each ion across the membrane. A central feature of the model is that ion channels are voltage dependent. The voltage dependence is such that sodium and potassium currents vary out of phase, creating a sustained switch between positive potential (when sodium channels are open and potassium channels are closed) and negative potential (in the opposite situation). The original model, not recalled here, consists of a state space model of dimension 4: one variable to describe the membrane potential, and three additional variables to describe the voltage dependent opening of the ion channels. More detailed models of action potentials take into account the effect of further ionic currents, increasing the dimension of the model up to 10 or 15 state variables.

The Hodgkin–Huxley model and its many variants exhibit sustained oscillations in numerical simulations, in good (quantitative) agreement with experimental data. Rigorous stability analysis of the limit cycle is usually restricted to two-dimensional simplifications of the model, such as the Fitzhugh–Nagumo model

$$\dot{V} = kV - V^3 + R$$
$$\tau\dot{R} = -R - V,$$

(13)

which qualitatively describes the limit cycle oscillation of the potential V with a single adaptation variable R to model the voltage dependence of the ion channels. We will come back on this example in the next section.

In order to study the dynamical behavior of large networks of interconnected neurons, the dynamics of each neuron are usually further simplified.

Two important models extensively studied in the literature are the models of Hopfield [8] and Kuramoto [12].

In Hopfield model, the dynamics of neuron k are described by a single variable x_k and the first-order equation

$$\tau \dot{x}_k = -x_k + S(u_k), \tag{14}$$

where $S(\cdot)$ is monotone and usually has a finite range (classical descriptions of this static nonlinearity include the sigmoid function [in computer science] or the Michaelis–Menton function [in reaction networks]). The state x_k has no correspondence with the physical variables of the Hodgkin–Huxley model but models an average activity of the neuron (it is often thought of as the averaging firing rate of the neuron). Hopfield studied the dynamics of N interconnected neurons with the interconnection determined by the linear coupling

$$u_k = \sum_{k=1}^{N} \Gamma_{kj} x_j.$$

The matrix Γ thus determines the network topology and affects the dynamical behavior of the network. Hopfield showed that symmetric network topologies $\Gamma = \Gamma^T$ result in gradient dynamics, in which case all solutions converge to critical points of a scalar potential. Hopfield models abound in neuroscience and have been used to describe the dynamics of a number of computational tasks (see, for instance, [34] for several illustrations in vision). In these examples, the oscillatory behavior of the neuron is unimportant. The state x_k only models the storage capacity of the neuron.

Storage models of oscillators neglect the phase variable of periodic solutions. As a consequence, they are inadequate for synchrony analysis. In contrast, phase models of oscillators disregard the dynamical behavior of the oscillator away from its limit cycle solution. The dynamics of neuron k are described by a single phase variable θ_k and the first-order equation

$$\dot{\theta}_k = \omega_k + u_k, \tag{15}$$

which is a state space equation on the circle S^1. In the absence of input, the phase variable travels on the circle at uniform speed ω_k. Kuramoto studied the dynamics of N interconnected phasors with the interconnection determined by the all-to-all coupling

$$u_k = -\frac{K}{N} \sum_{k=1}^{N} \sin(\theta_k - \theta_j), \quad K > 0.$$

If all the oscillators are identical ($\omega_k = \omega$ for all k), the dynamics of the interconnected system are

$$\dot{\theta}_k = \omega - \frac{K}{N} \sum_{k=1}^{N} \sin(\theta_k - \theta_j).$$

In a coordinate frame rotating at uniform speed ω, these are the gradient dynamics (9) discussed in Section 3 and the synchronized state is a stable equilibrium. The convergence analysis is much more involved when the oscillators are not identical (see [31] for a recent review), but a stable phase-locked equilibrium exists if the coupling is large enough. The centroid $\frac{1}{N}\sum_{j=1}^{N} e^{i\theta_j}$ of the oscillators plays an important role in the analysis of the Kuramoto model as a measure of synchrony. It coincides with the linear momentum of the group of particles in Section 3.

Several authors have studied how to reduce general models of oscillators to phase models of the type (15) in the limit of weak coupling, that is, when the coupling between the oscillators does not affect the convergence of each oscillator to a limit cycle solution. For more details, we refer the reader to the recent paper [2] and to references therein.

5 A dissipativity theory for oscillators

Dissipativity theory, introduced by Willems [33], is an interconnection theory for open systems described by state space models:

$$\begin{cases} \dot{x} = f(x) + g(x)u, & x \in \mathbb{R}^n, \quad u \in \mathbb{R}^m \\ y = h(x), \quad y \in \mathbb{R}^m. \end{cases} \tag{16}$$

We assume that the vector fields f and g and the function h are smooth, and that the origin $x = 0$ is an equilibrium point of the zero-input system, that is, $f(0) = 0$.

Lyapunov stability of the equilibrium $x = 0$ of the closed system $\dot{x} = f(x)$ is often characterized by a dissipation inequality $\dot{V} \leq 0$ where the scalar Lyapunov function $V(x) > 0$ has a strict minimum at $x = 0$. Dissipativity generalizes this concept to the open system (16): the system is dissipative if it satisfies a dissipation inequality

$$\dot{S} \leq w(u, y), \tag{17}$$

where the scalar *storage* function $S(x) \geq 0$ is analogous to the Lyapunov function of a closed system (with the physical interpretation of an internal energy). The scalar function $w(u, y)$ is called the supply rate. The dissipation inequality expresses that the rate of change of the internal energy (storage) of the system is bounded by the supply rate, that is, the rate at which the system can exchange energy with the external world through its external variables.

Dissipativity theory is fundamental to the stability analysis of interconnections. We here restrict the discussion to passivity theory, which is dissipativity theory for

$$w(u, y) = u^T y - d(y), \quad d(y) \geq 0, \tag{18}$$

a supply rate that prevails in physical models. The usual terminology is *passivity* when $d(y) = 0$ and *strict (output) passivity* when $d(y) > 0$ for $y \neq 0$. The (negative) feedback interconnection of two systems (16), labeled 1 and 2, respectively, defined by the interconnection rule $u_1 = u - y_2$, $u_2 = y_1 = y$, yields a new system (16) with external variables (u, y). The fundamental passivity theorem says that if both system 1 and system 2 are passive, then their feedback interconnection is also passive. Indeed, the storage function $S = S_1 + S_2$ satisfies the dissipation inequality $\dot{S} \leq u_1^T y_1 + u_2^T y_2 - d_1(y_1) - d_2(y_2) \leq u^T y - d_1(y)$. Moreover, the zero equilibrium of the (closed) system obtained by setting $u = 0$ is asymptotically stable modulo a detectability condition. The consequences of this basic theorem are far-reaching and have made dissipativity theory the central tool in nonlinear control theory for the stability analysis (and design) of equilibria; see [24, 32, 16] for illustrations.

In ongoing work [27, 29], we aim to show that dissipativity theory also provides an interconnection theory for oscillators. The main idea is very straightforward: we characterize an oscillator by a dissipation inequality with a supply rate

$$w(u, y) = u^T y + a_k(y) - d(y), \quad a_k(y) \geq 0, \quad d(y) \geq 0. \tag{19}$$

The supply rate (19) differs from the (strictly) passive supply rate (18) by the activation term $a_k(y)$. The passive oscillator is viewed as a system that passively exchanges its energy with the environment but that contains an active internal element. The competition between the internal elements that dissipate the storage and the active element that restores the storage is viewed as the basic oscillation mechanism. A necessary condition for sustained oscillations is that the system restores energy at low energy, that is, $a_k(y) - d(y) > 0$ when $|y|$ is small, and that the system dissipates energy at high energy, that is, $a_k(y) - d(y) < 0$ when $|y|$ is large.

A convenient way to obtain dissipativity with the supply rate (19) is to consider the feedback interconnection of a passive system with the static nonlinearity illustrated in Figure 4: $\phi_k(y) = -ky + \phi(y)$ where $\phi(\cdot)$ is a smooth sector nonlinearity in the sector $(0, \infty)$, which satisfies $\phi'(0) = \phi''(0) = 0$, $\phi'''(0) := \kappa > 0$ and $\lim_{|s| \to \infty} \frac{\phi(s)}{s} = +\infty$ ("stiffening" nonlinearity). The storage S of the passive system Σ satisfies

$$\dot{S} \leq ky^2 - y\phi(y) + uy, \tag{20}$$

which corresponds to $a_k(y) = ky^2$ and $d(y) = y\phi(y)$ in (19).

The parameter k regulates the level of activation near the equilibrium $x = 0$. When $k \leq 0$, the feedback system is strictly passive, and under a suitable detectability condition, the equilibrium is globally asymptotically stable. Stability of the equilibrium is lost at a critical value $k^* \geq 0$.

Stability

Two distinct bifurcation scenarios provide a stable oscillation mechanism for the feedback system in Figure 4. The first one corresponds to a supercritical Hopf bifurcation: two complex conjugate eigenvalues cross the imaginary axis

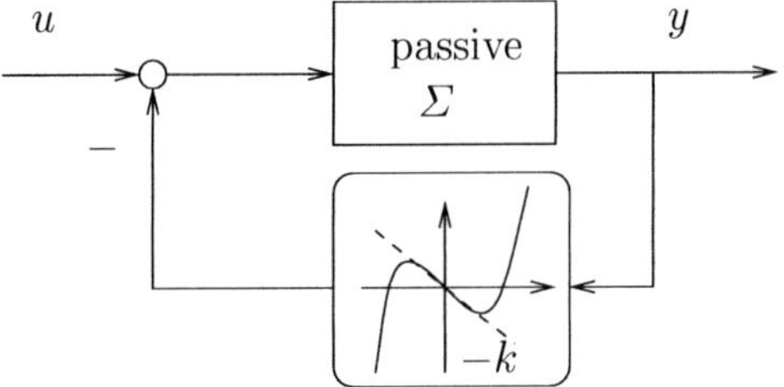

Fig. 4. Block diagram of a family of systems satisfying the dissipation inequality (20).

at $k = k^*$, giving rise to a stable limit cycle surrounding the unstable equilibrium $x = 0$ for $k > k^*$. The normal form of this bifurcation is obtained when the passive system Σ in the loop is a harmonic oscillator, characterized by the transfer function $H(s) = \frac{s}{s^2+\omega^2}$. The dynamics of the feedback system are then

$$\frac{d^2 y}{dt^2} + \omega^2 y + \frac{d}{dt}(ky - \phi(y)) = \dot{u}. \tag{21}$$

Equation (21) is the model of a van der Pol oscillator when $\phi(y) = y^3$: the circuit interpretation is that the sustained exchange of energy between a capacitor and an inductor is regulated by a static element (a tunnel-diode circuit) that dissipates energy when the current is high and restores energy when the current is low.

A second bifurcation scenario in the feedback system in Figure 4 is a supercritical pitchfork bifurcation: the stable equilibrium $x = 0$ becomes a saddle beyond the bifurcation value $k = k^*$ and two new stable equilibria appear for $k > k^*$. The normal form of this bifurcation is obtained when the passive system Σ in the loop is an integrator, characterized by the transfer function $H(s) = \frac{1}{s}$. The dynamics of the feedback system are then

$$\dot{y} = ky - \phi(y) + u. \tag{22}$$

When the input $u = 0$, the equilibrium $y = 0$ is stable for $k \leq 0$ and unstable for $k > 0$. Two other equilibria exist for $k > 0$ and create a bistable behavior: the positive (respectively, negative) equilibrium attracts all solutions with positive (respectively, negative) initial condition. For every $k > 0$, the bistable behavior persists over a range of (constant) inputs R, causing hysteresis in the static response of the system, as shown in Figure 5.

This hysteresis is turned into a relaxation oscillation when the input slowly *adapts* to follow the hysteresis loop, resulting in the closed-loop dynamics

$$\begin{aligned} \dot{y} &= ky - \phi(y) + R \\ \tau \dot{R} &= -R - y. \end{aligned} \tag{23}$$

Equation (23) is the Fitzhugh–Nagumo model recalled in the previous section. The same adaptation in the model of Figure 4 is illustrated in Figure 6.

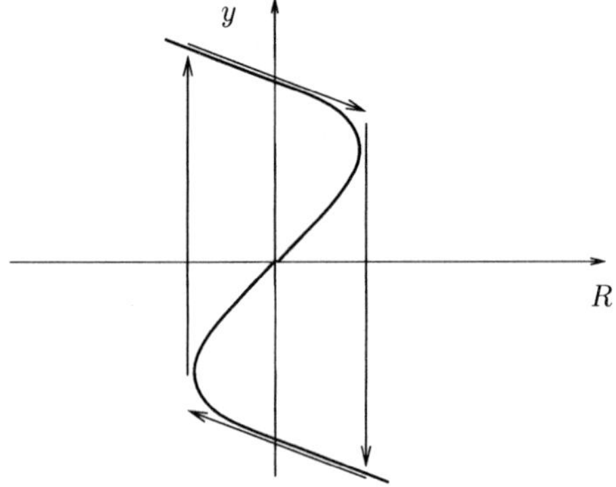

Fig. 5. The hysteresis associated to the bistable system (22) with constant input $u = R$.

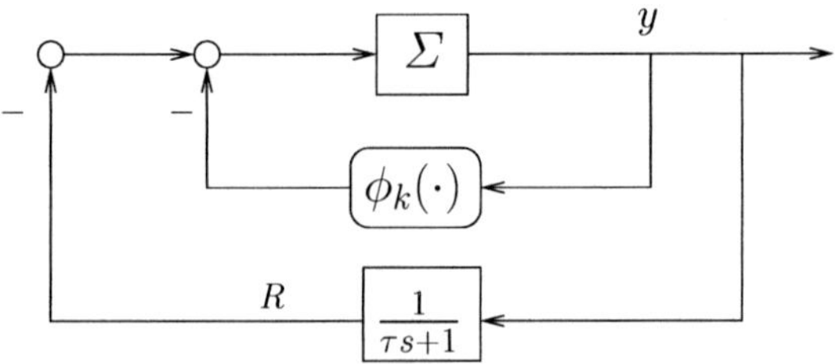

Fig. 6. Converting the pitchfork scenario into a relaxation oscillator with a slow adaptation mechanism ($\tau \gg 0$). The case $\Sigma = \frac{1}{s}$ corresponds to the Fitzhugh–Nagumo oscillator.

Oscillations resulting from slow adaptation in a bistable system seem to constitute a prevalent mechanism in models of biological oscillators (see, for instance, [4] for several illustrations).

The main result in [27] provides higher-dimensional generalization of the two above examples by showing that the dissipativity characterization of the feedback system forces one of the two bifurcation scenarios. The results are local in the parameter space (they hold for values of the parameter in the vicinity of the bifurcation), but they are global in the state space, that is, convergence to the stable limit cycle is proven for all initial conditions that do not belong to the stable manifold of the (unstable) equilibrium at the origin.

Interconnections

Dissipativity theory not only provides a stability theory for oscillators that admit the representation in Figure 4 but also an interconnection theory. As a direct consequence of passivity theory, any passive interconnection of dissipative systems with the supply rate (19) provides a new dissipative system with a supply rate of the same form.

For the sake of illustration, we only consider the interconnection of two identical passive oscillators, each characterized by a dissipation inequality

$$\dot{S}_i \leq k y_i^2 - \phi(y_i) y_i + u_i y_i, \quad i = 1, 2.$$

We consider the interconnections in Figure 7 and assume linear coupling

$$u = -\Gamma y + v,$$

using the vector notation $u = (u_1, u_2)^T$, $y = (y_1, y_2)^T$, and $v = (v_1, v_2)^T$.

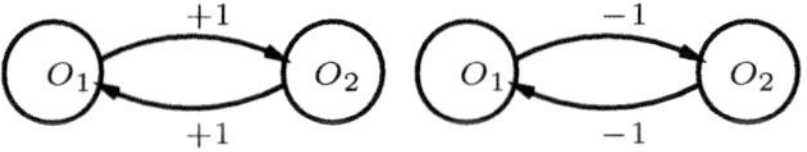

Fig. 7. Positive and negative feedback interconnection of two oscillators.

The interconnections in Figure 7 correspond to a symmetric matrix $\Gamma = \Gamma^T$ given by

$$\Gamma = \begin{bmatrix} 0 & \pm 1 \\ \pm 1 & 0 \end{bmatrix},$$

with $-y^T \Gamma y \leq y^T y$ for the two considered situations. The interconnection therefore satisfies the following dissipation inequality with storage $S = S_1 + S_2$:

$$\dot{S} \leq (k+1) y^T y - \Phi(y)^T y + v^T y,$$

which is dissipativity with the supply rate (19) for the system with input $v = (v_1, v_2)^T$ and output $y = (y_1, y_2)^T$. (We also use the notation $\Phi(y) = (\phi(y_1), \phi(y_2))^T$.) The bifurcation for the feedback interconnection of the two oscillators is the same as for each individual oscillator; only the bifurcation value $k = k^*$ is shifted by one for the interconnection. The general case of symmetric interconnections is treated in detail in [29]. The main observation is that any symmetric coupling $u = -\Gamma y + v$ can be written as $u = -(\Gamma + \lambda I) y + \lambda y + v$ with $\Gamma' = \Gamma + \lambda I$ a nonnegative symmetric matrix. The sum of the storages therefore satisfies

$$\dot{S} \leq (k + \lambda) y^T y - \Phi(y)^T y + v^T y,$$

which is dissipativity with the supply rate (19) for the system with input v and output y.

Synchronization

Dissipativity theory also provides a synchrony analysis for networks of oscillators. Synchrony is a convergence property for the difference between the solutions of different systems. Convergence properties for the *difference* between solutions of a closed system are characterized by notions of *incremental stability* [1, 14, 18]. For open systems, the corresponding notion is *incremental dissipativity*. Consider two different solutions $x_1(t)$ and $x_2(t)$ of the system (16) with inputs and outputs $(u_1(t), y_1(t))$ and $(u_2(t), y_2(t))$, respectively. Denote the incremental variables by $\delta x = x_1 - x_2$, $\delta u = u_1 - u_2$, and $\delta y = y_1 - y_2$. The system is *incrementally* dissipative if it satisfies a dissipation inequality

$$\dot{\delta S} \le w(\delta u, \delta y) \tag{24}$$

for the *incremental* scalar storage function $\delta S(\delta x) \ge 0$.

Consider two copies of the same system (16) with the difference coupling

$$u_i = -K(y_i - y_j), \ i, j = 1, 2,$$

which corresponds to the interconnection matrix

$$\Gamma = K \begin{bmatrix} -1 & 1 \\ 1 & -1 \end{bmatrix}.$$

Incremental dissipativity of the system (16) with the supply rate

$$w(\delta u, \delta y) = k\delta y^2 - \delta y \phi(\delta y) + \delta u \delta y \tag{25}$$

implies output synchronization for the interconnected system: substituting $\delta u = -2K\delta y$ in the inequality (24) yields

$$\dot{\delta S} \le (k - 2K)\delta y^2 - \delta y \phi(\delta y),$$

which implies asymptotic convergence of δy to zero, that is, output synchronization, when $2K > k$. State synchronization follows from output synchronization modulo a detectability condition.

In [29], we show that the implications of incremental dissipativity for synchronization extend to the interconnection of N identical systems with network topologies that include S_N symmetry (all-to-all topology), D_N symmetry (bidirectional ring topology), and Z_N symmetry (unidirectional ring topology). These results are closely related to other recent synchronization results in the literature [28, 19, 1], all based on incremental stability notions.

Synchronization of passive oscillators

We have shown that the Lure-type system in Figure 4 satisfies a dissipation inequality with the supply rate (19). We conclude this section by showing that the same system is also incrementally dissipative with a supply rate of the form (25) when the passive system Σ is linear and when the static nonlinearity ϕ is monotone.

For linear systems, dissipativity is equivalent to incremental dissipativity, that is, $S(x)$ satisfies the dissipation inequality $\dot{S} \leq w(u,y)$ if and only if the incremental storage $\delta S = S(\delta x)$ satisfies the incremental dissipation inequality $\dot{\delta S} \leq w(\delta u, \delta y)$.

The static nonlinearity $y = \phi(u)$ is passive if and only if $\phi(s)s \geq 0$ for all s. It is also *incrementally passive* if it satisfies the additional monotonicity property $(\phi(s_1) - \phi(s_2))(s_1 - s_2) = \delta\phi(s)\delta s \geq 0$ for all $\delta s = s_1 - s_2$.

If the passive system in Figure 4 is linear, it has a quadratic storage $S(x) = x^T P x$ and the feedback system satisfies the dissipation inequality

$$\dot{S} \leq ky^2 - y\phi(y) + uy.$$

The incremental storage is $\delta S = \delta x^T P \delta x$, which satisfies the incremental dissipation inequality

$$\dot{\delta S} \leq k\delta y^2 - \delta y \delta\phi(y) + \delta u \delta y.$$

If $\phi(\cdot)$ is monotone, then $\delta y \delta\phi(y) \geq \delta y \psi(\delta y) \geq 0$ for some static nonlinearity $\psi(\cdot)$ and the feedback system satisfies the incremental dissipation inequality

$$\dot{\delta S} \leq k\delta y^2 - \delta y \psi(\delta y) + \delta u \delta y.$$

Combining the global convergence result to a stable limit cycle for one system with the synchronization results for a network of interconnected identical systems, one obtains global convergence results to a synchrone oscillation for passive oscillators that admit the feedback representation in Figure 4.

6 Conclusion

Oscillators are important building blocks of dynamical systems. When suitably interconnected, they robustly produce ensemble phenomena with synchrony properties not encountered in equilibrium systems. The chapter has described two stabilization problems that illustrate the role of synchrony as a design principle: a rhythmic control task and the design of group motions for moving particles in a plane.

A system theory for oscillators requires an interconnection theory. The external characterization of oscillators adopted in this chapter follows the fundamental characterization of open systems by a dissipation inequality and proposes a supply rate that enables limit cycle oscillations in the isolated system. The proposed dissipation inequality has implications for the stability properties of the oscillator, both in isolation and when interconnected to other oscillators. In its incremental form, the same dissipation inequality has implications for the synchrony properties of networks of identical oscillators.

Acknowledgements. The author acknowledges several collaborators involved in the research projects reported in this paper: Manuel Gerard, Naomi Leonard, Derek Paley, Renaud Ronsse, and Guy-Bart Stan. He also wishes to acknowledge Jan Willems for several discussions and a continuing source of inspiration.

References

1. Angeli D (2002) A Lyapunov approach to incremental stability properties. IEEE Trans. on Automatic Control 47:410–422
2. Brown E, Moehlis J, Holmes P (2004) On phase reduction and response dynamics of neural oscillator populations. Neural Computation 16(4):673–715
3. Gérard M, Sepulchre R (2004) Stabilization through weak and occasional interactions: a billiard benchmark. In: 6th IFAC Symposium on Nonlinear Control Systems, Stuggart, Germany
4. Goldbeter A (1996) Biochemical oscillations and cellular rhythms. Cambridge University Press, Cambridge, UK
5. Hodgkin AL, Huxley AF (1952) A quantitative description of membrane current and its application to conduction and excitation in nerve. J. Physiology 117:500–544
6. Holmes PJ (1982) The dynamics of repeated impacts with a sinusoidally vibrating table. Journal of Sound and Vibration 84(2):173–189
7. Holmes PJ, Full RJ, Koditschek D, Guckenheimer J (2006) Dynamics of legged locomotion: Models, analyses and challenges. SIAM Review, to appear
8. Hopfield JJ (1982) Neural networks and physical systems with emergent collective computational abilities. Proceedings of the National Academy of Sciences USA 79:2554–2558
9. Justh EW, Krishnaprasad PS (2002) A simple control law for UAV formation flying Technical report Institute for Systems Research, University of Maryland
10. Justh EW, Krishnaprasad PS (2003) Steering laws and continuum models for planar formations. In: IEEE 42nd Conf. on Decision and Control, Maui, HI
11. Kumar V, Leonard N, Morse A (eds) (2004) Cooperative control, vol. 309 of Lecture Notes in Control and Information Sciences. Springer-Verlag, London
12. Kuramoto Y (1984) Chemical oscillations, waves, and turbulence. Springer-Verlag, London
13. Lehtihet BN, Miller HE (1986) Numerical study of a billiard in a gravitational field. Physica 21D:93–104
14. Lohmiller W, Slotine JJE (1998) On contraction analysis for nonlinear systems. Automatica 34(6):683–696
15. Ogren P, Fiorelli E, Leonard NE (2004) Cooperative control of mobile sensor networks: Adaptive gradient climbing in a distributed environment. IEEE Trans. on Automatic Control 49(8):1292–1302
16. Ortega R, Loria A, Nicklasson PJ, Sira-Ramirez H (1998) Passivity-based Control of Euler-Lagrange Systems. Springer-Verlag, London
17. Paley D, Leonard N, Sepulchre R (2004) Collective motion: bistability and trajectory tracking. In: IEEE 43rd Conf. on Decision and Control, Atlantis, Bahamas

18. Pavlov A, Pogromsky AY, van de Wouw N, Nijmeijer H (2004) Convergent dynamics, a tribute to Boris Pavlovich Demidovich. Systems & Control Letters 52:257–261

19. Pogromsky A (1998) Passivity based design of synchronizing systems. Int. J. Bifurcation and Chaos 8:295–319

20. Ronsse R, Lefevre P, Sepulchre R (2004a) Open-loop stabilization of 2D impact juggling. In: 6th IFAC Symposium on Nonlinear Control Systems, Stuggart, Germany

21. Ronsse R, Lefevre P, Sepulchre R (2004b) Sensorless stabilization of bounce juggling Technical report, Department of Electrical Engineering and Computer Science, University of Liège

22. Sastry S (1999) Nonlinear systems. Springer-Verlag, London

23. Sepulchre R, Gérard M (2003) Stabilization of periodic orbits in a wedge billiard. 1568–1573. In: IEEE 42nd Conf. on Decision and Control Maui, HI

24. Sepulchre R, Jankovic M, Kokotovic P (1997) Constructive nonlinear control. Springer-Verlag, London, UK

25. Sepulchre R, Paley D, Leonard N (2004a) Collective motion and oscillator synchronization. 189–205. In: Cooperative control, Kumar V, Leonard N, Morse A (eds), Springer-Verlag, London

26. Sepulchre R, Paley D, Leonard N (2004b) Collective stabilization of n steered particles in the plane, in preparation

27. Sepulchre R, Stan GB (2004) Feedback mechanisms for global oscillations in Lure systems. Systems & Control Letters 54(8):809–818

28. Slotine JJ, Wang W (2004) A study of synchronization and group cooperation using partial contraction theory. 207–228. In: Cooperative control, Kumar V, Leonard N, Morse A (eds) Springer-Verlag, London

29. Stan GB, Sepulchre R (2004) Dissipativity theory and global analysis of limit cycles, Technical report, Department of Electrical Engineering and Computer Science, University of Liège

30. Strogatz S (2003) Sync: the emerging science of spontaneous order. Hyperion

31. Strogatz SH (2000) From Kuramoto to Crawford: exploring the onset of synchronization in populations of coupled oscillators. Physica D 143:1–20

32. van der Schaft AJ (2000) L_2-gain and passivity techniques in nonlinear control. Springer-Verlag, London

33. Willems JC (1972) Dissipative dynamical systems. Arch. Rational Mechanics and Analysis 45:321–393

34. Wilson H (1999) Spikes, decisions, and actions. Oxford University Press, Oxford, UK

35. Winfree A (2000) The geometry of biological time, second edition. Springer-Verlag, London

36. Wojtkowski MP (1998) Hamiltonian systems with linear potential and elastic constraints. Communications in Mathematical Physics 194:47–60

Nonlinear Anti-windup for Exponentially Unstable Linear Plants*

Sergio Galeani,[1] Andrew R. Teel,[2] and Luca Zaccarian[1]

[1]Dipartimento di Informatica, Sistemi e Produzione, Università di Roma Tor Vergata, 00133 Roma, Italia
`[galeani,zack]@disp.uniroma2.it`
[2]Center for Control Engineering and Computation, Electrical and Computer Engineering, University of California, Santa Barbara, CA 93106, USA
`teel@ece.ucsb.edu`

Summary. In this chapter we discuss a constructive method for anti-windup design for general linear saturated plants with exponentially unstable modes. The constructive solution is independent of the controller dynamics, so the size of the (necessarily bounded) operating region in the exponentially unstable directions of the plant state space is large. We discuss the features of the anti-windup algorithm and illustrate its potential on several examples.

1 Introduction

Anti-windup compensation denotes an augmentation to an existing control scheme for a plant without input saturation aimed at recovering (as much as possible) the behavior of that existing control scheme on the same plant subject to input saturation. This control architecture, which appeared as early as the 1950s [19], was mainly motivated by industrial needs. Control designers were not able to directly apply to experimental devices the control laws (mainly PID ones) synthesized with linear design tools, which, of course, did not take input saturation into account.

Much progress has been made on anti-windup design, from ad hoc schemes for specific industrial devices to more systematic designs (see [16, 18] for surveys of some of these early schemes). Starting in the mid-1990s, modern developments in nonlinear control theory addressed the anti-windup construction problem in a more systematic way. The desirable nonlinear closed-loop properties that could be addressed and guaranteed via many modern anti-windup solutions ranged from stability to performance (see the references below).

*This work was supported in part by AFOSR grant number F49620-03-1-0203, NSF grant number ECS-0324679, ENEA-Euratom, ASI, and MIUR through PRIN and FIRB projects.

The underlying idea behind all anti-windup compensation schemes is that the original closed loop (herein called "unconstrained closed-loop system") is augmented with an extra (static or dynamic) filter. To detect the saturation activation, this filter is driven by the "excess of saturation," namely the signal $u - \text{sat}(u)$ that amounts to the quantity of commanded input that cannot reach the plant. On the other hand, to enforce a desirable closed-loop behavior, the filter injects modification signals into the unconstrained control scheme (to this aim, it may also have access to extra closed-loop signals available for measurement). With this architecture, provided that the filter produces a zero output as long as the input does not saturate, the preservation of the unconstrained behavior is guaranteed for all trajectories that stay within the saturation limits for all times. For all the other trajectories, modifications are required on the closed loop because the corresponding plant input is not achievable for all times.

Based on the above characterization, we can classify anti-windup compensators in two large families: the first one, called *essentially linear*, where the filter driven by the excess of saturation is linear; the second one, called *nonlinear*, where the filter driven by the excess of saturation is nonlinear. (In fairness, this classification is applicable in the continuous-time setting that we address here. However, in the discrete-time case alternative methods are available based on the so-called "reference governor" or "command governor scheme." See, e.g., [13] and references therein.)

Essentially linear anti-windup designs with useful closed loop guarantees have been widely used for control systems where the plant is exponentially stable (see, e.g., [29, 22, 20, 10, 21, 28, 14, 15]). Exponential stability of the plant is a key assumption when seeking global stability and performance results based on linear tools, such as sector bounds and quadratic stability results. Unfortunately, these tools become unusable when seeking global anti-windup solutions for plants that are not exponentially stable. It is well known that for these plants global stability can only be achieved if there are no exponentially unstable modes. In these cases nonlinear stabilizers are necessary in general (the corresponding anti-windup problem is solved using a nonlinear anti-windup scheme in [26]).

Furthermore, in the exponentially unstable case that we address here, things become even more complicated because the null controllability region of the saturated system is bounded in the exponentially unstable directions (see, e.g., [23]) and special care needs to be taken to keep the unstable part of the plant state within this safety region. Essentially linear anti-windup designs for exponentially unstable linear plants have been recently suggested in a number of papers. For example, in [6, 5, 9] recent methods for the characterization (and enlargement) of the stability domains for saturated feedback systems were employed to provide a systematic design for the selection of a static linear anti-windup gain. Moreover, in [27], the results of [14] were extended to the case of a narrowed sector bound, thus obtaining locally stabilizing anti-windup compensators for exponentially unstable linear plants.

Coprime factor based anti-windup of the type initially proposed in [20] was also extended to exponentially unstable plants in [8, 7].

In this chapter, we use the nonlinear anti-windup structure proposed in [12]. The architecture of the corresponding anti-windup compensator is the same as that first introduced in [26] and then further developed in [24, 3, 2] for exponentially unstable plants. One of the advantages of the scheme of [12] is that, unlike the previous approaches in [5, 6, 27, 8, 7, 9], the compensation structure is only dependent on the plant dynamics. Therefore, the boundaries of the operating region in the plant state space are independent of the unconstrained controller dynamics and are only dependent on the structural limitations of the saturated plant (whose null controllability region is bounded in the exponentially unstable direction). On the other hand, since the gains used in [5, 6, 27, 8, 7, 9] are linear functions also involving the unconstrained controller dynamics, when that controller is very aggressive, the corresponding constructions may lead to very small operating regions. Finally, an important advantage of this technique as compared to the previous ones is that we are able to guarantee bounded responses to references of arbitrarily large size, because the plant state is permanently monitored and kept within the null controllability region, thus preserving the overall stability property.

The chapter is structured as follows. In Section 2 we give the problem definition and then summarize and comment on the anti-windup construction of [12]. In Section 3 we illustrate the construction on several simulation examples.

2 Problem definition

2.1 Problem data

Consider a linear plant in the following form:

$$\begin{bmatrix} \dot{x}_s \\ \dot{x}_u \end{bmatrix} = Ax + Bu + B_d d$$
$$= \begin{bmatrix} A_s & A_{12} \\ 0 & A_u \end{bmatrix} \begin{bmatrix} x_s \\ x_u \end{bmatrix} + \begin{bmatrix} B_s \\ B_u \end{bmatrix} u + \begin{bmatrix} B_{ds} \\ 0 \end{bmatrix} d, \tag{1}$$
$$y = Cx + Du + D_d d,$$
$$z = C_z x + D_z u + D_{zd} d,$$

where A_s is a Hurwitz matrix, $x := [x_s^T \ x_u^T]^T \in \mathbb{R}^{n_s} \times \mathbb{R}^{n_u}$ is the plant state, $u \in \mathbb{R}^m$ is the control input, d is a disturbance input, y is the plant output available for measurement, and z is the performance output. Note that given a plant, the state partition in (1) is not unique. While fulfilling the requirement that A_s is Hurwitz, the partition should be carried out by inserting in A_u all the unsatisfactory open-loop modes (such as unstable or undesired slow modes); the effect of different choices of A_u is shown on a specific example in Section 3.

Assume that for the plant (1), a controller has been designed to guarantee desirable closed-loop behavior in terms of stability, performance, robustness, and convergence to a reference r[1]:

$$\begin{aligned}
\dot{x}_c &= A_c x_c + B_c u_c + B_r r, \\
y_c &= C_c x_c + D_c u_c + D_r r.
\end{aligned} \qquad (2)$$

We will denote the controller (2) as the *unconstrained controller* throughout the chapter, to emphasize the fact that its dynamics have been designed with the goal of guaranteeing desirable behavior when used in conjunction with the plant (1) through the following *unconstrained interconnection*:

$$u_c = y, \quad u = y_c. \qquad (3)$$

The closed-loop system (1), (2), (3) will be called *unconstrained closed loop* henceforth. Henceforth we will make use of the following assumption on the plant and on the unconstrained closed loop.

Assumption 1. *The state partition in (1) for the plant is such that the matrix A_s is Hurwitz and B_u has full column rank.*[2] *The unconstrained closed-loop system (1), (2), (3) is well posed and asymptotically stable.*

2.2 Input saturation and windup

We address in this chapter the problem that arises when the plant control input u is subject to input saturation. In particular, we consider a decentralized symmetric saturation function of the form $\mathrm{sat}(u) = [\sigma(u_1)\ \sigma(u_2)\ \cdots\ \sigma(u_m)]^T$, where $u_1, \ldots, u_m$ are the components of u and, without loss of generality, we can take the scalar saturation function $\sigma(\cdot)$ to be unitary: $\sigma(w) := \max\{-1, \min\{1, w\}\}$ (the approach easily extends to more general saturation functions by redefining the problem data in (1) and (2)).

When saturation is present at the plant input, the unconstrained interconnection (3) is no longer achievable on the closed loop system, and the following *saturated interconnection* is obtained instead:

$$u_c = y, \quad u = \mathrm{sat}(y_c). \qquad (4)$$

The corresponding *saturated closed-loop system* (1), (2), (4) typically exhibits undesirable behavior because the controller is designed for an unconstrained plant. In particular, especially for plants with exponentially unstable

---

[1]To simplify the exposition, we are only considering linear unconstrained controllers. However, the approach herein proposed can be extended to the case where (2) is a nonlinear controller.

[2]The assumption on the full column rank of B_u is required only for simplicity of exposition, and without any loss of generality: this assumption can be always satisfied by suitable selections of a minimal set of inputs for the dynamics of x_u.

modes (see the next section), the stability properties in Assumption 1 will be lost on the saturated closed loop and the closed-loop performance will deteriorate. In the next section we will give a formal mathematical definition of the performance and stability recovery goal underlying the anti-windup design used here.

2.3 Problem statement

When facing input saturation, the unconstrained controller (2) no longer guarantees the stability and performance properties of Assumption 1. It is therefore of interest to seek modifications of the saturated control scheme aimed at recovering (as much as possible) the unconstrained behavior characterizing the (ideal) interconnection (1), (2), (3) and, at the same time, guaranteeing suitable stability properties despite the presence of saturation.

To this aim, it is useful to define a compact and convex subset $\mathcal{X}_u \subset \mathbb{R}^{n_u}$ of the substate space of the plant (1) in the x_u direction, where we want the state x_u to be confined,[3] so that the overall plant state x will be confined in the set $\mathcal{X} := \mathbb{R}^{n_s} \times \mathcal{X}_u$. Note that, as commented in the following Remark 2, introducing this set is necessary to achieve the requirements of the next definition.

Once we define the region $\mathcal{X}_u$ where we want the state x_u to evolve, to suitably characterize the properties of the anti-windup closed loop, it is useful to define also a set of "feasible references" as the set of constant references leading to operating points within the set $\mathcal{X}_u$. This is done in the following definition.

Definition 1. *Given a set $\mathcal{X}_u$, a vector r_o is a* feasible reference *if the state response of the unconstrained closed loop (1), (2), (3) to the external inputs $(r(t), d(t)) = (r_\mathrm{o}, 0)$, $\forall t \geq 0$, converges to a steady-state value (x_s^*, x_u^*, x_c^*) with $x_u^* \in \mathcal{X}_u$ (note that by Assumption 1 this unconstrained response is always convergent).*

For the next definition, given certain selections of the external inputs $r(\cdot)$ and $d(\cdot)$ and initial conditions $x_s(0), x_u(0), x_c(0)$ for the plant and the unconstrained controller states, we will denote by subscripts ℓ all the responses arising from the unconstrained closed-loop system (1), (2), (3) (e.g., $x_\ell(\cdot), u_\ell(\cdot), z_\ell(\cdot)$). We will compare these responses to the responses of the compensated closed-loop system (or, equivalently, anti-windup closed-loop system), which are denoted without subscripts. The following problem has been addressed and solved in [12].

Definition 2. *Given a compact and convex set $\mathcal{X}_u \subset \mathbb{R}^{n_u}$, the* anti-windup problem for $\mathcal{X}_u$ *is to design an augmentation to the controller (2) such that*

---

[3]Actually, as it will be made clear, the state x_u will be confined in $\mathcal{X}_u$ only at the steady state, while it will be confined in a slightly inflated version $\mathcal{X}_u^+$ of $\mathcal{X}_u$ during transients.

for any initial condition $x_s(0), x_u(0), x_c(0)$ satisfying $x_u(0) \in \mathcal{X}_u$ and any selection of the external inputs $r(\cdot)$ and $d(\cdot)$, the corresponding responses of the unconstrained closed loop system and of the anti-windup closed-loop system satisfy the following properties:

1. *(**local preservation**) if $\mathrm{sat}(u_\ell(t)) = u_\ell(t)$, $x_{u\ell}(t) \in \mathcal{X}_u$, $\forall t \geq 0$, then $z(t) = z_\ell(t)$, $\forall t \geq 0$;*

2. *(**$\mathcal{L}_p$ recovery**) for all feasible references r_o, if $(r(\cdot) - r_\mathrm{o}, d(\cdot)) \in \mathcal{L}_p$ then $(z - z_\ell)(\cdot) \in \mathcal{L}_p$, $\forall p \in [1, \infty]$;*

3. *(**restricted tracking**) if $\lim\limits_{t \to +\infty} (x_{s\ell}(t), x_{u\ell}(t), x_{c\ell}(t)) = (\bar{x}_{s\ell}, \bar{x}_{u\ell}, \bar{x}_{c\ell})$ then $\lim\limits_{t \to +\infty} (x_s(t), x_u(t), x_c(t)) = (\bar{x}_s, \bar{x}_u, \bar{x}_c)$ with $\bar{x}_u \in \mathcal{X}_u$.*

Remark 1. It is worthwhile to comment on the three items stated in Definition 2, which represent three desirable properties guaranteed by the anti-windup construction. Item 1 (local preservation) guarantees that any trajectory generated by the unconstrained closed-loop system that never saturates and never violates the necessary constraint on the operating region $\mathcal{X}_u$ (therefore being safely reproducible on the saturated plant) will be preserved by the anti-windup compensation scheme. Item 2 ($\mathcal{L}_p$ recovery) guarantees that any unconstrained trajectory generated by a reference-disturbance pair converging (in an $\mathcal{L}_p$ sense) to a feasible reference selection will be asymptotically recovered (in an $\mathcal{L}_p$ sense). This property ensures that any unconstrained trajectory that converges to an admissible set point will be recovered by the anti-windup compensation scheme, even if some transient performance will be lost due to saturation. Note that this item evaluated for $p = \infty$ implies that the anti-windup closed loop is bounded input bounded state (BIBS) stable, that is: any (arbitrarily large) selection of the reference-disturbance pair will lead to a bounded response. Finally, item 3 (restricted tracking) guarantees that any converging unconstrained trajectory will correspond to a converging anti-windup trajectory. All trajectories that converge in forbidden regions for the saturated plant will be projected on a restricted set-point such that the unstable part of the plant state remains in $\mathcal{X}_u$. The selection of the projected set-point will be carried out via the projection function $\mathcal{P}(\cdot)$ (defined later in (13)).

Remark 2. Note that the main challenge in guaranteeing the anti-windup property of Definition 2 is that the null controllability region of the plant is bounded in the exponentially unstable directions (see, e.g., [23]). Therefore, special care by way of nonlinear functions needs to be taken to keep the state x_u within the set $\mathcal{X}_u$ at all times, otherwise stability could not be guaranteed. For this reason we will assume that the state x_u is exactly measured from the plant and that its dynamics are not affected by disturbances so that general results on the arising closed loop will be provable. On the other hand, for cases where an exact measurement of x_u is not available and/or disturbances affect the x_u component of the plant state equation, regional result can be obtained as commented in Remark 4.

Remark 3. Note that the maximum achievable set $\mathcal{X}_u$ is only dependent on the plant dynamics and not on the unconstrained controller dynamics. As a consequence, the operating region of the compensated closed-loop system is only restricted in the direction of the exponentially unstable states of the plant. This usually leads to extreme closed-loop performance as the transients commanded by any controller are preserved as much as possible until the safety boundary imposed by $\mathcal{X}_u$ is crossed. (On the contrary, approaches such as those introduced in [5, 6, 27, 8, 7, 9] only allow for operating regions that are restricted also in the controller state directions of the overall closed-loop state space.)

2.4 Anti-windup design

The anti-windup solution of [12] is based on an extension of the preliminary results of [25, 26] along the directions explored in [24, 2, 3] (see also [1] for an application of those preliminary schemes). The augmentation strategy is based on the introduction of the following model-based filter in the control scheme:

$$\begin{aligned}
\dot{\xi} &= A\xi + B(u - y_c), \\
v_2 &= -C\xi - D(u - y_c),
\end{aligned} \tag{5}$$

where $\xi = [\xi_s^T \ \xi_u^T]^T \in \mathbb{R}^{n_s} \times \mathbb{R}^{n_u}$, similar to the corresponding partition of the plant state in (1), and on the modification of the controller interconnection equations as follows:

$$u_c = y + v_2, \quad u = \mathrm{sat}(\alpha(x_u, x_u - \xi_u, y_c)), \tag{6}$$

where $\alpha(\cdot, \cdot, \cdot)$ is a function to be specified later. The corresponding scheme is represented in Figure 1.

The selection of $\alpha(\cdot, \cdot, \cdot)$ corresponds to the most involved aspect of this anti-windup construction. In [12], a general procedure was given for the design of the function $\alpha(\cdot, \cdot, \cdot)$, where the operating region $\mathcal{X}_u$ could be selected as an ellipsoid or as a polyhedral set. Moreover, the construction was based on the selection of a stabilizer guaranteeing suitable forward invariance properties for $\mathcal{X}_u$. Here, we rewrite the procedure in a more self-contained way, commenting and clarifying on the possible selections of the design parameters and on their impact on the arising anti-windup performance.

Procedure 1.(Anti-windup construction)

    Step 1. (Partition of the plant state in the exponentially stable part x_s and the remaining states x_u) Given a linear plant, select a coordinate representation such that its equations are in the form (1), where A_s is a Hurwitz matrix.

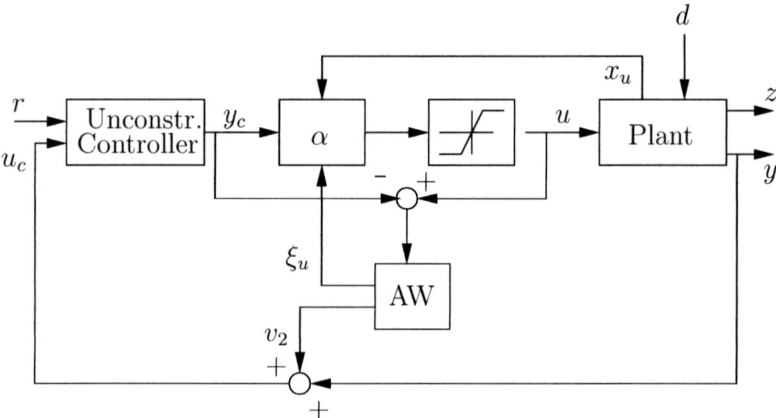

Fig. 1. The proposed anti-windup scheme.

Step 2. (Selection of the stabilizing gain F_u and the set $\mathcal{X}_u^+$) Define

$$F_u := GQ^{-1}, \qquad \mathcal{X}_u^+ := \{x_u : x_u^T Q^{-1} x_u \leq 1\}, \tag{7}$$

where Q and G are obtained from any solution of the following set of matrix inequalities in the variables $\lambda < 0$, γ, $Q = Q^T > 0 \in \mathbb{R}^{n_u \times n_u}$, $G \in \mathbb{R}^{m \times n_u}$ (where g_i denotes the ith row of G):

$$\begin{bmatrix} 1 & g_i \\ g_i^T & Q \end{bmatrix} \geq 0, \quad i = 1, \ldots, m,$$

$$\begin{bmatrix} \gamma I & I \\ I & Q \end{bmatrix} \geq 0, \tag{8}$$

$$QA_u^T + A_u Q + G^T B_u^T + B_u G < \lambda Q.$$

Step 3. (Definition of the set $\mathcal{X}_u$ and the bump function $\beta(\cdot)$) Select a (small) constant $\epsilon \in (0, 1)$ and define the set $\mathcal{X}_u := (1 - \epsilon)\mathcal{X}_u^+ = \{x_u : x_u^T Q_\epsilon^{-1} x_u \leq 1\}$ where $Q_\epsilon := (1 - \epsilon)^2 Q$. Define $\beta(\cdot)$ as

$$\beta(x_u) = \min\left\{1, \max\left\{0, \frac{1 - x_u^T Q^{-1} x_u}{1 - (1 - \epsilon)^2}\right\}\right\}. \tag{9}$$

Step 4. (Definition of $\mathcal{P}_u(\cdot)$ and $\gamma(\cdot, \cdot)$) Fix $\mathcal{P}_u(\cdot)$ as

$$\mathcal{P}_u(x_u) := -(B_u^T B_u)^{-1} B_u^T A_u x_u. \tag{10}$$

Following [4], the proposed set-point stabilizer $\gamma(\cdot, \cdot)$ for all $x_u \in \mathcal{X}_u^+$ and $x_u^* \in \mathcal{X}_u$ is given by

$$\gamma(x_u, x_u^*) = F_u x_u + (1 - \Psi_{\mathcal{X}_u^+}(x_u, x_u^*))(\mathcal{P}_u(x_u^*) - F_u x_u^*), \tag{11}$$

where, given a symmetric positive definite matrix $M \in \mathbb{R}^{m \times m}$, the ellipsoid $\mathcal{M} := \{x \in \mathbb{R}^m : x^T M x \leq 1\}$ and two vectors $x_1, x_2 \in \mathbb{R}^m$ with $v = x_1 - x_2$, the function $\Psi_{\mathcal{M}}(x_1, x_2)$ is defined as

$$\Psi_{\mathcal{M}}(x_1, x_2) = \frac{x_2^T M v + \sqrt{(x_2^T M v)^2 + (v^T M v)(1 - x_2^T M x_2)}}{1 - x_2^T M x_2}. \tag{12}$$

Step 5. (Definition of $\mathcal{P}(\cdot)$ and $\alpha(\cdot, \cdot, \cdot)$) Define the Lipschitz function $\mathcal{P}(\cdot)$ as

$$\mathcal{P}(x_u) := \frac{1}{\max\{1, \Psi_{\mathcal{X}_u}(x_u, 0)\}} x_u, \tag{13}$$

and the function $\alpha(\cdot, \cdot, \cdot)$ as

$$\alpha(x_u, x_u - \xi_u, y_c) := \gamma(x_u, \mathcal{P}(x_u - \xi_u)) + \beta(x_u)(y_c - \mathcal{P}_u(x_u - \xi_u)). \tag{14}$$

Step 6. The anti-windup compensator is given by the filter

$$\begin{aligned}
\dot{\xi} &= A\xi + B(u - y_c), \\
v_2 &= -C\xi - D(u - y_c),
\end{aligned} \tag{15a}$$

with the interconnection equations

$$u_c = y + v_2, \quad u = \operatorname{sat}(\alpha(x_u, x_u - \xi_u, y_c)). \tag{15b}$$

$$\star$$

The following theorem, proved in [12], states the effectiveness of the construction in Procedure 1.

Theorem 1. *Given a plant (1) and a controller (2) satisfying Assumption 1, the anti-windup construction of Procedure 1 solves the anti-windup problem of Definition 2 for the set $\mathcal{X}_u$.*

It is important to clarify the key features in the construction of Procedure 1, so that it also becomes clear how certain selections may impact the performance of the arising anti-windup solution. For example, different solutions that guarantee the $\mathcal{L}_p$ recovery property at item 2 of Definition 2 may be more or less desirable depending on the speed of convergence to zero of the performance output mismatch $(z - z_\ell)(\cdot)$ for certain converging signals of interest. This fact is commented below and also illustrated in the next section by some relevant examples.

Comments on Step 1. The first choice for the designer in Procedure 1 is the selection of the stable modes in x_s. Since the function $\alpha(\cdot, \cdot, \cdot)$ is selected only based on the remaining part of the state x_u, the larger x_s is, the simpler the anti-windup compensation law will be. Moreover, it is evident from

(15) that only measurement of the substate x_u is required by the anti-windup compensation law, therefore selecting x_s as large as possible greatly simplifies the arising anti-windup law. On the other hand, solutions that throw into x_s very slow modes (namely, modes very close to the imaginary axis) may result in very slow convergence. As a matter of fact, the stabilizing action effected through $\alpha(\cdot, \cdot, \cdot)$ would only monitor x_u and the slow modes would decay at their own rate, which might be unacceptable from a practical viewpoint. Therefore, the architectural complication arising from selecting a larger x_u (which includes the slow modes) may be worth the consequent performance advantage. In Section 3.3 we illustrate this fact with an example, where slow modes are first disregarded (so that x_u only contains the exponentially unstable parts of the system) and then are considered in x_u. The second solution, which is more involved, is shown to perform much more desirably than the first one. In Section 3.3 we also give some hints about possible modifications of the function $\alpha(\cdot, \cdot, \cdot)$ that would increase the speed of convergence of the slow modes even though these are not included in x_u.

Comments on Step 2. Once the state x_u has been decided, the next step is the selection of a feedback stabilizer for the corresponding saturated dynamics, through the solution of the matrix inequalities (8). Note that these inequalities are different from the linear matrix inequalities (LMIs) used in [12] because of the presence of the new term multiplied by the new variable λ on the right-hand side of the last inequality. This modified form of the equations in [12] is more appealing because it corresponds to a more advantageous way of carrying out high-performance selections of F_u guaranteeing fast convergence to zero. The LMIs (8) correspond to the LMI methods proposed in [17], which guarantee forward invariance of the region $\mathcal{X}_u^+$ in (7) through the saturated feedback $\mathrm{sat}(F_u x_u)$. In particular, by way of the second inequality of (8), a smaller γ corresponds to a larger Q, therefore to a larger set $\mathcal{X}_u^+$. On the other hand, a smaller λ (namely, a larger $|\lambda|$, because $\lambda < 0$) corresponds to a faster convergence rate to zero of the arising control law. It is evident from (8) that these two desirable properties are in contrast to each other (they both correspond to constraining Q to smaller and smaller sets). However, for any controllable matrix pair (A_u, B_u), the matrix inequalities (8) are feasible. Note also that when fixing λ, the inequalities (8) become linear and γ could be minimized by solving an eigenvalue problem. On the other hand, when fixing γ, minimizing λ in (8) corresponds to solving a generalized eigenvalue problem (which is, once again, convex). Both these eigenvalue problems can be addressed using standard LMI optimization tools such as [11]. Therefore, the selection of the solution to Step 2 should be carried out either with the goal of maximizing the size of the operating region with a certain guaranteed rate of convergence (therefore fixing λ and solving the arising eigenvalue problem that minimizes γ) or with the goal of maximizing the rate of convergence with a certain guaranteed size of the operating region (therefore fixing γ and solving the arising generalized eigenvalue problem that minimizes λ). In Section 3 we make extensive use of this strategy.

Comments on Step 3. The selection of $\mathcal{X}_u$ and $\beta(\cdot)$ in Step 3 is quite straightforward and does not typically impact the anti-windup performance. The function $\beta(\cdot)$ is a Lipschitz indicator function for $\mathcal{X}_u$ (where $\mathcal{X}_u$ is a deflated version of the forward invariant set $\mathcal{X}_u^+$). The role of $\beta(\cdot)$ is to assign full authority to the feedback stabilizer when the unstable state x_u is on the boundary of the set $\mathcal{X}_u^+$. This ensures boundedness of the x_u states. In general, the rule of thumb is to choose $\mathcal{X}_u$ as an ϵ contraction of $\mathcal{X}_u^+$ and to define $\beta(\cdot)$ as a Lipschitz function that is zero outside of $\mathcal{X}_u^+$ (and on its boundary), one inside $\mathcal{X}_u$ (and on its boundary), and varies continuously in the annulus $\mathcal{X}_u^+ - \mathcal{X}_u$. Note that by Definition 1, the set $\mathcal{X}_u$ characterizes the set of feasible references (namely the references for which the $\mathcal{L}_p$ recovery property at item 2 of Definition 2 is guaranteed). It is therefore desirable to keep ϵ small, so that the set $\mathcal{X}_u$ is large, and a larger number of references become feasible. Besides this fact, in general, the specific selection of $\beta(\cdot)$ does not impact noticeably the anti-windup performance.

Comments on Step 4. The functions $\mathcal{P}_u(\cdot)$ and $\gamma(\cdot, \cdot)$ defined at Step 4 are quite straightforward. Note that the proposed choice of $\gamma(\cdot, \cdot)$ is fully constructive, and so is $\mathcal{P}_u(\cdot)$ if B_u has full column rank. In cases where this rank assumption (see Assumption 1) is removed, the construction still applies but the definition of $\mathcal{P}_u(\cdot)$ is more involved because it is mandatory that at the steady state the unconstrained controller and $\gamma(\cdot, \cdot)$ make use of the same control allocation (which in this special case becomes nonunique).

Comments on Step 5. In Step 5, the proposed choice of the function $\alpha(\cdot, \cdot, \cdot)$ is once again fully constructive. As far as $\mathcal{P}(\cdot)$ is concerned, this function defines the location where infeasible steady states are projected by the anti-windup laws. It may be desirable in some cases to select different functions so that desirable restricted tracking properties are achieved; however, in most situations, the function (13) will be a satisfactory selection.

Remark 4. Note that we are assuming to have full measurement of the substate x_u of the plant and that the x_u dynamics are not affected by disturbances. This assumption is necessary to be able to prove clean results that are applicable to any selection of (r, d). Indeed, if any of these two assumptions were violated, it would be easy to build examples where stability is lost due to a large enough disturbance or a large enough initial condition for the estimation error of the unmeasurable part of the state x_u.

It is, however, possible to extend our results to the case where both these assumptions are not satisfied. The availability of x_u could be relaxed by designing an observer and stating regional results that rely on suitable bounds on the size of the estimation error at the initial time. Similarly, we could allow some components of the disturbance input d to affect the dynamics x_u provided that those components were suitably bounded so that "bad" selections that could push the state x_u outside the null controllability region are ruled out. In both cases, the region $\mathcal{X}_u^+$ would be restricted to allow for a suitable safety margin to tolerate the estimation error and the disturbance action.

3 Simulation examples

The three examples considered in this section will point out the flexibility offered by the proposed design procedure and the related trade-offs in anti-windup synthesis. In all the examples, an unconstrained controller is given, designed disregarding saturation and described by (2), which ensures asymptotic set-point tracking of step references and has been obtained by suitably connecting a reference prefilter and a LQG controller stabilizing the plant (in the last example, the plant augmented with an internal model of the constant references to be tracked).

It is worth remarking that the anti-windup design is fully independent of the controller (2), and so the same anti-windup compensator will work for any "reasonable" (in the sense of Assumption 1) controller; in particular, the forward invariant domain inside which the unconstrained closed-loop response is preserved (as long as saturation does not occur) will be as large as possible and independent of the specific controller, as required in our problem statement. On the other hand, by considering different anti-windup compensators, it will be apparent that the above-mentioned forward invariant domain will depend on the performance level imposed to the anti-windup compensator in terms of fast recovery after saturation.

3.1 Trading trackable signals for fast convergence

In this section we will illustrate how speed of convergence of the trajectories and size of the operating region are two performance features that arise from a trade-off corresponding to different selections of the variables λ and γ in the matrix inequalities (8). Consider a double integrator

$$\begin{bmatrix} \dot{x}_1 \\ \dot{x}_2 \end{bmatrix} = \begin{bmatrix} 0 & 1 \\ 0 & 0 \end{bmatrix} \begin{bmatrix} x_1 \\ x_2 \end{bmatrix} + \begin{bmatrix} 0 \\ 1 \end{bmatrix} u, \tag{16}$$

$$y = \begin{bmatrix} 1 & 0 \end{bmatrix} \begin{bmatrix} x_1 \\ x_2 \end{bmatrix} \tag{17}$$

This system has several interesting peculiarities. First, the null controllability region is unbounded in all directions; however, only semiglobal results are achievable when relying on quadratic stability results (such as the ones underlying the construction in [17], which we rely upon here). Moreover, it is interesting to note that for any step reference value, the control input will always be zero at the steady-state and that there is no need to add an internal model to guarantee zero steady-state error. For this plant, we design an unconstrained controller based on the paradigm discussed above, where the prefilter is $F(s) = \dfrac{100}{s + 100}$ and the LQG part is designed by solving a control Riccati equation for $(A + 0.1I, B)$ with state weight $C'C$ and input weight 0.1, and a filtering Riccati equation $(A+0.3I, C)$ with state and output disturbance covariance matrices both equal to I.

For the unconstrained closed-loop system formed by the plant (16) and the above controller, we compare the simulation results arising from three different anti-windup designs, all of them arising from applying Procedure 1 with x_u being the whole state (so that A_s is an empty matrix). The three designs differ from each other based on the type of optimization carried out in the gain selection at Step 2 of the procedure. In particular, we have

1. Anti-windup 1, where the matrix inequalities (8) are solved by fixing the minimum required convergence rate as $\lambda = -0.1$ and minimizing γ (resulting in $\gamma^* = 0.0405$), so that the size of the ellipsoid $\mathcal{X}_u^+$ is maximized;
2. Anti-windup 2, where the matrix inequalities (8) are solved by fixing the minimum required convergence rate as $\lambda = -0.2$ and minimizing γ (resulting in $\gamma^* = 0.1675$), so that the size of the ellipsoid $\mathcal{X}_u^+$ is maximized;
3. Anti-windup 3, where the matrix inequalities (8) are solved by fixing a minimum required size of the ellipsoid $\mathcal{X}_u^+$ by selecting $\gamma = 1$ and minimizing λ (through a generalized eigenvalue problem, obtaining $\lambda = -0.4485$) so that the convergence rate is maximized.

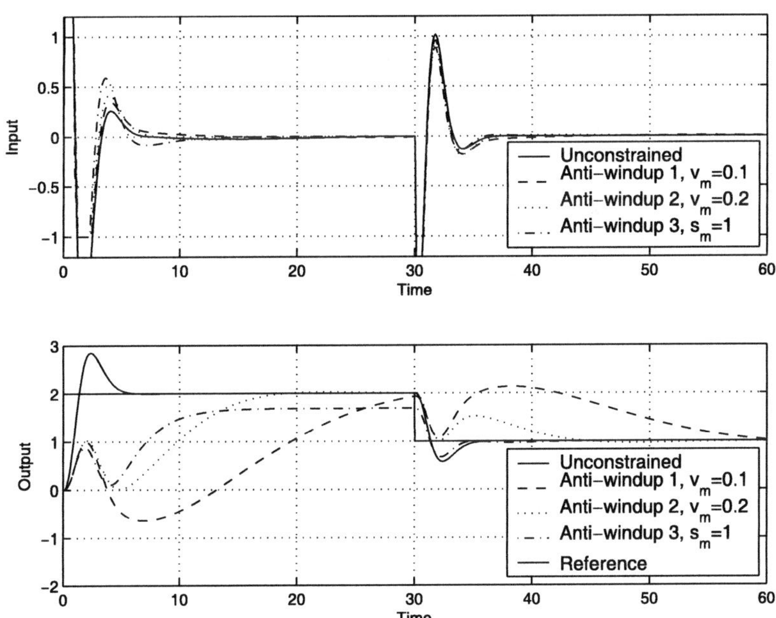

Fig. 2. Comparison among input and output closed-loop responses for the example in Section 3.1.

The three responses corresponding to the step reference $r(t) = 2$ for $t \in [0, 30)$, $r(t) = 1$ for $t \in [30, +\infty)$, are reported in Figure 2, where several interesting conclusions can be drawn. First, notice that the first solution (dashed) provides a poor output response with severe undershoot (although a larger operating region is guaranteed); the second solution (dotted) achieves a better response because a faster speed of convergence is enforced through the selection of λ. Finally, the last response (dash-dotted) corresponds to a faster convergence rate, although the operating region becomes too small and the anti-windup compensator enforces restricted tracking.

3.2 Enforcing state constraints

In this second example we consider the same double integrator introduced in the previous section and the same unconstrained controller, apart from the time constant of the prefilter, which is taken equal to 0.1 while preserving the same static gain $F(0) = 1$; the purpose of this choice is to avoid saturation in the subsequent simulation and show only a "state-constraining" feature of the proposed anti-windup compensation scheme. Using the same prefilter as in Section 3.1, the anti-windup compensation would still work and assure both anti-windup and state-constraining features, but the simultaneous appearance of both effects on would be harder to follow on the arising simulations. We design the anti-windup compensator following a different design goal from before: namely, we want to enforce certain constraints on the values of the plant states. Indeed, since the anti-windup law guarantees that the states x_u are confined within the set $\mathcal{X}_u^+$, then by enforcing a suitable bound on the size of Q in (8), we can guarantee that certain preassigned limits will be never exceeded by any trajectory.

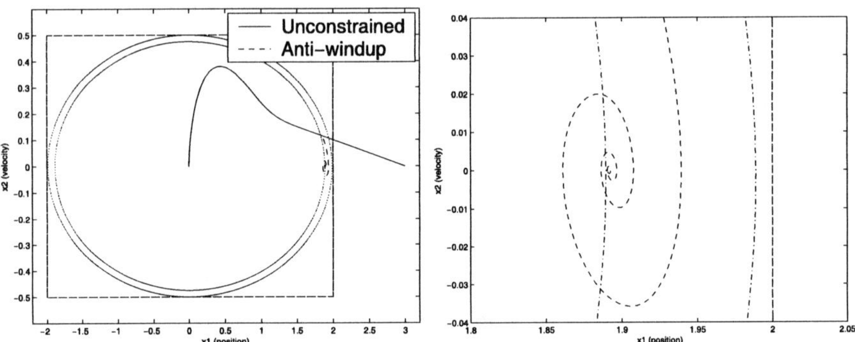

Fig. 3. State trajectories for the closed-loop systems of the example of Section 3.2 (left); convergence to the border of $\mathcal{X}_u$ (right).

For illustration purposes, we choose to constrain the state inside the region $|x_1| \leq 2$, $|x_2| \leq 0.5$ (whose border is shown as a dashed thin line in Figure 3), the region $\mathcal{X}_u^+$ has been determined by maximizing the sum of the lengths of the semiaxes of an invariant ellipsoid contained inside the above given region (the borders of $\mathcal{X}_u^+$ and $\mathcal{X}_u$ are shown as dash-dotted thin lines in Figure 3). Note that an alternative approach that would lead to a larger operating region could be undertaken by selecting polyhedral invariant sets in the construction of [12] so that, in principle, $\mathcal{X}_u^+$ could correspond exactly to the solid box in Figure 3.[4]

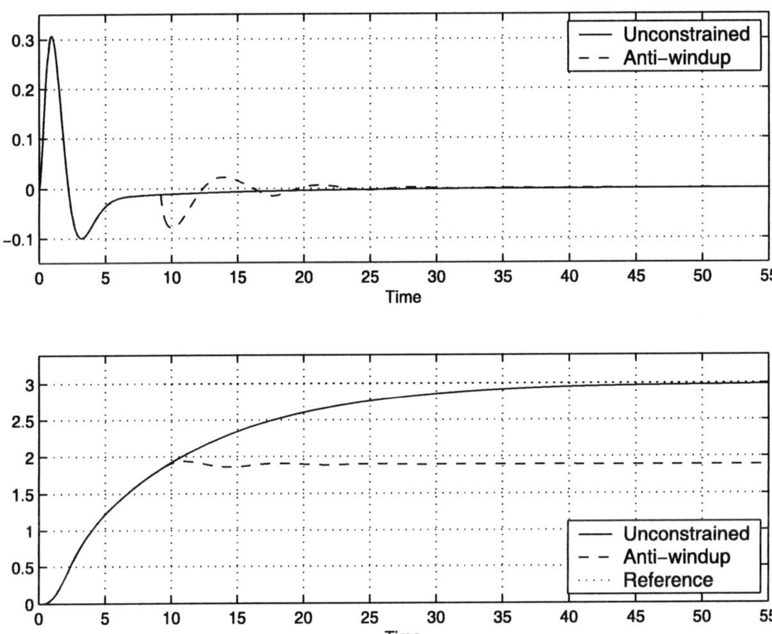

Fig. 4. Input and output closed-loop responses for the example of Section 3.2.

The response to a reference signal $r(t) \equiv 3$, which is infeasible at the steady state, is considered. The unconstrained and the anti-windup closed-loop responses are shown by solid and dashed curves, respectively, in Figure 3 (where the trajectories on the phase plane are reported) and Figure 4 (where the input and output time histories are reported). As expected, the state of the anti-windup closed-loop system converges to the projection on the border of $\mathcal{X}_u$ of the infeasible steady state reached by the unconstrained closed loop: this fact can be better appreciated by looking at the right plot of Figure 3, which is a blown-up detail of the left plot in that same figure.

---

[4]However, in this case the LMI technique of [17] cannot be used to determine a stabilizing law that makes that region forward invariant, and alternative techniques may be needed for the selection of F_u or suitable nonlinear generalizations of it (see, e.g., [4]).

3.3 A MIMO system

The last example corresponds to a MIMO control system where we want to illustrate the effects of making different selections of A_u at the first step of Procedure 1 on the overall performance. Consider the MIMO plant described by (1) with

$$A = \begin{bmatrix} 1 & 1 & 0 & 0 & 0 \\ -1 & 1 & 0 & 0 & 0 \\ 1 & 0 & -0.15 & -1 & 0 \\ 0 & 1 & 1 & -0.15 & 0 \\ 1 & 0 & -1 & 0 & -10 \end{bmatrix}, \qquad B = \begin{bmatrix} 10 & 0 \\ 0 & 0 \\ 0 & -10 \\ 10 & 0 \\ 10 & 10 \end{bmatrix}, \qquad B_d = 0,$$

$$C = \begin{bmatrix} 0.1 & 0 & 1 & 0 & 0 \\ 0 & 0.1 & 0 & 1 & 0 \end{bmatrix}, \qquad D = 0, \qquad D_d = 0,$$

and $z = y$. Both the inputs to the plant are subject to saturation so that $\|u(t)\|_\infty \leq 1, \forall t \geq 0$. Note that for this example, the first two states (x_1, x_2) correspond to exponentially unstable modes, the state x_5 corresponds to a fast exponentially stable mode, and the states (x_3, x_4) correspond to weakly damped oscillatory modes. The unconstrained controller for this plant is designed following the same controller architecture adopted in the previous examples, selecting the prefilter as $F(s) = \dfrac{100}{s + 100} I$ and designing the LQG part by solving a control Riccati equation for $(A_{aug} + 2I, B_{aug})$ (where A_{aug} and B_{aug} denote the A and B matrices of the cascade connection of the controlled plant and the two integrators constituting the internal model of the reference signals) with state weight I and input weight $0.1I$, and a filtering Riccati equation $(A + 5I, C)$ with state and output disturbance weights both equal to I.

For the unconstrained closed-loop system designed above, three different anti-windup compensators are designed and compared, all of them corresponding to applying Procedure 1 as follows:

1. Anti-windup 1, where at Step 1 the matrix A_u is selected as the upper left 4×4 submatrix of A, therefore the slow modes are considered in the anti-windup design. The gain $F_u \in \mathbb{R}^{2 \times 4}$ and the operating region $\mathcal{X}_u^+ \subset \mathbb{R}^4$ are determined based on the matrix inequalities (8) by fixing $\lambda = -0.5$ and minimizing γ (resulting in $\gamma^* = 0.4739$) so that the operating region is maximized.

2. Anti-windup 2, where at Step 1 the matrix A_u is selected as the upper left 2×2 submatrix of A, therefore the slow modes are *not* considered in the anti-windup design. The gain $F_u \in \mathbb{R}^{2 \times 2}$ and the operating region $\mathcal{X}_u^+ \subset \mathbb{R}^2$ are determined based on the matrix inequalities (8) by fixing $\lambda = -0.5$ and minimizing γ (resulting in $\gamma^* = 0.4737$) so that the operating region is maximized.

3. Anti-windup 3, where the same gain as in the previous step is used but the function α in (14) is replaced by the following modified version:

$$\alpha := \gamma(x_u, \mathcal{P}(x_u - \xi_u)) + \beta(x_u)(y_c + F_s\xi_s - \mathcal{P}_u(x_u - \xi_u)), \qquad (18)$$

where the extra stabilizing signal $F_s\xi_s$ is aimed at improving the performance recovery and is designed on a modified version of the matrix inequalities (8).

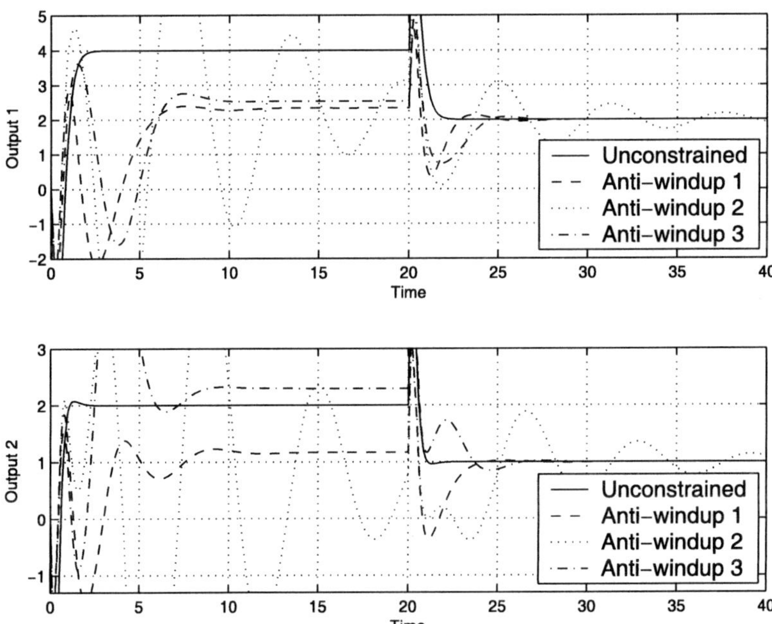

Fig. 5. Output closed-loop responses for the example of Section 3.3.

For our simulations we consider the reference signal $r(t) = \begin{bmatrix} 4 & 2 \end{bmatrix}^T$ for $t \in [0, 20)$, $r(t) = \begin{bmatrix} 2 & 1 \end{bmatrix}^T$ for $t \in [30, +\infty)$, which is not steady-state feasible[5] for $t \in [0, 20)$. The closed-loop responses are shown in Figures 5 and 6, which represent the output and input responses, respectively. Moreover, in Figure 7, we represent the plant state trajectories in the (x_1, x_2) and (x_3, x_4) directions, respectively.

Although the minimum required convergence rate is the same for both the first and the second designs, it is clear from Figure 5 that overlooking the slowly convergent modes can lead to extremely poor performance in convergence rate. On the other hand, the price to be paid for the improved performance of the first design is that the anti-windup compensator is more

---

[5] Namely, in the first time interval, it is not compatible with the required minimum convergence rates–in the sense that the corresponding matrix inequalities become infeasible when requiring such a large operating region.

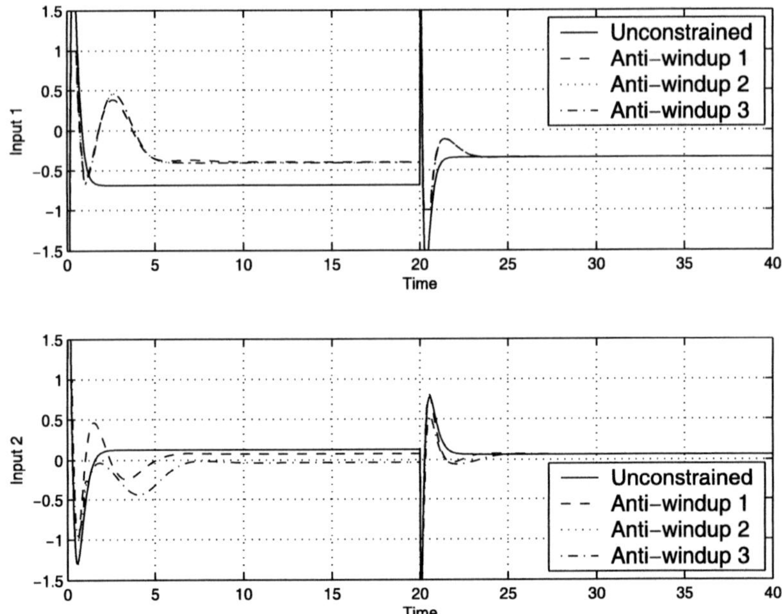

Fig. 6. Input closed-loop responses for the example of Section 3.3.

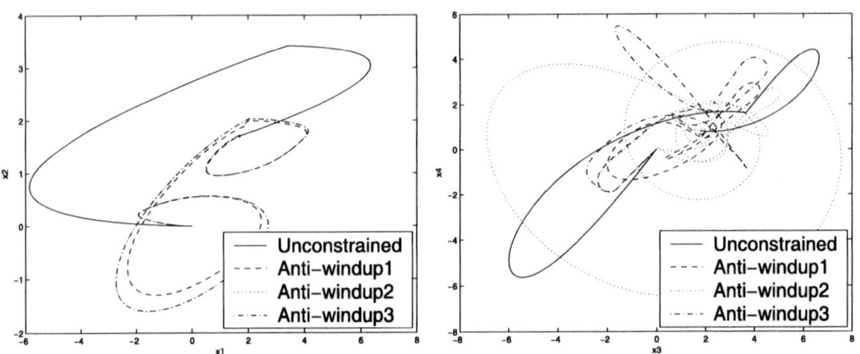

Fig. 7. Plant state trajectories for the example of Section 3.3: (x_1, x_2) plane (left) and (x_3, x_4) plane (right).

involved as the stabilizing function requires measurement of four states instead of only two. This problem is partially solved when using the third construction, which goes slightly beyond the theory developed in [12] and introduces an extra stabilizing term in the function α. This term uses the information captured by the subvector ξ_s of the state of the anti-windup compensator in a similar fashion to what was suggested in [24], thereby forcing the response toward the unconstrained one also in the slow-mode directions whenever the unstable state is not close to the boundary of $\mathcal{X}_u^+$ (namely, when $\beta(x_u) = 1$). The beneficial action of this extra compensation can be appreciated on the output responses of Figure 5. Moreover, from Figure 7, it is apparent that the poor response experienced at the plant output in the second case is mainly due to the large oscillations in the (x_3, x_4) directions of the right plot: as a matter of fact, the state trajectory in the (x_1, x_2) directions represented in the left plot remains relatively unchanged in the three anti-windup cases.

4 Conclusions

In this chapter we illustrated several peculiarities of the nonlinear anti-windup technique presented in [12] for exponentially unstable linear plants. Some important aspects related to the selection of the free parameters in the anti-windup design procedure were first discussed and then illustrated on several case studies.

References

1. Barbu C, Reginatto R, Teel A, Zaccarian L (1999) Anti-windup design for manual flight control. 3186–90. In: American Control Conference, San Diego, CA
2. Barbu C, Reginatto R, Teel A, Zaccarian L (2000) Anti-windup for exponentially unstable linear systems with inputs limited in magnitude and rate. 1230–1234. In: American Control Conference, Chicago, IL
3. Barbu C, Reginatto R, Teel A, Zaccarian L (2002) Anti-windup for exponentially unstable linear systems with rate and magnitude limits In: Kapila V, Grigoriadis K (eds), Actuator saturation control, 1–31. Marcel Dekker, New York
4. Blanchini F, Miani S (2000) Any domain of attraction for a linear constrained system is a tracking domain of attraction. SIAM J. Contr. Opt. 38(3):971–994
5. Cao Y, Lin Z, Ward D (2002a) Antiwindup design for linear systems subject to input saturation. Journal of Guidance Navigation and Control 25(3):455–463
6. Cao Y, Lin Z, Ward D (2002b) An antiwindup approach to enlarging domain of attraction for linear systems subject to actuator saturation. IEEE Trans. Aut. Cont. 47(1):140–145
7. Crawshaw S (2003) Global and local analysis of coprime factor-based anti-windup for stable and unstable plants. In: European Control Conference, Cambridge, UK
8. Crawshaw S, Vinnicombe G (2002) Anti-windup for local stability of unstable plants. 645–650. In: American Control Conference, Anchorage, AK
9. Da Silva Jr JG, Tarbouriech S (2003) Anti-windup design with guaranteed regions of stability: an LMI-based approach. 4451–4456. In: Conference on Decision and Control, Maui, HI
10. Edwards C, Postlethwaite I (1999) An anti-windup scheme with closed-loop stability considerations. Automatica 35(4):761–765

11. Gahinet P, Nemirovski A, Laub A, Chilali M (1995) LMI Control Toolbox. MathWorks, Natick, MA
12. Galeani S, Teel A, Zaccarian L (2004) Constructive nonlinear anti-windup design for exponentially unstable linear plants. In: Conference on Decision and Control, Atlantis, Bahamas
13. Gilbert E, Kolmanovsky I (1999) Fast reference governors for systems with state and control constraints and disturbance inputs. Internat. J. Robust Nonlinear Control 9(15):1117–1141
14. Grimm G, Hatfield J, Postlethwaite I, Teel A, Turner M, Zaccarian L (2003) Antiwindup for stable linear systems with input saturation: an LMI-based synthesis. IEEE Trans. Aut. Cont. (A) 48(9):1509–1525
15. Grimm G, Teel A, Zaccarian L (2004) Linear LMI-based external anti-windup augmentation for stable linear systems. Automatica (B) 40(11):1987–1996
16. Hanus R (1988) Antiwindup and bumpless transfer: a survey. 59–65. In: Proceedings of the 12th IMACS World Congress, Paris, France
17. Hu T, Lin Z, Chen B (2002) An analysis and design method for linear systems subject to actuator saturation and disturbance. Automatica 38(2):351–359
18. Kothare M, Campo P, Morari M, Nett N (1994) A unified framework for the study of anti-windup designs. Automatica 30(12):1869–1883
19. Lozier J (1956) A steady-state approach to the theory of saturable servo systems. IRE Transactions on Automatic Control 1:19–39
20. Miyamoto S, Vinnicombe G (1996) Robust control of plants with saturation nonlinearity based on coprime factor representation. 2838–2840. In: Conference on Decision and Control, Kobe, Japan
21. Mulder E, Kothare M, Morari M (2001) Multivariable anti-windup controller synthesis using linear matrix inequalities. Automatica 37(9):1407–1416
22. Park J, Choi C (1995) Dynamic compensation method for multivariable control systems with saturating actuators. IEEE Trans. Aut. Cont. 40(9):1635–1640
23. Sontag E (1984) An algebraic approach to bounded controllability of linear systems. Int. Journal of Control 39(1):181–188
24. Teel A (1999) Anti-windup for exponentially unstable linear systems. Int. J. Robust and Nonlinear Control 9:701–716
25. Teel A, Kapoor N (1997a) Uniting local and global controllers. In: Proc. 4th ECC, Brussels, Belgium
26. Teel A, Kapoor N (1997b) The $\mathcal{L}_2$ anti-windup problem: Its definition and solution. In: Proc. 4th ECC, Brussels, Belgium
27. Wu F, Lu B (2004) Anti-windup control design for exponentially unstable LTI systems with actuator saturation. Systems and Control Letters 52(3-4):304–322
28. Zaccarian L, Teel A (2002) A common framework for anti-windup, bumpless transfer and reliable designs. Automatica (B) 38(10):1735–1744
29. Zheng A, Kothare MV, Morari M (1994) Anti-windup design for internal model control. Int. J. of Control 60(5):1015–1024

Constrained Pole Assignment Control*

Mikuláš Huba

University of Technology in Bratislava
Faculty of Electrical Engineering and Information Technology
Ilkovičova 3
812 19 Bratislava, Slovak Republic
Mikulas.Huba@stuba.sk

Summary. This chapter gives an overview of simple controllers for SISO systems based on the generalization of the linear pole assignment method for constrained systems with dynamics ranging from relay minimum time systems to linear pole assignment systems. The design is based on splitting the nth-order system dynamics into n first-order ones, which can be constrained without any problems with stability and overshooting. It requires a successive decrease of the distance of the representative point from the next invariant set with lower dimension. Since the distance of the representative point to such invariant set can be defined in many ways, the construction of the constrained controllers is not unique. The controllers derived from the second-order integrator are simple, appropriate also for extremely fast application, and easy to tune by a procedure that generalizes the well-known methods by Ziegler and Nichols, or Åström and Hägglund, respectively.

1 Introduction

In control design, the task is usually formulated as: for a given system let us find a control law (controller) satisfying chosen design criteria. Depending on the criteria used, the basic approaches are the (linear) pole assignment control, the linear quadratic control, or the (linear) predictive control. In reality, the control signal is always constrained, which can be expressed as

$$u_r \in [U_1, U_2]; \quad U_1 < 0 < U_2. \tag{1}$$

So, despite the fact that the control signal is usually generated by linear controllers, the input to the plant u_r is modified by a nonlinear function that can be denoted as the saturation function:

$$u_r(t) = sat\{u(t)\}; \qquad sat\{u\} = \begin{cases} U_2\,; & u > U_2 > 0 \\ u\,; & U_1 \le u \le U_2 \\ U_1\,; & u < U_1 < 0 \end{cases}. \tag{2}$$

*This work has been partially supported by the Slovak Scientific Grant Agency, Grant No. 1/7621/20.

In the classical period of control (up to the late 1960s), the constraints in control were discussed by a huge amount of researchers. However, in the subsequent decades, the problem of constraints practically disappeared in the mainstream of control theory. A limited number of authors dealing with anti-windup design continued to investigate this important feature. Today, the problem of constraints is again in the focus of research activities, and it is impossible to give just a brief overview of the different approaches (see, e.g., [2], [4]). However, they mostly have one common feature—trying to be general and dealing with MIMO systems and within the robust theory framework, they are mostly complex, difficult to understand and to use. Does it mean that the problem of constrained control cannot be treated by a simple and physically transparent theory? Or is it a consequence of the well-known fact that academicians do not like simple problems and prefer to follow the mainstream approaches (so, today, everybody is dealing with the predictive control)? Or is it because control theory is dominated by mathematicians who are following the priorities of mathematics, while the engineering solutions to practical problems are mostly kept secret by corporations?

1.1 Simple solutions for simple problems

In this chapter we show that constrained controllers can be treated, understood, implemented, and tuned in a simple way—even simpler and more reliable than in the linear case. The pioneering times of 1960s and the simple solutions already forgotten are reconsidered. Minimum time and linear pole assignment control are shown to be not antagonistic solutions, but limit features of a broad range of physically efficient solutions.

Results of this chapter partially summarize and extend results of the first volume of *Constrained Systems Design* ([10]), published in Slovak and so not accessible to a broader community. For the sake of brevity, the content is limited to continuous-time systems. Some new ideas related to the design via modes decomposition and time-delayed systems are added.

1.2 Dynamical classes of control

Dynamical classes of control can be considered an extension of the Feldbaum's theorem on n intervals of optimal control published in 1949 to more general transients. Despite the fact that in controlling oscillatory systems the transients corresponding to larger initial deviations were shown to consist of more than n control intervals, this theorem represents very important physical control features, later practically forgotten within linear control theory. We will extend this theorem to deal with constrained pole assignment control, considering more general control sequences than the piecewise constant ones used under relay minimum time control. Instead of n intervals of time optimal control we will consider the nth dynamical class of control typical with n control pulses of a more general shape. The natural dynamical class is given by the system's relative degree. In general, it can be decreased by appropriate feedback—mostly to increase the closed-loop robustness and to decrease the control sensitivity. Such controllers are often denoted as *compensating*. While the problems of dynamical classes 0 and 1 can be successfully treated by linear design complemented by the classical anti-windup approaches, it is not the case in dealing with the dynamical class 2. Here, when the control signal attacks both the upper and the lower control limits, the classical anti-windup approaches fail (see, e.g., [15], [5]). For the sake of brevity, but also for practical reasons (the rapidly increasing complexity and sensitivity), we will focus our considerations to this dynamical class. Higher dynamical classes will not be mentioned here.

2 Control of first-order systems

For the sake of brevity, we consider just the I_1-plant

$$\frac{d\bar{y}(t)}{dt} = K_s \bar{u}(t), \tag{3}$$

with $u(t) = K_s \bar{u}(t)$ being the normalized control signal constrained to (1). Let $y = \bar{y} - r$ be the transformed output shifted by the reference signal r so that the new reference signal is $y_r = 0$.

Minimum time control can be simply implemented using an on-off (bang-bang) relay control. This approach, however, does not solve maintenance of the steady state in the vicinity of the output reference value y_r, which is possible just by oscillation, or by use of the relay with dead zone.

The *(linear) pole assignment control* of this plant corresponds to the output decrease carried out proportionally to the output value described by

$$\frac{dy(t)}{dt} = \alpha\, y(t). \tag{4}$$

The quotient of such a decrease α (when $\alpha < 0$) is usually denoted as the *closed-loop pole*. The control algorithm satisfying (4) follows from comparison of the right-hand sides of (4) and (3):

$$u(t) = -K_R y(t)\,; \; K_R = -\alpha\,; \; \bar{u} = u/K_s. \tag{5}$$

However, the regular output decrease (4) can only be achieved on a restricted interval containing $y_r = 0$. Outside of this interval the output decrease corresponding to the limit control value is time optimal, but slower as in the case of the linear pole assignment control (4). All points in which u satisfies (1) will be denoted as the *proportional band of control* (P_b). In the proportional band the pole assignment decrease (4) is slower than the minimum time decrease corresponding to the limit control signal value. For a properly tuned P-controller the control saturation causes neither overshooting nor instability.

Due to this, the saturating P-controller (5) (with the subsequent control signal limiter (2)) has properties required very often by practice (Figure 1) and will be denoted as the *constrained pole assignment controller*. It is a minimum time controller with an additional constraint

$$\frac{-dy(t)/dt}{y(t)} \leq |\alpha| \, . \tag{6}$$

It could be shown (e.g., choosing the Lyapunov function $V = y^2$, or by other techniques like Popov and circle criterion and hyperstability) that the closed control loop remains stable for all constraints (2) containing the steady state control value. This fact will be used later in designing controllers for higher-order systems.

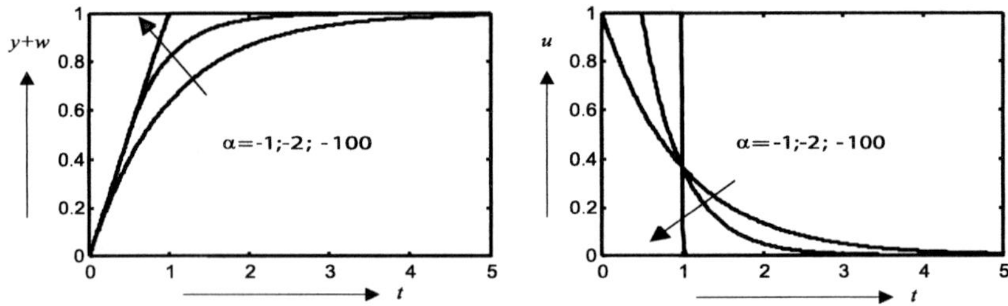

Fig. 1. Output (left) and control signal (right) for the constrained pole assignment controller for several values of the closed-loop pole α. In limit cases, the closed loop can be either fully linear or closed to the relay minimum time control. Well-designed control usually requires mixed behavior with both constrained transients and exponential damping close to the steady states.

3 Control of second-order systems

For the sake of brevity, we again consider just the double integrator

$$\frac{d^2\bar{y}(t)}{dt^2} = K_s\bar{u}\,(t)\,. \tag{7}$$

For $y = \bar{y} - r$; r being the reference input, $\mathbf{x}(t) = [y(t), d(t)]$; $d = dy/dt$ the system state and

$$u = K_s\bar{u} \; ; \quad U_i = K_s\bar{U}_i \; ; \quad i = 1,\, 2 \tag{8}$$

being the normalized control, it can be described in the state space as

$$\frac{d\mathbf{x}\,(t)}{dt} = \mathbf{A}\mathbf{x}\,(t) + \mathbf{b}u\,(t) \; ; \quad \mathbf{A} = \begin{bmatrix} 0 & 1 \\ 0 & 0 \end{bmatrix} \; ; \quad \mathbf{b} = \begin{bmatrix} 0 \\ 1 \end{bmatrix}. \tag{9}$$

3.1 Linear pole assignment control: real poles

The linear pole assignment PD-controller is given as

$$u = \mathbf{r}^t \mathbf{x} \; ; \quad \mathbf{r}^t = [r_0; r_1] \; ; \quad \bar{u} = u/K_s. \tag{10}$$

In several papers ([6], [8]) it was shown that the controller (10) fulfills two tasks:

a) It regularly decreases the (oriented) distance[1]

$$\rho_1 \left\{ \mathbf{x} \left(t \right) \right\} = \mathbf{a}^t \mathbf{x} \tag{11}$$

of the representative point $\mathbf{x}(t)$ from a line

$$L = \left\{ \mathbf{x} \mid \mathbf{a}^t \mathbf{x} = 0 \; ; \; \mathbf{a}^t = [a_0; a_1] \right\} \tag{12}$$

with the rate proportional to the pole α_2

$$\frac{d\rho_1 \left\{ \mathbf{x} \left(t \right) \right\}}{dt} = \alpha_2 \, \rho_1 \left\{ \mathbf{x} \left(t \right) \right\} \tag{13}$$

$$\rho_1 \left\{ \mathbf{x} \left(t \right) \right\} \neq 0 \, .$$

From

$$\frac{d\mathbf{x} \left(t \right)}{dt} = \mathbf{A}_R \mathbf{x} \left(t \right) \; ; \; \mathbf{A}_R = \mathbf{A} + \mathbf{b} \mathbf{r}^t, \tag{14}$$

whereby $\mathbf{A}_R$ is the closed-loop matrix, and from (11) and (13) it follows that $\mathbf{a}^t$ is the left eigenvector of $\mathbf{A}_R$ corresponding to α_2, since

$$\mathbf{a}^t \left[\alpha_2 \mathbf{I} - \mathbf{A}_R \right] \mathbf{x} = 0 \, . \tag{15}$$

After a slight modification we get the pole assignment control algorithm

$$u = \frac{\mathbf{a}^t \left[\alpha_2 \mathbf{I} - \mathbf{A} \right]}{\mathbf{a}^t \mathbf{b}} \mathbf{x} \, . \tag{16}$$

b) At the points of L (satisfying $\rho_1 \left\{ \mathbf{x} \right\} = 0$), the pole assignment controller guarantees a regular decrease of the (oriented) distance $\rho_0 \left\{ \mathbf{x} \right\}$ from the origin with the rate proportional to the closed-loop pole α_1:

$$\frac{d}{dt} \rho_0 \left\{ \mathbf{x} \left(t \right) \right\} = \alpha_1 \, \rho_0 \left\{ \mathbf{x} \left(t \right) \right\} \tag{17}$$

$$\rho_1 \left\{ \mathbf{x} \left(t \right) \right\} = \mathbf{a}^t \mathbf{x}(\mathbf{t}) = 0 \, .$$

Due to the motion along the line, in the case of Euclidean metric the last equation can be simply rewritten into

---

[1]This corresponds to the (Euclidean) distance measured in the direction of the normal line vector. But it could also be measured in another direction specified by a chosen vector not parallel to the line.

$$\frac{d\mathbf{x}\left(t\right)}{dt} = \alpha_1 \mathbf{x}\left(t\right) , \tag{18}$$

when also

$$\left[\alpha_1 \mathbf{I} - \mathbf{A}_R\right]\mathbf{x} = 0. \tag{19}$$

The above equation is fulfilled for points $\mathbf{x}$ of L, which are traced out by the right eigenvector $\mathbf{v}$ of $\mathbf{A}_R$ corresponding to α_1, i.e., for $\mathbf{x} = q\mathbf{v}$; $q \in (-\infty, \infty)$. It is known (see, e.g., [3]) that $\mathbf{a}$ and $\mathbf{v}$ are orthogonal to each other, i.e., $\mathbf{a}^t\mathbf{v} = 0$. For $\mathbf{v}$, it follows from (19) that

$$\left[\alpha_1 \mathbf{I} - \mathbf{A}\right]\mathbf{v} = \mathbf{b}\mathbf{r}^t\mathbf{v} .$$

The length of $\mathbf{v}$ can be specified arbitrarily different from zero, e.g., by $\mathbf{r}^t\mathbf{v} = 1$, which yields

$$\mathbf{v} = -\left[\alpha_1 \mathbf{I} - \mathbf{A}\right]^{-1}\mathbf{b} = \begin{bmatrix} 1/\alpha_1^2 \\ 1/\alpha_1 \end{bmatrix} . \tag{20}$$

The normal vector can be specified from (20) and from the orthogonality condition $\mathbf{a}^t\mathbf{v} = 0$, e.g., as

$$\mathbf{a}^t = \left[2 \ -2/\alpha_1\right] . \tag{21}$$

Two ordered combinations of the closed-loop poles $[\alpha_1, \alpha_2]$ yield two eigenvectors

$$\mathbf{v}_i = \left[\alpha_i \mathbf{I} - \mathbf{A}\right]^{-1}\mathbf{b} \ ; \ i = 1, 2 \tag{22}$$

and two lines $L_i, i = 1, 2$. For the poles denoted with respect to

$$\alpha_1 < \alpha_2 < 0, \tag{23}$$

the deeper slope in the phase plane (y, d) corresponds to L_1. Finally, after substituting (21) into (16) it is possible to derive the controller vector $[r_0; r_1]$. For the double integrator, it is given by the closed-loop poles vector $[\alpha_1, \alpha_2]$ as

$$r_0 = -\alpha_1\alpha_2 \ ; \ r_1 = \alpha_1 + \alpha_2. \tag{24}$$

It is obviously invariant to the order of poles in the closed-loop poles vector. It means that it is not important which of the two lines traced out by the eigenvector will be chosen as the reference braking line L for the braking phase.

When the corresponding eigenvectors $\mathbf{v}_i$ are not collinear and form a basis, any state can be expressed as a sum of two modes

$$\mathbf{x} = c_1\mathbf{v}_1 + c_2\mathbf{v}_2. \tag{25}$$

After denoting $u = u_1 + u_2$ and due to $\mathbf{r}^t\mathbf{v}_i = 1 \ ; \ i = 1, 2,$

$$\dot{\mathbf{x}} = \dot{\mathbf{x}}_1 + \dot{\mathbf{x}}_2 = \mathbf{A}\left(\mathbf{x}_1 + \mathbf{x}_2\right) + \mathbf{b}\left(u_1 + u_2\right) .$$

Let each mode correspond to the elementary first-order solution

$$\dot{\mathbf{x}}_i = \mathbf{A}\mathbf{x}_i + \mathbf{b}u_i = \alpha_i \mathbf{x}_i \; ; \; \mathbf{x}_i = c_i \mathbf{v}_i,$$

$$\begin{bmatrix} 0 & 1 \\ 0 & 0 \end{bmatrix} c_i \begin{bmatrix} v_{1i} \\ v_{2i} \end{bmatrix} + \begin{bmatrix} 0 \\ 1 \end{bmatrix} u_i = \alpha_i c_i \begin{bmatrix} v_{1i} \\ v_{2i} \end{bmatrix}.$$

Then, from the second row of the last equation it follows that

$$u_i = c_i,$$

i.e., by expressing the actual state in the basis formed by the eigenvectors, we have control values for both particular modes. Because it holds that

$$\mathbf{x} = \begin{bmatrix} 1/\alpha_1^2 & 1/\alpha_2^2 \\ 1/\alpha_1 & 1/\alpha_2 \end{bmatrix} \begin{bmatrix} c_1 \\ c_2 \end{bmatrix} = \mathbf{M} \begin{bmatrix} c_1 \\ c_2 \end{bmatrix},$$

$$\begin{bmatrix} c_1 \\ c_2 \end{bmatrix} = \mathbf{M}^{-1}\mathbf{x} = \begin{bmatrix} \left(-\alpha_1^2 \alpha_2 y + \alpha_1^2 d\right)/\left(\alpha_1 - \alpha_2\right) \\ \left(\alpha_1 \alpha_2^2 y - \alpha_2^2 d\right)/\left(\alpha_1 - \alpha_2\right) \end{bmatrix}. \tag{26}$$

After expressing $u = u_1 + u_2 = c_1 + c_2$ we again get the controller (24). This even holds for $\alpha_2 \to \alpha_1$, when the mode decomposition cannot be used (division by zero in (26)).

3.2 Saturating pole assignment control

Putting limiter (2) into the second-order loop, its behavior can become useless or unstable. The dynamics specified by the closed-loop poles can be guaranteed just over the invariant set of linear control. This is a subset of the proportional band of control P_b, in which the control signal does not saturate. Points of P_b can be expressed as

$$P_b : \mathbf{x} = x_v \mathbf{v} + x_z \mathbf{z} \; ; \; x_v \in [U_1, U_2] \; ; \; x_z \in (-\infty, \infty) \; ; \; \mathbf{r}^t \mathbf{z} = 0. \tag{27}$$

Due to this, the two lines B_j corresponding to $\mathbf{r}^t \mathbf{x} = U_j$ parallel to vector $\mathbf{z}$ limit the strip-like zone of P_b. For the double integrator and real closed-loop poles

$$B_j : y - \left(\frac{1}{\alpha_1} + \frac{1}{\alpha_2}\right) d + \frac{U_j}{\alpha_1 \alpha_2} = 0 \; ; \; j = 1, 2. \tag{28}$$

Just a line segment of the lines L lying in P_b traced out by vertices

$$X_0^j = \mathbf{v} U_j \tag{29}$$

can be considered as the reference braking trajectory and as the target for the first phase of control.

In defining the shape of the invariant set of linear control, it is important to identify vertices of B_j in which the trajectories of the closed-loop system have tangents parallel to $\mathbf{z}$. These points P_0^j can be defined by

$$\mathbf{r}^t P_0^j = U_j \ ; \quad \mathbf{r}^t \frac{d\mathbf{x}}{dt} = \mathbf{r}^t \left(\mathbf{A} + \mathbf{b}\mathbf{r}^t \right) P_0^j = 0. \tag{30}$$

The invariant set of linear control is then limited by the closed-loop trajectories crossing the vertices P_0^j:

$$P_0^j = \begin{bmatrix} \frac{\alpha_j^2 + \alpha_2^2 + \alpha_1 \alpha_2}{\alpha_j^2 \alpha_2^2} \\ \frac{\alpha_1 + \alpha_2}{\alpha_1 \alpha_2} \end{bmatrix}. \tag{31}$$

For the relatively fast poles, the invariant set of linear control can shrink to just a fraction of the working range of control. Then constrained controllers have to be used.

3.3 Constrained pole assignment control

One way to deal with constrained pole assignment control is the invariant set-based approach ([7], [6], [8], [9]) based on a distance decrease.

Reference braking curve (RBC)

Looking for analogies with first order saturating control, it will be further assumed that the control signal was set to the limit value U_j during the first part of a braking phase ending by the movement along the line segment $X_0^j 0$; $j = 1, 2$. Starting backward from X_0^j with $u = U_j$ and the time increasing in the inverse direction, we get corresponding points

$$\mathbf{x}_b = e^{-\mathbf{A}\tau} X_0^j + U_j \int_0^{-\tau} e^{\mathbf{A}\xi} \mathbf{b} d\xi \ ; \quad j = 1, 2 \ ; \quad \tau > 0. \tag{32}$$

For

$$j = \left(3 + sign\left(y \right) \right) /2, \tag{33}$$

the braking line segments $X_0^j 0$ are complemented by parabolic reference braking curve RBC_j that can be (after eliminating time τ) expressed as

$$y_b = \frac{d}{\alpha_1} \ ; \qquad d \in \left[d_0^2, d_0^1 \right] \tag{34}$$

$$y_b^j = \frac{d^2 - \left(d_0^j \right)^2}{2U_j} + y_0^j \ ; \quad d \notin \left[d_0^2, d_0^1 \right].$$

The same can also be expressed in the form

$$d_b = \alpha_1 y \ ; \qquad y \in \left[y_0^1, y_0^2 \right] \tag{35}$$

$$d_b^j = -sign(y) \sqrt{ \left(d_0^j \right)^2 + 2U_j \left(y - y_0^j \right) } \ ; \quad y \notin \left[y_0^1, y_0^2 \right].$$

Measuring the distance to RBC as a position difference

We now propose a control algorithm to decrease the distance ρ of $\mathbf{x}$ to RBC_j according to (4) with the quotient α_2. In defining the (oriented) distance of $\mathbf{x}$ from RBC_j, there are infinitely many possibilities. To illustrate, we introduce three basic solutions, when the distance is measured in direction of a chosen vector $\mathbf{w}$. The first one is given by a measurement specified by a "horizontal" phase plane vector $\mathbf{w} = (w, 0)$, when

$$\rho_y\left(\mathbf{x}\right) = y - y_b. \tag{36}$$

For $d \in \left[d_0^2, d_0^1\right]$ the corresponding controller is again (10) with limiter (2). For $\rho_1(\mathbf{x}) = \rho_y(\mathbf{x})$ and higher velocities d the controller can be derived from

$$\frac{d}{dt}\rho_y\left(\mathbf{x}\right) = \alpha_2\,\rho_y\left(\mathbf{x}\right), \tag{37}$$

when for the double integrator

$$\frac{d}{dt}\rho_y\left(\mathbf{x}\right) = \frac{d}{dy}\rho_y\left(\mathbf{x}\right)\cdot d + \frac{d}{dd}\rho_y\left(\mathbf{x}\right)\cdot u \tag{38}$$

$$u = \left[1 - \alpha_2\frac{y - y_b}{d}\right] U_j \tag{39}$$

$$\bar{u}_r = sat\left\{u/K_s\right\}. \tag{40}$$

Measuring the distance to RBC as a velocity difference

By specifying the distance to RBC as the velocity difference, i.e., by a vertical phase plane vector $\mathbf{w} = (0, w)$ with

$$\rho_d\left(\mathbf{x}\right) = d - d_b \tag{41}$$

for $y \in \left[y_0^1, y_0^2\right]$, the control is again computed according to (10) and (2). Outside of this region, from (41) and following equations it follows that

$$u = \alpha_2 d + \sqrt{p}\left[\alpha_2 - U_j d/p\right]\ ;\ \ p = 2U_j y - U_j^2/\alpha_1^2. \tag{42}$$

It is interesting to note that (1.) the resulting control quality depends on the order of the closed-loop poles; (2.) the algorithms and transients corresponding to different direction of the distance measurement (or, more generally, to different distance definitions) are not identical (just similar); (3.) for the relatively "slow" closed-loop poles the transients can be fully linear. By shifting $a_{1,2} \to -\infty$ the proportional band becomes narrower and the transients are similar to those of the relay minimum time systems.

Measuring the distance to RBC along the complementary eigenvector: mode decomposition

In looking for some specific solutions, we can find another interesting property in the case of the distance measurement specified by the eigenvector

$$\mathbf{v}_2 = -\left[\alpha_2 \mathbf{I} - \mathbf{A}\right]^{-1} \mathbf{b} = \begin{bmatrix} 1/\alpha_2^2 \\ 1/\alpha_2 \end{bmatrix},$$

which is complementary to the eigenvector specifying RBC.

The control signal dynamics will now be decomposed into two eventually constrained exponentials. If u satisfies to (2) and the control and state are decomposed into particular modes, whereby

$$\mathbf{x} = \mathbf{x}_1 + \mathbf{x}_2 \; ; \; u = u_1 + u_2$$

for $u_1 \in [0, U_j]$ denoting the mode control corresponding to braking the mode of control corresponding to acceleration must fulfill $u_2 = u - u_1 \in [0, U_{3-j} - U_j]$. The total transient will be given as the sum of the particular modes in the time domain (Figure 2). Again, the dynamic is specified by the order of poles. Since the exponential corresponding to the "faster" pole is damped better than that one corresponding to the "slower" pole, just the transients corresponding to $\alpha_2 < \alpha_1 < 0$ finish without overshoot (an overshoot can also be found in Figure 2a). For closed-loop control, the control input computation is based on expressing the position of the actual point to RBC ([13]). By adding an unconstrained value of $c_2\mathbf{v}_2$ to its point representing the first mode, we can reach any point $\mathbf{x} = [y, d]^t$ of the double integrator phase plane as

$$y = \begin{cases} d_1/\alpha_1 + c_2/\alpha_2^2 \; ; \; d_1 \in [0, U_j/\alpha_1] \\ \left[d_1^2 + (U_j/\alpha_1)^2\right]/(2U_j) + c_2/\alpha_2^2 \; ; \; d_1 \notin [0, U_j/\alpha_1] \end{cases} \tag{43}$$

$$d = d_1 + c_2/\alpha_2.$$

By substituting for c_2

$$u_2 = c_2 = \alpha_2 d_2 = \alpha_2(d - d_1), \tag{44}$$

the solution of (43) for the unknown d_1 can be derived as

$$d_1 = \begin{cases} \alpha_1\left(\alpha_2 y - d\right)/\left(\alpha_2 - \alpha_1\right) \\ U_j\left[1/\alpha_2 \pm \sqrt{2\left(\alpha_2 y - d\right)/\left(\alpha_2 U_j\right) + 1/\alpha_2^2 - 1/\alpha_1^2}\right]. \end{cases}$$

After substituting into (44) and after simplifying, we get

$$u = u_1 + u_2 = \alpha_1 d_1 + \alpha_2(d - d_1) = \tag{45}$$

$$= \begin{cases} -\alpha_1\alpha_2 y + (\alpha_1 + \alpha_2)\, d \; ; \; u_1 \in [0, U_j] \; ; \; d_1 \in [0, U_j/\alpha_1] \\ \alpha_2 d + U_j\alpha_2\sqrt{2\left(\alpha_2 y - d\right)/\left(\alpha_2 U_j\right) + 1/\alpha_2^2 - 1/\alpha_1^2} \; ; \; u_1 = U_j \; ; \\ \qquad\qquad\qquad\qquad\qquad\qquad\qquad\qquad\qquad d_1 \notin [0, U_j/\alpha_1]. \end{cases}$$

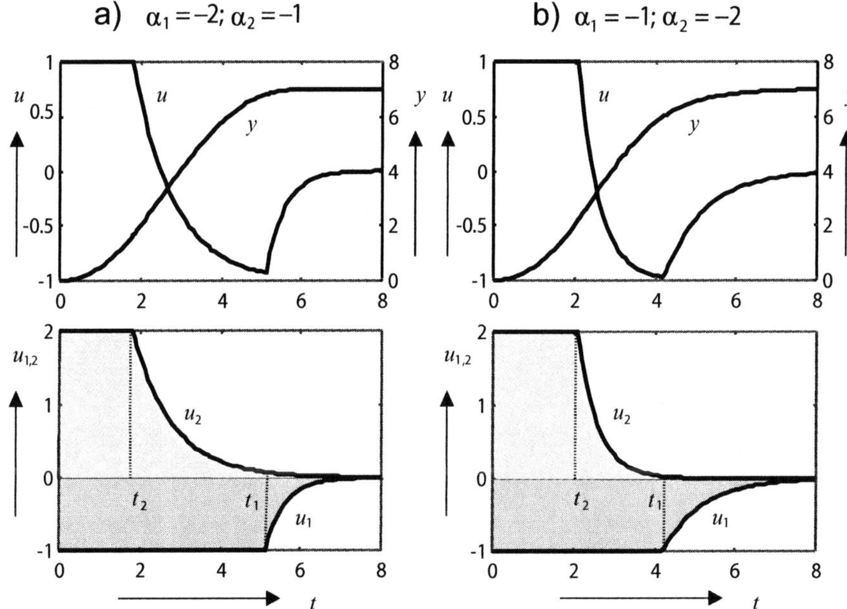

Fig. 2. Decomposition of the second-order dynamics into two first-order constrained exponentials (modes) for different pole orders.

This has yet to be limited to the admissible range (2). The value of U_j has to be chosen in such a way that

$$j = \left[3 + sign(y - d/\alpha_2)\right]/2.$$

For $2\left(\alpha_2 y - d\right)/\left(\alpha_2 U_j\right) + 1/\alpha_2^2 - 1/\alpha_1^2 < 0$, $u = U_j$ is set.

Finally, for $\alpha_1 = \alpha_2 = \alpha$ the linear segment of control disappears and

$$u = sat\left\{\alpha d + U_j \alpha \sqrt{2\left(y - d/\alpha\right)/U_j}\right\} \ ; \quad j = \left[3 + sign(y - d/\alpha)\right]/2. \quad (46)$$

Besides the three properties mentioned in the paragraph above it is interesting to note that by specifying the poles, we simultaneously restrict the rate of the control signal changes (see Figure 4) to

$$\frac{du}{dt} \in \left[(U_2 - U_1)/\alpha_2, U_1/\alpha_1\right].$$

4 Time-delayed systems

One of the basic shortcomings of minimum time control is its high sensitivity to time delays. The methods for its compensation were based on shifting the switching curves (surfaces) against the motion of the representative point. This, however, does not solve the situation around the reference state. Constrained pole assignment control enables us to take into account the time delays by modifying the closed-loop poles. Detailed study of time delay compensation in constrained pole assignment control design can be found in [10].

4.1 Dominant first-order dynamics

Let us consider the single integrator with elementary time delays: the dead time $e^{-T_d s}$, or the time lag $1/\left(1 + T_1 s\right)$ controlled by the P-controller (5).

Double real pole setting

In the first case with

$$I_1 T_d \; : \; S\left(s\right) = \frac{K_s e^{-T_d s}}{s} = \frac{\bar{Y}\left(s\right)}{\bar{U}\left(s\right)} \tag{47}$$

and $u = K_s \bar{u}$, the characteristic polynomial

$$A\left(s\right) = s + K_R e^{-T_d s} \tag{48}$$

has infinitely many poles. The fastest monotonous transients correspond to the double real dominant closed-loop pole s_0 that satisfies also equation

$$\dot{A}\left(s_0\right) = 1 - K_R T_d e^{-T_d s_0} = 0. \tag{49}$$

It can be determined as

$$s_0 = -\left[\ln\left(K_R T_d\right)\right]/T_d.$$

From these last two equations it follows that the corresponding controller gain is

$$K_R = 1/\left(e T_d\right). \tag{50}$$

By comparing this value with the pole assignment controller gain $K_R = -\alpha$, it can be found that the required controller gain can be achieved by working with *equivalent pole*

$$\alpha_e = -1/\left(e T_d\right) = -0.368/T_d. \tag{51}$$

Similarly, for the $I_1 T_1$-system

$$I_1 T_1 \; : \; S\left(s\right) = \frac{K_s}{s\left(T_1 s + 1\right)} = \frac{\bar{Y}\left(s\right)}{\bar{U}\left(s\right)}, \tag{52}$$

we can derive

$$\alpha_e = -1/\left(4 T_1\right) = -0.25/T_1. \tag{53}$$

4.2 Dominant second-order dynamics

Controller tuning motivated by the triple real closed-loop pole

As a generalization of the previous procedure, the first criterion used for the constrained controller tuning of the time-delayed second order systems [7] was to find the r_{00}, r_{10} that will guarantee a triple real closed-loop pole α_0. We will explain the procedure by considering an additional time constant T_1, when

$$I_2 T_1 : \quad S(s) = \frac{K_s}{s^2 (T_1 s + 1)}. \tag{54}$$

The plant gain can again be treated by considering $u = K_s \bar{u}$. For an input delay, the system is described by the state equations

$$\dot{\mathbf{x}} = \mathbf{A}\mathbf{x} + \mathbf{b}u \; ; \quad \mathbf{A} = \begin{bmatrix} -1/T_1 & 0 & 0 \\ 1 & 0 & 0 \\ 0 & 1 & 0 \end{bmatrix} \; ; \quad \mathbf{b} = \begin{bmatrix} 1/T_1 \\ 0 \\ 0 \end{bmatrix}. \tag{55}$$

The controller equation is

$$u = \mathbf{r}^t \mathbf{x} = \begin{bmatrix} 0 & r_1 & r_0 \end{bmatrix} \mathbf{x}.$$

The closed-loop characteristic polynomial is

$$A(s) = \det(s\mathbf{I} - \mathbf{A}_R) = \begin{vmatrix} s + 1/T_1 & -R_1 & -R_0 \\ -1 & s & 0 \\ 0 & -1 & s \end{vmatrix} = s^3 + s^2/T_1 - R_1 s - R_0,$$

whereby

$$R_0 = r_0/T_1 \; ; \quad R_1 = r_1/T_1.$$

In general, this polynomial has three different roots, which correspond to the three different modes of the control error. Since control quality is dominated by the slowest mode, the aim of the control design is to make this mode as fast as possible. A compromise solution is then given by a triple real closed-loop pole corresponding to equations

$$A(s_o) = 0 \; ; \; \dot{A}(s_o) = 0 \; ; \; \ddot{A}(s_o) = 0.$$

By solving these equations for the double integrator we get

$$s_0 = -1/(3T_1) \; ; \; R_1 = -1/(3T_1^2) \; ; \; R_0 = -1/(27T_1^3), \tag{56}$$

when

$$r_{10} = R_1 T_1 = -1/(3T_1) \; ; \; r_{00} = R_0 T_1 = -1/(27T_1^2). \tag{57}$$

By comparing these values with (24), we get the equivalent poles

$$\alpha_{1,2} = (-1/2 \pm j\sqrt{3}/6)/(3T_1) = (-.1666666666 \pm j0.09622504486)/T_1. \tag{58}$$

Similarly, for the I_2T_d system

$$I_2T_d: \quad S(s) = \frac{K_s e^{-T_d s}}{s^2}, \tag{59}$$

we can derive controller parameters and equivalent poles

$$r_{10} = -2(\sqrt{2}-1)e^{(-2+\sqrt{2})}/T_d = -0.4611587914/T_d; \tag{60}$$

$$r_{00} = -2(7-5\sqrt{2})e^{(-2+\sqrt{2})}/T_d^2 = -0.07912234014/T_d^2;$$

$$\alpha_{1,2} = (-.2305793957 \pm j0.1611070527)/T_d. \tag{61}$$

The problem is that the equivalent poles are complex conjugate numbers, while the previous constrained design has considered just real closed-loop poles. Does it mean that it is not appropriate for time-delayed systems? This question was partially answered already in [7]. By approximating the complex pair of equivalent poles by their modules or real parts, when

$$\alpha_e = -0.167/T_1 \text{ (real part), or } \alpha_e = -0.192/T_1 \text{ (module)} \tag{62}$$

$$\alpha_e = -0.231/T_d \text{ (real part), or } \alpha_e = -0.281/T_d \text{ (module)},$$

and using such poles in the previously derived constrained controllers, it is possible to get solutions that in the linear range of operation yield monotonous responses close to those with complex poles, fully compensating the influence of constraints for large disturbances.

Design of constrained pole assignment control derived for complex poles considered recently in [12], [14] showed improved dynamics of the final phase of control. It avoids the necessity of approximating the complex poles by real ones. However, the used gain scheduling solutions are not naturally valid for any initial conditions. So, by their simplicity and global validity the approximative solutions based on (62) do not lose their importance.

What are the links between first-order and second-order design?

New motivation for research on constrained time-delayed systems was given by the design based on mode decomposition [13]. If it is possible to split the total transient into particular first-order modes, why it is not possible to base the time-delayed system design on solutions derived for the first-order time-delayed systems and work with the closed-loop poles (51, 53)? Figure 3 shows application of the controller (46) to (55) for two different time delays and the settings corresponding to (53) and (62). For the relatively small time delay the transients corresponding to (53) seem to be the fastest ones without oscillations for the given time delay. Already a small increase of the absolute value of the poles leads to oscillatory behavior. The tuning according to (62) is too conservative. However, for the larger delay the control signal (that does not reach saturation limits) already tends to oscillate.

In the case of a system involving dead time (Figure 4) the situation is much worse. What is the cause of oscillations? Stability of the closed loop with the nonlinear controller (46) can be analyzed using the Lyapunov linearization method, when for differences from the steady state the controller is approximated by a PD-controller with gains

$$r_0 = \left[\frac{\partial u}{\partial y}\right]_{y=0;d=0} = \lim_{y,d\to 0} \frac{\alpha}{\sqrt{\frac{2(y-d/\alpha)}{U_j}}} \to -\infty \tag{63}$$

$$r_1 = \left[\frac{\partial u}{\partial d}\right]_{y=0;d=0} = \alpha - \lim_{y,d\to 0} \frac{1}{\sqrt{\frac{2(y-d/\alpha)}{U_j}}} \to -\infty.$$

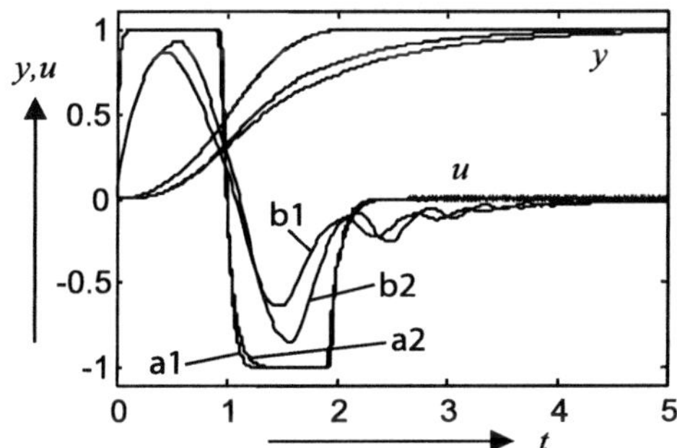

Fig. 3. Application of the controller (46) to a system with an inertial actuator (55) for the time delays a) $T_1 = 0.02$ and b) $T_1 = 0.2$ and the settings corresponding to 1) $\alpha_e = -0.25/T_1$ and 2) $\alpha_e = -0.192/T_1$.

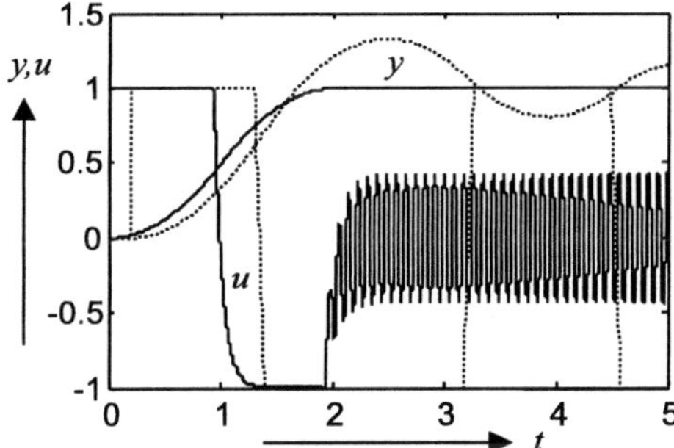

Fig. 4. Application of the controller (46) to a system with actuator having dead time a) $T_d = 0.02$ (full curves) and b) $T_d = 0.2$ (dotted) and the settings corresponding to $\alpha_e = -0.281/T_d$.

Obviously, this controller is singular at origin, which in the presence of dead time leads to limit cycles. With the controller gains (63) the open-loop Nyquist curve must circle around the critical point, which is different from the system with the time lag T_1, when it is possible to derive stability conditions. However, in reality each control loop involves some dead time. So, in the simulation of (55) some invisible oscillations always occur because of the delay introduced by the finite integration step. Therefore, the constrained decomposition with equal modes has a theoretical value and is not recommended to be used in practice. Instead, it is possible to use a setting with different poles, when α_2 is given by (51), (53) and

$$\alpha_1 = \alpha_2/c \; ; \; c > 1 \,.$$

Experiments show that the value of c, which is an indicator of different directions of the eigenvectors, should be at least $c > 1.2$. This, however, still enables $c < 1.7151$ to achieve higher dynamics of transients, as well as higher controller gains than in the approach base on approximating the triple real pole setting, when it holds that

$$r_0 = \alpha_1\alpha_2 = (0.368/T_d)^2 \, /c > (0.281/T_d)^2 \,.$$

This can be useful in decreasing the steady state error for acting disturbances (Figure 5). Increased values of c slow down transients, but contribute to increased system robustness.

Use of "slower" poles for acceleration $\alpha_2 = \alpha_1/c \; ; \; c > 1$ would lead to transients with overshooting.

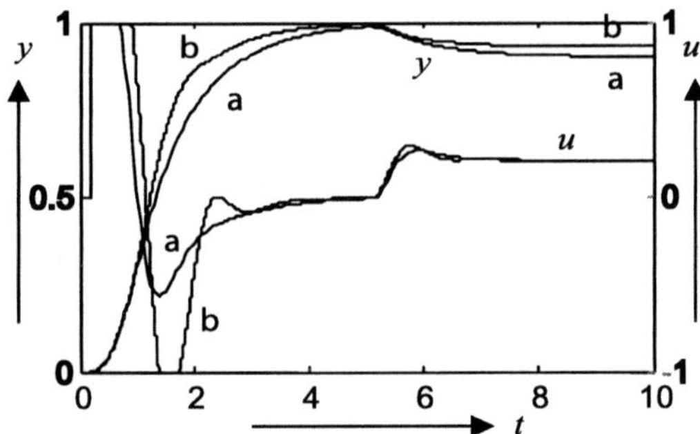

Fig. 5. Double integrator control for $T_d = 0.2$. a) Controller base on the pole decomposition for $c = 1.2$; b) controller (39) with $\alpha_e = -0.281/T_d$. For $t = 5$ a constant input disturbance $v = -0.2$ occurred.

Another interesting feature is that the controllers based on decreasing the distance from RBC measured in velocity are faster than those based on difference in the position (Figure 6).

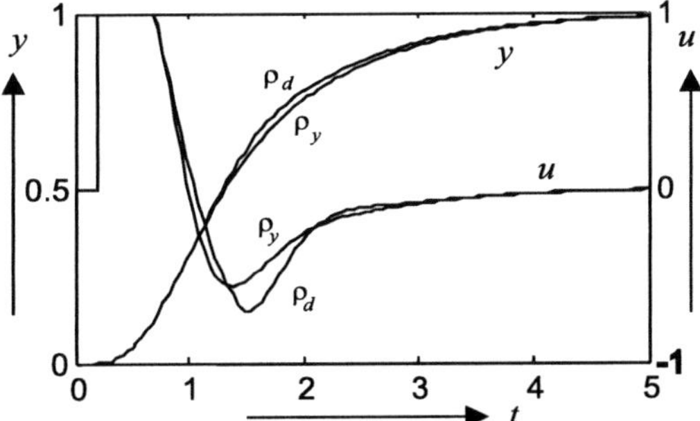

Fig. 6. Comparing controllers based on decreasing the distance $\rho_y = y - y_b$ and $\rho_d = d - d_b$ with equivalent poles $\alpha_e = -0.281/T_d$; $T_d = 0.2$.

Coming back to the question—why it is not possible to work with the equivalent poles derived for the first-order time-delayed systems, when the constrained controller is based on splitting the second-order dynamics into two first-order ones—we tried to rethink this point. Use of both controllers based on the distance decrease (Figure 7) shows that by applying the first-order setting (51) the number of the control pulses (dynamical class of control) increases. However, the output remains monotonous. Considering the fact that a dead time brings theoretically infinite number of modes into control, the increase of the number of control intervals seems to be acceptable. Both solutions can preserve specific control features. In order to come to more general conclusions, we will analyze in detail control of higher-order dynamical systems, which could show some regularities in the nature of these processes. So, at this moment it is possible to conclude that the problem of the optimal distance choice and of the optimal tuning of constrained controllers for time delayed systems is not definitely closed.

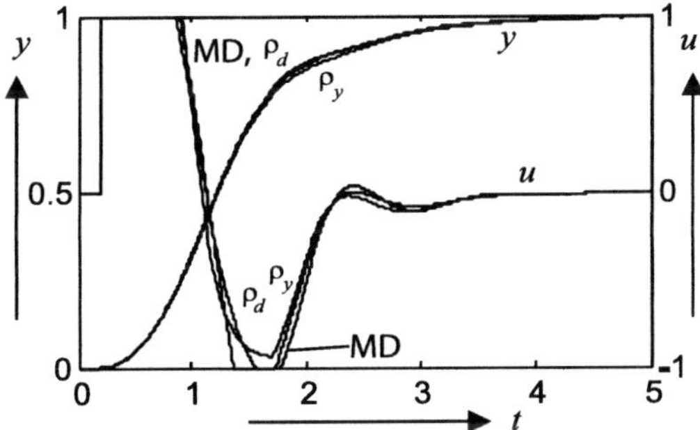

Fig. 7. Comparing controllers based on decreasing the distance $\rho_y = y - y_b$ and $\rho_d = d - d_b$ with equivalent poles $\alpha_e = -0.386/T_d$ and with the controller based on the mode decomposition (MD) with poles $\alpha_2 = -0.386/T_d$, $c = 1.2$; $T_d = 0.2$.

5 Controller tuning in a general case

Paper [7] already showed how it is possible to apply algorithms derived for the double integrator for controlling time-delayed and higher-order systems by tuning such controllers by step response-based experiments. The controller tuning requires determination of the plant gain K_s and of one of the time delays T_d, or T_1. Both can be determined according to the approximation of the measured process reaction curve by the I_2T_d-model or by the I_2T_1-model. Together with the "ultimate sensitivity method" based on identification of the approximative (I_2T_d or I_2T_1) using the measurement at the stability border [10], these approaches can be considered as an extension of the well-known method by Ziegler and Nichols [16]. Despite the fact that it was derived experimentally, it can be explained by considering the I_1T_d and I_1T_1-approximations of the controlled processes. In many situations it is also possible to use the relay identification method popularized by [1], which can be modified also to work with the second-order time-delayed models (see, e.g., [11]).

6 Illustrative example

Although the best demonstration of the effectiveness of the new approach would be to control a real plant, we will do so in a simulation example that can be easily repeated by the reader. Let the problem be to design a constrained controller with $u_r \in [U_{r1}, U_{r2}] = [0, 1.5]$ for the plant

$$S(s) = \frac{1}{\left(s^2 + 0.1s + 1\right)\left(0.2s + 1\right)}, \tag{64}$$

given just by the step response. Approximating its initial phase by the I_2T_d-model we get, e.g., $K_s = 0.76$; $T_d = 0.1$ (Figure 8). It is also important to measure the static gain $K = 1$ by evaluating the steady state input-output relation.

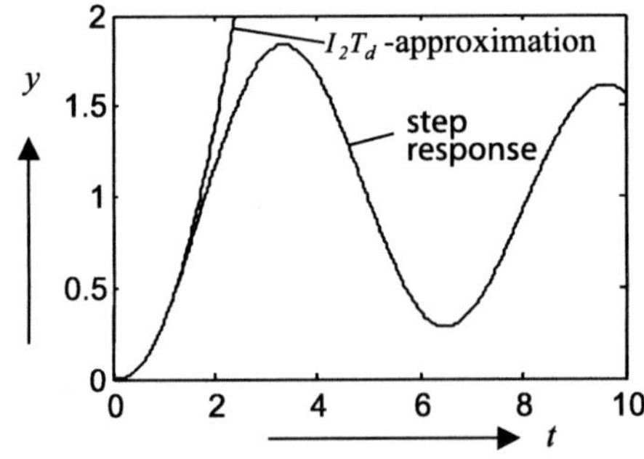

Fig. 8. Approximation of the plant step response by the I_2T_d-model.

To get a zero steady state error in controlling static plants it is necessary to add a feedforward component part of control u_f, when

$$u_{tot} = u/K_s + u_f \; ; \;\; u_f = r/K \in [U_{r1}, U_{r2}] \, , \tag{65}$$

whereby r is a permissible piecewise constant reference signal fulfilling (65). Use of a feedforward part, however, changes the limit control values available to the controller for the transients to

$$U_{ci} = K_s \left(U_{ri} - u_f \right); \; i = 1, 2 \, . \tag{66}$$

Transients corresponding to the controller (39) for both settings (62) and (51) are shown in Figure 9. It is shown that the output responses remain monotonous while the control signal actively compensates the oscillatory plant behaviour.

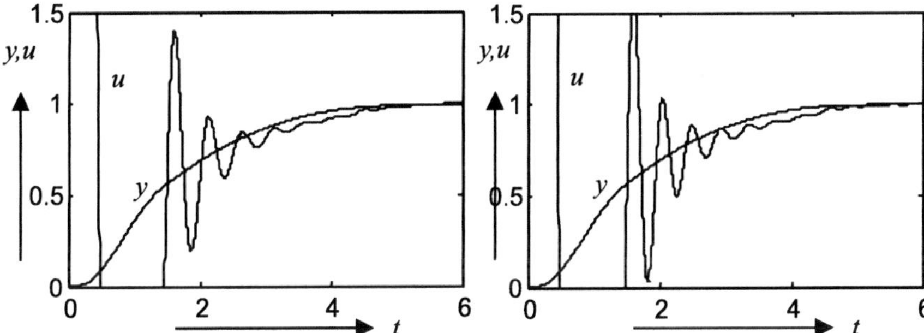

Fig. 9. Closed-loop control of the plant (64) approximated according to Figure 8 corresponding to the controller (65) with the closed-loop poles (62) (left) and (51) (right): $u_{tot} \in [0, 1.5]$.

7 Conclusions

The presented approach to constrained control design based on generalization of the linear pole assignment control is powerful and efficient because of its simplicity and possible ability to deal with time-delayed systems. Furthermore, the results derived for the time-delayed single and double integrators can be successfully applied to a broad class of linear and nonlinear higher-order systems by procedures that can be considered as a modification or extension of the well-known method by Ziegler and Nichols. Similarly, simple and easy applicable procedures are derived for the plant approximations using the relay experiments.

This chapter brings, however, an introduction to constrained pole assignment control design. The topics not considered here due to the limited space include the case of complex poles, introduction of windupless I-action, nonlinear constrained control design, active compensations of time delays, and discrete-time controllers with both pulse amplitude and pulse width modulation. With the exception of the complex closed-loop poles that were treated in [14], [12] and systems with pulse width modulation, these topics can be found in [10].

References

1. Aström, KJ, Hägglund T (1984). Automatic tuning of simple regulators with specification of phase and amplitude margins. Automatica 20(5):645–651
2. Bemporad A, Morari M, Dua V, Pistikopoulos EN (2002). The explicit linear quadratic regulator for constrained systems. Automatica 38:3–20
3. Föllinger O (1994). Regelungstechnik. 8. Auflage, Hüthig Buch Verlag Heidelberg
4. Henrion D, Tarbouriech S, Kučera V (2001). Control of linear systems subject to input constraints: a polynomial approach. Automatica 37:597–604
5. Hippe P (2003). A new windup prevention scheme for stable and unstable systems. In: Proceedings of 2nd IFAC Conference on Control Systems Design '03, Bratislava, Slovak Republic
6. Huba M (1999). Dynamical classes in the minimum time pole assignment control. In: Computing anticipatory systems–CASYS '98. American Institute of Physics, Woodbury, NY
7. Huba M, Bisták P, Skachová Z, Žáková K (1998). P- and PD-controllers for I1 and I2 models with dead time. 6th IEEE Mediterranean Conference on Control and Systems, Alghero, Italy 514–519
8. Huba M, Bisták P (1999). Dynamic classes in the PID control. In: Proceedings of the 1999 American Control Conference, San Diego, CA
9. Huba M, Sovišová D, Oravec I (1999). Invariant sets based concept of the pole assignment control. In: European Control Conference, VDI/VDE Düsseldorf, Germany
10. Huba M (2003). Constrained systems design. Vol. 1, Basic controllers. Vol. 2, Basic structures. STU Bratislava (in Slovak)
11. Huba M (2003). The helicopter rack control, 11th Mediterranean Conference on Control and Automation, Rhodes, Greece
12. Huba M. (2004). Gain scheduled constrained controller for SISO plants. In: Southeastern Europe, USA, Japan and EC Workshop on Research and Education in Control and Signal Processing REDISCOVER 2004, Cavtat, Croatia, 1:104–107
13. Huba M, Bisták P (2004). Constrained controllers based on the dynamics decomposition. In: Southeastern Europe, USA, Japan and EC Workshop on Research and Education in Control and Signal Processing REDISCOVER 2004, Cavtat, Croatia, 1:49–52
14. Huba M (2004). Gain scheduled constrained controller for SISO plants. IFAC Conf. NOLCOS 2004, Stuttgart
15. Rönnbäck S (1996). Nonlinear dynamic windup detection in anti-windup compensators. Preprints CESA '96, Lille, 1014–1019

16. Ziegler JG, Nichols NB (1942). Optimum settings for automatic controllers. Trans. ASME, 759–768

An Overview of Finite-Time Stability

Peter Dorato

Department of Electrical and Computer Engineering
University of New Mexico
Albuquerque, NM 87131, USA
peter@ece.unm.edu

Summary. Finite-time stability (FTS) is a concept that was first introduced in the 1950s. The FTS concept differs from classical stability in two important ways. First, it deals with systems whose operation is limited to a *fixed finite* interval of time. Second, FTS requires *prescribed bounds* on system variables. For systems that are known to operate only over a finite interval of time and whenever, from practical considerations, the systems' variables must lie within specific bounds, FTS is the only meaningful definition of stability. This overview will first present a short history of the development of the concept of FTS. Then it will present some important analysis and design results for linear, nonlinear, and stochastic systems. Finally some applications of the theory will be presented.

1 Introduction

Classical stability concepts, e.g., Lyapunov stability, asymptotic stability, bounded-input-bounded-output (BIBO) stability, all deal with systems operating over an *infinite* interval of time. In addition, while classical stability concepts require that system variables be bounded, the values of the bounds are not prescribed. The term *practical stability* was introduced for systems operating over an infinite time interval with *prescribed* bounds, in the text of LaSalle and Lefschetz [37], published in 1961. Earlier, however (1953–1954), papers were published in the Russian literature, e.g., [29], [38], dealing with both prescribed bounds and finite-time intervals, under the title of *finite-time stability*. The term *short-time stability* has also been used for finite-time stability, see, for example, [11], but the more common terminology in the literature remains *finite-time stability*. It should be noted that the term *finite-time stability*, with a very different meaning than that considered here, has also appeared in the literature, e.g., [26], [7]. In these latter publications, the term FTS is used to describe systems whose state approaches zero in a finite time.

We next summarize some key classical and finite-time stability definitions.
Definition 1 [25]:
The equilibrium point, $x_e = 0$, of the system, $\dot{x} = f(x,t)$, is said to be

Lyapunov stable if given a $\beta > 0$, there exists an $\alpha > 0$, which in general depends on β and t_0, such that

$$||x(t_0)|| < \alpha \Rightarrow ||x(t)|| < \beta, \quad \text{for all } t \geq t_0 \,.$$

▲

Definition 2 [37]:
The equilibrium point, $x_e = 0$, of the system, $\dot{x} = f(x,t) + p(x,t)$, is said to be *practically stable*, with respect to sets Q_0, Q and real variable $\delta > 0$, if

$$x(t_0) \in Q_0 \text{ and } ||p(x,t|| \leq \delta \Rightarrow \text{ for all } x(t) \in Q, \quad t \geq t_0 \,.$$

Note that in contrast to Lyapunov stability, practical stability requires that sets Q_0 and Q be independently prespecified. These sets are generally determined from practical considerations.   ▲

Definition 3 [29]:
The nonlinear system, $\dot{x} = f(x,t)$, is said to be *finite-time stable (FTS)* with respect to given positive real numbers (α, β, T), $\alpha < \beta$, if

$$||x(t_0)|| < \alpha \Rightarrow ||x(t)|| < \beta, \quad \text{for all } t \in [t_0, t_0 + T) \,.$$

Note that in contrast to practical stability, FTS is defined over a specified finite interval of time. Also FTS is defined with respect to a specific norm, and initial time. As mentioned earlier, the term *short-time stability* has also been used for FTS, see, for example, Dorato [11], D'Angelo [10], and Richards [44]. Since all real systems operate over finite intervals of time, however long, the term "short-time" helps distinguish between classical and FTS. Finally, note that for FTS, the set of initial states may include equilibrium points other than zero.   ▲

Definition 4 [53]:
The nonlinear system, $\dot{x} = f(x,t)$, is said to be *(finite-time) contractively stable* with respect to positive real numbers $(\alpha, \beta, \gamma, T)$, if it is FTS with respect to (α, γ, T), and there exists a $t_1 \in (t_0, t_0 + T)$ such that

$$||x(t)|| < \beta, \quad t_1 < t < t_0 + T$$

with $\beta < \alpha$. A simple interpretation of t_1 is that of a *settling time* over the finite interval $(t_0, t_0 + T)$.   ▲

Definition 5 [52]:
The nonlinear system, $\dot{x} = f(x,t) + u(x,t)$, is said to be *stable under perturbing forces*, or *finite-time bounded (FTB)* [6], with respect to positive real numbers $(\alpha, \beta, \epsilon, T)$ if

$$||x(t_0)|| < \alpha \text{ and } ||u(x,t)|| \leq \epsilon \Rightarrow ||x(t)|| < \beta, \quad \text{for all } t \in [t_0, t_0 + T) \,.$$

In D'Angelo [10], the term *short-time nonresonance* is used for FTB in linear systems.   ▲

One variation in Definition 5, with respect to the definition given in [52], is that the perturbation term, $u(x,t)$, is assumed to be uniformly bounded by ϵ for all x and t. This assumption makes the definitions in [52] and [6] compatible. Finally it should be noted that *practical stability* and *stability under perturbation forces* concepts are very similar, the distinction being that practical stability is defined over an infinite time interval, rather than a finite time interval.

Definition 6 [34]:
The stochastic nonlinear system, $dx = f(x)dt + \sigma(x)dz$, where z is a Wiener process, is said to be *finite-time stochastically stable (FTSS)*, with respect to $(Q_1, Q_2, 1 - \lambda, T)$, where Q_1 and Q_2 are given sets, and λ and T are given positive real numbers, with $\lambda < 1$, if

$$x(0) \in Q_1 \Rightarrow Pr\{x(t) \in Q_2, 0 \le t \le T\} \ge 1 - \lambda.$$

▲

The probability that $x(t) \in Q_2$ over the interval $[0, T]$ has several names; it is referred to as *inclusion probability* by Van Mellaert [47] and *containment probability* by Wonham (Section VII-B of [54]).

2 A short history of FTS

In 1953, the concept of FTS was first introduced by Kamenkov [29] in an article published in the Russian journal, PMM (*Journal of Applied Mathematics and Mechanics*). Other related articles appeared shortly thereafter in the same journal, such as the articles by Lebedev [30] [38] [39] in 1954 and Chzhan-Sy-In [8] [9] in 1959. However, it was not until a discussion of FTS appeared in the translation of Hahn's text *Theorie und Anwendung der Direkten Methode von Ljapunov*, published in 1963, that the results of Kamenkov and Lebedev became available in English. The translation of PMM from Russian to English initiated in 1958 made available the 1959 results of Chzhan-Sy-In to English readers. These early Russian papers on FTS dealt with both linear and nonlinear systems.

In 1961 several studies were reported on FTS for linear time-varying systems, such as those by Dorato [11] [12] [13], under the title, *short-time stability*. The term "short-time stability" is also used in the texts of D'Angelo [10] and Richards [44] for stability over finite time. In 1961, the related concept of *practical stability* was introduced in the text of LaSalle and Lefschetz [37]. The two concepts, *finite-time stability* and *practical stability*, have in common the specification of bounds, but differ in the size of the time interval of interest. The text of Lakshmikantham, Leela, and Martynyuk [35] deals in detail with practical stability, while articles have been published that combine a discussion of FTS and practical stability. See, for example, [41], [42], [19], and [20].

In 1965, Weiss and Infante [53] published a very complete analysis of FTS for nonlinear systems, including the concept of finite-time *contractive stability*. Shortly thereafter Weiss and Infante published an extension of FTS to nonlinear systems in the presence of perturbation signals [52], leading to the concept of finite-time BIBO stability, now commonly referred to as *finite-time bounded (FTB) stability* [6]. In 1966, Kushner [34] investigated the concept of FTS for stochastic systems, with more details presented in Chapter III of his text on *Stochastic Stability and Control* [33]. Finally, in 1969, Michel and Wu [43] extended many of the existing analysis results on FTS for continuous-time systems to discrete-time systems. During the period 1965–1975, numerous other results were reported on FTS, e.g., [28], [32], [49], [24], [23], [51], [50], [27], [31], [36]. All of these results, however, were limited to the *analysis* of given systems, rather than *control design/synthesis*.

In 1969, Garrard [17] presented a paper at the 4th IFAC Congress on finite-time control system synthesis for nonlinear systems, which was further expanded into a journal article [18] in 1972. In 1972, another journal article by Van Mellaert and Dorato [48] appeared on FTS design. This article dealt with designs for stochastic systems, based on the doctoral thesis of the first author, which appeared in 1967 [47]. Finally, during this period, a study was published by San Filippo and Dorato [45] on the robust design of linear systems with a linear-quadratic performance and FTS, as applied to a flight control problem. This study was based on the doctoral thesis of the first author, San Filippo, published in 1973 [46]. In a conference paper, Grujic [21] applied FTS concepts to the design of adaptive systems (see also Grujic [22]). Most of the design techniques presented during the 1969–1976 period were computationally very intensive. It was not until recently that computable design algorithms were reported, at least for linear systems.

In 1997, Dorato, Abdallah, and Famularo [16] presented a study at the 36th IEEE CDC, on the robust FTS design for linear systems, using linear matrix inequalities (LMIs) to compute state-feedback control laws. Further LMI-based design results for linear systems were presented in Amato et al. [5], [2], [3], and [6]. In 2002, Abdallah et al. presented statistical-learning techniques for FTS design with static output-feedback. Recently, discrete-time FTS design techniques have been applied to the control of ATM networks and to network controlled systems (see Amato et al. [1] and Mastellone [40]).

3 Some Finite-time Analysis Results

In this section we summarize a few key results for FTS analysis.

Theorem 1 (Dorato [11]): The linear time-varying system, $\dot{x} = A(t)x$, is FTS

(Definition 3), with respect to (α, β, T), if

$$\int_{t_0}^{t} \Lambda(\tau)d\tau \leq ln\frac{\beta}{\alpha}$$

for $t \in [t_0, t_0 + T)$, where $\Lambda(t)$ is the maximum eigenvalue of $\frac{1}{2}(A'(t) + A(t))$.
∎

Theorem 2 (Weiss and Infante [52]):
A nonlinear system, $\dot{x} = f(x,t) + u(x,t)$, is *stable under perturbing forces*
(Definition 5), with respect to $(\alpha, \beta, \epsilon, T)$, if there exists a real-valued function
$V(x,t)$ and real-valued functions $\phi(t), \rho(t)$ integrable on $\mathcal{J}$ such that

1. $||\frac{\partial V(x,t)}{\partial x}|| \leq \rho(t)$ for all $x \in \mathcal{R}, t \in \mathcal{J}$
2. $\dot{V}_f(x,t) < \phi(t)$ for all $x \in \mathcal{R}, t \in \mathcal{J}$
3. $\int_{t_1}^{t_2}[\phi(t) + \epsilon\rho(t)]dt \leq V_m^{\beta}(t_2) - V_M^{\alpha}(t_1)$, $t_1, t_2 \in \mathcal{J}$

where
$\mathcal{J} = [t_0, t_0 + T)$
$\mathcal{R} = \{x; \alpha < ||x|| \leq \beta\}$
$V_f(x,t) = \frac{\partial V(x,t)}{\partial x}f(x,t) + \frac{\partial V}{\partial t}$
and

$$V_M^{\alpha}(t) = \max_{||x||=\alpha} V(x,t), \quad V_m^{\beta}(t) = \min_{||x||=\beta} V(x,t).$$

∎

If ϵ is set equal to zero, the above theorem provides a theorem for FTS (Definition 3). Note that the function $V(x,t)$ in this theorem, unlike classical Lyapunov functions, is not required to be positive definite. In addition, $\dot{V}$ is not required to be negative semi-definite. A theorem similar to Theorem 2 is given in [43] for discrete-time systems. Also, in [14], Theorem 2 is applied to linear time-varying systems of the form, $\dot{x} = A(t)x + b(t)u(t)$, where $u(t)$ is a scalar perturbation term, with magnitude bounded by ϵ. There is, however, an error in the results stated in [14]: the function $\rho(t)$ should be 2β, and the term $\epsilon\rho(t)$ in Part 3 of Theorem 2 should be replaced by $\epsilon\Lambda_b^{1/2}(t)(2\beta)$. The condition for stability (equation (3) in [14]) should read

$$\int_{t_1}^{t_2}(\Lambda_A(t)\beta^2 + 2\epsilon\Lambda_b^{1/2}(t)\beta)dt \leq (\beta^2 - \alpha^2).$$

Theorem 3 (Kushner [34]):
The stochastic nonlinear system, $dx = f(x)dt + \sigma(x)dz$, is *FTSS* (Definition 6) with respect to $(Q_1, Q_2, 1 - \lambda, T)$, if there exists a continuous non-negative function $V(x)$, such that

$$\mathcal{A}V(x) \leq \phi(t), \text{ for all } x \in Q_2$$

and

$$\frac{V(x) + \int_0^T \phi(\tau)d\tau}{q_2} \leq \lambda, \text{ for all } x \in Q_1$$

where $\mathcal{A}$ is the infinitesimal operator,

$$\mathcal{A} = \sum f_i \frac{\partial}{\partial x_i} + \frac{1}{2} \sum \sum S_{ij} \frac{\partial^2}{\partial x_i \partial x_j}$$

where $Q_1 = \{x; V(x) < q_1\}$, $Q_2 = \{x; V(x) < q_2\}$ and

$$S_{ij} = \sum_k \sigma_{ik} \sigma_{jk}.$$

$\blacksquare$

In [34] conditions are also given for discrete-time systems to be *stochastically finite-time stable*.

The above theorems are *analysis* results for FTS, which basically require the bounding of nonlinear functions over a given set. For *control design*, a design signal must be introduced in the system equations; for example, a nonlinear system needs to be written, $\dot{x} = f(x,t) + B(x,t)u$ where u is a control signal (not to be confused with the perturbation term $u(x,t)$ in Definition 5). In the next section some design results will be presented.

4 Some Finite-Time Design Results

In this section we summarize some FTS *design/synthesis* results.

Theorem 4 (Garrard [18]):
A control signal u in a nonlinear system of the form, $\dot{x} = f(x,t) + Bu$, will result in an FTS system (Definition 3) with respect to (α, β, T), if u satisfies the inequality

$$\sum_{i=1}^n x_i f_i + \sum_{i=1}^n x_i \sum_{j=1}^m b_{ij} u_i - \frac{1}{T} ln\left(\frac{\beta}{\alpha}\right) \sum_{i=1}^n x_i^2 \leq 0, \text{ for all } x \in \mathcal{R},$$

where n is the dimension of the state x, m is the dimension of the control vector u, and $\mathcal{R}$ is as defined in Theorem 2. $\blacksquare$

It is interesting to note that this design result follows from a V function of the form, $V(x) = ln(\frac{||x||}{\alpha})^2$.

To design a control signal from Theorem 4 is, in general, computationally very difficult. If the nonlinearities $f_i(x,t)$ and the function $V(x)$ are multivariate polynomial functions, the design problem is then reduced to quantifier-elimination for multivariate polynomial inequalities (MPIs). The objective is to eliminate the logic quantifier "for all" in the inequality of Theorem 4, to obtain a Boolean function involving only the control variable u. While still a difficult problem, some symbolic and numerical techniques are available to solve this quantifier elimination problem (see, for example, [15]).

Theorem 5 (Amato, Ariola, and Dorato [6]):
If for a given $a > 0$, there exits a matrix L and a symmetric matrix Q such that the LMIs

$$A(p_{(i)})Q + QA^T(p_{(i)}) + B(p_{(i)})L + L^T B^T(p_{(i)}) - aQ < 0$$

$$\frac{\alpha^2}{\beta^2}e^{aT}I < Q < I, \ \ I = \text{identity matrix}$$

have a feasible solution, then the control law, $u = LQ^{-1}x$, will make the linear uncertain system, $\dot{x} = A(p)x + B(p)u$, FTS (Definition 3), with respect to (α, β, T). ∎

In Theorem 5, the uncertainty in the A and B system matrices is represented by the parameter vector p, and the uncertainty is assumed to be multi-affine in p, with $p_{(i)}$ representing the corner points of the uncertainty set. Theorem 5 is taken from Corollary 9 in [6], with some change in notation, i.e., α, R, c_1, and c_2 in Corollary 9, replaced, respectively, by a, I, α^2, and β^2, to make the results compatible with the notation used here.

The reduction of the design problem to an LMI feasibility problem is computationally significant. However, it should be noted that Theorem 5, like all other theorems cited here, provide only *sufficient* conditions, so that the results may be overly conservative. For linear systems some *necessary and sufficient* conditions for FTS are given in [4]; however, these results are *analysis* results only.

5 Conclusions

While most systems must operate satisfactorily over arbitrarily large intervals of time, some systems are required to operate satisfactorily only over fixed finite intervals of time, e.g., missile systems, certain aircraft maneuvers, communication networks, robotic maneuvers. For the latter situation, the only meaningful concept of stability is FTS. We now have available many results for the *analysis* of finite-time stable systems. It is a bit surprising that more of these results have not been incorporated in recent textbooks on linear and nonlinear systems. There is a need, however, for more results in the *design* of finite-time stable systems, especially for nonlinear and stochastic systems. It would also be of interest to see more applications of FTS concepts to practical design problems.

References

1. Amato F, Ariola M, Abdallah C, Cosentino C (2002) Application of finite-time stability concepts to control of ATM networks. In: 40th Allerton Conf. on Communication, Control and Computers, Allerton, IL
2. Amato F, Ariola M, Abdallah C, Dorato P (1999a) Dynamic output feedback finite-time control of LTI systems subject to parametric uncertainties and disturbances. In: European Control Conference, Karlsruhe, Germany
3. Amato F, Ariola M, Abdallah C, Dorato P (1999b) Finite-time control for uncertain linear systems with disturbance inputs. 1776–1780. In: Proc. American Control Conf., San Diego, CA
4. Amato F, Ariola M, Cosentino C, Abdallah C, Dorato P (2003) Necessary and sufficient conditions for finite-time stability of linear systems. 4452–4456. In: Proc. American Control Conf., Denver, CO
5. Amato F, Ariola M, Dorato P (1998) Robust finite-time stabilization of linear systems depending on parameter uncertainties. 1207–1208. In: Proc. IEEE Conf. on Decision and Control, Tampa, FL
6. Amato F, Ariola M, Dorato P (2001) Finite-time control of linear systems subject to parametric uncertainties and disturbances. Automatica 37:1459–1463
7. Bhat S, Bernstein D (1998) Continuous finite-time stabilization of the translational and rotational double integrators. IEEE Trans. Automat. Contr. 43:678–682
8. Chzhan-Sy-In (1959a) On stability of motion for a finite interval of time. Journal of Applied Math. and Mechanics (PMM) 23:333–344
9. Chzhan-Sy-In (1959b) On estimates of solutions of systems of differential equations, accumulation of perturbation, and stability of motion during a finite time interval. Journal of Applied Math. and Mechanics 23:920–933
10. D'Angelo H (1970) Time-varying systems: analysis and design. Allyn and Bacon, Boston, MA
11. Dorato P (1961a) Short-time stability in linear time-varying systems. 83–87. In: IRE International Convention Record, Part IV
12. Dorato P (1961b) Short-time stability in linear time-varying systems, PhD thesis, Polytechnic Institute of Brooklyn
13. Dorato P (1961c) Short-time stability. IRE Trans. Automat. Contr. 6:86
14. Dorato P (1967) Comment on finite-time stability under perturbing forces and on product spaces. IEEE Trans. Automat. Contr. 12:340
15. Dorato P (2000) Quantified multivariate polynomial inequalities. IEEE Control Systems Magazine 20:48–58
16. Dorato P, Abdallah C, Famularo D (1997) Robust finite-time stability design via linear matrix inequalities. In: Proc. IEEE Conf. on Decision and Control, San Diego, CA
17. Garrard W (1969) Finite-time stability in control system synthesis. 21–31. In: Proc. 4th IFAC Congress, Warsaw, Poland
18. Garrard W (1972) Further results on the synthesis of finite-time stable systems. IEEE Trans. Automat. Contr. 17:142–144
19. Grujic L (1973) On practical stability. Int. J. Control 17:881–887
20. Grujic L (1975) Uniform practical and finite-time stability of large-scale systems. Int. J. Systems Sci. 6:181–195

21. Grujic L (1976) Finite-time adaptive control. In: Proc. 1976 JACC, Purdue University

22. Grujic L (1977) Finite time noninertial adaptive control. AIAA Journal 15:354–359

23. Gunderson R (1967a) Qualitative solution behavior on a finite time interval, PhD thesis, University of Alabama

24. Gunderson R (1967b) On stability over a finite interval. IEEE Trans. Automat. Contr. AC-12:634–635

25. Hahn W (1963) Theory and applications of Liapunov's direct method. Prentice-Hall, Englewood Cliffs, NJ

26. Haimo V (1986) Finite time controllers. SIAM J. Control and Optimization 24:760–770

27. Hallam T, Komkov V (1969) Application of Liapunov's functions to finite time stability. Revue Roumaine de Mathematiques Pure et Appliquees 14:495–501

28. Heinen J, Wu S (1969) Further results concerning finite-time stability. IEEE Trans. Automat. Contr. AC-14:211–212

29. Kamenkov G (1953) On stability of motion over a finite interval of time [in Russian]. Journal of Applied Math. and Mechanics (PMM) 17:529–540

30. Kamenkov G, Lebedev A (1954) Remarks on the paper on stability in finite time interval [in Russian]. Journal of Applied Math. and Mechanics (PMM) 18:512

31. Kayande A (1971) A theorem on contractive stability. SIAM J. Appl. Math. 21:601–604

32. Kayande A, Wong J (1968) Finite time stability and comparison principles. Proc. Cambridge Philosophical Society 64:749–756

33. Kushner H (1967) Stochastic stability and control. Academic Press, New York, NY

34. Kushner H (1966) Finite-time stochastic stability and the analysis of tracking systems. IEEE Trans. Automat. Contr. AC-11:219–227

35. Lakshmikantham V, Leela S, Martynyuk A (1990) Practical stability of nonlinear systems. World Scientific, Singapore

36. Lam L, Weiss L (1974) Finite time stability with respect to time-varying sets. J. Franklin Inst. 298:425–421

37. LaSalle J, Lefschetz S (1961) Stability by Liapunov's direct method. Academic Press, New York, NY

38. Lebedev A (1954a) The problem of stability in a finite interval of time [in Russian]. Journal of Applied Math. and Mechanics (PMM) 18:75–94

39. Lebedev A (1954b) On stability of motion during a given interval of time [in Russian]. Journal of Applied Math. and Mechanics (PMM) 18:139–148

40. Mastellone S (2004) Finite-time stability of nonlinear networked control systems, Master's thesis, University of New Mexico

41. Michel A (1970) Quantitative analysis of simple and interconnected systems: Stability, boundedness, and trajectory behavior. IEEE Trans. Circuit Theory CT-17:292–301

42. Michel A, Porter D (1972) Practical stability and finite-time stability of discontinuous systems. IEEE Trans. Circuit Theory CT-19:123–129

43. Michel A, Wu S (1969) Stability of discrete-time systems over a finite interval of time. Int. J. Control 9:679–694

44. Richards J (1983) Analysis of periodically time-varying systems. Springer-Verlag, Berlin

45. San Filippo F, Dorato P (1974) Short-time parameter optimization with flight control applications. Automatica 10:425–430

46. San Fillipo F (1973) Short time optimization of parametrically disturbed linear control system, PhD thesis, Polytechnic Institute of Brooklyn

47. Van Mellaert L (1967) Inclusion-probability-optimal control, PhD thesis, Polytechnic Institute of Brooklyn

48. Van Mellaert L, Dorato P (1972) Numerical solution of an optimal control problem with a probability criterion. IEEE Trans. Automat. Contr. AC-17:543–546

49. Watson J, Stubberud A (1967) Stability of systems operating in a finite time interval. IEEE Trans. Automat. Contr. AC-12:116

50. Weiss L (1969) On uniform and nonuniform finite time stability. IEEE Trans. Automat. Contr. AC-14:313–314

51. Weiss L (1968) Converse theorems for finite-time stability. SIAM J. Appl. Math. 16:1319–1324

52. Weiss L, Infante E (1967) Finite time stability under perturbing forces and on product spaces. IEEE Trans. Automat. Contr. AC-12:54–59

53. Weiss L, Infante E (1965) On the stability of systems defined over a finite time interval. Proc. of the National Academy of Sciences 54:440–448

54. Wonham W (1970) Random differential equations in control theory In: Bharucha-Reid A (ed), Probabilistic methods in applied mathematics, Volume 2, 132–208. Academic Press, New York

Finite-Time Control of Linear Systems:
A Survey

Francesco Amato,[1] Marco Ariola,[2] Marco Carbone,[3] and Carlo Cosentino[2]

[1]Corso di Laurea in Ingegneria Informatica e Biomedica, Dip. di Medicina
Sperimentale e Clinica, Università Magna Græcia, Via T. Campanella 115,
88100 Catanzaro, Italia. `amato@unicz.it`
[2]Dipartimento di Informatica e Sistemistica, Università degli Studi di Napoli
Federico II, Via Claudio 21, 80125 Napoli, Italia. `{ariola,carcosen}@unina.it`
[3]Dipartimento di Informatica, Matematica, Elettronica e Trasporti, Università
degli Studi Mediterranea di Reggio Calabria, Via Graziella, Loc. Feo di Vito,
89100 Reggio Calabria, Italia. `mcarbone@ing.unirc.it`

Summary. This chapter illustrates various finite-time analysis and design problems for linear systems. Most of this work deals with continuous-time systems. First, some conditions for finite-time stability and boundedness are presented; then we turn to the design problem. In this context, we consider both the state feedback and the output feedback synthesis. For both cases, we end up with some sufficient conditions involving linear matrix inequalities (both algebraic and differential). The last section of the chapter extends the previous results to discrete-time systems.

1 Introduction

The concept of finite-time control dates back to the 1960s, when the idea of finite-time stability (FTS) was introduced in the control literature [11], [12], [14]. A system is said to be finite-time stable if, given a bound on the initial condition, its state does not exceed a certain threshold during a specified time interval.

It is important to recall that FTS and Lyapunov asymptotic stability (LAS) are independent concepts; indeed, a system can be FTS but not LAS, and vice versa. While LAS deals with the behaviour of a system within a sufficiently long (in principle, infinite) time interval, FTS is a more practical concept, useful to study the behaviour of the system within a finite (possibly short) interval, and therefore it finds application whenever it is desired that the state variables do not exceed a given threshold (for example, to avoid saturations or the excitation of nonlinear dynamics) during the transients.

FTS in the presence of exogenous inputs leads to the concept of finite-time boundedness (FTB). In other words, a system is said to be FTB if, given a bound on the initial condition and a characterization of the set of admissible inputs, the state variables remain below the prescribed limit for all inputs in the set. It is clear that FTB implies FTS but the converse is not true.

FTS and FTB are open-loop concepts. The finite-time control problem concerns the design of a linear controller, which ensures the FTS or the FTB of the closed-loop system.

In this chapter we present some results concerning finite-time analysis and control of linear systems. Most of the chapter is devoted to continuous-time systems, while discrete-time systems are discussed in the last section.

Most of the conditions for FTS, FTB, and finite-time control involve linear matrix inequalities (LMIs) or differential linear matrix inequalities (DLMIs).

The chapter is divided as follows: in Section 2 we give some definitions of FTS and FTB for the continuous-time case and we state the problems that we want to solve. Section 3 is devoted to the analysis conditions for FTS and FTB, whereas in Section 4 and Section 5 we present some sufficient conditions for the design of state feedback and output feedback controllers, respectively, guaranteeing FTS and/or FTB. Finally in Section 6 we consider discrete-time systems.

2 Definitions and Problem Statement

In the following we will give the basic definitions concerning FTS and FTB; these definitions are from [4], [5], [8]. The definitions will refer to the time-varying, continuous-time case; they generalize to time-invariant systems in an obvious way. The corresponding definitions for discrete-time systems are given in Section 5. All matrices and vectors are of suitable dimensions.

Definition 1 (FTS). *Given three positive scalars c_1, c_2, T, with $c_1 < c_2$, and a positive definite symmetric matrix function $\Gamma(t)$ defined over $[0, T]$, the time-varying linear system*

$$\dot{x}(t) = A(t)x(t)\,, \quad x(0) = x_0 \tag{1}$$

is said to be FTS with respect to $(c_1, c_2, T, \Gamma(t))$, if

$$x_0^T \Gamma(0) x_0 \leq c_1 \Rightarrow x(t)^T \Gamma(t) x(t) < c_2 \quad \forall t \in [0, T]\,. \tag{2}$$

$\Diamond$

Remark 1. Lyapunov Asymptotic Stability (LAS) and FTS are independent concepts: a system that is FTS may be not LAS; conversely a LAS system could be not FTS if, during the transients, its state exceeds the prescribed bounds. $\triangle$

Definition 2 (FTB). *Given three positive scalars c_1, c_2, T, with $c_1 < c_2$, a positive definite symmetric matrix function $\Gamma(t)$ defined over $[0, T]$, and a class of signals $\mathcal{W}$, the time-varying linear system*

$$\dot{x}(t) = A(t)x(t) + G(t)w(t), \quad x(0) = x_0 \tag{3}$$

is said to be FTB with respect to $(c_1, c_2, \mathcal{W}, T, \Gamma(t))$ if

$$x_0^T \Gamma(0)x_0 \le c_1 \Rightarrow x(t)^T \Gamma(t)x(t) < c_2 \quad \forall t \in [0, T],$$

for all $w(\cdot) \in \mathcal{W}$. ◇

Remark 2 (Time-invariant case). In the time-invariant case, we will consider a *constant* matrix Γ. This will allow us to give some conditions, both for FTS and for FTB, expressed in terms of LMIs. △

Remark 3. An important difference between LAS and FTB is that, for a given set of initial states, LAS is a structural property of the system, depending only on the system matrix, while FTB also depends on the kind and amplitude of the inputs acting on the system. △

Note that FTS and FTB refer to open-loop systems. The next problem puts FTS and FTB in the design context.

Problem 1 (Finite-Time Control via State Feedback). Consider the time-varying linear system

$$\dot{x}(t) = A(t)x(t) + B(t)u(t) + G(t)w(t), \quad x(0) = x_0, \tag{4}$$

where $u(t)$ is the control input and $w(t)$ is the disturbance. Then, given three positive scalars c_1, c_2, T, with $c_1 < c_2$, a positive definite symmetric matrix function $\Gamma(t)$ defined over $[0, T]$, and the class of signals $\mathcal{W}$, find a state feedback controller in the form

$$u(t) = K(t)x(t), \tag{5}$$

such that the closed-loop system obtained by the connection of (4) and (5), namely

$$\dot{x}(t) = (A(t) + B(t)K(t)) x(t) + G(t)w(t), \quad x(0) = x_0,$$

is FTB with respect to $(c_1, c_2, \mathcal{W}, T, \Gamma(t))$. ◇

Problem 2 (Finite-Time Control via Output Feedback). Consider the time-varying linear system

$$\dot{x}(t) = A(t)x(t) + B(t)u(t) + G(t)w(t), \quad x(0) = x_0 \tag{6a}$$
$$y(t) = C(t)x(t) + H(t)w(t), \tag{6b}$$

where $u(t)$ is the control input, $w(t)$ is the disturbance, and $y(t)$ is the output. Then, given three positive scalars c_1, c_2, T, with $c_1 < c_2$, two positive definite symmetric matrix functions $\Gamma(t)$, $\Gamma_K(t)$ defined over $[0, T]$, and a class of signals $\mathcal{W}$, find a dynamical output feedback controller in the form

$$\dot{x}_c(t) = A_K(t)x_c(t) + B_K(t)y(t) \tag{7a}$$
$$u(t) = C_K(t)x_c(t) + D_K(t)y(t), \tag{7b}$$

where $x_c(t)$ has the same dimension of $x(t)$, such that the closed-loop system obtained by the connection of (6) and (7) is FTB with respect to $(c_1, c_2, \mathcal{W}, T, \text{blockdiag}(\Gamma(t), \Gamma_K(t)))$. $\diamondsuit$

Note that we assume, for the sake of simplicity, that the weighting matrix does not contain cross-coupling terms between the system state and the controller state. Moreover, the definition of c_1 and c_2 in Problem 2 must take into account the augmented state of the closed-loop system.

Obviously, by letting $w(t) = 0$ in Problems 1 and 2, we obtain the corresponding finite-time stabilization problems.

3 Main Results: Analysis

In this section we will give some conditions for FTS and FTB of a continuous-time linear system. We will first present some conditions, both necessary and sufficient and just sufficient, expressed in form of DLMIs. Then, for the time-invariant case, we will present some sufficient conditions that can be expressed in form of LMIs.

3.1 Analysis conditions expressed in form of DLMIs

The following theorem gives three necessary and sufficient conditions for FTS of system (1) and one sufficient condition.

Theorem 1 ([4]). *The following statements are equivalent:*

i) System (1) is FTS wrt $(c_1, c_2, T, \Gamma(t))$.
ii) For all $t \in [0, T]$

$$\Phi(t, 0)^T \Gamma(t) \Phi(t, 0) < \frac{c_2}{c_1} \Gamma(0),$$

where $\Phi(t, 0)$ is the state transition matrix.
iii) For all $t \in [0, T]$ the differential Lyapunov inequality with terminal and initial conditions

$$\dot{P}(\tau) + A(\tau)^T P(\tau) + P(\tau)A(\tau) < 0, \quad \tau \in]0, t]$$
$$P(t) \geq \Gamma(t)$$
$$P(0) < \frac{c_2}{c_1} \Gamma(0),$$

admits a symmetric solution $P(\cdot)$.

Moreover, the following condition is sufficient for FTS.

iv) The differential Lyapunov inequality

$$\dot{P}(t) + A(t)^T P(t) + P(t)A(t) < 0$$
$$P(t) \geq \Gamma(t), \quad \forall t \in [0, T]$$
$$P(0) < \frac{c_2}{c_1}\Gamma(0)$$

admits a symmetric solution $P(\cdot)$.

Remark 4. Condition iii) is proven by imposing that the norm of the operator mapping the initial condition to the state at time t is less than the ratio c_2/c_1. Moreover, condition iv) is readily seen to imply condition iii).   △

Remark 5. Note that condition ii) is useful for the analysis; however, it cannot be used for design purposes. On the other hand condition iii) requires us to check infinitely many DLMIs. Therefore, the starting point for the solution of the design problems will be condition iv).   △

The following result deals with FTB in the presence of square integrable inputs.

Theorem 2 ([4]). *Consider the following class of signals:*

$$\mathcal{W} := \left\{ w(\cdot) \mid w(\cdot) \in \mathcal{L}^2([0, T]), \ \int_0^T w(\tau)^T w(\tau)\, d\tau \leq d \right\},$$

where $\mathcal{L}^2([0, T])$ is the set of square integrable vector-valued functions in $[0, T]$ and d is a positive scalar. Then system (3) is FTB wrt to $(c_1, c_2, \mathcal{W}, T, \Gamma(t))$ if there exists a symmetric matrix-valued function $P(\cdot)$ such that

$$\dot{P}(t) + A(t)^T P(t) + P(t)A(t) + \frac{c_1 + d}{c_2}P(t)G(t)G(t)^T P(t) < 0$$
$$P(t) \geq \Gamma(t), \quad \forall t \in [0, T]$$
$$P(0) < \frac{c_2}{c_1 + d}\Gamma(0).$$

Note that the differential Riccati inequality in Theorem 2 can be converted to a DLMI by using Schur complements.

3.2 Analysis conditions expressed in form of LMIs

The following theorems state *sufficient* conditions for the FTS and the FTB of a linear time-invariant system in the form

$$\dot{x}(t) = Ax(t) + Gw, \tag{8}$$

where, in this case, w is a constant input.

The approach makes use of Lyapunov-type functions which, in the case of finite-time boundedness, also depend on the disturbance vector. Roughly speaking, the Lyapunov function is derived along the system trajectories and a bound is found for the state at time t.

Theorem 3 ([8]). *Let Γ a positive definite matrix. System (8) with $w \equiv 0$ is FTS with respect to (c_1, c_2, T, Γ) if, letting $\tilde{Q} = \Gamma^{-\frac{1}{2}} Q \Gamma^{-\frac{1}{2}}$, there exist a nonnegative scalar α and a symmetric positive definite matrix Q such that*

$$A\tilde{Q} + \tilde{Q}A^T - \alpha\tilde{Q} < 0 \tag{9a}$$

$$\mathrm{cond}(Q) < \frac{c_2}{c_1}e^{-\alpha T}, \tag{9b}$$

where $\mathrm{cond}(Q) = \frac{\lambda_{\max}(Q)}{\lambda_{\min}(Q)}$ *indicates the condition number of Q.*

Theorem 4 ([8]). *Consider the following class of constant signals:*

$$\mathcal{W} := \left\{ w \,|\, w^T w \le d \right\},$$

where d is a nonnegative scalar. Then system (8) is FTB with respect to $(c_1, c_2, \mathcal{W}, T, \Gamma)$ if, letting $\tilde{Q}_1 = \Gamma^{-\frac{1}{2}} Q_1 \Gamma^{-\frac{1}{2}}$, there exist a positive scalar α and two symmetric positive definite matrices Q_1 and Q_2 such that

$$\begin{pmatrix} A\tilde{Q}_1 + \tilde{Q}_1 A^T - \alpha\tilde{Q}_1 & GQ_2 \\ Q_2 G^T & -\alpha Q_2 \end{pmatrix} < 0 \tag{10a}$$

$$\frac{c_1}{\lambda_{\min}(Q_1)} + \frac{d}{\lambda_{\min}(Q_2)} < \frac{c_2 e^{-\alpha T}}{\lambda_{\max}(Q_1)}, \tag{10b}$$

where $\lambda_{\max}(\cdot)$ and $\lambda_{\min}(\cdot)$ indicate the maximum and minimum eigenvalue of the argument, respectively.

Remark 6. In Theorem 3 we allow α to be zero; this is not permitted in Theorem 4. It can be easily seen that if we can guarantee that the conditions in Theorem 3 are satisfied for $\alpha = 0$, then the system $\dot{x} = Ax$ is *also* asymptotically stable. In Theorem 4, instead, if $\alpha = 0$, we cannot guarantee the negative definiteness of (10a). $\triangle$

It is easy to check that condition (9b) is guaranteed by imposing the condition

$$I < Q < \frac{c_2}{c_1}e^{-\alpha T},$$

whereas condition (10b) is guaranteed by imposing the conditions

$$\lambda_1 I < Q_1 < I \tag{11a}$$

$$\lambda_2 I < Q_2 < \lambda_3 I \tag{11b}$$

$$c_1/\lambda_1 + d/\lambda_2 < c_2 e^{-\alpha T} \tag{11c}$$

for some positive numbers λ_1, λ_2, and λ_3. Inequality (11c) can be converted to an LMI using Schur complements. It is equivalent to

$$\begin{pmatrix} c_2 e^{-\alpha T} & \sqrt{c_1} & \sqrt{d} \\ \sqrt{c_1} & \lambda_1 & 0 \\ \sqrt{d} & 0 & \lambda_2 \end{pmatrix} > 0.$$

Therefore, from a computational point of view, once we have fixed a value for α, the feasibility of the conditions stated in Theorems 3 and 4 can be turned into LMI-based feasibility problems [10].

4 State Feedback Controller Design

In this section we will provide some sufficient conditions that allow us to design state feedback controllers that render a continuous-time linear system either FTS or FTB. As in Section 3, we will first present conditions expressed in form of DLMIs and, then, for the time-invariant case, some conditions that can be expressed in form of LMIs.

The proofs are obtained by considering the corresponding analysis conditions, in which $A + BK$ replaces A, and using the change of variable, first suggested in [9], $KQ = L$.

4.1 State feedback design conditions expressed in form of DLMIs

The first result deals with finite-time stabilization.

Theorem 5 ([4]). *System* (4) *with* $w(t) \equiv 0$ *is finite-time stabilizable via state feedback with respect to* $(c_1, c_2, T, \Gamma(t))$ *if there exist a symmetric matrix-valued function* $Q(\cdot)$ *and a matrix function* $L(\cdot)$ *such that*

$$- \dot{Q}(t) + A(t)Q(t) + Q(t)A(t)^T + L(t)^T B(t)^T + B(t)L(t) < 0$$

$$Q(t) \leq \Gamma^{-1}(t), \quad \forall t \in [0, T]$$

$$Q(0) > \frac{c_1}{c_2} \Gamma^{-1}(0).$$

In this case a controller gain that solves Problem 1 with $w(t) \equiv 0$ *is* $K(t) = L(t)Q^{-1}(t)$.

Next we consider the solution of the more general Problem 1 with $w(t) \neq 0$.

Theorem 6 ([4]). *Consider the following class of signals:*

$$\mathcal{W} := \left\{ w(\cdot) \mid w(\cdot) \in \mathcal{L}^2([0, T]), \ \int_0^T w(\tau)^T w(\tau) \, d\tau \leq d \right\}.$$

Then Problem 1 is solvable if there exist a symmetric matrix-valued function $Q(\cdot)$ *and a matrix function* $L(\cdot)$ *such that*

$$- \dot{Q}(t) + A(t)Q(t) + Q(t)A(t)^T$$

$$+ L(t)^T B(t)^T + B(t)L(t) + \frac{c_1 + d}{c_2} G(t)G(t)^T < 0$$

$$Q(t) \leq \Gamma^{-1}(t), \quad \forall t \in [0, T]$$

$$Q(0) > \frac{c_1 + d}{c_2} \Gamma^{-1}(0).$$

In this case a controller gain that solves Problem 1 is $K(t) = L(t)Q^{-1}(t)$.

4.2 State feedback design conditions expressed in form of LMIs

In this section we will consider the solution to Problem 1 when all the system matrices are time-invariant. Therefore, we will consider the following system:

$$\dot{x}(t) = Ax(t) + Bu(t) + Gw. \tag{12}$$

As usual, we first consider finite-time stabilization.

Theorem 7 ([8]). *System (12) with $w \equiv 0$ is finite-time stabilizable via state feedback with respect to (c_1, c_2, T, Γ) if, letting $\tilde{Q} = \Gamma^{-\frac{1}{2}} Q \Gamma^{-\frac{1}{2}}$, there exist a nonnegative scalar α, a symmetric positive definite matrix Q, and a matrix L such that*

$$A\tilde{Q} + \tilde{Q}A^T + BL + L^T B^T - \alpha\tilde{Q} < 0$$

$$\operatorname{cond}(Q) < \frac{c_2}{c_1} e^{-\alpha T}.$$

In this case a controller gain that solves Problem 1 with $w \equiv 0$ is $K = L\tilde{Q}^{-1}$.

Then we consider the case in which disturbances are present.

Theorem 8 ([8]). *Problem 1 for system (12) is solvable if, letting $\tilde{Q}_1 = \Gamma^{-\frac{1}{2}} Q_1 \Gamma^{-\frac{1}{2}}$, there exist a positive scalar α, two symmetric positive definite matrices Q_1 and Q_2, and a matrix L such that*

$$\begin{pmatrix} A\tilde{Q}_1 + \tilde{Q}_1 A^T + BL + L^T B^T - \alpha\tilde{Q}_1 & GQ_2 \\ Q_2 G^T & -\alpha Q_2 \end{pmatrix} < 0$$

$$\frac{c_1}{\lambda_{\min}(Q_1)} + \frac{d}{\lambda_{\min}(Q_2)} < \frac{c_2 e^{-\alpha T}}{\lambda_{\max}(Q_1)}.$$

In this case a controller that solves Problem 1 is given by $K = L\tilde{Q}_1^{-1}$.

Reduction to LMIs

The feasibility of the conditions stated in Theorems 7 and 8 can be turned into the LMI-based feasibility problems as shown below.

LMI Feasibility Problem (from Theorem 7)

Given system (12) with $w \equiv 0$ and (c_1, c_2, T, Γ), letting $\tilde{Q} = \Gamma^{-\frac{1}{2}} Q \Gamma^{-\frac{1}{2}}$, fix an $\alpha \geq 0$ and find L, Q satisfying the LMIs

$$\frac{c_1}{c_2} e^{\alpha T} I < Q < I$$

$$A\tilde{Q} + \tilde{Q}A^T + BL + L^T B^T - \alpha\tilde{Q} < 0.$$

If the problem is feasible, the controller $K = L\tilde{Q}^{-1}$ solves Problem 1.   ◇

In a similar way, we can derive the following feasibility problem from the conditions stated in Theorem 8.

LMI Feasibility Problem (from Theorem 8)
Given system (12), (c_1, c_2, T, Γ), and $d > 0$ letting $\tilde{Q}_1 = \Gamma^{-\frac{1}{2}} Q_1 \Gamma^{-\frac{1}{2}}$, fix an $\alpha > 0$ and find L, Q_1, Q_2, λ_1, λ_2, and λ_3 satisfying the LMIs

$$\lambda_1 > 0, \quad \lambda_2 > 0, \quad \lambda_3 > 0$$

$$\lambda_1 I < Q_1 < I, \quad \lambda_2 I < Q_2 < \lambda_3 I$$

$$\begin{pmatrix} A\tilde{Q}_1 + \tilde{Q}_1 A^T + BL + L^T B^T - \alpha \tilde{Q}_1 & GQ_2 \\ Q_2 G^T & -\alpha Q_2 \end{pmatrix} < 0$$

$$\begin{pmatrix} c_2 e^{-\alpha T} & \sqrt{c_1} & \sqrt{d} \\ \sqrt{c_1} & \lambda_1 & 0 \\ \sqrt{d} & 0 & \lambda_2 \end{pmatrix} > 0.$$

If the problem is feasible, the controller $K = L\tilde{Q}_1^{-1}$ solves Problem 1.   ◇

5 Output Feedback Controller Design

This section deals with the design of a dynamic output feedback controller, which guarantees that the closed-loop system is FTS (or FTB) with respect to some design parameters. Also in this case, two different approaches can be used, which yield DLMI and LMI conditions, respectively.

5.1 Output feedback design conditions expressed in form of DLMIs

To derive the next results, the starting point is again the analysis conditions of Section 3, in which the system matrix $A(\cdot)$ is replaced by the closed-loop matrix. Then the linearizing change of variables proposed in [13] is adapted to deal with the time-varying case.

Theorem 9 ([4]). *System (6) with $w = 0$ is finite-time stabilizable via dynamic output feedback wrt $(c_1, c_2, T, \mathrm{blockdiag}(\Gamma(t), \Gamma_K(t)))$, if there exist symmetric matrix-valued functions $Q(\cdot)$, $S(\cdot)$, an invertible matrix function $N(\cdot)$, and matrix-valued functions $\hat{A}_K(\cdot)$, $\hat{B}_K(\cdot)$, $\hat{C}_K(\cdot)$, and $D_K(\cdot)$ such that*

$$\begin{pmatrix} \hat{\Theta}_{11}(t) & \hat{\Theta}_{12}(t) \\ \hat{\Theta}_{12}^{T}(t) & \hat{\Theta}_{22}(t) \end{pmatrix} < 0 \tag{13a}$$

$$\begin{pmatrix} Q(t) & \hat{\Psi}_{12}(t) & \hat{\Psi}_{13}(t) & \hat{\Psi}_{14}(t) \\ \hat{\Psi}_{12}^{T}(t) & \hat{\Psi}_{22}(t) & 0 & 0 \\ \hat{\Psi}_{13}^{T}(t) & 0 & I & 0 \\ \hat{\Psi}_{14}^{T}(t) & 0 & 0 & I \end{pmatrix} \geq 0, \quad t \in [0, T] \tag{13b}$$

$$\begin{pmatrix} Q(0) & I \\ I & S(0) \end{pmatrix} \leq \frac{c_2}{c_1} \begin{pmatrix} \hat{\Delta}_{11} & Q(0)\Gamma(0) \\ \Gamma(0)Q(0) & \Gamma(0) \end{pmatrix}, \tag{13c}$$

where

$$\hat{\Theta}_{11}(t) = \dot{Q}(t) + A(t)Q(t) + Q(t)A^{T}(t) + B(t)\hat{C}_K(t) + \hat{C}_K^{T}(t)B^{T}(t)$$

$$\hat{\Theta}_{12}(t) = A(t) + \hat{A}_K^{T}(t) + B(t)D_K(t)C(t)$$

$$\hat{\Theta}_{22}(t) = \dot{S}(t) + S(t)A(t) + A^{T}(t)S(t) + \hat{B}_K(t)C(t) + C^{T}(t)\hat{B}_K^{T}(t)$$

$$\hat{\Psi}_{12}(t) = I - Q(t)\Gamma(t)$$

$$\hat{\Psi}_{13}(t) = Q(t)\Gamma^{1/2}$$

$$\hat{\Psi}_{14}(t) = N(t)\Gamma_K^{1/2}(t)$$

$$\hat{\Psi}_{22}(t) = S(t) - \Gamma(t)$$

$$\hat{\Delta}_{11} = Q(0)\Gamma(0)Q(0) + N(0)\Gamma_K(0)N^{T}(0).$$

Remark 7 (Change of variables). Note that (13a) and (13b) are DLMIs; the initial condition (13c) has to be a posteriori checked. To obtain a linear problem, the following change of variables has been used in the proof of the theorem (the time argument is omitted for brevity):

$$\hat{B}_K = MB_K + SBD_K \tag{14a}$$

$$\hat{C}_K = C_K N^{T} + D_K CQ \tag{14b}$$

$$\hat{A}_K = \dot{S}Q + \dot{M}N^{T} + MA_K N^{T} + SBC_K N^{T} + MB_K CQ + S(A + BD_K C)Q. \tag{14c}$$

$\triangle$

Remark 8 (Controller design). Assume now that the hypotheses of Theorem 9 are satisfied; to design the controller the following steps have to be followed:

i) Find $Q(\cdot)$, $S(\cdot)$, $N(\cdot)$, $\hat{A}_K(\cdot)$, $\hat{B}_K(\cdot)$, $\hat{C}_K(\cdot)$, and $D_K(\cdot)$ such that (13) is satisfied.

ii) Find matrix function $M(\cdot)$ such that $M(t) = (I - S(t)Q(t))N^{-T}(t)$.

iii) Obtain $A_K(\cdot)$, $B_K(\cdot)$, $C_K(\cdot)$, and $D_K(\cdot)$ by inverting (14).

$\triangle$

To compute $M(\cdot)$, the matrix-valued function $N(\cdot)$ has to be invertible. To this end we can force, without loss of generality, the condition $N(t) > 0$ for all $t \in [0, T]$.

Next, we consider the case where $w \neq 0$.

Theorem 10 ([4]). *Problem 2 is solvable if there exist symmetric matrix-valued functions $Q(\cdot)$ and $S(\cdot)$, an invertible matrix-valued function $N(\cdot)$, and matrix-valued functions $\hat{A}_K(\cdot)$, $\hat{B}_K(\cdot)$, $\hat{C}_K(\cdot)$, and $D_K(\cdot)$ such that*

$$
\begin{pmatrix}
\Theta_{11}(t) & \Theta_{12}(t) & \Theta_{13}(t) \\
\Theta_{12}^T(t) & \Theta_{22}(t) & \Theta_{23}(t) \\
\Theta_{13}^T(t) & \Theta_{23}^T(t) & -I
\end{pmatrix} < 0
$$

$$
\begin{pmatrix}
Q(t) & \Psi_{12}(t) & \Psi_{13}(t) & \Psi_{14}(t) \\
\Psi_{12}^T(t) & \Psi_{22}(t) & 0 & 0 \\
\Psi_{13}^T(t) & 0 & I & 0 \\
\Psi_{14}^T(t) & 0 & 0 & I
\end{pmatrix} \geq 0, \quad t \in [0, T]
$$

$$
\begin{pmatrix} Q(0) & I \\ I & S(0) \end{pmatrix} \leq \frac{c_2}{c_1 + d} \begin{pmatrix} \Delta_{11} & Q(0)\Gamma(0) \\ \Gamma(0)Q(0) & \Gamma(0) \end{pmatrix},
$$

where

$$
\Theta_{11}(t) = -\dot{Q}(t) + A(t)Q(t) + Q(t)A^T(t) + B(t)\hat{C}_K(t) + \hat{C}_K^T(t)B^T(t)
$$

$$
\Theta_{12}(t) = A(t) + \hat{A}_K^T(t) + B(t)D_K(t)C(t)
$$

$$
\Theta_{13}(t) = \alpha^{1/2}(G(t) + B(t)D_K(t)H(t))
$$

$$
\Theta_{22}(t) = \dot{S}(t) + S(t)A(t) + A^T(t)S(t) + \hat{B}_K(t)C(t) + C^T(t)\hat{B}_K^T(t)
$$

$$
\Theta_{23}(t) = \alpha^{1/2}(S(t)G(t) + \hat{B}_K(t)H(t))
$$

$$
\Psi_{12}(t) = I - Q(t)\Gamma(t)
$$

$$
\Psi_{13}(t) = Q(t)\Gamma^{1/2}(t)
$$

$$
\Psi_{14}(t) = N(t)\Gamma_K^{1/2}(t)
$$

$$
\Psi_{22}(t) = S(t) - \Gamma(t)
$$

$$
\Delta_{11} = Q(0)\Gamma(0)Q(0) + N(0)\Gamma_K(0)N^T(0).
$$

For a given α, similar comments to those ones of the finite-time stabilization case apply.

5.2 Output feedback design conditions expressed in form of LMIs

In this section we consider the time-invariant system

$$
\dot{x}(t) = Ax(t) + Bu(t), \quad x(0) = x_0 \tag{15a}
$$

$$
y(t) = Cx(t) \tag{15b}
$$

and focus on the finite-time stabilization problem.

Let us consider the output feedback controller of the form

$$\dot{x}_c(t) = Ax_c(t) + Bu(t) + L\left(Cx_c(t) - y(t)\right)$$
$$= Ax_c(t) + Bu(t) + LC\left(x_c(t) - x(t)\right), \quad x_c(0) = 0 \tag{16a}$$
$$u(t) = Kx_c(t). \tag{16b}$$

To obtain a computationally tractable condition for finite-time stabilization, we also modify the corresponding definition. Such a modified definition takes into account the fact that, from a practical point of view, only the finite-time stabilization of the physical state of the system is of interest. Concerning the controller, rather than bounding its state vector x_c, it is important to limit the control variable $u(t)$, which can be obtained by imposing further LMI constraints in the design phase (see [10]).

To this end, note that the closed-loop connection between system (15) and controller (16) is well posed. Therefore, given controller (16), for any initial condition x_0 with $x_0^T \Gamma x_0 \leq c_1$ and $x_c(0) = 0$, signal $x_c(\cdot)$ is unique and it makes sense to define $\mathcal{W}_K$ as the set composed of all trajectories $x_c(\cdot)$ of the closed-loop system with x_0 such that $x_0^T \Gamma x_0 \leq c_1$ and $x_c(0) = 0$.

Definition 3. *Given three positive scalars c_1, c_2, T, with $c_1 < c_2$, a positive definite symmetric matrix Γ, system (15) is said to be finite-time stabilizable via output feedback with respect to (c_1, c_2, T, Γ) if there exists a dynamic controller in the form (16) such that system (15) under the input (16b), namely*

$$\dot{x}(t) = Ax(t) + BKx_c(t), \quad x(0) = x_0, \tag{17}$$

is FTB with respect to $(c_1, c_2, \mathcal{W}_K, T, \Gamma)$.   ◇

The above definition reduces finite-time stabilization via output feedback to a finite-time boundedness problem. This key idea will allow us to establish a procedure for controller design.

Note that the controller (16) consists of a classical Luenberger observer, plus a static feedback of the estimated state. To find such a controller, we make the following assumption: a state feedback controller $u(t) = Kx(t)$, which finite-time stabilizes system (15a) via state feedback, exists and has been designed using the results of Theorem 7.

On the basis of this assumption, it should be clear that the finite-time control problem reduces to finding a suitable observer gain L.

To better investigate this point let us consider the closed-loop system obtained by combining (16) with (15). It is well known that, by using a state transformation, it is possible to bring the system in the state-estimation error base, that is,

$$\dot{x}(t) = (A + BK)x(t) - BKe(t), \quad x(0) = x_0 \tag{18a}$$
$$\dot{e}(t) = (A + LC)e(t), \quad e(0) = x_0, \tag{18b}$$

where $e(t) := x(t) - x_c(t)$.

It follows that the system state evolution is determined by the closed-loop matrix $A + BK$ and by the behavior of the exogenous input $e(t)$. By virtue of our assumption, for $e(t) = 0$ system (18a) is FTS, while the presence of a nonzero $e(t)$ could bring the norm of the state $x(t)$ outside the prespecified bound c_2.

On the basis of the above discussion, the goal is to design an observer gain L in (16a) such that the FTS property of the system $\dot{x} = (A + BK)\,x$ is not lost in presence of the estimation error; this is equivalent to requiring that system (18a) is FTB with respect to all admissible time behaviors of the estimation error $e(t)$.

To sum up, system (15) is finite-time stabilizable via output feedback if the following problem admits a solution.

Problem 3 (FTS via Observer Based Output Feedback). Given K such that system $\dot{x}(t) = (A + BK)\,x(t)$ is FTS wrt (c_1, c_2, T, Γ), find an observer gain L such that system (18a) is FTB wrt $(c_1, c_2, \mathcal{W}_L, T, \Gamma)$ where $\mathcal{W}_L$ is the set

$$\mathcal{W}_L := \left\{ e(\cdot) \, : \, \dot{e}(t) = \left(A + LC\right)e(t), \, e(0) = x_0 \, , \, x_0^T \Gamma x_0 \leq c_1 \right\} \, .$$

$$\diamondsuit$$

Theorem 11 ([2]). *System (12) with $w = 0$ is finite-time stabilizable via output feedback wrt (c_1, c_2, T, Γ) if, letting $\tilde{Q}_1 = \Gamma^{1/2}Q_1\Gamma^{1/2}$, $\tilde{Q}_2 = \Gamma^{1/2}Q_2\Gamma^{1/2}$, there exist symmetric matrices Q_1, Q_2, a matrix M, a nonnegative scalar α, and positive scalars λ_k, $k = 1, 2, 3$, such that*

$$\begin{pmatrix} (A + BK)^T\tilde{Q}_1 + \tilde{Q}_1(A + BK) - \alpha\tilde{Q}_1 & -\tilde{Q}_1 BK \\ K^T B^T \tilde{Q}_1 & A^T\tilde{Q}_2 + \tilde{Q}_2 A + C^T M^T + MC - \alpha\tilde{Q}_2 \end{pmatrix} < 0$$

$$\lambda_3 I < Q_1 < \lambda_1 I$$
$$0 < Q_2 < \lambda_2 I$$
$$c_1(\lambda_1 + \lambda_2) \leq c_2 e^{-\alpha T}\lambda_3 \, .$$

In this case a dynamic controller that makes system (15) FTS has the structure (16), with $L = \tilde{Q}_2^{-1}M$.

6 Finite-Time Control of Discrete-Time Systems

In this section we consider the finite-time control problem for linear discrete-time systems. For the sake of simplicity we i) generalize only the conditions based on DLMIs and ii) focus on the finite-time stabilization problem.

6.1 Analysis

First, we generalize to discrete-time systems the definition of FTS.

Definition 4 (FTS). *The discrete-time linear system*

$$x(k+1) = A(k)x(k) \qquad k \geq 0 \tag{19}$$

is said to be FTS with respect to $(c_1, c_2, \Gamma(k), N)$, *where* $\Gamma(\cdot)$ *is a positive definite matrix function,* $0 < c_1 < c_2$, *and* $N > 0$, *if*

$$x^T(0)\Gamma(0)x(0) \leq c_1 \Rightarrow x^T(k)\Gamma(k)x(k) < c_2 \quad \forall k \in \{1, \ldots, N\} \ .$$

$$\diamond$$

As in the continuous-time case, the following theorem can be proven by imposing that the ratio between the state at the time k and the state at the time 0 is less than c_2/c_1.

Theorem 12 (Necessary and Sufficient Conditions for FTS [6]). *The following statements are equivalent:*

i) System (19) *is FTS with respect to* $(c_1, c_2, \Gamma(k), N)$.
ii) $\Phi(k,0)^T \Gamma(k) \Phi(k,0) < \frac{c_2}{c_1} \Gamma(k)$ *for all* $k \in \{1, \ldots, N\}$, *where* $\Phi(\cdot, \cdot)$ *denotes the state transition matrix.*
iii) For each $k \in \{1, \ldots, N\}$ *let*

$$P_k(k) = \Gamma(k)$$
$$P_k(h) = A(k)^T P_k(h+1) A(k), \quad h \in \{0, 1, \ldots, k-1\},$$

then $P_k(0) < \frac{c_2}{c_1} \Gamma(0)$.
iv) For each $k \in \{1, \ldots, N\}$ *there exists a symmetric matrix-valued function* $P_k(\cdot), \ h \in \{0, 1, \ldots, k\}$ *such that*

$$A(k)^T P_k(h+1) A(k) - P_k(h) < 0, \quad h \in \{0, 1, \ldots, k-1\}$$
$$P_k(k) \geq \Gamma(k)$$
$$P_k(0) < \frac{c_2}{c_1} \Gamma(0) \ .$$

Moreover, each one of the above conditions is implied by the following.

v) There exists a symmetric matrix-valued function $P(\cdot), \ k \in \{0, 1, \ldots, N\}$, *such that*

$$A(k)^T P(k+1) A(k) - P(k) < 0, \quad k \in \{0, 1, \ldots, N-1\}$$
$$P(k) \geq \Gamma(k), \qquad k \in \{1, \ldots, N\}$$
$$P(0) < \frac{c_2}{c_1} \Gamma(0) \ .$$

vi) There exists a symmetric matrix-valued function $Q(\cdot)$, $k \in \{0, 1, \ldots, N\}$ such that

$$A(k)Q(k)A(k)^T - Q(k+1) < 0, \quad k \in \{0, 1, \ldots, N-1\}$$
$$Q(k) \leq \Gamma(k)^{-1}, \quad k \in \{1, \ldots, N\}$$
$$Q(0) > \frac{c_1}{c_2}\Gamma(0)^{-1}.$$

We have some remarks about the use of the results contained in Theorem 12.

Remark 9. Statements ii) and iii) are very useful to test the FTS of a given system. However, they cannot be used for design purposes. In the same way, condition iv) is not useful from a practical point of view, since it requires the study of the feasibility of N difference inequalities, if $[1, N]$ is the time interval of interest.

Conversely the sufficient condition v) and vi) require us to check only *one* difference inequality. $\triangle$

Remark 10. Note that a matrix function $P(\cdot)$ satisfying condition v) in Theorem 12 can be found, if one exists, by solving recursively an LMI feasibility problem. $\triangle$

Remark 11. Condition vi) will be used for the state feedback design. $\triangle$

6.2 Design

Given the system

$$x(k+1) = A(k)x(k) + B(k)u(k), \tag{20}$$

we consider the time-varying state feedback controller

$$u(k) = G(k)x(k). \tag{21}$$

One of the goal of this section is to find some sufficient conditions that guarantee that the state of the system given by the interconnection of system (20) with the controller (21) is stable over a finite-time interval.

Problem 4. Given system (20), find a state feedback controller (21) such that the closed-loop system

$$x(k+1) = (A(k) + B(k)G(k))x(k) \tag{22}$$

is FTS with respect to $(c_1, c_2, \Gamma(k), N)$. $\Diamond$

Theorem 13 (FTS via State Feedback [6]). *Problem 4 is solvable if there exists a positive definite matrix-valued function $Q(\cdot)$ and a matrix-valued function $L(\cdot)$ such that*

$$\begin{pmatrix} -Q(k+1) & A(k)Q(k) + B(k)L(k) \\ Q(k)A(k)^T + L(k)^T B(k)^T & -Q(k) \end{pmatrix} < 0$$

$$k \in \{0, 1, \ldots, N-1\} \qquad (23\text{a})$$

$$Q(k) \leq \Gamma(k)^{-1}, \qquad k \in \{1, \ldots, N\} \quad (23\text{b})$$

$$Q(0) > \frac{c_1}{c_2} \Gamma(0)^{-1}. \qquad (23\text{c})$$

In this case the gain of a state feedback controller solving Problem 4 is given by $G(k) = L(k)Q(k)^{-1}$.

Remark 12. To find a numerical solution to Problem 4, i.e., to compute the matrix-valued functions $Q(\cdot)$ and $K(\cdot)$, a back-stepping algorithm can be used for conditions (23). In the first step, inequalities (23a) and (23b) can be solved, obtaining the matrices $Q(N)$, $Q(N-1)$, $K(N-1)$. Given $Q(N-1)$, in the next step, (23a) and (23b) can be solved for $k = N-2$, finding $Q(N-2)$, $K(N-2)$, and so on. The final step consists of solving (23a) and (23c) together for $k = 0$. To find the smallest value for c_2, a further condition can be added to the various steps, which imposes the maximization of the smallest eigenvalue of $Q(k)$ at each step. $\triangle$

Now let us consider the system

$$x(k+1) = A(k)x(k) + B(k)u(k) \qquad (24\text{a})$$

$$y(k) = C(k)x(k). \qquad (24\text{b})$$

Then with respect to system (24), we consider the following dynamic output feedback controller:

$$x_c(k+1) = A_K(k)x_c(k) + B_K(k)y(k) \qquad (25\text{a})$$

$$u(k) = C_K(k)x_c(k) + D_K(k)y(k), \qquad (25\text{b})$$

where the controller state vector $x_c(k)$ has the same dimension of $x(k)$.

The main goal of this section is to find some sufficient conditions that guarantee the existence of an output feedback controller that finite-time stabilizes the overall closed-loop system, as stated in the following problem.

Problem 5. Let us denote by $\Gamma_K(\cdot)$ a positive definite matrix function weighting the controller state. Then, given system (24), find an output feedback controller (25) such that the corresponding closed-loop system is FTS with respect to $(c_1, c_2, \text{blockdiag}(\Gamma(k), \Gamma_K(k)), N)$. $\triangle$

Theorem 14 (Solution of Problem 5 [7]). *Problem 5 is solvable if there exist positive definite matrix-valued functions $Q(\cdot)$, $S(\cdot)$, an invertible matrix $N(\cdot)$, matrix-valued functions $\hat{A}_K(\cdot)$, $\hat{B}_K(\cdot)$, $\hat{C}_K(\cdot)$, and $D_K(\cdot)$ such that*

$$\begin{pmatrix} \Theta_{11}(k) & \Theta_{12}(k) \\ \Theta_{12}^T(k) & \Theta_{22}(k) \end{pmatrix} < 0, \quad k \in \{0, 1, \ldots, N-1\} \tag{26a}$$

$$\begin{pmatrix} Q(k) & \Psi_{12}(k) & \Psi_{13}(k) & \Psi_{14}(k) \\ \Psi_{12}^T(k) & \Psi_{22}(k) & 0 & 0 \\ \Psi_{13}^T(k) & 0 & I & 0 \\ \Psi_{14}^T(k) & 0 & 0 & I \end{pmatrix} \geq 0, \quad k \in \{1, 2, \ldots, N\} \tag{26b}$$

$$\begin{pmatrix} Q(0) & I \\ I & S(0) \end{pmatrix} \leq \frac{\epsilon^2}{\delta^2} \begin{pmatrix} \Delta_{11} & Q(0)\Gamma(0) \\ \Gamma(0)Q(0) & Q(0) \end{pmatrix}, \tag{26c}$$

where

$$\Theta_{11}(k) = -\begin{pmatrix} Q(k) & I \\ I & S(k) \end{pmatrix}$$

$$\Theta_{12}(k) = \begin{pmatrix} Q(k)A(k)^T + \hat{C}_K^T(k)B(k)^T & \hat{A}_K^T(k) \\ A(k)^T + C(k)^T D_K^T(k)B(k)^T & A^T S(k+1) + C^T \hat{B}_K^T(k) \end{pmatrix}$$

$$\Theta_{22}(k) = -\begin{pmatrix} Q(k+1) & I \\ I & S(k+1) \end{pmatrix}$$

$$\Psi_{12}(k) = I - Q(k)\Gamma(k)$$

$$\Psi_{13}(k) = Q(k)\Gamma(k)^{1/2}$$

$$\Psi_{14}(k) = N(k)\Gamma_K(k)^{1/2}$$

$$\Psi_{22}(k) = S(k) - \Gamma(k)$$

$$\Delta_{11} = Q(0)N(0)Q(0) + N(0)\Gamma_K(0)N^T(0).$$

Remark 13 (Change of variables). To obtain a linear problem, the following change of variables has been used in the proof of the theorem:

$$\hat{B}_K(k) = M(k+1)B_K(k) + S(k+1)B(k)D_K(k) \tag{27a}$$

$$\hat{C}_K(k) = C_K(k)N^T(k) + D_K(k)C(k)Q(k) \tag{27b}$$

$$\hat{A}_K(k) = M(k+1)A_K(k)N^T(k) + S(k+1)B(k)C_K(k)N^T(k)$$
$$+ M(k+1)B_K(k)C(k)Q(k)$$
$$+ S(k+1)\big(A(k) + B(k)D_K(k)C(k)\big)Q(k). \tag{27c}$$

$$\triangle$$

Remark 14 (Controller design). Assume now that the hypothesis of Theorem 14 are satisfied; to design the controller the following steps have to be followed:

i) Find $Q(\cdot)$, $S(\cdot)$, $N(\cdot)$, $\hat{A}_K(\cdot)$, $\hat{B}_K(\cdot)$, $\hat{C}_K(\cdot)$, and $D_K(\cdot)$ such that (26) are satisfied.

ii) Find matrix function $M(\cdot)$ such that $M(k) = (I - S(k)Q(k))N^{-T}(k)$.
iii) Obtain $A_K(\cdot)$, $B_K(\cdot)$ and $C_K(\cdot)$ by inverting (27).

$\triangle$

Note that (26a) and (26b) are linear difference matrix inequalities. Concerning the initial condition (26c), it has to be a posteriori checked; alternatively it can be taken into account within the design cycle by solving a quadratic optimization problem for $k = 0$ over $Q(0)$, $S(0)$, and $N(0)$. Concerning the nonsingularity of $N(\cdot)$, the same considerations of the continuous-time case apply.

7 Conclusions

In this chapter sufficient conditions as well as necessary and sufficient conditions for FTS and sufficient conditions for finite-time boundedness of linear systems have been provided. These conditions, which involve the feasibility of either LMI or DLMI problems, have then been used in the design context for controller synthesis. The extension to discrete-time systems has been considered for what concerns the FTS context. It is interesting to note that there is no conceptual difficulty in extending these results to uncertain systems along the guidelines of [1] and [3].

References

1. Abdallah CT, Amato F, Ariola M, Dorato P, Koltchinski V (2002) Statistical learning methods in linear algebra and control problems: the example of finite-time control of uncertain linear systems. Linear Algebra and its Applications 351–352: 11–26
2. Amato F, Ariola M, Cosentino C (2003a) Finite-time control with pole placement. Proc. of the 2003 European Control Conference, Cambridge
3. Amato F, Ariola M, Cosentino C (2003b) Robust finite-time stabilization via dynamic output feedback: an LMI approach. Proc. of IFAC ROCOND 2003, Milan
4. Amato F, Ariola M, Cosentino C (2005) Finite-time control of linear time-varying systems via output feedback. Submitted
5. Amato F, Ariola M, Cosentino C, Abdallah C, Dorato P (2003) Necessary and sufficient conditions for finite-time stability of linear systems. Proc. of the 2003 American Control Conference, Denver, CO, 4452–4456
6. Amato F, Ariola M, Cosentino C, Carbone M (2004a) Finite-time stability of discrete-time systems. Proc. of the 2004 American Control Conference, Boston, MA, 1440–1444
7. Amato F, Ariola M, Cosentino C, Carbone M (2004b) Finite-time output feedback control of discrete-time systems. Submitted

8. Amato F, Ariola M, Dorato P (2001) Finite-time control of linear systems subject to parametric uncertainities and disturbances. Automatica 37:1459–1463

9. Bernussou J, Peres PLD, Geromel JC (1989) A linear programming oriented procedure for quadratic stabilization of uncertain systems. Systems & Control Letters 13: 65–72

10. Boyd S, El Ghaoui L, Feron E, Balakrishnan V (1994) Linear matrix inequalities in system and control theory. SIAM, Philadelphia, PA

11. D'Angelo H (1970) Linear time-varying systems: analysis and synthesis. Allyn and Bacon, Boston, MA

12. Dorato P (1961) Short time stability in linear time-varying systems. Proc. IRE International Convention Record Part 4, 83–87

13. Gahinet P (1996) Explicit controller formulas for LMI-based H_∞ synthesis. Automatica 32:1007–1014

14. Weiss L, Infante EF (1967) Finite time stability under perturbing forces and on product spaces. IEEE Transactions Automatic Control 12:54–59

Part III

Robotics

An Application of Iterative Identification and Control in the Robotics Field

Pedro Albertos,[1] Angel Valera,[1] Julio A. Romero,[2] and Alicia Esparza[1]

[1]Department of Systems Engineering and Control
University of Valencia, C/Vera s/n Valencia, 46021 Spain
`pedro@aii.upv.es,{giuprog,alespei}@isa.upv.es`
[2]Department of Technology, Riu Sec Campus, University of Jaume I, 12071
Castellón de la Plana, Spain
`romeroj@tec.uji.es`

Summary. The plant model appropriate for designing the control strongly depends on the requirements. Simple models are enough to compute nondemanding controls. The parameters of well-defined structural models of flexible robot manipulators are difficult to determine because their effect is only visible if the manipulator is under strong actions or with high-frequency excitation. Thus, in this chapter, an iterative approach is suggested. This approach is applied to a one-degree-of-freedom flexible robot manipulator, first using some well-known models and then controlling a lab prototype. This approach can be used with a variety of control design and/or identification techniques.

1 Introduction

In model-based control, the process model accuracy determines the achievable performances. For low-demand control, a simple model may be good enough. A more accurate model is necessary for a high-performance control system. The modeling of processes that are open loop unstable or with different time scale modes is difficult. High-frequency modes can be masked by measurement noise, being difficult to locate under relatively low gain inputs. Also, fast modes require strong or high-frequency actions to be excited and only appear if fast response of the controlled system is required.

This is the case of robot arms where the open-loop behavior includes one or two integrators in the chain. If flexible joints or flexible arms are considered, resonant modes appear, which should be taken into account in the controller design to achieve high performances.

The modeling and control of flexible robot manipulators has attracted the interest of many researchers (see, for instance, [10] for a review of this topic), as has the transfer function of a single flexible robot [11]. For the sake of simplicity, some important phenomena such as joint flexibility, actuator dynamics and friction were neglected in basic control algorithms. However, experimental results have shown that joint elasticity should be taken into account in the modeling of robotic manipulators. The elasticity in the joints may be caused by the harmonic drives, which are a special type of gear mechanism having high transmission ratio, low weight, and small size.

To get a reliable model, these systems should be identified when operating under control, using closed loop identification techniques, [8]. Another option is to perform these phases in an iterative framework [7], as discussed in [1]. This is the proposed approach in this chapter.

The problem of designing a control system for a plant whose initial model is not well known, with unbounded uncertainties, and where the main goal is the maximization of a kind of performance index under some constraints, can be approached following an iterative schema.

Assume an initial model of the plant, design the most performing controller, and iteratively repeat the following:

- Estimate a more accurate model, until required.
- Design a suitable controller improving the performances, until required.

Specifications may also be upgraded for a given model, as the iterations proceed, if there is no constraint violation in the actual controlled plant. Under the current operating conditions, that is, on the limit of the constraints, get new data from the process and estimate a more accurate model. The correct operation of the plant is experimentally checked, as the required performances are increased. The model uncertainty is not estimated and, thus, stability and performance degrading issues are not theoretically studied. This decoupling of the identification and control design activities has been also studied in different frameworks [12, 9].

The approach followed in this chapter is also related to the features of the frequency domain response of the plant. The manipulator model may be decomposed into different subsystems, with different frequency ranges, and only those subsystems excited by the controller action, which depends on the required performances, should be considered in the design of the appropriate controller. Given an initial low-frequency range model, the goal consists of iteratively designing a controller, which stabilizes the manipulator while increasing the closed-loop bandwidth [3]. In the next section, the basic models of the one-degree-of-freedom flexible manipulator used in this work are summarized. Then, a direct application of frequency scale decomposition is shown. As an alternative, a generalized predictive controller (GPC) [5] design approach can be implemented, and by reducing the control weight, faster response is achieved. Finally, these designs are applied to a laboratory flexible join manipulator. A progressive reduction of the control weight in the GPC setting can be considered, as far as the improvement on the process model allows for stronger control actions.

This methodology is rather general and different identification, analysis, design, and iteration algorithms may be applied. In this chapter, based on the theoretical models of the robot manipulators, the attention is focused on a simple set of such algorithms, in order to explore the main issues of this approach, but other more sophisticated identification and/or control algorithms can be foreseen. In the next section, first principle models of one-degree-of-freedom flexible arms point out the presence of resonance modes. These models are well suited to apply the proposed methodology, based on the iterative control design, which is summarized in Section 3. Basic control design techniques and identification approaches are used in Section 4 to illustrate the procedure on simulated models. A practical experiment on a laboratory manipulator is reported in Section 5, and some conclusions are drafted in the last section.

2 Structural models

In this section, the theoretical models of one-degree-of-freedom flexible robot manipulator are summarized. A rigid rotational body, with flexible joint, is assumed.

2.1 Flexible joint

For modeling purposes, the manipulator consists of an actuator whose rotor axis, with inertia J and viscous damping B, is directly connected to an axis with inertia D. The elasticity in the joint is modeled as a torsion spring with known characteristics (a linear spring with stiffness K is considered) [10], as shown in Figure 1.

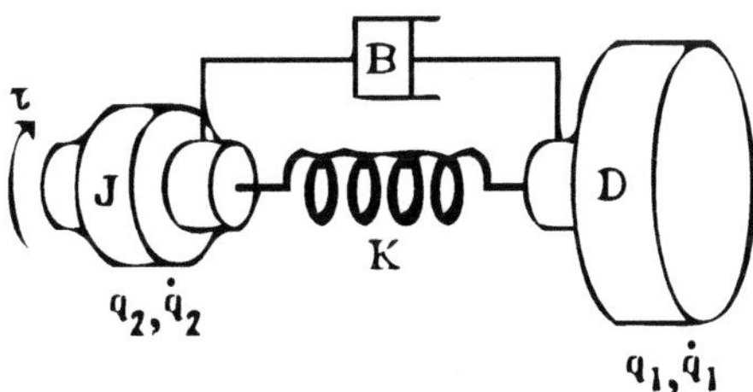

Fig. 1. Flexible robot.

The motor shaft angle q_2 and the link angle q_1 are taken as the generalized coordinates. Thus, the dynamic equations are

$$J\ddot{q}_2 + B\left(\dot{q}_2 - \dot{q}_1\right) + K\left(q_2 - q_1\right) = \tau \tag{1}$$
$$D\ddot{q}_1 + B\left(\dot{q}_1 - \dot{q}_2\right) + K\left(q_1 - q_2\right) = 0. \tag{2}$$

From (1) and (2) the transfer functions of the robot link angle (end effector) and the motor shaft angle, with respect to the generalized force applied to the motor shaft by the actuator, are

$$\frac{q_1}{\tau} = \frac{Bs + K}{I_R I_F s^4 + I_R B s^3 + I_R K s^2} = \frac{1/I_R}{s^2} \frac{\frac{B}{I_F}s + \frac{K}{I_F}}{s^2 + \frac{B}{I_F}s + \frac{K}{I_F}} \tag{3}$$

$$\frac{q_2}{\tau} = \frac{Ds^2 + Bs + K}{(Js^2 + Bs + K)(Ds^2 + Bs + K) - (Bs + K)^2} =$$

$$\frac{(I_R/J)\,s^2 + R/I_F s + K/I_F}{I_R s^2 \left(s^2 + R/I_F s + K/I_F\right)}\,, \tag{4}$$

where $\frac{1}{I_F} = \frac{1}{J} + \frac{1}{D}$ and $I_R = J + D$.

The basic model of the robot (a double integrator) appears explicitly. If the stiffness is null, $K = 0$, the model is reduced to a double integrator plus a time constant (due to the damping B).

A similar model can be derived for a rotary manipulator on a fixed table, like the Rotary Flexible Joint (RFJ), from Quanser Consulting [4]. The RFJ is attached to a servo plant in such a way that joint flexibility is attained via two identical springs anchored to the body and to the load. This manipulator is shown in Figure 2.

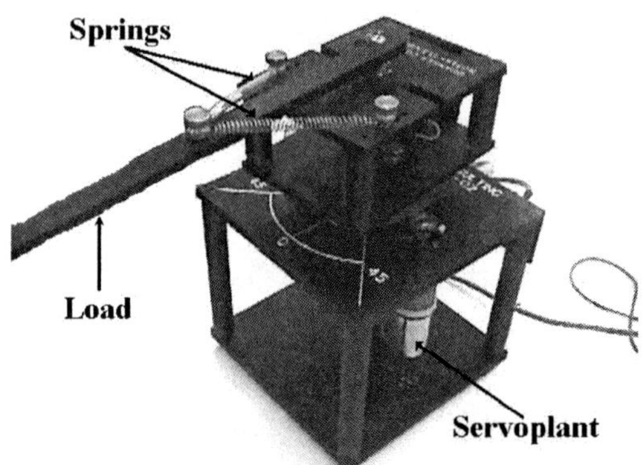

Fig. 2. Rotational actuator with flexible joint.

This mechanical configuration changes the dynamic equations of the previous robot. In this case the torque balance yields

$$\begin{aligned} J\ddot{q}_2 - K q_1 &= \tau \\ D\ddot{q}_1 + D\ddot{q}_2 + K q_1 &= 0\,. \end{aligned} \tag{5}$$

If the voltage applied to the robot motor generates the generalized torque

$$\tau = K_e I = \frac{K_e}{R}V - \frac{K_e^2}{R}\dot{q}_2\,, \tag{6}$$

the full model can be expressed in state space representation:

$$
\left\{\begin{array}{c} \dot{q}_2(t) \\ \dot{q}_1(t) \\ \ddot{q}_2(t) \\ \ddot{q}_1(t) \end{array}\right\} = \begin{bmatrix} 0 & 0 & 1 & 0 \\ 0 & 0 & 0 & 1 \\ 0 & \frac{K}{J} & \frac{-K_e^2}{JR} & 0 \\ 0 & \frac{-K(J+D)}{JD} & \frac{K_e^2}{JR} & 0 \end{bmatrix} \left\{\begin{array}{c} q_2(t) \\ q_1(t) \\ \dot{q}_2(t) \\ \dot{q}_1(t) \end{array}\right\} + \begin{bmatrix} 0 \\ 0 \\ \frac{K_e}{JR} \\ \frac{-K_e}{JR} \end{bmatrix} V(t); \qquad (7)
$$

$$
y(t) = \begin{bmatrix} 1 & 0 & 0 & 0 \end{bmatrix} \left\{\begin{array}{c} q_2(t) \\ q_1(t) \\ \dot{q}_2(t) \\ \dot{q}_1(t) \end{array}\right\} + 0 V(t). \qquad (8)
$$

Assuming the following parameters: $D = 0.0059$, $J = 0.0021$, $K = 1.61$, $K_e = 0.5353$, and $R = 2.573$, the state equation (7) is

$$
\left\{\begin{array}{c} \dot{q}_2(t) \\ \dot{q}_1(t) \\ \ddot{q}_2(t) \\ \ddot{q}_1(t) \end{array}\right\} = \begin{bmatrix} 0 & 0 & 1 & 0 \\ 0 & 0 & 0 & 1 \\ 0 & 766 & -53 & 0 \\ 0 & -1040 & 53 & 0 \end{bmatrix} \left\{\begin{array}{c} q_2(t) \\ q_1(t) \\ \dot{q}_2(t) \\ \dot{q}_1(t) \end{array}\right\} + \begin{bmatrix} 0 \\ 0 \\ 99 \\ -99 \end{bmatrix} V(t) \qquad (9)
$$

and the manipulator transfer function is

$$
G_{FR}(s) = \frac{q_1}{\tau} = \frac{27126}{s^4 + 53s^3 + 1040s^2 + 14522s} = \frac{1}{s\,(s + 35.1)} \frac{27126}{s^2 + 17.83s + 412.91}. \qquad (10)
$$

In this case, the basic model is composed of an integrator, a low time constant, and a high-frequency resonant mode.

Other than the flexible joint manipulator, the flexibility can be also assumed in the link, leading to the flexible-link robot [11]. In this case small elastic deformations in the robot links are assumed. One example can be found in spatial robots, where the links are very long and thin. So, for high velocities the deformational behavior of the flexible links must be considered. To find the model of flexible link robots some approaches can be used. Equations similar to (9) and (10) can be obtained with the resulting transfer function composed of a double integrator and a number of resonant modes and zeros.

3 Iterative control design

The general control design methodology is described as follows: given a process and some control goals and constraints, by an iterative approach, design the highest performing control system without violating the constraints. The proposed procedure is

1. Get a raw model of the manipulator to be controlled. Usually, this model explains its low-frequency behavior.

2. For the available model, design a controller. Any model-based design approach can be used.
3. Apply the designed controller to the actual manipulator and check the constraint's fulfillment. Due to the mismatch between the robot and its model, conservative performances are required.
4. If the constraints are fulfilled:
   a) stop if a satisfactory control is achieved, or
   b) redesign or retune the controller to enhance the controlled manipulator performances, and go to step 3. Otherwise, if there is a constraint violation, use the last controller.
5. Carry on an identification experiment:
   a) if a better model is obtained, go to step 2, otherwise,
   b) stop and keep the achieved controller.

A detailed procedure is described in [2].

This approach seems to be appropriate for the control design of the one-degree-of-freedom flexible manipulator for the following reasons:

- A rough mode is available based on structural properties, as described in the previous section. It could be one or two integrators, as well as an additional time constant.
- The model mismatching is due to high-frequency, almost resonant, modes.
- It is important to achieve a fast response.
- There is always some high-frequency noise in the measurements.
- The plant is available and some experiments can be easily carried out.
- The constraints, such as a maximum overshoot or an excessive ripple, can be checked.

For the sake of simplicity, the control design approaches to be considered are very simple: cancellation controllers and GPC. The requirements are expressed either as a suitable model for the controlled plant or as a cost index, and the constraints are reduced to keep the stability of the controlled robot.

The parameter estimation methods used in the experiments are also very simple. Step response model matching or physical reasoning will provide for low-frequency models. Model updating will be achieved by closed loop least squares parameter estimation.

4 Applications of the iterative control design

First, the approach is illustrated by designing the control for a simulated manipulator, assuming a transfer function like (10), with an output measurement Gaussian noise, with zero mean and variance $\sigma = 0.001$.

4.1 Cancellation controller and step response

To design a cancellation controller, the reference model is chosen to represent a stable and damped plant with a progressively faster closed-loop step response. The reference model is thus based on a multipole transfer function, whose poles will be faster at each iteration. The constraint is to limit the overshoot to a maximum of 10%.

It is known that the process includes an integrator, as shown in the transfer function (10). So a closed-loop experiment is carried out to determine the gain constant, $k = 1.88\frac{rad}{sec^2V}$, very similar to the theoretical one. Then, the algorithm follows the steps above:

Step 1: The assumed initial model is $\hat{G}_0 = \frac{k}{s}$. The desired behavior of the closed loop is expressed by $M_0 = \frac{\lambda}{s+\lambda}$.

Step 2: The cancellation controller is given by $C_0 = \frac{1}{\hat{G}_0} \cdot \frac{M_0}{1-M_0} = \frac{\lambda}{k}$. Assume conservative initial requirements expressed by $\lambda = 1$ rad/sec.

Step 3: The step response of the manipulator is too slow (see Figure 3).

Step 4: There is no sign of instability or overshoot. Thus λ is cautiously increased. Some different responses are plotted in Figure 3.

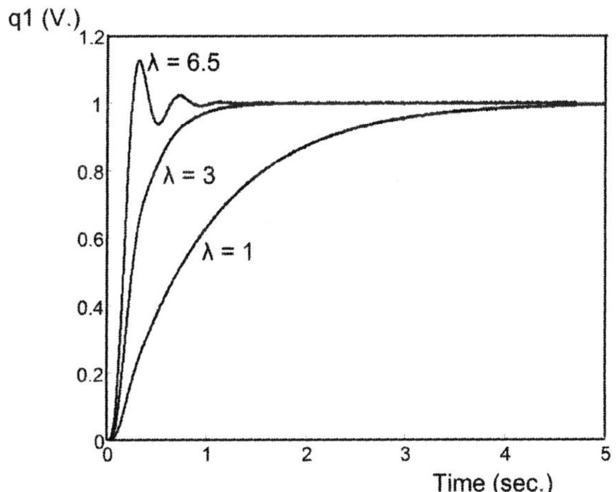

Fig. 3. Step response of the manipulator, (9), (10), for $\lambda = 1, 3$ and 6.5 rad/sec.

Step 4b: Validity check. For $\lambda = 6.5$ rad/sec the constraint is violated, as the overshoot is greater than 10%. It is thus necessary to look for a new model able to explain the step response oscillations.

Step 5: Run a new identification test. We assume a multiplicative uncertainty. The plant model, $\hat{G}_1 = \hat{G}_0 \cdot \hat{G}_1^*$, is composed by two parts, one of them is known, $\hat{G}_0$, and the other one is unknown, $\hat{G}_1^*$. In this case, a least squares algorithm is applied using the model output prediction error. A ± 5 V pseudorandom signal is used as excitation, and a normally distributed output measurement noise is considered with zero mean and a variance of 0.001. A third-order model, $\hat{G}_1^*$, is estimated, leading to

$$\hat{G}_1 = \frac{1.88}{s} \cdot \frac{0.006031(s^2 - 2967s + 2.95 \cdot 10^6)}{(s + 48.87)(s^2 + 17.81s + 365.3)}$$

$$\bar{G}_1 = \frac{1.88}{s} \cdot \frac{17791}{(s + 48.87)(s^2 + 17.81s + 365.3)}$$

$$\hat{G}_1 = \bar{G}_1 \cdot (1 - 0.001s + 3.39 \cdot 10^{-7} s^2)$$

where a pair of very fast nonminimum phase zeros appeared.

The new control requirements should be expressed by means of a fourth-order reference model. The cancellation controller must avoid dealing with the unstable poles. Thus, the proposed cancellation controller is $C_1 = \frac{1}{\bar{G}_1} \cdot \frac{M_1}{1-M_1}$ and the reference is taken as $M_1 = \frac{\lambda^4(1-0.001s+3.39\cdot10^{-7}s^2)}{(s+\lambda)^4}$. It has been checked that the current closed-loop model is not much disturbed by the addition of these zeros. Now, the design procedure is repeated, by cautiously increasing the performances (λ). The first value of λ is a bit smaller than the last one reached in the previous iteration, in order to assure the loop stability. With this model, λ is increased until $\lambda = 1000$. The iterations finish when the step response is good enough, as Figure 4 shows. In this case, a small overshoot appears but it does not violate the constraint.

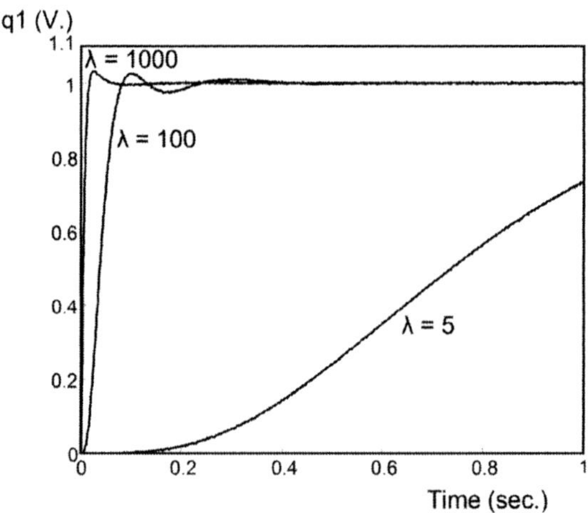

Fig. 4. Step response of the manipulator controlled by C_1, for $\lambda = 5, 100$, and 1000.

Remark 1. Looking at the controlled plant performances, it should be emphasized that the settling time was reduced from around 4 sec (first design, with $\lambda = 1$), as shown in Figure 3, to $1/10$ sec, in the last design.

Remark 2. The modeling and control design are very simple. Other identification and controller design techniques may be used. In this example the main goal was to illustrate the procedure.

4.2 GPC controller

The design framework is now as follows:

- The simulated plant is as given by (10).
- The simplified model of the manipulator, including an estimated time constant, is

$$\hat{G}_0(s) = \frac{65}{s(s+35)}.$$

- The control design is discretized. A sampling period $T = 0.01$ sec is assumed:

$$\hat{G}_0(z) = 10^{-3} \cdot \frac{2.933z + 2.61}{z^2 - 1.7047z + 0.7047}.$$

- The design criterion is a generalized predictive controller (GPC), resulting in a control scheme as shown in Figure 5. In the cost index

$$J = \sum_{i=N_1}^{N_2} \alpha_i [y(k+i|k) - w(k+i)]^2 + \sum_{j=N_1}^{Nu} \lambda_j [\Delta u(k+j-1)]^2,$$

where $w(k)$ represents the output reference, different weights λ will be considered for the control action. The lower the λ weight is, the stronger control actions are allowed, leading to sharper controlled plant response to a step change in the reference. Some conventional parameters are assumed in the GPC setting [5].

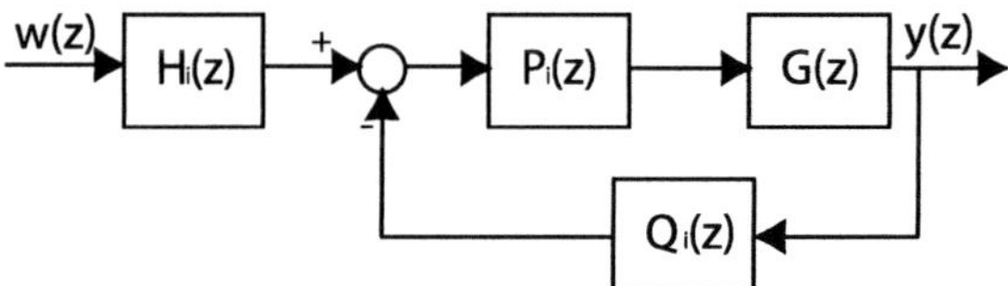

Fig. 5. GPC-controlled manipulator.

- In this case, the constraint is the absence of oscillations (but an integral of the time absolute error criteria could be used).
- Least squares identification techniques will be applied to obtain a new refined model.

The initial GPC parameters are $N_1 = 1$, $N_2 = 40$, $Nu = 4$, $\alpha = 1$. The assumed initial control weight is $\lambda_0 = 10$. The resulting controller blocks are

$$H_0 = 0.2396; \quad P_0(z) = \frac{z-0.9}{z^2-1.65z+0.653}; \quad Q_0(z) = \frac{5.5107z^2-9.3116z+3.8249}{z-0.9}.$$

The simulated step responses (step reference applied at 1 sec) for different values of λ are plotted in Figure 6.

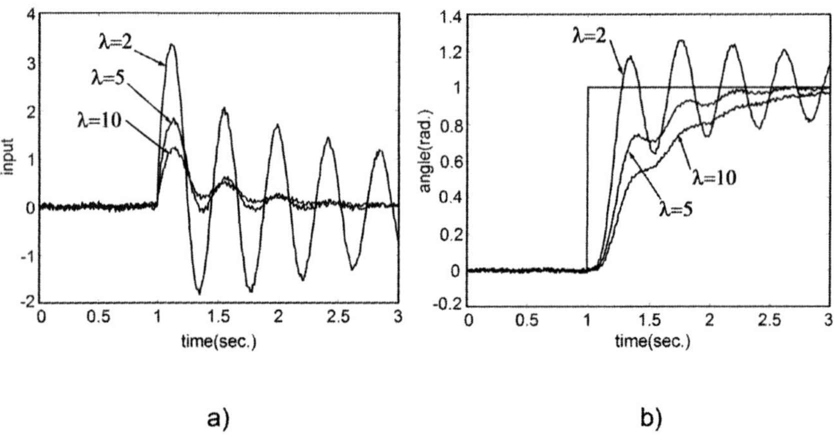

Fig. 6. Iterative designs of GPC for $\hat{G}_0$ and different weighting factor λ, applied to the model (10); a) Control input; b) Manipulator output, y.

The controlled plant time response is improved as far as the control weight is reduced. For $\lambda = 2$, the manipulator response becomes oscillatory. Also, the control action increases. At this point, a new model must be obtained if better performances are required. Thus, a new model for the plant is estimated and, in this case, the current model is not maintained:

$$\hat{G}_1(z) = 10^{-3} \cdot \frac{2.73z^2 - 7.067z + 5.04}{z^3 - 2.775z^2 + 2.59z - 0.815}.$$

The design leads to a new controller. The response is plotted in Figure 7. Furthermore, the control weight can be reduced to obtain better performances. The response is also shown for $\lambda = 2, 1$ and $\lambda = 0.5$.

The procedure can be repeated again, until the response of the manipulator is considered to be appropriate. Observe that, starting from different raw models of the manipulator and enhancing the models in different directions, both design techniques allow for similar results in the controlled manipulator behavior.

5 Experimental results

A similar approach is followed by using the laboratory RFJ system as the process to be controlled. For the corresponding GPC design, the same raw

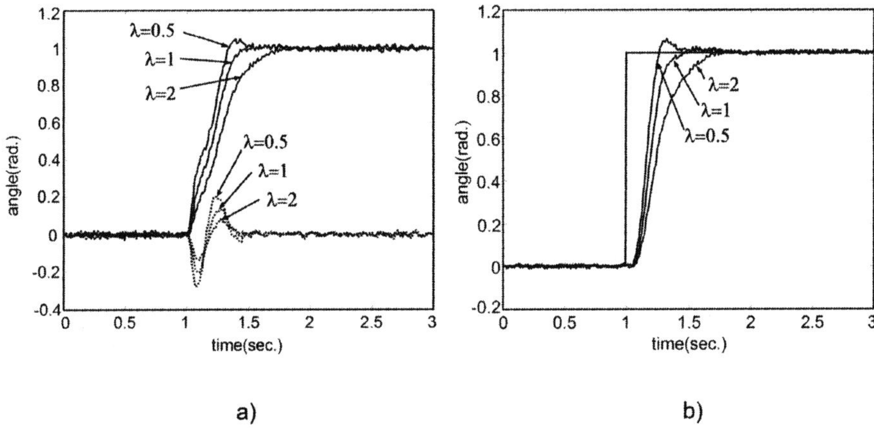

Fig. 7. Final GPC controlled designs for $\hat{G}_1(z)$. a) Solid line: motor shaft angle, q_2; Dotted line: link angle, q_1. b) Manipulator output, y.

plant model was initially assumed and the same steps were followed. This manipulator also presents oscillatory behavior if the control weight is less than $\lambda = 2$, as shown in Figure 8. It should be noticed that in the experimental setting the measurement noise is larger than the previously simulated one.

The improvement in the closed-loop response can be recognized in Figure 8. The oscillations have been suppressed, but the effects of both the noise and the rest of unmodeled dynamics more clearly appear.

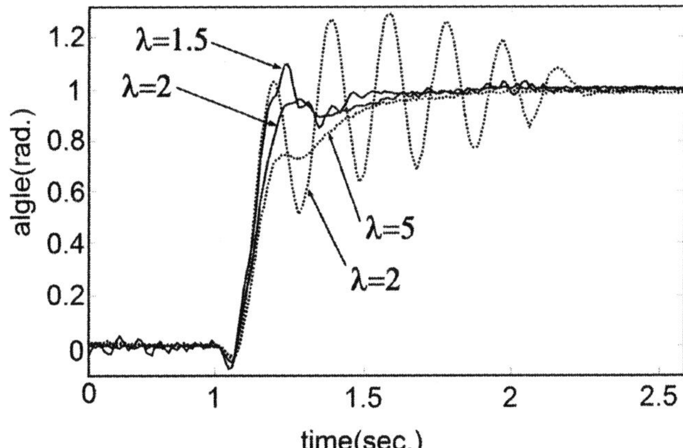

Fig. 8. Step responses of the controlled manipulator. Dashed line: initial model. Solid line: refined model.

When the controller gain is increased, the ratio noise/signal at the controller output is bigger. The noise can even be due to the quantization process, always present in the implementation of digital controllers. To reduce the noise effect, a filter has been considered in the GPC design. In particular, a filter transfer function $T(z) = 1 - 0.9z^{-1}$ has been assumed.

The experimental results are in agreement with those obtained using the simulated manipulator, although additional disturbances appear in the implementation.

6 Conclusions

In mechatronic designs it is normal to find mechanical systems with flexible components and joints. The effect of these elements mainly appears under high-performing operating conditions, being quite difficult if the system is softly activated or without any control. In these cases, rather than carrying out sophisticated identification procedures, an iterative approach has been suggested.

The iterative procedure is based on a model-based control design and a closed loop identification approach. Starting with a simple model, a controller is designed satisfying a set of requirements. If the requirements are increased, the designed controller will excite the hidden modes of the flexible system. The actual controlled system will violate some restrictions. It is under these conditions that a new and more accurate model of the process can be estimated. The control design step is repeated. The procedure ends when the global behavior is accepted or when no further improvement in the model or the control can be achieved.

The procedure has been applied to one-degree-of-freedom flexible robot manipulators, both simulated and lab prototypes. Although the control design methods, as well as the parameter estimation techniques, used in these experiments are very simple, they illustrate the proposal. Any other more appropriate technique can be used. For more complicated manipulators, the approach could be valid as far as the dynamic behavior can be approximately captured by a linear model.

References

1. Albertos P (2002) Iterative identification and control. In: Albertos P, Sala A (eds) Advances in Theory and Applications. Springer Verlag, Berlin
2. Albertos P, Esparza A, Romero J (2000) Model-based iterative control design. In: American Control Conference, 2578–2582. Chicago, IL
3. Albertos P, Pico J (1993) Iterative controller design by frequency scale experimental decomposition. In: IEEE CDC, 2828–2832. Tucson, AZ
4. Apkarian J (1995) Quanser Consulting handbook

5. Clarck DW, Mohtadi C, Tuffs PS (1987) Generalised predictive control. Automatica, 23(2):137–148
6. Franklin GF, Powell JD, Emami-Naeini A (1991) Feedback control of dynamic systems. Addison-Wesley, Reading, MA
7. Gevers M (1996) Modeling, identification and control. In: Communications, computing control and signal processing 2000, 1–15. Kluwer Academic Publishers, Dordrecht
8. Landau ID (2001) Identification in closed loop: a powerful design tool (better models, simple controllers). Control Engineering Practice, 9(1):51–65
9. Lee WS, Anderson BDO, Kosut RL, Mareels IM (1993) On robust performance improvement through the windsurfer approach to adaptive robust control. In: IEEE CDC, 2821–2827. Tucson, AZ
10. Spong MW (1987) Modelling and control of elastic joint robots. ASME J. of Dynamic, Systems, Measurement and Control, 109:310–319
11. Wang D, Vidyasagar M (1991) Transfer functions for a single flexible link. The International Journal of Robotics Research, 10(5):540–549
12. Zang Z, Bitmead RR, Gevers M (1991) Iterative model refinement and control robustness enhancement. In: IEEE CDC, 279–284. Brighton, UK

Friction Identification and Model-Based Digital Control of a Direct-Drive Manipulator

Basilio Bona, Marina Indri, and Nicola Smaldone

Dipartimento di Automatica e Informatica, Politecnico di Torino
Corso Duca degli Abruzzi 24, 10129 Torino, Italia,
{basilio.bona,marina.indri,nicola.smaldone}@polito.it

Summary. Several tasks of the most recent robotics applications require high control performances, which cannot be achieved by the classical joint independent control schemes widely used in the industrial field. The necessity to directly take into account parasitic phenomena affecting motion control, such as friction, often leads to the development of model-based control schemes. The actual effectiveness of such schemes is strongly dependent on the accuracy with which the robot dynamics and the friction effects are compensated by the identified models, and it must be assessed by suitable experimental tests. In this chapter, different solutions are investigated for the development of a model-based control scheme, including joint friction compensation, for a two-links, planar, direct-drive manipulator. In particular, the use of available nominal robot inertial parameters for the identification of a nonlinear friction function, based on the well-known LuGre model, is compared with a complete dynamic calibration of the manipulator, including the estimation of both the robot dynamics and the parameters of a polynomial friction function. The identification results are discussed in the two cases, and inverse dynamics control schemes, based on the identified models, are experimentally applied to the manipulator for the execution of different trajectories, which allow the evaluation of the control performances in different conditions.

1 Introduction

To attain accurate geometric motion at very low velocities, parasitic phenomena, such as nonlinear friction and stiction acting at joints, must be investigated, understood, and explicitly taken into account in robotic systems.

Traditional controllers, based on linear PID algorithms, are unable to achieve good geometric performances, both on transient regimes and in static conditions, when friction is nonnegligible. Often, the conspicuous integrative term, necessary to reduce the steady-state error, may excite the neglected elastic dynamics. This is particularly true in direct-drive robots, where the beneficial effects of gearboxes are absent. In other circumstances, poorly tuned PID gains may produce sustained limit cycles, negatively affecting the robot behavior. Conversely, model-based controllers, embedding physically based models of the friction phenomena, have significantly improved the general performances of the manipulators [2, 3, 4, 5, 7, 8, 15, 18, 19].

The incorporation of friction terms in the manipulator model requires the solution of theoretical issues, such as the structure identification and parameter estimation problems, as well as practical ones connected to the real-time implementation of model-based control algorithms, such as computation time constraints and measurements availability.

The material presented in this chapter investigates these aspects in detail and experimentally tests the proposed solutions on a real robotic equipment, making use of a rapid prototyping architecture for real-time control design, available at our laboratory.[1]

This chapter is organized as follows: Section 2 describes the robotic plant used as a test case for the model-based control algorithms, as well as the rapid prototyping control environment, based on the QNX real-time operating system. Section 3 illustrates the dynamic model of the manipulator, including nonlinear terms taking care of friction and stiction phenomena, and the related identification process, reporting the results obtained under different simplifying assumptions. Section 4 accounts, compares, and discusses a number of experimental tests where the model-based controllers, whose parameters were estimated in the previous section, are applied to the robotic system. Section 5 draws the conclusions and outlines some possible lines of further investigation.

2 The robotic system

A two-revolute-link IMI planar manipulator, moving in a horizontal plane as sketched in Figure 1, has been used for the experimental tests. The link lengths are $L_1 = 0.3683$ m and $L_2 = 0.2413$ m; the angular limits are ± 2.15 rad for both joints; and the tip height from the robot base is 0.45 m.

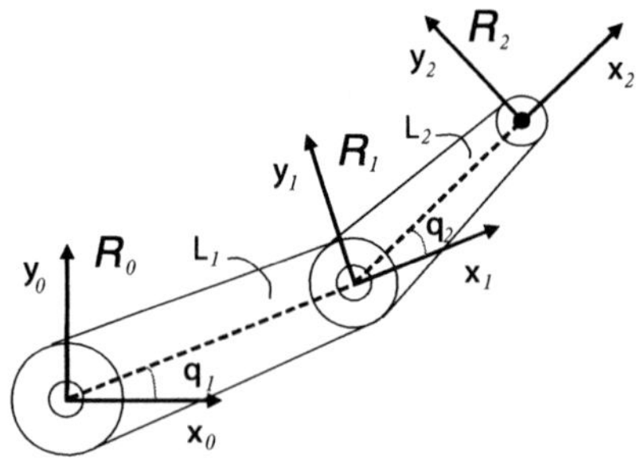

Fig. 1. Sketch of the IMI planar manipulator.

---

[1] See www.ladispe.polito.it/robotica/Activities/activities.htm.

Each joint is moved by a direct-drive (i.e., no reduction gears) brushless NSK Megatorque motor, actuated by a power drive that handles all the various functions of the motor itself (e.g., the digital input/output signals interchange and the application of analog command inputs) and makes available the angular measurements from the resolvers.

The drive cabinets contain the electronic boards for PWM motor command and the 16-bit A/D converters, which transform analog resolver signals into digital signals.

The signals coming from the controller are treated as torque or velocity reference commands, according to a selectable operating mode, respectively called *torque mode* or *velocity mode*. Torque mode is the default mode that will be used in the actual testing phase; when selected the actuator model can be reduced to a proportional gain $K_{V\tau}$ between the command input voltage, V_m, and the torque τ_c supplied by the motor. Velocity mode, in which an additional velocity loop is present, is used only in emergency situations, when the pressing of the emergency stop button must instantaneously stop the manipulator motion. The stopping phase is executed as specified by the internal velocity control algorithm.

The implementation and testing of model-based controllers require fast and systematic interactions between the algorithmic design phase and the experimental testing on the real plant, together with the possibility of acquiring all the data necessary for the model identification and for the evaluation of the control performances. It is therefore essential to use a PC-based rapid prototyping environment, where different controller blocks, each one with an easily modifiable structure and parametrization, may be designed and tested on a simulated plant. Subsequently the candidate controller solution must be translated in a suitable computer code and downloaded on the target hardware platform for real-time validation.

The prototyping architecture used for the experiments reported in the next sections is based on the RT-Lab environment by Opal-RT, which allows the distribution of the simulation from a Simulink project on a multiprocessor hardware architecture. In particular, the SOLO release, working with a Command Station and a unique Master target, has been used. The target runs the QNX real-time operating system, and hence a "hardware-in-the-loop simulation" can be easily performed, in which the controller is defined by a Simulink model, interacting with the controlled plant through I/O boards.

RT-Lab exploits the automatic code generation features supplied by Mathworks Real-Time Workshop (RTW) and Stateflow coder toolboxes, so it is possible to completely avoid the manual coding phase. Not only is the control scheme defined by a high-abstraction Simulink model, but the overall logic governing the plant working operations is also represented in the same environment by a finite state machine using the Stateflow tool, and then automatically coded in C. The C files are downloaded to the Master Node for the compilation phase, producing the final QNX executable file. Real-time constraints are automatically granted by the Target QNX machine.

Different control-loop algorithms, stored in a library of Simulink blocks, can be easily modified by the user, and data files can be saved on the Target hard disk. Due to the standard PC hardware requirements of the RT-Lab Host/Target architecture, no consistent memory or mass storage limitations are present. Sustained data acquisitions can be stored on the Target hard disk and retrieved by a simple FTP client for further analysis.

3 Identification of the robot dynamic model including friction

A model-based controller guarantees satisfactory performances, only if the robot dynamic model used in the control loop is sufficiently accurate [16].

The dynamic behavior of the considered manipulator can be described by the following second-order nonlinear differential equation:

$$M(q)\ddot{q} + C(q,\dot{q})\dot{q} + \tau_f(q,\dot{q}) = \tau_c \,, \tag{1}$$

where $q, \dot{q}$, and $\ddot{q}$ are the vectors of joint angles, angular velocities, and angular accelerations, $M(q)$ is the configuration-dependent inertia matrix, including both links and motors inertia, $C(q,\dot{q})\dot{q}$ is the term containing Coriolis and centrifugal torques, τ_f is the friction torque vector, and τ_c is the command torque vector. The gravitation effects are negligible since the motion plane is horizontal.

The precise knowledge of each term in (1) is necessary to implement model-based controllers, such as the inverse dynamics control scheme, which will be considered in the next section.

Two main basic issues define such an identification problem: the knowledge of the robot inertial parameters, which are present in $M(q)$ and $C(q,\dot{q})\dot{q}$, and the choice of a proper friction model to represent $\tau_f(q,\dot{q})$.

The structure of the inertial part of the robot dynamic model (i.e., the expressions of $M(q)$ and $C(q,\dot{q})$) is assumed to be correctly known, as it often happens in case of direct-drive motors, since the uncertainties related to the presence of elasticity and backlash in the gearboxes, are not present in such actuators.

The dynamic model (1) can be rewritten in the following form:

$$D_d(q,\dot{q},\ddot{q})\theta_d + \tau_f(q,\dot{q}) = \tau_c \,, \tag{2}$$

where the contributions of the inertial, centrifugal, and Coriolis torques have been regrouped in the term $D_d(q,\dot{q},\ddot{q})\theta_d$, which is linear with respect to the vector of the identifiable inertial parameters θ_d [6], [17], [12].

In particular, for the considered manipulator, it results $D_d(q,\dot{q},\ddot{q}) = [D_d(i,j)] \in \mathbb{R}^{2\times4}$, $\theta_d \in \mathbb{R}^4$, with

$$
\begin{aligned}
D_d(1,1) &= \ddot{q}_1 \\
D_d(1,2) &= L_1[(2\ddot{q}_1 + \ddot{q}_2)\cos q_2 - (2\dot{q}_1 + \dot{q}_2)\dot{q}_2 \sin q_2] \\
D_d(1,3) &= L_1[-(2\ddot{q}_1 + \ddot{q}_2)\sin q_2 - (2\dot{q}_1 + \dot{q}_2)\dot{q}_2 \cos q_2] \\
D_d(1,4) &= \ddot{q}_1 + \ddot{q}_2 \\
D_d(2,1) &= 0 \\
D_d(2,2) &= L_1(\ddot{q}_1 \cos q_2 + \dot{q}_1^2 \sin q_2) \\
D_d(2,3) &= L_1(-\ddot{q}_1 \sin q_2 + \dot{q}_1^2 \cos q_2) \\
D_d(2,4) &= \ddot{q}_1 + \ddot{q}_2
\end{aligned}
\tag{3a}
$$

$$
\theta_d = \big[\, (\Gamma_{1z} + m_2 l_1^2) \quad m_2 s_{2x} \quad m_2 s_{2y} \quad \Gamma_{2z} \,\big]^T ,
\tag{3b}
$$

where $(\Gamma_{1z} + m_2 l_1^2)$ and Γ_{2z} are the overall inertia moments of the two links with respect to the axis perpendicular to the motion plane, and $m_2 s_{2x}$, $m_2 s_{2y}$ are the first-order moments of the second link.

Nominal values of the inertial parameters, coming from data sheets and/or direct measurements, can be used both in the friction identification phase (if needed), and in the expression of the model-based control law. The differences between the actual inertial, centrifugal, and Coriolis torques and the computed ones will affect both the friction torques data, considered to identify the friction parameters, and the performances of the implemented model-based controller. Otherwise a complete identification of the robot dynamic parameters can be performed, including both the inertial part of the model and the friction torques. Both procedures will be investigated in the remainder of the chapter.

Several choices are possible for the representation of the friction torques $\tau_f(q, \dot{q})$, as different models have been proposed in literature, see, e.g., [1, 2, 7, 8, 9, 10, 13, 18], to describe in a more or less accurate way all the friction components. A basic classification of the friction models is relative to their *static* or *dynamic* characteristics [15].

The classical *static* models describe friction as a function of the *current* relative velocity of the bodies in contact, taking into account some of the various aspects of the friction force, such as Coulomb friction, viscous friction, stiction, and the Stribeck effect, which is relative to the low-velocities region, in which friction decreases as velocity increases. Different nonlinear functions have been proposed to describe such aspects, the simplest choice being a polynomial function of a sufficiently high order.

The model proposed in [2] can be considered as an intermediate step toward the actual *dynamic* models, in which presliding displacement is handled and addressed by a proper dynamic friction model, dealing with the behavior of the microscopical contact points between the surfaces.

The well-known *LuGre* model [8, 15] (with some recent modifications in [11], [14]) takes into account both the steady-state friction curve and the presliding phase by means of flexible bristles, representing the contact points of the moving surfaces. According to such a model, the behavior of the friction torque τ_{fi} on the ith joint of a manipulator can be described by the following equations:

$$\dot{z}_i = \dot{q}_i - \frac{|\dot{q}_i|}{g_i(\dot{q}_i)}\sigma_{0i}z_i \tag{4}$$

$$\tau_{fi} = \sigma_{0i}z_i + \sigma_{1i}\dot{z}_i + f_i(\dot{q}_i), \tag{5}$$

where $\dot{q}_i$ is the angular velocity of the joint, z_i is a state variable representing the average bristle deflection for joint i, σ_{0i} and σ_{1i} are model parameters assumed to be constant, and $g_i(\dot{q}_i)$ and $f_i(\dot{q}_i)$ model the Stribeck effect and the viscous friction, respectively. For constant velocity, the steady-state friction torque is then given by

$$\tau_{fi_{ss}} = g_i(\dot{q}_i)\,\mathrm{sgn}(\dot{q}_i) + f_i(\dot{q}_i). \tag{6}$$

Different parameterizations are possible for $g_i(\dot{q}_i)$ and $f_i(\dot{q}_i)$: the first one is a nonlinear function of velocity, generally expressed by means of exponential terms, while the second one can be given by a simple linear viscous function or by a higher-order polynomial function.

In the next subsections two different approaches have been followed for the identification of the complete robot dynamic model. In the first one, inertial parameters are defined by their nominal values, whereas the static, nonlinear part of the LuGre model is used to compute the friction torques, assuming different parameters when negative or positive velocity values are involved. In the second one, a simpler polynomial function of the joint velocity is considered to approximate each friction torque (independently of the velocity sign), but its parameters are identified together with the inertial ones, on the basis of the data acquired while the robot executes a trajectory that "optimizes" in some sense the excitation of the manipulator dynamics. No dynamic component is considered for friction, since the identification of its parameters would require high-precision sensors, which are not available in the present experimental set-up. Only joint position measurements are available, whereas velocities and accelerations are computed using low-pass numerical filters.

3.1 Identification of the static part of the LuGre friction model

Different expressions can be used for $g_i(\dot{q}_i)$ and $f_i(\dot{q}_i)$ to obtain a better fitting with the acquired experimental data. The *standard* ones, considered in [8, 15], are given by

$$g_i(\dot{q}_i) = \left(\alpha_{0i} + \alpha_{1i}\,e^{-\left(\frac{\dot{q}_i}{\omega_{s,i}}\right)^2}\right)\mathrm{sgn}(\dot{q}_i) \tag{7}$$

$$f(\dot{q}_i) = \alpha_{2i}\dot{q}_i. \tag{8}$$

The parameters of such functions must be identified by collecting data, while joints are moved at constant velocity values by means of some kind of joint-independent (not model-based) control law. A PD control law, with sampling time of 1 ms, has been actually considered, and large data session have been acquired to derive friction torque samples from the manipulator dynamic equations (1), as

$$\tau_f(\dot{q}) + \tau_{err} = \tau_c - D_d(q, \dot{q}, \ddot{q})\theta_d,\tag{9}$$

where τ_{err} is a torque vector that contains all neglected modeling errors including dynamic friction phenomenon. The following nominal inertial parameters values (in the proper SI units) were used:

$$\theta_d = \theta_{d,nom} = \begin{bmatrix} 3.69 & 0.97 & 0 & 0.27 \end{bmatrix}^T.\tag{10}$$

Experimental friction data, collected both at low and high velocities, and indicated by stars in Figures 2 and 3, have been fitted using expressions (7), (8) for $g_i(\dot{q}_i)$ and $f_i(\dot{q}_i)$, by considering, for each joint, different values for parameters $\omega_{s,i}$, and $\alpha_{0i} \ldots \alpha_{2i}$ when positive or negative velocities are involved.

A least-squares (LS) algorithm has been applied to estimate the α parameters for each joint by considering, in subsequent iterations, tentative values of $\omega_{s,i}$ between 0.10 and 0.35 rad/s. This range has been chosen by inferring from the acquired data the Stribeck region, in which friction decreases as velocity increases. An alternative solution can be found using a nonlinear estimation technique, for the simultaneous identification of all the parameters. The estimated values are reported in Table 1.

Table 1. Estimated static parameters of the LuGre friction model.

	Joint 1 $\omega > 0$	Joint 1 $\omega < 0$	Joint 2 $\omega > 0$	Joint 2 $\omega < 0$
α_0	7.78	7.78	2.42	2.29
α_1	-0.60	0.63	0.54	0.72
α_2	1.24	1.62	0.27	0.43
ω_s	0.31	0.10	0.15	0.15

Figures 2 and 3 compare the friction torques (solid line), computed by (6), (7), (8) using the values reported in Table 1, with the acquired friction data indicated by stars.

As the figures show, whereas satisfying results have been obtained for the first joint (Figure 2), an adequate data fitting has not been achieved for the second joint (Figure 3). It must be considered that the *acquired* friction torque data do not come from direct measurements, but are actually derived from the manipulator dynamic model, as illustrated before. Such data are affected: i) by errors in the computation of matrix $D_d(q, \dot{q}, \ddot{q})$, due to parameter uncertainties, measurement noise in the joint positions, and computational approximations in the determination of the joint velocities and accelerations, and ii) by modeling errors and disturbances, generically collected in the τ_{err} term in (9). It is evident from Figure 3 that the samples obtained when the velocity of the second joint is greater than 4 rad/s (in absolute value) are affected by relevant errors and/or disturbances.

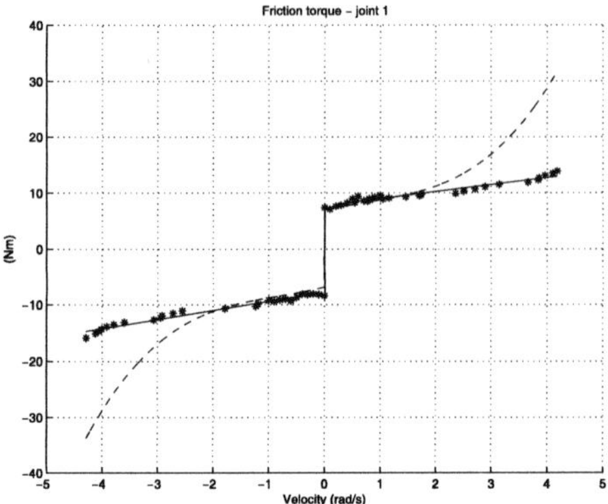

Fig. 2. Friction torque (Nm) on joint 1: acquired (stars), static LuGre model (solid), and third-order polynomial function (dashed).

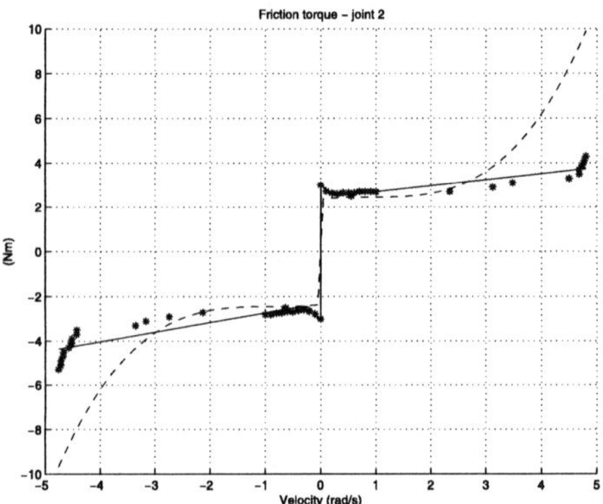

Fig. 3. Friction torque (Nm) on joint 2: acquired (stars), static LuGre model (solid), and third-order polynomial function (dashed).

Different strategies could be adopted to obtain better performances, such as introducing a more accurate robot dynamic model (e.g., including parasitic elastic terms for the second joint, which is smaller and lighter than the first one), estimating the robot inertial parameters (to eliminate or reduce the influence of parameter uncertainties in the torque computation), and/or considering a different friction model, possibly on the basis of a different data acquisition procedure.

In the following subsection, a *complete* robot dynamic parameter estimation is implemented, by considering matrix $D_d(q, \dot{q}, \ddot{q})$ as in (3a) (i.e., without introducing other dynamic contributions), and representing the friction torques by a third-order velocity polynomial function.

3.2 Complete dynamic model identification with a polynomial friction function

Let the friction torque on the ith joint be described by the following third-order polynomial function:

$$\tau_{fi} = a_{0i}\text{sign}(\dot{q}_i) + a_{1i}\dot{q}_i + a_{2i}\text{sign}(\dot{q}_i)\dot{q}_i^2 + a_{3i}\dot{q}_i^3 . \tag{11}$$

It is then possible to rewrite the manipulator dynamic model (1) in the following form:

$$\tau_c = D(q, \dot{q}, \ddot{q})\theta , \tag{12}$$

where $D(q, \dot{q}, \ddot{q}) = [D_d(q, \dot{q}, \ddot{q})\, D_f(\dot{q})] \in \mathbb{R}^{2 \times 12}$, $\theta = [\theta_d\, \theta_f]^T \in \mathbb{R}^{12}$, being $D_f(\dot{q}) = [D_f(j, k)] \in \mathbb{R}^{2 \times 8}$, $\theta_f \in \mathbb{R}^8$, with

$$
\begin{aligned}
&D_f(1,1) = \text{sgn}(\dot{q}_1) && D_f(1,2) = \dot{q}_1 \\
&D_f(1,3) = \text{sgn}(\dot{q}_1)\dot{q}_1^2 && D_f(1,4) = \dot{q}_1^3 \\
&D_f(1,5) = D_f(1,6) = 0 && D_f(1,7) = D_f(1,8) = 0 \\
&D_f(2,1) = D_f(2,2) = 0 && D_f(2,3) = D_f(2,4) = 0 \\
&D_f(2,5) = \text{sgn}(\dot{q}_2) && D_f(2,6) = \dot{q}_2 \\
&D_f(2,7) = \text{sgn}(\dot{q}_2)\dot{q}_2^2 && D_f(2,8) = \dot{q}_2^3
\end{aligned}
\tag{13a}
$$

$$\theta_f = \begin{bmatrix} a_{01} & a_{11} & a_{21} & a_{31} & a_{02} & a_{12} & a_{22} & a_{32} \end{bmatrix}^T . \tag{13b}$$

Equation (12) is linear with respect to vector θ, which contains all the parameters defining the manipulator model.

According to the method developed in [6], parameters have been estimated by a least-squares algorithm, collecting data while following an "optimal" trajectory (i.e., a trajectory that optimally excites the robot dynamics described by model (12)) of the following type:

$$q_i(t) = \sum_{j=1}^{n_a} \alpha_{j,i} \sin(\omega_{j,i}t) \tag{14}$$

for the ith joint. Different optimality criteria can be applied, such as the minimization of the estimates bias due to unmodeled dynamics errors, and their sensitivity to variations of the measured torque vector, or the minimization of the uncertainty bound of the parameter estimates. The first goal can be obtained minimizing the so-called J_k index, whereas the second one is performed by the maximization of the so-called J_d index (details can be found in [6] and in the references therein). As a somehow simplified model has been used (especially for friction), and no deeper investigation has been performed on possible unmodeled dynamics, the J_k optimality criterion has been considered as prior, thus obtaining a four-harmonics "optimal" trajectory function, defined in (14) with $n_a = 4$, $\alpha_{11} = -0.4355$, $\alpha_{21} = -0.4032$, $\alpha_{31} = -0.5$, $\alpha_{41} = -0.371$, $\alpha_{12} = 0.371$, $\alpha_{22} = 0.2097$, $\alpha_{32} = 0.4677$, $\alpha_{42} = 0.3387$,

$\omega_{11} = 0.7213$, $\omega_{21} = 0.8656$, $\omega_{31} = 1.082$, $\omega_{41} = 1.8754$, $\omega_{12} = 1.8754$, $\omega_{22} = 2.0918$, $\omega_{32} = 0.8656$, $\omega_{42} = 1.3705$ (the $\omega_{i,j}$ terms are expressed in rad/s). Such function was supplied as reference trajectory to the manipulator, for a time period of 15 s, and repeated ten times, to investigate the system repeatability and to estimate the measurement noise. Joint position and torque data have been collected, whereas velocities and accelerations have been computed numerically, with the insertion of a proper filtering action, as before.

The LS algorithm described in [6] has been used for the off-line parameter estimation, collecting more than 2500 equally spaced samples for each trajectory repetition. The LS algorithm yields the following parameters:

$$\hat{\theta} = [3.65 \quad 0.98 \quad -0.0012 \quad 0.29 \quad 6.84 \quad 2.62$$
$$-1.16 \quad 0.47 \quad 2.35 \quad 0.31 \quad -0.34 \quad 0.12]^T . \tag{15}$$

The order of magnitude of their variances is between 10^{-7} and 10^{-5} for the estimates of the inertial parameters, and between 10^{-4} and 10^{-2} for the friction ones. The comparison with the nominal values (10) of the inertial parameters confirms the validity of the estimates, at least for the inertial part of the model. The estimated friction parameters have been used to compute the friction torques on the joint according to (11). Figures 2 and 3 show such torques (dashed line); the velocity range is the same that was considered in the static term identification of the LuGre model in Section 3.1.

It is interesting to note that a quite satisfactory correspondence with the previously acquired friction data (stars) and the LuGre static model (solid line) is obtained only at low velocities. This is because the optimal trajectory corresponds to a joint velocity range of about $-1.5 \div 1$ rad/s, and it is just within this range that the estimated polynomial friction model can be considered as valid. Again the estimate for the second joint (Figure 3) is more critical, suggesting the possible presence of significant modeling errors (i.e., the considered "friction" term is not given by pure friction).

It must be stressed that, since our main goal is the development of a manipulator dynamic model suitable for control purposes, the actual validity of the identified models must be judged by evaluating their capacity to approximate the overall robot dynamics, given by the torque vector τ_c in (12).

The estimated parameters have been used to compute the model torques $\hat{\tau} = D\hat{\theta}$ from (12), which are compared with the measured ones, considering two cases: a) the execution of the four-harmonic trajectory (14), and b) a Cartesian circular trajectory, centered in $x = 0.4$ m, $y = 0.2$ m, with radius 0.12 m. The relative estimation errors, defined as the ratio between the rms estimation error and the rms value of the measured torques, are given as $\varepsilon_{r,1} = 31.8\,\%$, $\varepsilon_{r,2} = 31.7\,\%$ for case a), and $\varepsilon_{r,1} = 33.1\,\%$, $\varepsilon_{r,2} = 39.9\,\%$ for case b), when performed with joint velocities not greater than 2 rad/s. In the latter case, similar or marginally worse results are obtained using the nominal values of the inertial parameters together with the friction static LuGre model (identified in Section 3.1). Figures 4 and 5 compare the performed torque reconstructions with the acquired data for such a circular trajectory, and show the time history of the corresponding torque errors with the two identified models. It appears that the relative estimation errors are significantly worsened by local peaks of the torque errors, corresponding to changes of the velocity sign, especially for the second joint. Satisfying performances could, however, be obtained in spite of the high values of $\varepsilon_{r,2}$, embedding such models for torque compensation within a control scheme. It is interesting to note also that the use of the more accurate static LuGre model seems to give no particular advantage in our case. In the next section the two solutions will be compared from the control point of view, by applying them into an inverse dynamics control scheme for the execution of different trajectories (and in particular of the considered circular one).

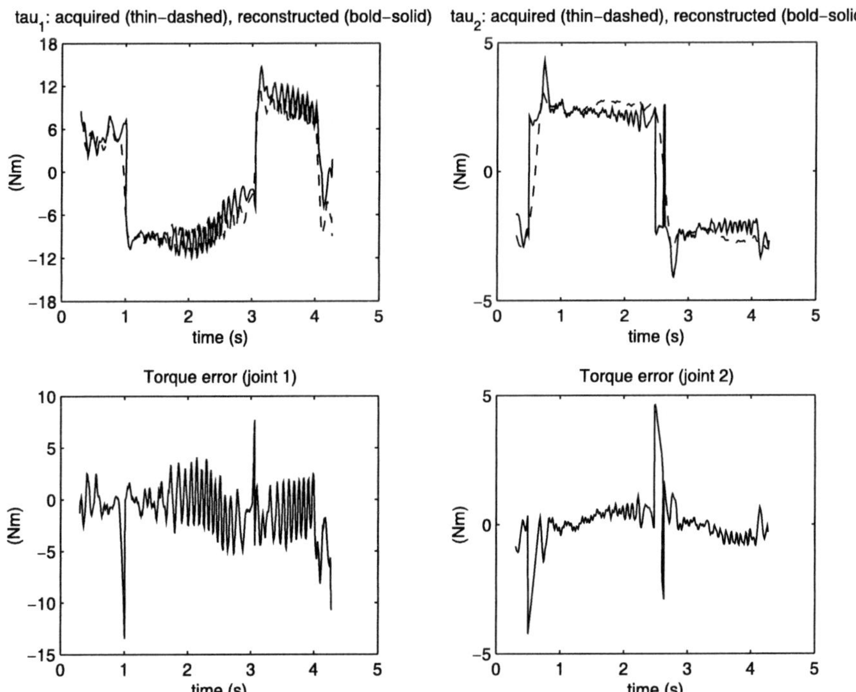

Fig. 4. Torque reconstruction for the circular trajectory with the polynomial friction function.

4 Experimental application of model-based controllers

Inverse dynamics control schemes, embedding the above identified models, are applied to the manipulator for the execution of trajectories at different

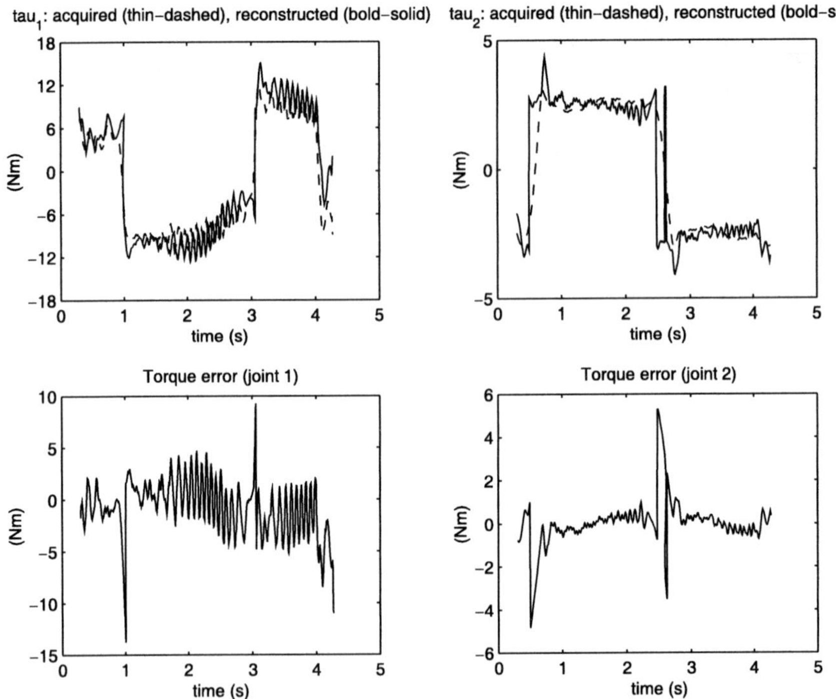

Fig. 5. Torque reconstruction for the circular trajectory with the static LuGre friction model.

velocities, to investigate the performances of different friction models in the control loop.

The model-based control scheme tested on the real plant is given by the following inverse dynamics control law:

$$\tau_c = D_d(q, \dot{q}, \ddot{q}_r - v_c)\hat{\theta}_d + \hat{\tau}_f(q, \dot{q}), \tag{16}$$

where $\ddot{q}_r$ is the reference acceleration vector, $\hat{\theta}_d$ and $\hat{\tau}_f(q, \dot{q})$ are estimates of the inertial parameter vector θ_d and of the friction torques $\tau_f(q, \dot{q})$, respectively. The term v_c represents the command vector of the outer loop, obtained by a standard PID control algorithm, identical in all the tests. Three different solutions have been considered for $\hat{\theta}_d$ and $\hat{\tau}_f(q, \dot{q})$:

C1) $\hat{\theta}_d = \theta_{d,nom}$, given in (10), and $\hat{\tau}_f(q, \dot{q}) = 0$. Only the inertial part of the robot dynamics is compensated by the inner loop, on the basis of the nominal parameter values, without any friction compensation;

C2) $\hat{\theta}_d = \theta_{d,nom}$ and the friction torques $\hat{\tau}_{f,i}$ on each joint given by the static LuGre friction model (6), (7), (8), using the estimated values reported in Table 1;

C3) $\hat{\theta}_d$ given by the first four entries of the estimated vector $\hat{\theta}$ in (15), and the friction torques $\hat{\tau}_{f,i}$ on each joint computed from the third-order polynomial (11), using the estimated parameter values (the last eight entries of $\hat{\theta}$ in (15)).

The effectiveness of the different inverse dynamics controllers in following an assigned geometric Cartesian path has been tested by considering the circular trajectory previously defined, cyclically repeated. The circle has been tracked at different velocities to evaluate the friction compensation performances. The results obtained in two "extreme" cases are reported and discussed: the first one, denoted as LV, corresponds to *low* joint velocities (less than 2 rad/s), while the second one, denoted as HV, corresponds to *high* joint velocities (values greater than 4 rad/s).

The results obtained in the LV test are reported in Figures 6–8 and summarized in Table 2, which shows the maximum and the rms values of the joint and Cartesian errors.

Table 2. LV test results: joint and Cartesian errors.

LV Test	Errors					
	Joint 1 (mrad)		Joint 2 (mrad)		Cartesian (mm)	
	max	rms	max	rms	max	rms
C1	4.7	3.1	38.1	23.1	8.6	5.0
C2	2.8	0.8	3.2	1.8	1.1	0.5
C3	3.7	2.3	16.4	8.8	3.3	1.8

The second joint reveals stick-slip phenomena, due to stiction at low velocities. Figure 6, which compares the time history of joint velocities with the three control solutions, shows that when no friction compensation is inserted (solid line), the second joint briefly stops when the velocity changes sign. Such a behaviour is completely eliminated by the C2 solution (bold, dashed line), and strongly reduced by C3 (thin, dash-dotted line). The best friction reconstruction at very low velocities was in fact provided by the static LuGre friction model, as was shown in Figure 3.

The torque reconstruction performed by C2 leads to the best control performances. As shown in Table 2 and Figure 7, a strong reduction of the tracking errors is obtained (an entire order of magnitude for the second joint) with respect to C1, so that the circular trajectory is much better than that obtained with C1 (see Figure 8). The performances obtained by C3 are only marginally worse than those of C2 (see Table 2), so that a satisfactory tracking of the circular trajectory is obtained. As shown by Figure 8, the difference between the executed trajectory and the reference one can be appreciated only in some limited regions of the circumference, where the joint velocities change sign.

The test has been repeated also by considering an inner *feedforward* compensation loop, i.e., by using the reference position and velocity values q_r and

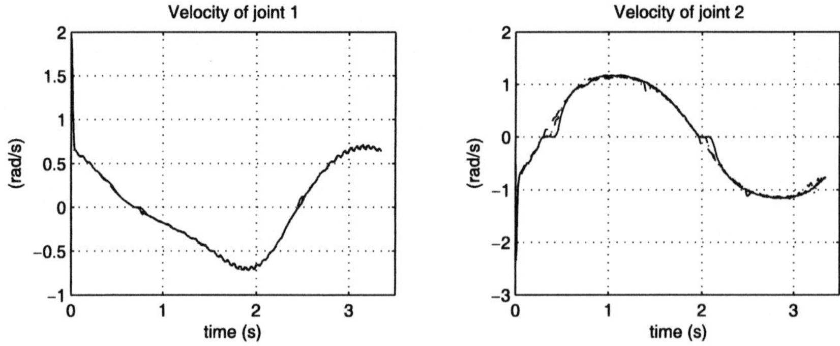

Fig. 6. LV test: joint velocities with C1 (solid), C2 (bold, dashed), and C3 (thin, dash-dotted) model-based control schemes.

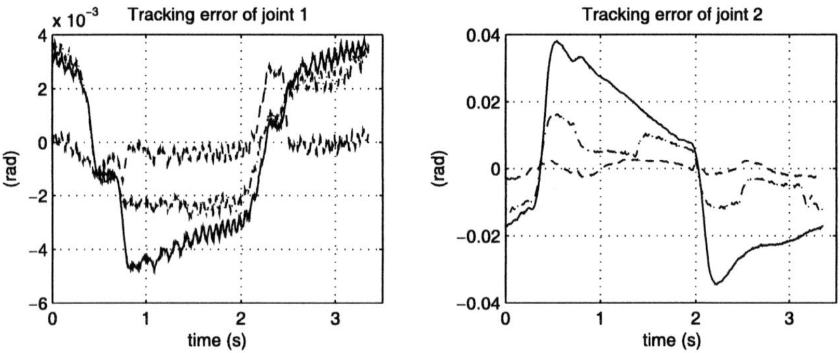

Fig. 7. LV test: joint tracking errors with C1 (solid), C2 (bold, dashed) and C3 (thin, dash-dotted) model-based control schemes.

$\dot{q}_r$, instead of q and $\dot{q}$, respectively, in $D_d(q, \dot{q}, \ddot{q}_r - v_c)$ and $\hat{\tau}_f(q, \dot{q})$ in (16). While the control performances obtained by C1 and C3 do not significantly change, the maximum error of the second joint becomes a little worse for C2, passing from 3.2 mrad to 4.7 mrad, which does not lead to an appreciable worsening of the Cartesian error, as reported in Table 3.

Table 3. LV-feedforward test results: joint and Cartesian errors.

LV-FF Test	Errors					
	Joint 1 (mrad)		Joint 2 (mrad)		Cartesian (mm)	
	max	rms	max	rms	max	rms
C1	4.5	3.0	39.5	22.9	8.5	4.9
C2	3.0	0.9	4.7	1.7	1.0	0.5
C3	3.7	2.3	16.6	8.8	3.2	1.8

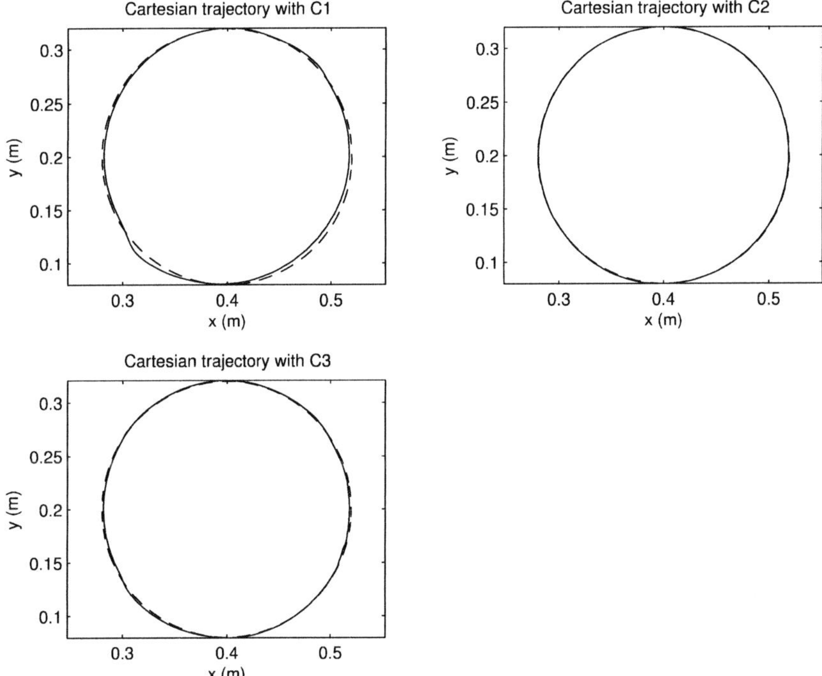

Fig. 8. LV test: Cartesian reference (dashed) and executed (solid) trajectory with the three model-based control schemes.

For the trajectory of the LV test, feedforward compensation can then be successfully applied whenever needed to reduce real-time computational burden, by precalculating off-line the torque compensation terms on the basis of one of the identified models, and by inserting them in the control software by means of a *look-up table* procedure.

The results obtained in the HV test are reported in Figures 9 and 10 and summarized in Table 4, which shows the maximum and the rms values of the joint and Cartesian errors.

Table 4. HV test results: Joint and Cartesian errors.

HV Test	Errors					
	Joint 1 (mrad)		Joint 2 (mrad)		Cartesian (mm)	
	max	rms	max	rms	max	rms
C1	6.2	2.7	37.3	21.6	8.4	5.2
C2	6.4	3.0	26.1	16.5	5.9	3.5
C3	10.1	4.7	27.3	14.1	4.9	2.9

Due to the higher joint velocities involved in this test, stick-slip phenomena are not present. Nevertheless, the trajectory executed by C1 presents large errors (see the first plot of Figure 10), as a result of the significant joint errors reported in Figure 9 by a solid line. Comparing the Cartesian errors in Tables 2 and 4, it is shown that the Cartesian error obtained by C1 is a little bit worse in this "high-velocity" test. Friction compensation is then fundamental in this case.

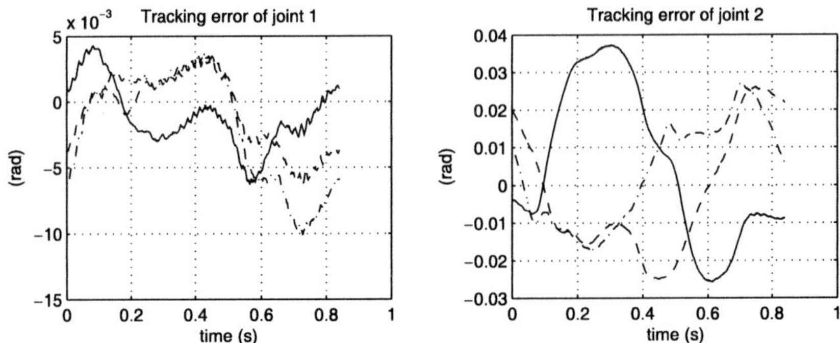

Fig. 9. HV test: joint tracking errors with C1 (solid), C2 (bold, dashed), and C3 (thin, dash-dotted) model-based control schemes.

The application of the C2 control scheme, including the friction compensation provided by the identified LuGre static model, does not lead to the satisfactory results that were achieved in the LV test. Only a small reduction of the second joint error (see Figure 9; bold, dashed line), and consequently of the Cartesian one (see Table 4), is obtained. The execution of the circular trajectory results to be quite poor, as shown by the second plot of Figure 10. The application of the C3 control solution determines a little degradation of the tracking error of the first joint, but the reduction of the rms value of the second joint error allows a reduction of the Cartesian error, as reported in Table 4. In the velocity range relative to the HV test, the friction torque computed from (11) gives better results than the LuGre model, at least for the second joint. The resulting circular trajectory, executed by the robot with the C3 control solution, is the best one for the HV test, as is clearly shown in Figure 10.

The results obtained by repeating the HV test with the three control solutions, with a *feedforward* inner compensation loop, are similar to those obtained when feedback is used. There is some little further reduction of the errors in the C3 case, probably due to the use of "better," noiseless signals q_r and $\dot{q}_r$ instead of the actual q and $\dot{q}$. In particular, the resulting Cartesian error shows a maximum value of 4.3 mm (instead of 4.9 mm), and a rms

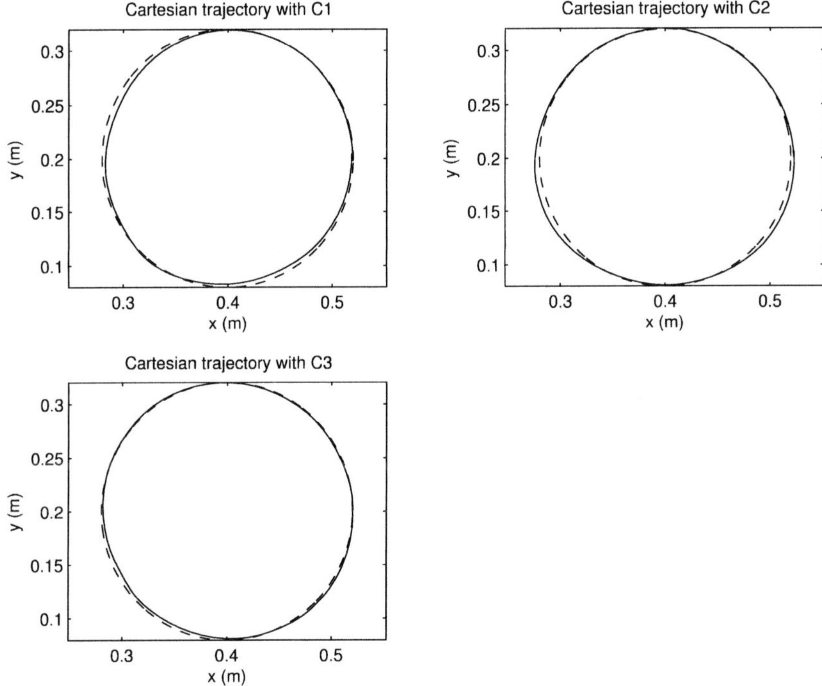

Fig. 10. HV test: Cartesian reference (dashed) and executed (solid) trajectory with the three model-based control schemes.

value of 2.7 mm (instead of 2.9 mm), as reported in Table 5. Feedforward compensation is then not only applicable in this case, but even preferable.

Table 5. HV-feedforward test results: joint and Cartesian errors.

HV-FF Test	Errors					
	Joint 1 (mrad)		Joint 2 (mrad)		Cartesian (mm)	
	max	rms	max	rms	max	rms
C1	5.9	2.2	35.2	17.4	8.0	4.4
C2	5.5	2.9	32.3	14.7	6.5	3.3
C3	9.2	4.5	25.3	13.7	4.3	2.7

A final test has been performed to check the effectiveness of the three control solution in a *regulation* task: starting from an initial condition corresponding to $q_1(0) = -2$ rad, $q_2(0) = -2$ rad, the same kind of reference trajectory has been assigned to both joints to lead their final positions to $q_1(t_f) = q_2(t_f) = 2$ rad. In particular, the reference trajectory has been computed as a sequence of two bang-bang profiles, from -2 to 0 and from 0 to 2 rad, crossing the zero position without a complete stop. No velocity sign change is present, and a maximum joint velocity of 3 rad/s is reached. Table 6 summarizes the obtained results, showing that the C3 solution allows the best tracking results for the second joint and the smallest final errors for both joints.

Table 6. Regulation test results: joint errors.

Reg. test	Errors					
	Joint 1 (mrad)			Joint 2 (mrad)		
	max	rms	final	max	rms	final
C1	8.0	1.8	1.0	38.7	21.3	15.5
C2	8.9	2.0	0.65	24.4	11.3	4.3
C3	9.0	2.2	0.1	14.0	6.5	0.85

It is interesting to note from the second plot of Figure 11 that for C1 and C2 negative peaks are reached, before the second joint error settles down to its final value. The desired final position is then significantly exceeded, before the final settling. When C3 is applied, no negative peak error is present for the second joint (thus reducing the corresponding Cartesian peak error, if the imposed final joint position corresponds to a desired Cartesian position for the end-effector), whereas the negative peak of the first joint error remains quite unchanged with the three control solutions, but its amplitude is limited to about 7 mrad.

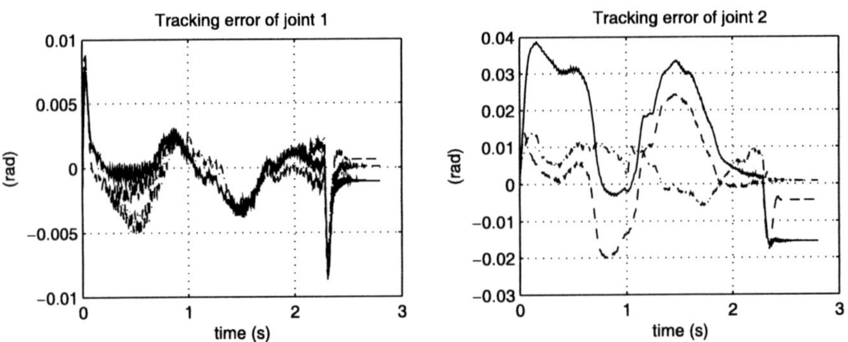

Fig. 11. Regulation test: joint tracking errors with C1 (solid), C2 (bold, dashed), and C3 (thin, dash-dotted) model-based control schemes.

The best regulation results, obtained by the third-order polynomial friction compensation with respect to the LuGre model ones, could be explained by analyzing the different procedures followed to collect data for friction identification in the two cases. As explained in Sections 3.1 and 3.2, the friction torque data were collected for the LuGre static model by moving the joints at assigned constant velocity values, whereas the data for the identification of the parameters of the third-order polynomial function (together with the inertial one) were acquired during the execution of an assigned, optimal trajectory. The procedure followed in the first case could be more sensitive to hysteresis effects not included in the model, i.e., different friction torque values could be evaluated at the same joint velocity, if such a velocity is reached starting from different conditions (increasing or decreasing the velocity). The hysteresis effects, if present, are particularly critical at low velocities, so that the friction compensation performed by C2 when the joints are going to stop could be inadequate, because the data used in the identification of the included friction model could be inconsistent with the condition occurring in the performed test. The compensation obtained by C3 seems to be more robust, since the data used in the identification were collected during the execution of a trajectory of the type given in (14), which includes various passages through zero velocity, at different conditions.

The execution of the regulation test with the three control solutions with a *feedforward* inner compensation loop has led to small variations of the final errors with respect to the feedback implementation. Final errors lower than 1 mrad have been obtained with the C3 solution for both joints also in this case (even if the final value of the first one has passed from 0.1 to 0.43 mrad), whereas the final error of the second joint has shown a more significant worsening when C2 is applied, passing from 4.3 to 7.2 mrad, confirming a larger sensitivity of this solution.

5 Conclusions

When the effectiveness of a model-based controller must be experimentally executed on different manipulator tasks, it is fundamental to pass rapidly from the design to the experimental testing phase, with the possibility of acquiring large amounts of data. The choice of the dynamic model to be used in a model-based control scheme is crucial for the attainable performances of the particular tasks to be executed and is greatly affected by the actual identification stage based on the available sensors. As the experimental results have shown, the choice of an accurate friction model (as in the C2 control scheme) has led to the best results at low velocities (LV test), when friction is particularly significant, but not in the high-velocity range (HV test) and in the regulation one. These results make us conclude that the tracking performances are strongly dependent on the executed task. Then, from the control point of view, the polynomial friction model could be preferable, since the quality of the achieved tracking results are comparable in all the tests. In this case it becomes possible to define some general accuracy performances that can be guaranteed independently of the particular trajectory executed, inserting the compensation term in feedback or in feedforward without significant differences.

References

1. Armstrong-Hélouvry B (1991) Control of machines with friction. Kluwer Academic Publishers, Boston, MA
2. Armstrong-Hélouvry B, Dupont P, Canudas de Wit C (1994) A survey of models, analysis tools and compensation methods for the control of machines with friction. Automatica 30(7):1083–1138
3. Bona B, Indri M, Smaldone N (2002) An experimental setup for modelling, simulation and fast prototyping of mechanical arms. 207–212. In: IEEE Conf. on Computer-Aided Control Systems Design, Glasgow, UK
4. Bona B, Indri M, Smaldone N (2003a) Nonlinear friction phenomena in direct-drive robotic arms: An experimental set-up for rapid modelling and control prototyping. 59–64. In: 7th IFAC 2003 Symposium on Robot Control, Wroclaw, Poland
5. Bona B, Indri M, Smaldone N (2003b) Nonlinear friction estimation for digital control of direct-drive manipulators. In: European Control Conference (ECC'03), Cambridge, UK
6. Calafiore G, Indri M, Bona B (2001) Robot dynamic calibration: optimal excitation trajectories and experimental parameter estimation. J. Robotic Systems 18(2):55–68
7. Canudas de Wit C, Noël P, Aubin A, Brogliato B (1991) Adaptive friction compensation in robot manipulators: Low velocities. Int. J. of Robotic Research 10(3):189–199
8. Canudas de Wit C, Olsson H, Åström K, Lischinsky P (1995) A new model for control of systems with friction. IEEE Trans. on Automatic Control 40(3):419–425
9. Chen YY, Huang PY, Yen JY (2002) Frequency-domain identification algorithms for servo systems with friction. IEEE Trans. on Control Systems Technology 10(5):654–665
10. Dupont P, Armstrong B, Hayward V (2000) Elasto-plastic friction model: contact compliance and stiction. 1072–1077. In: 2000 American Control Conference, Chicago, IL
11. Dupont P, Hayward V, Armstrong B, Altpeter F (2002) Single state elasto-plastic friction models. IEEE Trans. on Automatic Control 47(5):787–792
12. Gautier M, Khalil W (1990) Direct calculation of minimum set of inertial parameters of serial robots. IEEE Trans. on Robotics and Automation 6(3):368–373
13. Hensen R, van de Molengraft M, Steinbuch M (2002) Frequency domain identification of dynamic friction model parameters. IEEE Trans. on Control Systems Technology 10(2):191–196
14. Lampaert V, Swevers J, Al-Bender F (2002) Modification of the leuven integrated friction model structure. IEEE Trans. on Automatic Control 47(4):683–687
15. Olsson H, Åström K, Canudas de Wit C, Gäfvert M, Lischinsky P (1998) Friction models and friction compensation. European Journal of Control 4:176–195
16. Sciavicco L, Siciliano B (2000) Modelling and control of robot manipulators, 2nd edition. Springer, Berlin
17. Sheu SY, Walker M (1989) Estimating the essential parameter space of the robot manipulator dynamics. 2135–2140. In: 28th Conference on Decision and Control, Tampa, FL

18. Swevers J, Al-Bender F, Ganseman C, Prajogo T (2000) An integrated friction model structure with improved presliding behaviour for accurate friction compensation. IEEE Trans. on Automatic Control 45(4):675–686
19. Tataryn P, Sepehri N, Strong D (1996) Experimental comparison of some compensation techniques for the control of manipulators with stick-slip friction. Control Engineering Practice 4(9):1209–1219

A Singular Perturbation Approach to Control of Flexible Arms in Compliant Motion

Bruno Siciliano and Luigi Villani

Dipartimento di Informatica e Sistemistica
Università degli Studi di Napoli Federico II
Via Claudio 21, 80125 Napoli, Italia
{siciliano,lvillani}@unina.it

Summary. The problem of controlling the interaction of a flexible link arm with a compliant environment is considered. The arm's tip is required to keep contact with a surface by applying a constant force and maintaining a prescribed position or following a desired path on the surface. Using singular perturbation theory, the system is decomposed into a slow subsystem associated with rigid motion and a fast subsystem associated with link flexible dynamics. A parallel force and position control developed for rigid robots is adopted for the slow subsystem, while a fast control action is employed to stabilize the link deflections. Simulation results are presented for a two-link planar arm under gravity in contact with an elastically compliant surface.

1 Introduction

Lightweight flexible robots offer many advantages over conventional industrial robots, like high speed, large workspace, and high payload-to-arm weight [1]. In fact, they are conveniently employed in a large variety of fields including teleoperation, space robotics, and nuclear waste manipulation.

The dynamics of multilink flexible arms are, however, much more complex than rigid robot dynamics, due to the distributed flexibility of the links [2]. As a consequence, several challenging problems are still open, regarding both modeling and control aspects.

From the modeling standpoint, the dynamics of a flexible structure are described by an infinite-dimensional model. Various techniques have been proposed to achieve approximate finite-dimensional models, e.g., the assumed modes method, the finite elements method, and the Ritz-Kantorovich expansion. In the case of multilink flexible arms, a recursive procedure can be set up for dynamic model computation by using a Lagrangian formulation in conjunction with the assumed mode technique [3].

The inherent difficulty of the control problem can be ascribed to the fact that the number of controlled variables is strictly less than the number of mechanical degrees of freedom. Moreover, the dynamic relation between the input torques of the joint actuators and the tip position reveals a behavior that is the nonlinear counterpart of the nonminimum phase phenomenon of linear systems. Hence, inversion-based control strategies would normally lead to instability in the closed loop. See [4] and the references therein for further discussion about modeling and control problems for flexible link arms.

An effective approach to motion control design is based on singular perturbation theory [5]. When the link stiffness is large, a two-time scale model of the flexible arm can be derived [6], consisting of a slow subsystem corresponding to the rigid body motion and a fast subsystem describing the flexible motion. A composite control strategy can then be applied, based on a slow control designed for the equivalent rigid arm and a fast control that stabilizes the fast subsystem. Further developments of perturbation techniques for flexible arms can be found in [7, 8, 9, 10].

When the arm interacts with an external environment, suitable strategies have to be adopted to control both the tip position and the contact force. In fact, during the interaction, the environment sets constraints on the geometric path that can be followed by the end effector, and high contact forces may arise if purely motion strategies are adopted. The higher the environment stiffness and position control accuracy are, the easier an unstable behavior with damage to the robot or to the environment may occur. Hence, interaction control should ensure a suitable compliant behavior to the robotic arm to limit the contact forces.

Notice that the intrinsic compliance of a flexible link arm may reduce the value of the forces that can be generated when the interaction task is executed by a rigid robot. This means that by using flexible robots to perform interaction tasks some benefits may arise, even though the distributed flexibility of the links makes the interaction control problem more complex than for rigid robots.

The most common solution to interaction control is the use of a force/torque sensor, mounted between the last link and the end effector, which provides force measurements that can be suitably exploited by the robot controller.

While several control schemes have been proposed to force and position control of rigid robot manipulators [11], only few papers on interaction control of flexible arms have been published so far.

Early works addressing stability problems in force-controlled flexible manipulators are [12, 13]. Models for multilink constrained flexible robots have been developed in [14, 15] where a hybrid position and force control approach is adopted. Hybrid control is used in [16] and [17] to design robust and adaptive control strategies, respectively, as well as in [18, 19] to control a flexible macro manipulator carrying a rigid micro. In most of these papers ([13, 15, 17]), singular perturbation techniques are exploited to cope with link flexibility; a singular perturbed model for a constrained multilink flexible arm was developed in [20].

The singular perturbation method is adopted in this chapter to design a force and position control for flexible arms based on the parallel approach developed in [21, 22] for rigid robots in contact with compliant environments. As opposed to the hybrid control strategies where force and position are controlled in reciprocal subspaces [23, 24], both force and position variables are used in each subspace without any selection mechanism. This makes parallel controllers suitable to manage contacts with nonperfectly known environments and unplanned collisions, which represent a drawback for hybrid controllers. Moreover, differently from previous works tackling the problem of force and position control of flexible manipulators (e.g., [13, 15, 17, 20]), the equations of the constraint environment do not have to be taken into account for control design.

The proposed control scheme guarantees force regulation and position tracking for the slow dynamics. An additional control action is required in both cases to stabilize the fast dynamics related to link flexibility.

Simulation results have been carried out on the model of a two-link planar arm developed in [25]; interaction with an elastically compliant plane have been considered. The numerical case study confirms the results anticipated in theory.

2 Modeling

Consider a robot arm composed of a serial chain of n flexible links connected by rigid revolute joints subject only to bending deformations in the plane of motion, without torsional effects. A sketch of a two-link arm is shown in Figure 1 with coordinate frame assignment; the tip of the robotic arm is assumed to be in contact with a planar surface. The rigid motion is described by the joint angles ϑ_i, while $w_i(x_i)$ denotes the transversal deflection of link i at x_i with $0 \leq x_i \leq \ell_i$, being ℓ_i the link length.

2.1 Kinematics

Let $\boldsymbol{p}_i^i(x_i) = [x_i \; w_i(x_i)]^\mathrm{T}$ be the position of a point along the deflected link i with respect to frame (X_i, Y_i) and $\boldsymbol{p}_i$ be the position of the same point in the base frame. Also let $\boldsymbol{r}_{i+1}^i = \boldsymbol{p}_i^i(\ell_i)$ be the position of the origin of frame (X_{i+1}, Y_{i+1}) with respect to frame (X_i, Y_i), and $\boldsymbol{r}_{i+1}$ its position in the base frame.

The joint (rigid) rotation matrix $\boldsymbol{R}_i$ and the rotation matrix $\boldsymbol{E}_i$ of the (flexible) link at the end point are, respectively,

$$\boldsymbol{R}_i = \begin{bmatrix} \cos \vartheta_i & -\sin \vartheta_i \\ \sin \vartheta_i & \cos \vartheta_i \end{bmatrix} \tag{1}$$

and

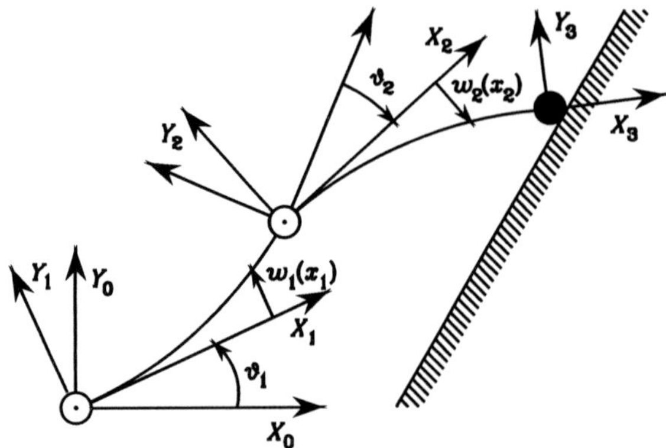

Fig. 1. Planar two-link flexible arm in contact with a planar surface.

$$E_i = \begin{bmatrix} 1 & -w'_{ie} \\ w'_{ie} & 1 \end{bmatrix}, \tag{2}$$

where $w'_{ie} = (\partial w_i / \partial x_i)|_{x_i = \ell_i}$, and the approximation $\arctan w'_{ie} \simeq w'_{ie}$, in the hypothesis of small deflections, has been made. Hence the above absolute position vectors can be expressed as

$$\boldsymbol{p}_i = \boldsymbol{r}_i + \boldsymbol{W}_i \boldsymbol{p}_i^i \tag{3}$$

and

$$\boldsymbol{r}_{i+1} = \boldsymbol{r}_i + \boldsymbol{W}_i \boldsymbol{r}_{i+1}^i, \tag{4}$$

where $\boldsymbol{W}_i$ is the global transformation matrix from the base frame to (X_i, Y_i) given by the recursive equation

$$\boldsymbol{W}_i = \boldsymbol{W}_{i-1} \boldsymbol{E}_{i-1} \boldsymbol{R}_i = \widehat{\boldsymbol{W}}_{i-1} \boldsymbol{R}_i, \tag{5}$$

with

$$\widehat{\boldsymbol{W}}_0 = \boldsymbol{I}. \tag{6}$$

On the basis of the above relations, the kinematics of any point along the arm are completely specified as a function of joint angles and link deflections.

A finite-dimensional model (of order m_i) of link flexibility can be obtained by the assumed mode technique [2]. Links are modeled as Euler-Bernoulli beams of uniform density ρ_i and constant flexural rigidity $(EI)_i$, with deflection $w_i(x_i, t)$ satisfying the partial differential equation

$$(EI)_i \frac{\partial^4 w_i(x_i, t)}{\partial x_i^4} + \rho_i \frac{\partial^2 w_i(x_i, t)}{\partial t^2} = 0, \qquad i = 1, \ldots, n. \tag{7}$$

Exploiting separability in time and space of solutions of (7), the link deflection $w_i(x_i, t)$ can be expressed as the sum of a finite number of modes

$$w_i(x_i, t) = \sum_{i=1}^{m_i} \phi_{ij}(x_i)\delta_{ij}(t), \tag{8}$$

where $\phi_{ij}(x)$ is the shape assumed for the jth mode of link i, and $\delta_{ij}(t)$ is its time-varying amplitude. The mode shapes have to satisfy proper boundary conditions at the base (clamped) and at the end of each link (mass).

In view of (8), a direct kinematics equation can be derived expressing the (2×1) position vector $\boldsymbol{p}$ of the arm tip point as a function of the $(n \times 1)$ joint variable vector $\boldsymbol{\vartheta} = [\vartheta_1 \dots \vartheta_n]^{\mathrm{T}}$ and the $(m \times 1)$ deflection variable vector $\boldsymbol{\delta} = [\delta_{11} \dots \delta_{1m_1} \dots \delta_{n1} \dots \delta_{nm_n}]^{\mathrm{T}}$ [3, 25], i.e.,

$$\boldsymbol{p} = \boldsymbol{k}(\boldsymbol{\vartheta}, \boldsymbol{\delta}). \tag{9}$$

For later use, the differential kinematics is also needed. The absolute linear velocity of an arm point is

$$\dot{\boldsymbol{p}}_i = \dot{\boldsymbol{r}}_i + \dot{\boldsymbol{W}}_i \boldsymbol{p}_i^i + \boldsymbol{W}_i \dot{\boldsymbol{p}}_i^i, \tag{10}$$

with $\dot{\boldsymbol{r}}_{i+1}^i = \dot{\boldsymbol{p}}_i^i(\ell_i)$. Since the links are assumed inextensible ($\dot{x}_i = 0$), then $\dot{\boldsymbol{p}}_i^i(x_i) = [0 \; \dot{w}_i(x_i)]^{\mathrm{T}}$. The computation of (10) takes advantage of the recursion

$$\dot{\boldsymbol{W}}_i = \widehat{\dot{\boldsymbol{W}}}_{i-1}\boldsymbol{R}_i + \widehat{\boldsymbol{W}}_{i-1}\dot{\boldsymbol{R}}_i \tag{11}$$

with

$$\widehat{\dot{\boldsymbol{W}}}_i = \dot{\boldsymbol{W}}_i\boldsymbol{E}_i + \boldsymbol{W}_i\dot{\boldsymbol{E}}_i. \tag{12}$$

Also, note that

$$\dot{\boldsymbol{R}}_i = \boldsymbol{S}\boldsymbol{R}_i\dot{\vartheta}_i, \quad \dot{\boldsymbol{E}}_i = \boldsymbol{S}\dot{w}'_{ie} \tag{13}$$

with

$$\boldsymbol{S} = \begin{bmatrix} 0 & -1 \\ 1 & 0 \end{bmatrix}. \tag{14}$$

In view of (9)–(14), it is not difficult to show that the differential kinematics equation expressing the tip velocity $\dot{\boldsymbol{p}}$ as a function of $\dot{\boldsymbol{\vartheta}}$ and $\dot{\boldsymbol{\delta}}$ can be written in the form

$$\dot{\boldsymbol{p}} = \boldsymbol{J}_\vartheta(\boldsymbol{\vartheta}, \boldsymbol{\delta})\dot{\boldsymbol{\vartheta}} + \boldsymbol{J}_\delta(\boldsymbol{\vartheta}, \boldsymbol{\delta})\dot{\boldsymbol{\delta}}, \tag{15}$$

where $\boldsymbol{J}_\vartheta = \partial\boldsymbol{k}/\partial\boldsymbol{\vartheta}$ and $\boldsymbol{J}_\delta = \partial\boldsymbol{k}/\partial\boldsymbol{\delta}$.

2.2 Dynamics

Using the assumed modes link approximation (8), a finite-dimensional Lagrangian dynamic model of the planar arm can be obtained as a function of the $n + m$ vector of generalized coordinates $\boldsymbol{q} = [\boldsymbol{\vartheta}^{\mathrm{T}} \; \boldsymbol{\delta}^{\mathrm{T}}]^{\mathrm{T}}$ in the form [3, 25]

$$\boldsymbol{B}(\boldsymbol{q})\ddot{\boldsymbol{q}} + \boldsymbol{c}(\boldsymbol{q}, \dot{\boldsymbol{q}}) + \boldsymbol{g}(\boldsymbol{q}) + \begin{bmatrix} \boldsymbol{0} \\ \boldsymbol{D}\dot{\boldsymbol{\delta}} + \boldsymbol{K}\boldsymbol{\delta} \end{bmatrix} = \begin{bmatrix} \boldsymbol{\tau} \\ \boldsymbol{0} \end{bmatrix}, \tag{16}$$

where $\boldsymbol{B}$ is the positive definite symmetric inertia matrix, $\boldsymbol{c}$ is the vector of Coriolis and centrifugal torques, $\boldsymbol{g}$ is the vector of gravitational torques, $\boldsymbol{K}$ is the diagonal and positive definite link stiffness matrix, $\boldsymbol{D}$ is the diagonal and positive semidefinite link damping matrix, and $\boldsymbol{\tau}$ is the vector of the input joint torques.

In the case that the arm's tip is in contact with the environment, by virtue of the virtual work principle, the vector $\boldsymbol{f}$ of the forces exerted by the arm on the environment performing work on $\boldsymbol{p}$ has to be related to the $(n \times 1)$ vector $\boldsymbol{J}_{\vartheta}^{\mathrm{T}} \boldsymbol{f}$ of joint torques performing work on $\boldsymbol{\vartheta}$ and to the $(m \times 1)$ vector $\boldsymbol{J}_{\delta}^{\mathrm{T}} \boldsymbol{f}$ of the elastic reaction forces performing work on $\boldsymbol{\delta}$. Hence, the dynamic model (16) can be rewritten in the form:

$$\begin{bmatrix} \boldsymbol{B}_{\vartheta\vartheta}(\boldsymbol{\vartheta},\boldsymbol{\delta}) & \boldsymbol{B}_{\vartheta\delta}(\boldsymbol{\vartheta},\boldsymbol{\delta}) \\ \boldsymbol{B}_{\vartheta\delta}^{\mathrm{T}}(\boldsymbol{\vartheta},\boldsymbol{\delta}) & \boldsymbol{B}_{\delta\delta}(\boldsymbol{\vartheta},\boldsymbol{\delta}) \end{bmatrix} \begin{bmatrix} \ddot{\boldsymbol{\vartheta}} \\ \ddot{\boldsymbol{\delta}} \end{bmatrix} + \begin{bmatrix} \boldsymbol{c}_{\vartheta}(\boldsymbol{\vartheta},\boldsymbol{\delta},\dot{\boldsymbol{\vartheta}},\dot{\boldsymbol{\delta}}) \\ \boldsymbol{c}_{\delta}(\boldsymbol{\vartheta},\boldsymbol{\delta},\dot{\boldsymbol{\vartheta}},\dot{\boldsymbol{\delta}}) \end{bmatrix}$$

$$+ \begin{bmatrix} \boldsymbol{g}_{\vartheta}(\boldsymbol{\vartheta},\boldsymbol{\delta}) \\ \boldsymbol{g}_{\delta}(\boldsymbol{\vartheta},\boldsymbol{\delta}) \end{bmatrix} + \begin{bmatrix} \boldsymbol{0} \\ \boldsymbol{D}\dot{\boldsymbol{\delta}} + \boldsymbol{K}\boldsymbol{\delta} \end{bmatrix} = \begin{bmatrix} \boldsymbol{\tau} \\ \boldsymbol{0} \end{bmatrix} - \begin{bmatrix} \boldsymbol{J}_{\vartheta}^{\mathrm{T}}(\boldsymbol{\vartheta},\boldsymbol{\delta})\boldsymbol{f} \\ \boldsymbol{J}_{\delta}^{\mathrm{T}}(\boldsymbol{\vartheta},\boldsymbol{\delta})\boldsymbol{f} \end{bmatrix}, \qquad (17)$$

where the matrix and vectors have been partitioned in blocks according to the rigid and flexible components.

3 Singularly perturbed model

When the link stiffness is large, it is reasonable to expect that the dynamics related to link flexibility are much faster than the dynamics associated with the rigid motion of the robot so that the system naturally exhibits a two-time scale dynamic behaviour in terms of rigid and flexible variables. This feature can be conveniently exploited for control design.

Following the approach proposed in [6], the system can be decomposed in a slow and a fast subsystem by using singular perturbation theory; this leads to a composite control strategy for the full system based on separate control designs for the two reduced-order subsystems.

3.1 Unconstrained motion

In the absence of contact with the environment, assuming that full-state measurements are available, the joint torques can be conveniently chosen as

$$\boldsymbol{\tau} = \boldsymbol{g}_{\vartheta}(\boldsymbol{\vartheta},\boldsymbol{\delta}) + \boldsymbol{u}, \qquad (18)$$

in order to cancel out the effects of the static torques acting on the rigid part of the arm dynamics. The vector $\boldsymbol{u}$ is the new control input to be designed on the basis of the singular perturbation approach.

The time scale separation between the slow and fast dynamics can be determined by defining the singular perturbation parameter $\epsilon = 1/\sqrt{k_m}$, where k_m is the smallest coefficient of the diagonal stiffness matrix $\boldsymbol{K}$, and the new variable

$$\boldsymbol{z} = \bar{\boldsymbol{K}}\boldsymbol{\delta} = \frac{1}{\epsilon^2}\hat{\boldsymbol{K}}\boldsymbol{\delta} \tag{19}$$

corresponds to the elastic force, where $\boldsymbol{K} = k_m\hat{\boldsymbol{K}}$. Considering the inverse $\boldsymbol{H}$ of the inertia matrix $\boldsymbol{B}$, the dynamic model (17), with control law (18), can be rewritten in terms of the new variable $\boldsymbol{z}$ as

$$\ddot{\boldsymbol{\vartheta}} = \boldsymbol{H}_{\vartheta\vartheta}(\boldsymbol{\vartheta}, \epsilon^2\boldsymbol{z}) \left(\boldsymbol{u} - \boldsymbol{c}_\vartheta(\boldsymbol{\vartheta}, \epsilon^2\boldsymbol{z}, \dot{\boldsymbol{\vartheta}}, \epsilon^2\dot{\boldsymbol{z}})) \right)$$

$$-\boldsymbol{H}_{\vartheta\delta}(\boldsymbol{\vartheta}, \epsilon^2\boldsymbol{z}) \left(\boldsymbol{c}_\delta(\boldsymbol{\vartheta}, \epsilon^2\boldsymbol{z}, \dot{\boldsymbol{\vartheta}}, \epsilon^2\dot{\boldsymbol{z}}) + \boldsymbol{g}_\delta(\boldsymbol{\vartheta}, \boldsymbol{\delta}) \right.$$

$$\left. +\epsilon^2\boldsymbol{D}\hat{\boldsymbol{K}}^{-1}\dot{\boldsymbol{z}} + \boldsymbol{z} \right) \tag{20}$$

$$\epsilon^2\ddot{\boldsymbol{z}} = \hat{\boldsymbol{K}}\boldsymbol{H}_{\vartheta\delta}^{\mathrm{T}}(\boldsymbol{\vartheta}, \epsilon^2\boldsymbol{z}) \left(\boldsymbol{u} - \boldsymbol{c}_\vartheta(\boldsymbol{\vartheta}, \epsilon^2\boldsymbol{z}, \dot{\boldsymbol{\vartheta}}, \epsilon^2\dot{\boldsymbol{z}}) \right)$$

$$-\hat{\boldsymbol{K}}\boldsymbol{H}_{\delta\delta}(\boldsymbol{\vartheta}, \epsilon^2\boldsymbol{z}) \left(\boldsymbol{c}_\delta(\boldsymbol{\vartheta}, \epsilon^2\boldsymbol{z}, \dot{\boldsymbol{\vartheta}}, \epsilon^2\dot{\boldsymbol{z}}) + \boldsymbol{g}_\delta(\boldsymbol{\vartheta}, \boldsymbol{\delta}) \right.$$

$$\left. +\epsilon^2\boldsymbol{D}\hat{\boldsymbol{K}}^{-1}\dot{\boldsymbol{z}} + \boldsymbol{z} \right), \tag{21}$$

where a suitable partition of $\boldsymbol{H}$ has been considered

$$\boldsymbol{H} = \boldsymbol{B}^{-1} = \begin{bmatrix} \boldsymbol{H}_{\vartheta\vartheta} & \boldsymbol{H}_{\vartheta\delta} \\ \boldsymbol{H}_{\vartheta\delta}^{\mathrm{T}} & \boldsymbol{H}_{\delta\delta} \end{bmatrix}. \tag{22}$$

Equations (20) and (21) represent a singularly perturbed form of the flexible arm model; when $\epsilon \to 0$, the model of an equivalent rigid arm is recovered. In fact, setting $\epsilon = 0$ and solving for $\boldsymbol{z}$ in (21) gives

$$\boldsymbol{z}_s = \bar{\boldsymbol{H}}_{\delta\delta}^{-1}(\boldsymbol{\vartheta}_s)\bar{\boldsymbol{H}}_{\vartheta\delta}^{\mathrm{T}}(\boldsymbol{\vartheta}_s) \left(\boldsymbol{u}_s - \bar{\boldsymbol{c}}_\vartheta(\boldsymbol{\vartheta}_s, \dot{\boldsymbol{\vartheta}}_s) \right) - \bar{\boldsymbol{c}}_\delta(\boldsymbol{\vartheta}_s, \dot{\boldsymbol{\vartheta}}_s) - \bar{\boldsymbol{g}}_\delta(\boldsymbol{\vartheta}_s), \tag{23}$$

where the subscript s indicates that the system is considered in the *slow* time scale and the overbar denotes that a quantity is computed with $\epsilon = 0$. Plugging (23) into (20) with $\epsilon = 0$ yields

$$\ddot{\boldsymbol{\vartheta}}_s = \bar{\boldsymbol{B}}_{\vartheta,\vartheta}^{-1}(\boldsymbol{\vartheta}_s) \left(\boldsymbol{u}_s - \bar{\boldsymbol{c}}_\vartheta(\boldsymbol{\vartheta}_s, \dot{\boldsymbol{\vartheta}}_s) \right), \tag{24}$$

where the equality

$$\bar{\boldsymbol{B}}_{\vartheta,\vartheta}^{-1}(\boldsymbol{\vartheta}_s) = \left(\bar{\boldsymbol{H}}_{\vartheta\vartheta}(\boldsymbol{\vartheta}_s) - \bar{\boldsymbol{H}}_{\vartheta\delta}(\boldsymbol{\vartheta}_s)\bar{\boldsymbol{H}}_{\delta\delta}^{-1}(\boldsymbol{\vartheta}_s)\bar{\boldsymbol{H}}_{\vartheta\delta}^{\mathrm{T}}(\boldsymbol{\vartheta}_s) \right) \tag{25}$$

has been exploited, $\bar{\boldsymbol{B}}_{\vartheta\vartheta}(\boldsymbol{\vartheta}_s)$ being the inertia matrix of the equivalent rigid arm and $\bar{\boldsymbol{c}}_\vartheta(\boldsymbol{\vartheta}_s, \dot{\boldsymbol{\vartheta}}_s)$ the vector of the corresponding Coriolis and centrifugal torques.

The dynamics of the system in the *fast* time scale can be obtained by setting $t_f = t/\epsilon$, treating the slow variables as constants in the fast time scale, and introducing the fast variables $z_f = z - z_s$. Thus, the fast system of (21) is

$$\frac{d^2 z_f}{dt_f^2} = -\hat{K}\bar{H}_{\delta\delta}(\vartheta_s)z_f + \hat{K}\bar{H}_{\vartheta\delta}^{\mathrm{T}}(\vartheta_s)u_f, \tag{26}$$

where the fast control $u_f = u - u_s$ has been introduced accordingly.

On the basis of the above two-time scale model, the design of a feedback controller for the system (20) and (21) can be performed according to a composite control strategy, i.e.,

$$u = u_s(\vartheta_s, \dot{\vartheta}_s) + u_f(z_f, dz_f/dt_f), \tag{27}$$

with the constraint that $u_f(0,0) = 0$, so that u_f is inactive along the equilibrium manifold specified by (23).

Notice that the fast system (26) is a marginally stable linear slowly time-varying system that can be stabilized to the equilibrium manifold $\dot{z}_f = 0$ ($\dot{z} = 0$) and $z_f = 0$ ($z = z_s$) by a proper choice of the control input u_f. A reasonable way to achieve this goal is to design a state space control law of the form

$$u_f = K_1 \dot{z}_f + K_2 z_f, \tag{28}$$

where, in principle, the matrices K_1 and K_2 should be tuned for every configuration ϑ_s. However, the computational burden necessary to perform this strategy can be avoided by using constant matrix gains tuned with reference to a given robot configuration [6]; any state space technique can be used, e.g., based on classical pole placement algorithms.

3.2 Constrained motion

When the arm's tip is constrained by the environment, a similar model can be derived provided that the contact force is measured by using a force sensor mounted on the tip.

In detail, in lieu of (18), the joint torques can be chosen as

$$\tau = g_\vartheta(\vartheta, \delta) + J_\vartheta^{\mathrm{T}}(\vartheta, \delta)f + u, \tag{29}$$

in order to cancel out also the effects of the contact force f acting on the rigid part of the arm dynamics.

By introducing the variable z as in (19) and following the same procedure as in the unconstrained motion case, the rigid robot dynamics for the slow time scale can be achieved in the same form (24).

As for the fast dynamics, the same expression (26) holds by defining the fast variable as $z_f = z - z_s'$, where

$$z_s' = z_s - \bar{J}_\delta^{\mathrm{T}}(\vartheta_s)f_s \tag{30}$$

with z_s defined in (23). Hence the same control law (28) can be adopted to stabilize the fast dynamics.

4 Compliant motion control

Control of the interaction between a robotic arm and the environment by using a pure motion control strategy is a candidate to fail if the task is not accurately planned. In practice, planning errors may give rise to a contact force causing a deviation of the tip from the desired trajectory. Since motion control reacts to reduce such deviation, the contact force may reach high values that can lead to saturation of joint actuators or breakage of the parts in contact.

The higher environment stiffness and position control accuracy are, the easier a situation like the one just described can occur. This drawback can be overcome if a compliant behavior is ensured during the interaction. This is partially achieved in a passive fashion, due to the flexibility of the arm, but it can be enhanced in active fashion by adopting a suitable compliant control strategy.

In this section, a compliant control strategy that offers the possibility of controlling the contact force to a desired value, as well as the tip position to a desired trajectory assigned along the unconstrained directions, is presented. This strategy is based on the parallel force/position control approach [21], which is especially effective in the case of inaccurate contact modeling. The key feature is to have a force control loop working in parallel to a position control loop. The logical conflict between the two loops is managed by imposing a dominance of the force control action over the position one, i.e., force regulation is always guaranteed at the expense of a position error along the constrained directions.

4.1 Position control

To gain insight into parallel force/position control applied to a flexible arm, the position control loop is first designed. To this purpose, it is useful to derive the slow dynamics corresponding to the tip position. Differentiating (15) gives the tip acceleration

$$\ddot{\boldsymbol{p}} = \boldsymbol{J}_\vartheta(\boldsymbol{\vartheta},\boldsymbol{\delta})\ddot{\boldsymbol{\vartheta}} + \boldsymbol{J}_\delta(\boldsymbol{\vartheta},\boldsymbol{\delta})\ddot{\boldsymbol{\delta}} + \boldsymbol{h}(\boldsymbol{\vartheta},\boldsymbol{\delta},\dot{\boldsymbol{\vartheta}},\dot{\boldsymbol{\delta}}),\tag{31}$$

where $\boldsymbol{h} = \dot{\boldsymbol{J}}_\vartheta\dot{\boldsymbol{\vartheta}} + \dot{\boldsymbol{J}}_\delta\dot{\boldsymbol{\delta}}$; hence the corresponding slow system is

$$\ddot{\boldsymbol{p}}_s = \bar{\boldsymbol{J}}_\vartheta(\boldsymbol{\vartheta}_s)\bar{\boldsymbol{B}}_{\vartheta,\vartheta}^{-1}(\boldsymbol{\vartheta}_s)\left(\boldsymbol{u}_s - \bar{\boldsymbol{c}}_\vartheta(\boldsymbol{\vartheta}_s,\dot{\boldsymbol{\vartheta}}_s)\right) + \bar{\boldsymbol{h}}(\boldsymbol{\vartheta}_s,\dot{\boldsymbol{\vartheta}}_s),\tag{32}$$

where (24) has been used. The slow dynamic models (24) and (32) enjoy the same notable properties of the rigid robot dynamic models [4], hence the control strategies used for rigid arms can be adopted.

If tracking of a time-varying position $\boldsymbol{p}_r(t)$ is desired (with an order ϵ approximation), an inverse dynamics motion scheme can be adopted for the slow system, i.e.,

$$u_s = \bar{B}_{\vartheta,\vartheta}(\vartheta_s)\bar{J}_\vartheta^{-1}(\vartheta_s)\left(a_s - \bar{h}(\vartheta_s,\dot{\vartheta}_s)\right) + \bar{c}_\vartheta(\vartheta_s,\dot{\vartheta}_s),\qquad(33)$$

where a_s is a new control input and the Jacobian matrix is assumed to be nonsingular.

Folding (33) into (32) gives

$$\ddot{p}_s = a_s;\qquad(34)$$

hence the control input a_s can be chosen as

$$a_s = \ddot{p}_r + k_D(\dot{p}_r - \dot{p}_s) + k_P(p_r - p_s),\qquad(35)$$

giving the closed loop equation for the slow subsystem

$$\ddot{p}_r - \ddot{p}_s + k_D(\dot{p}_r - \dot{p}_s) + k_P(p_r - p_s) = 0.\qquad(36)$$

The system (36) is exponentially stable for any choice of the positive gains k_D and k_P and thus tracking of p_r and $\dot{p}_r$ is ensured for the slow subsystem.

As a further step, the full-order system (16) with the composite control law (27), (33), (35), and (28) have to be analyzed. By virtue of Tikhonov's theorem, it can be shown that tracking of the reference position $p_r(t)$ is achieved with an order ϵ approximation.

4.2 Parallel force/position control

The interaction of a flexible arm with a compliant environment can be managed by controlling both the contact force and the tip position.

A better insight into the behaviour of the system during the interaction can be achieved by considering a model of the compliant environment. To this purpose, a planar surface is considered, which is locally a good approximation of surfaces of regular curvature, and the model of the contact force is given by

$$f = k_e nn^{\mathrm{T}}(p - p_o),\qquad(37)$$

where p_o represents the position of any point on the undeformed plane, n is the unit vector along the normal to the plane, and $k_e > 0$ is the contact stiffness coefficient. For the purpose of this work, it is assumed that the same equation can be established in terms of the slow variables. Such a model shows that the contact force is normal to the plane, and thus a null force error can be obtained only if the desired force f_d is aligned with n. Also, null position errors can be obtained only on the contact plane while the component of the position along n has to accommodate the force requirement specified by f_d.

The parallel force/position controller is based on the inverse dynamics law (33) and (35), where p_r is chosen as

$$p_r = p_d + p_c,\qquad(38)$$

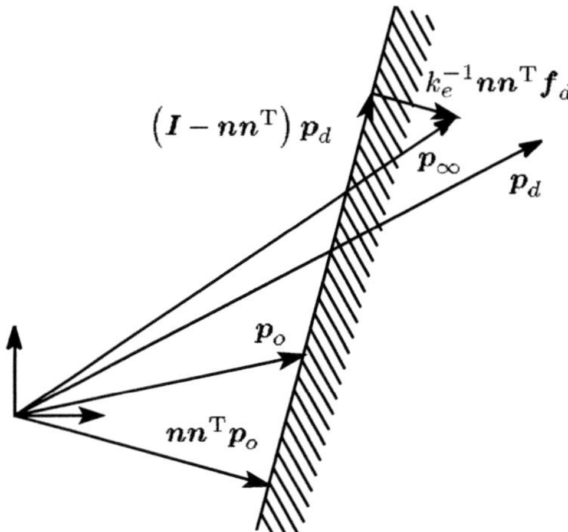

Fig. 2. Equilibrium position with parallel force and position control.

and p_c is the solution of the differential equation

$$k_A\ddot{p}_c + k_V\dot{p}_c = f_d - f_s; \tag{39}$$

$k_P, k_D, k_A, k_V > 0$ are suitable feedback gains. It is worth pointing out that p_c resulting from integration of (39) provides an integral control action on the force error.

The stability analysis for the slow system (32) with the control law (33), (35), (38), and (39) can be carried out with the same arguments used in [11] for the case of rigid robots. In particular, the force/position parallel control scheme ensures regulation of the contact force to the desired set-point f_d and tracking of the time-varying component of the desired position on the contact plane $\left(I - nn^{\mathrm{T}}\right)p_d(t)$.

To better understand the compliant behavior ensured by parallel control, consider for simplicity the case that p_d is constant. It can be shown that the closed loop system has an exponentially stable equilibrium at

$$p_\infty = \left(I - nn^{\mathrm{T}}\right)p_d + nn^{\mathrm{T}}\left(k_e^{-1}f_d + p_o\right) \tag{40}$$

$$f_\infty = k_e nn^{\mathrm{T}}(p_\infty - p_o) = f_d, \tag{41}$$

where the matrix $\left(I - nn^{\mathrm{T}}\right)$ projects the vectors on the contact plane.

The equilibrium position is depicted in Figure 2. It can be recognized that p_∞ differs from p_d by a vector aligned along the normal to the contact plane whose magnitude is that necessary to guarantee $f_\infty = f_d$ in view of (41). Therefore (for the slow system) force regulation is ensured while a null position error is achieved only for the component parallel to the contact plane.

If $\boldsymbol{f}_d$ is not aligned with $\boldsymbol{n}$, then it can be found that a drift motion of the arm's tip is generated along the plane; for this reason, if the contact geometry is unknown, it is advisable to set $\boldsymbol{f}_d = \boldsymbol{0}$.

As before, Tikhonov's theorem has to be applied to the full-order system (17) with the composite control law (27), (33), (35)–(39), and (28). It can be shown that that force regulation and position tracking are achieved with an order ϵ approximation.

5 Simulation

To illustrate the effectiveness of the proposed strategy, a planar two-link flexible arm (Figure 1) is considered:

$$\boldsymbol{\vartheta} = [\vartheta_1 \ \vartheta_2]^{\mathrm{T}}$$

and an expansion with two clamped-mass assumed modes is taken for each link:

$$\boldsymbol{\delta} = [\delta_{11} \ \delta_{12} \ \delta_{21} \ \delta_{22}]^{\mathrm{T}}.$$

The following parameters are set up for the links and a payload is assumed to be placed at the arm's tip:

$\rho_1 = \rho_2 = 1.0$ kg/m (link uniform density)

$\ell_1 = \ell_2 = 0.5$ m (link length)

$d_1 = d_2 = 0.25$ m (link center of mass)

$m_1 = m_2 = 0.5$ m (link mass)

$m_{h1} = m_{h2} = 1$ kg (hub mass)

$m_p = 0.1$ kg (payload mass)

$(EI)_1 = (EI)_2 = 10$ N m^2 (flexural link rigidity).

The stiffness coefficients of the diagonal matrix K are

$k_{11} = 38.79$ N $k_{12} = 513.37$ N

$k_{21} = 536.09$ N $k_{22} = 20792.09$ N.

The dynamic model of the arm and the missing numerical data can be found in [25], while the direct and differential kinematics equations are reported in [26].

The contact surface is a vertical plane, thus the normal vector in (37) is $\boldsymbol{n} = [1 \ 0]^{\mathrm{T}}$; a point of the undeformed plane is

$$\boldsymbol{p}_o = [0.55 \ 0]^{\mathrm{T}} \text{ m}$$

and the contact stiffness is $k_e = 50$ N/m.

The arm was initially placed with the tip in contact with the undeformed plane in the position

$$\boldsymbol{p}(0) = [0.55 \ -0.55]^{\mathrm{T}} \text{ m}$$

with null contact force; the corresponding generalized coordinates of the arm are

$\boldsymbol{\vartheta} = [-1.396\ 1.462]^{\mathrm{T}}$ rad

$\boldsymbol{\delta} = [-0.106\ 0.001\ -0.009\ -0.0001]^{\mathrm{T}}$ m.

It is desired to reach the tip position

$\boldsymbol{p}_d = [0.55\ -0.35]^{\mathrm{T}}$ m

and a fifth-order polynomial trajectory with null initial and final velocity and acceleration is imposed from the initial to the final position with a duration of 5 s.

The desired force is taken from zero to the desired value

$\boldsymbol{f}_d = [5\ 0]^{\mathrm{T}}$ N,

according to a fifth-order polynomial trajectory with null initial and final first and second derivative and a duration of 1 s.

The fast control law $\boldsymbol{u}_f$ has been implemented with $\epsilon = 0.1606$. The matrix gains in (28) have been tuned by solving an LQ problem for the system (26) with the configuration dependent terms computed in the initial configuration of the arm. The matrix weights of the index performance have been chosen so as to preserve the time-scale separation between slow and fast dynamics for both the control schemes. The resulting matrix gains are

$$\boldsymbol{K}_1 = \begin{bmatrix} -0.0372 & -0.0204 & -0.0375 & 0.1495 \\ 0.0573 & 0.0903 & 0.0080 & -0.7856 \end{bmatrix}$$

$$\boldsymbol{K}_2 = \begin{bmatrix} -0.1033 & -0.0132 & -0.0059 & -0.0053 \\ -0.0882 & 0.0327 & -0.0537 & -0.0217 \end{bmatrix}.$$

Numerical simulations have been performed via MATLAB/Simulink. To reproduce a real situation of a continuous-time system with a digital controller, the control laws are discretized with 5 ms sampling time, while the equations of motion are integrated using a variable step Runge-Kutta method with a minimum step size of 1 ms.

The slow controller (33), (35)–(39) has been used in the composite control law (27). The actual force $\boldsymbol{f}$ and position $\boldsymbol{p}$ are used in the slow control law instead of the corresponding slow values, assuming that direct force measurement is available and that the tip position is computed from joint angles and link deflection measurements via the direct kinematics equation (9). The control gains have been set to $k_P = 100$, $k_D = 22$, $k_A = 0.7813$, $k_V = 13.75$.

In Figure 3 the time histories of the desired (dashed) and actual (solid) contact force are reported, together with the position error. It is easy to see that the contact force remains close to the desired value during the tip motion (notice that the commanded position trajectory has a 5 s duration) and reaches the desired set-point after about 3 s, before the completion of the tip motion. Tracking of the y-component of the position is ensured, while a significant error occurs for the x-component. Its (constant) value at steady state is exactly that required to achieve null force error along the same axis, according to the equilibrium equations (40) and (41).

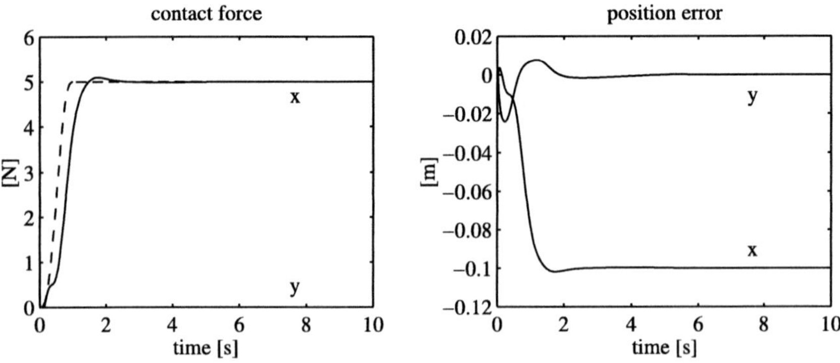

Fig. 3. Time histories of contact force and position error.

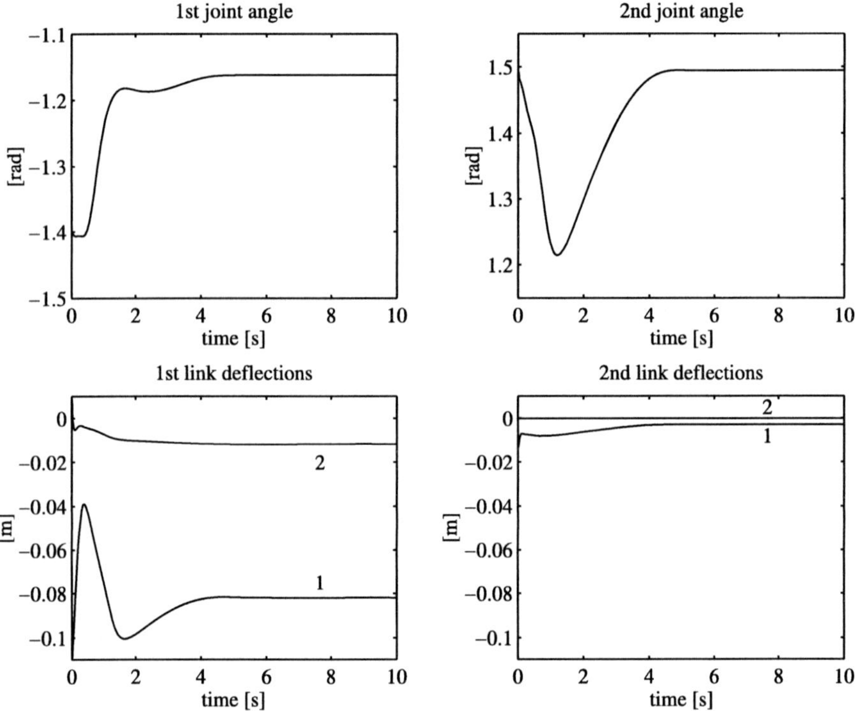

Fig. 4. Time histories of contact force and position error.

The time histories of the joint angles and link deflections are reported in Figure 4. It can be recognized that the oscillations of the link deflections are well damped; moreover, because of gravity and contact force, the arm has to bend to reach the desired tip position with the desired contact force.

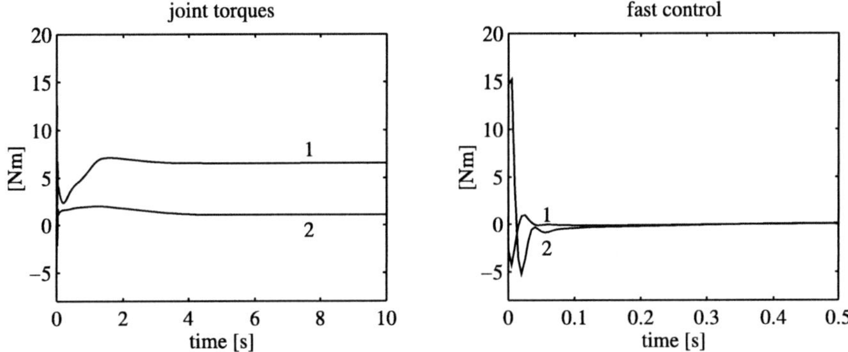

Fig. 5. Time histories of contact force and position error.

Figure 5 shows the time history of the joint torque u and the first 0.5 s of the time history of the fast torque u_f. It can be observed that the control effort keeps limited values during task execution; remarkably, the control torque u_f converges to zero with a transient much faster than the transient of u, as expected.

It is worth pointing out that the simulation of both slow control laws without the fast control action (28) has revealed an unstable behaviour; the results have not been reported here for brevity.

6 Conclusion

The problem of force and position control for flexible link arms has been considered in this chapter. Because of the presence of structural link flexibility, the additional objective of damping the vibrations that are naturally excited during task execution was considered. By using singular perturbation theory, under the reasonable hypothesis that link stiffness be large, the system has been split into a slow subsystem describing the rigid motion dynamics and a fast subsystem describing the flexible dynamics. Then a force and position parallel control has been adopted for the slow subsystem, while a fast action has been designed for vibration damping. Simulation results have confirmed the feasibility of the proposed approach.

References

1. Book WJ (1993) Controlled motion in an elastic world. ASME Journal of Dynamic Systems, Measurement, and Control 115:252–261
2. Meirovitch L (1967) Analytical methods in vibrations. Macmillan, New York, NY

3. Book WJ (1984) Recursive Lagrangian dynamics of flexible manipulator arms. International Journal of Robotics Research 3:(3)87–101
4. Canudas De Wit C, Siciliano B, Bastin G (Eds) (1996) Theory of robot control. Springer-Verlag, London
5. Kokotovic P, Khalil HK, O'Reilly J (1986) Singular perturbation methods in control: analysis and design. Academic Press, New York
6. Siciliano B, Book WJ (1988) A singular perturbation approach to control of lightweight flexible manipulators. International Journal of Robotics Research 7(4):79–90
7. Fraser AR, Daniel RW (1991) Perturbation techniques for flexible manipulators. Kluwer Academic Publishers, Boston, MA
8. Siciliano B, Prasad JVR, Calise AJ (1992) Output feedback two-time scale control of multi-link flexible arms. ASME Journal of Dynamic Systems, Measurement, and Control 114:70–77
9. Vandergrift MW, Lewis FL, Zhu SQ (1994) Flexible-link robot arm control by a feedback linearization/singular perturbation approach. Journal of Robotic Systems 11:591–603
10. Moallem M, Khorasani K, Patel RV (1997) An integral manifold approach for tip-position tracking of flexible multi-link manipulators. IEEE Transactions on Robotics and Automation 13:823–837
11. Siciliano B, Villani L (2000) Robot force control. Kluwer Academic Publishers, Boston, MA
12. Chiou BC, Shahinpoor M (1990) Dynamic stability analysis of a two-link force-controlled flexible manipulator. ASME Journal of Dynamic Systems, Measurement, and Control 112:661–666
13. Mills JK (1992) Stability and control aspects of flexible link robot manipulators during constrained motion tasks. Journal of Robotic Systems 9:933–953
14. Matsuno F, Asano T, Sakawa Y (1994) Modeling and quasi-static hybrid position/force control of constrained planar two-link flexible manipulators. IEEE Transactions on Robotics and Automation 10:287–297
15. Matsuno F, Yamamoto K (1994) Dynamic hybrid position/force control of a two degree-of-freedom flexible manipulator. Journal of Robotic Systems 11:355–366
16. Hu FL, Ulsoy AG (1994) Force and motion control of a constrained flexible robot arm. ASME Journal of Dynamic Systems, Measurement, and Control 116:336–343
17. Yang JH, Lian FL, Fu LC (1995) Adaptive hybrid position/force control for robot manipulators with compliant links. Proceedings 1995 IEEE International Conference on Robotics and Automation, Nagoya, Japan, 603–608
18. Lew JY, Book WJ (1993) Hybrid control of flexible manipulators with multiple contact. Proceedings 1993 IEEE International Conference on Robotics and Automation, Atlanta, GA, 2:242–247
19. Yoshikawa T, Harada K, Matsumoto A (1996) Hybrid position/force control of flexible-macro/rigid-micro manipulator systems. IEEE Transactions on Robotics and Automation 12:633–640
20. Rocco P, Book WJ (1996) Modelling for two-time scale force/position control of flexible robots. Proceedings 1996 IEEE International Conference on Robotics and Automation, Minneapolis, MN, 1941–1946
21. Chiaverini S, Sciavicco L (1993) The parallel approach to force/position control of robotic manipulators. IEEE Transactions on Robotics and Automation 9:361–373

22. Chiaverini S, Siciliano B, Villani L (1994) Force/position regulation of compliant robot manipulators. IEEE Transactions on Automatic Control 39:647–652
23. Raibert MH, Craig JJ (1981) Hybrid position/force control of manipulators. ASME Journal of Dynamic Systems, Measurement, and Control 103:126–133
24. Yoshikawa T (1987) Dynamic hybrid position/force control of robot manipulators—Description of hand constraints and calculation of joint driving force. IEEE Journal of Robotics and Automation 3:386–392
25. De Luca A, Siciliano B (1991) Closed-form dynamic model of planar multi-link lightweight robots. IEEE Transactions on Systems, Man, and Cybernetics 21:826–839
26. Siciliano B (1999) Closed-loop inverse kinematics algorithms for constrained flexible manipulators under gravity. Journal of Robotic Systems 16:353–362

Fault Tolerant Tracking of a Robot Manipulator: An Internal Model Based Approach*

Claudio Bonivento, Luca Gentili, and Andrea Paoli

CASY-DEIS
Università di Bologna
Via Risorgimento 2
40136 Bologna, Italia
{cbonivento,lgentili,apaoli}@deis.unibo.it

Summary. In this paper an implicit fault tolerant control scheme is specialized for an n-degrees-of-freedom fully actuated mechanical manipulator subject to sinusoidal torque disturbances acting on joints. We show in detail how a standard tracking controller can be "augmented" with an internal model unit designed to compensate the unknown spurious torque harmonics. In this way the controller is proved to be *global implicitly fault tolerant* to all the faults belonging to the model embedded in the regulator. Moreover, by simply testing the state of the internal model we will show how to perform fault detection and isolation.

1 Introduction

Fault tolerant control (FTC) systems are able, on the one hand, to detect incipient faults in sensors and/or actuators and, on the other, to promptly adapt the control law in such a way as to preserve prespecified performances in terms of quality of the production, safety, and so on.

The common approach in dealing with such a problem (see [18], [8], [2], [13] and the references therein) is to split the overall design into two distinct phases. In the first phase, the fault detection and isolation (FDI) problem is addressed. This part consists of designing a dynamical system (filter) that, by processing input/output data, is able to detect the presence of an incipient fault and to isolate it from other faults and/or disturbances. The second phase usually consists of the design of a supervisory unit that, on the basis of the information provided by the FDI filter, reconfigures the control to compensate

---

*A preliminary version of this work was presented in [3]. This work was supported by MIUR and EC-Project IFATIS partly funded by the European Commission in the IST program 2001 of the 5th EC framework programme (IST-2001-32122).

for the effect of the fault and to fulfill performance constraints, i.e., by means of a parameterized controller that is suitably updated.

In [4] a different approach to FTC was discussed. Specifically, the case was addressed in which the faults affecting the controlled system can be modeled as functions (of time) within a finitely parameterized family of such functions. Then a controller that embeds an *internal model* of this family is designed to generate a supplementary control action compensating for the presence of any of such faults, regardless of their entity. The idea is pursued using the theoretical machinery of the (nonlinear) output regulation theory (see [6], [11]) under the assumption that the side effects generated by the occurrence of the fault can be modeled as an exogenous signal generated by an autonomous "neutrally stable" system (the so-called "exosystem"). In this framework, the FDI phase is postponed to that of control reconfiguration since it can be carried out by testing the state of the internal model unit, which automatically activates to offset the presence of the fault.

In this chapter the approach outlined above is specialized to the design of a fault tolerant control system for a n-degrees-of-freedom (dof) fully actuated mechanical robot subject to constant and sinusoidal torque disturbances acting on joints (see [14], [7]). We show how this framework can be cast as an output regulation problem. More in detail we show how a standard tracking robot control (see [20], [21], [9]) can be "augmented" with an internal model unit designed to compensate the unknown spurious torque harmonics. In this way the controller is proved to be *global implicitly fault tolerant* to all the faults belonging to the model embedded in the regulator. It is worth remarking that, with respect to the design solution presented in [3], the regulator discussed here is also able to take into account constant torque disturbances.

In Section 2 the problem is introduced and some preliminary positions are given to show how the problem can be cast in the framework illustrated in [4]. In Section 3 the control design of a canonical internal model unit able to achieve the implicit fault tolerance of the robot controlled is presented. In Section 4 the main result regarding the implicit fault tolerance of the robot controlled with the given algorithm, augmented with an adaptive mechanism, is given. This result is proved to be robust with respect to the uncertainty of the characteristic frequencies of the sinusoidal disturbances. In Section 5 some simulation tests are provided to show the effectiveness of the proposed controller, while Section 6 concludes the chapter with some final remarks.

2 Problem statement and preliminary positions

In this section we introduce the model of a n-dof fully actuated robot manipulator and state the FTC-FDI problem. Usually the joint actuators are modeled as pure torque sources; however they can be subject to some asymmetries (e.g., due to some electrical or mechanical faults) that comport the rise of spurious harmonics in the electrical variables and then in the generated torques. Hence, in the following we will model these effects as sinusoidal signals superimposed on the controlled torque signals. We will then show how it is possible to cast this problem in the framework illustrated in [4].To point out that a preexisting control can be augmented without modification with the FTC-FDI module designed (*internal model unit*) able to overcome the disturbance and, moreover, to isolate it, in this section a simple tracking controller is also considered.

The regulation scheme developed is depicted in Figure 1. Consider an

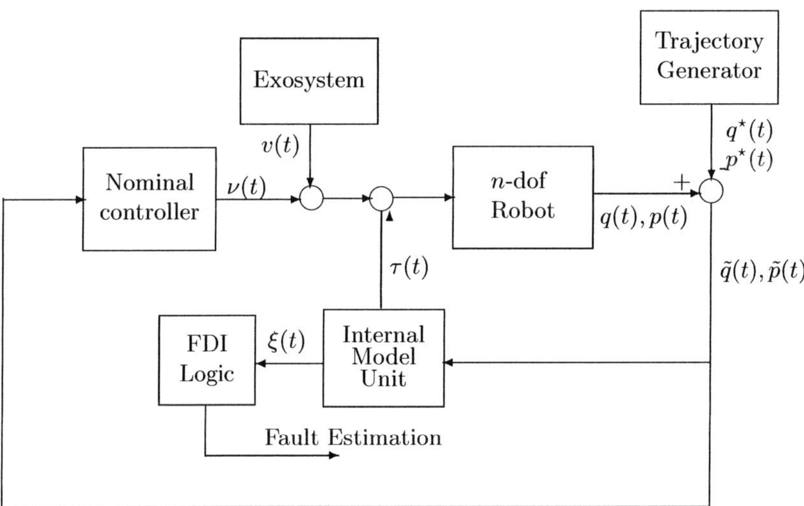

Fig. 1. FTC controller scheme.

n-dof fully actuated robot manipulator with generalized coordinates $q = (q_1, \cdots, q_n)^{\mathrm{T}}$. If $p = M(q)\dot{q} = (p_1, \cdots, p_n)^{\mathrm{T}}$ are the generalized momenta, with $M(q)$ the inertia matrix, symmetric and positive definite for all q, an explicit port-Hamiltonian representation of this system can be obtained defining the whole state $x := (q, p)^{\mathrm{T}}$, the Hamiltonian function as the total energy of the system (sum of kinetic energy and potential energy)

$$H(q, p) := \frac{1}{2} p^{\mathrm{T}} M^{-1}(q) p + P(q),$$

and, finally, the matrices

$$J = \begin{pmatrix} 0 & I_n \\ -I_n & 0 \end{pmatrix}, \quad R = \begin{pmatrix} 0 & 0 \\ 0 & D(q) \end{pmatrix}, \quad G = \begin{pmatrix} 0 \\ I_n \end{pmatrix},$$

with $D(q) = D^T(q) \geq 0$ taking into account the dissipation effects (see [15], [21] and references therein for an exhaustive insight on port-Hamiltonian formalism). The input is an effort representing the actuation torques and the output is a flow representing the joint velocities. These considerations lead to the following typical port-Hamiltonian model:

$$\begin{bmatrix} \dot{q} \\ \dot{p} \end{bmatrix} = [J - R] \begin{bmatrix} \dfrac{\partial H}{\partial q} \\ \dfrac{\partial H}{\partial p} \end{bmatrix} + Gv$$

$$y = G^{\mathrm{T}} \begin{bmatrix} \dfrac{\partial H}{\partial q} \\ \dfrac{\partial H}{\partial p} \end{bmatrix}.$$

This system will be affected by an external torque ripple $v(t)$ acting through the control input channel (i.e., actually the torque applied to the system will be the sum of the control torque and the external disturbance $v + v(t)$). The problem addressed in this chapter is to compensate this disturbance, detecting and isolating in the meanwhile the entity of this (unknown) disturbance.

It is worth pointing out again that the design of the internal model unit does not affect a previous regulator designed to carry out a particular task. To explain this feature, in the following we will introduce a control scheme whose aim is to make the manipulator track a known trajectory.

Remark 1. The tracking control is developed following [9], but the same results can be obtained using a simpler controller. This feature will be explained in the following, pointing out that the framework presented is suitable for dealing with exogenous disturbances acting on a generic mechanical system already regulated to accomplish a certain task with a classical control strategy.

2.1 Tracking control

First a preliminary torque input able to compensate potential energies (as gravity) is designed:

$$\nu = \frac{\partial P(q)}{\partial q} + \nu'.$$

Let us define the desired trajectory for the generalized coordinates and the generalized momenta as $(q^\star(t),\, p^\star(t))$. This trajectory, to be realizable, has to satisfy $p^\star(t) = M(q^\star)\dot{q}^\star(t)$. To define new error variables, let us consider the following change of coordinates:

$$\tilde{q} = q - q^\star(t)$$

$$\tilde{p} = p - M(q)\dot{q}^\star(t).$$

Deriving the new error coordinates we obtain

$$\dot{\tilde{q}} = M^{-1}(q)\tilde{p}$$

$$
\begin{aligned}
\dot{\tilde{p}} &= -\frac{\partial H}{\partial q} - D(q)\frac{\partial H}{\partial p} + \nu' - \frac{d}{dt}(M(q)\dot{q}^{\star}(t)) = \\
&= -\frac{1}{2}p^{\mathrm{T}}\frac{\partial M^{-1}(q)}{\partial q}p - DM^{-1}(q)p + \nu' - \frac{d}{dt}(M\dot{q}^{\star}(t)) = \\
&= -\frac{1}{2}(\tilde{p} + M\dot{q}^{\star}(t))^{\mathrm{T}}\frac{\partial M^{-1}(q)}{\partial q}(\tilde{p} + M\dot{q}^{\star}(t)) + \\
&\quad -DM^{-1}(\tilde{p} + M\dot{q}^{\star}(t)) + \nu' - \frac{d}{dt}(M\dot{q}^{\star}(t)) = \\
&= -\frac{1}{2}\tilde{p}^{\mathrm{T}}\frac{\partial M^{-1}(q)}{\partial q}\tilde{p} - D(q)M^{-1}(q)\tilde{p} + \nu' - \Pi(q,\dot{q}^{\star}(t),\ddot{q}^{\star}(t))
\end{aligned}
\tag{1}
$$

for a suitably defined vector $\Pi(q,\dot{q}^{\star}(t),\ddot{q}^{\star}(t))$.

Defining a new Hamiltonian function as

$$H' = \frac{1}{2}\tilde{p}^{\mathrm{T}}M^{-1}(q)\tilde{p},$$

it is possible to write (1) again as a port-Hamiltonian system:

$$
\begin{aligned}
\dot{\tilde{q}} &= \frac{\partial H'}{\partial \tilde{p}} \\
\dot{\tilde{p}} &= -\frac{\partial H'}{\partial \tilde{q}} - D(q)\frac{\partial H'}{\partial \tilde{p}} + \nu' - \Pi(q,\dot{q}^{\star}(t),\ddot{q}^{\star}(t)).
\end{aligned}
\tag{2}
$$

It is now possible to obtain a perfect asymptotic tracking designing the control torque to delete the "bad" term $\Pi(\cdot)$, to shape the energy of the error system to have a minimum in the origin,[1] and to add some damping action to have this minimum globally attractive:

$$\nu' = \Pi(q,\dot{q}^{\star}(t),\ddot{q}^{\star}(t)) + DM^{-1}(q)\tilde{p} - \tilde{q} - k_p M^{-1}(q)\tilde{p} + \tau, \tag{3}$$

where k_p is a symmetric design matrix ($-k_p$ is Hurwitz) and τ is an additional control torque that will be used in the following section to compensate the presence of additional torque disturbances.

The whole error system (2) with the controller (3) writes as

$$
\begin{aligned}
\dot{\tilde{q}} &= \frac{\partial \tilde{H}}{\partial \tilde{p}} \\
\dot{\tilde{p}} &= -\frac{\partial \tilde{H}}{\partial \tilde{q}} - k_p\frac{\partial \tilde{H}}{\partial \tilde{p}} + \tau,
\end{aligned}
\tag{4}
$$

where the new Hamiltonian is defined by

---

[1] Note that $\tilde{q} = 0$ means that the tracking is achieved as $q \to q^{\star}(t)$.

$$\tilde{H} = \frac{1}{2}\tilde{p}^{\mathrm{T}} M^{-1}(q)\tilde{p} + \frac{1}{2}\tilde{q}^{\mathrm{T}}\tilde{q} \,. \tag{5}$$

It is easy to realize that the tracking objective is globally asymptotically achieved: it is, in fact, straightforward to choose the Hamiltonian $\tilde{H}$ as a Lyapunov function and to state, exploiting the La Salle's invariant principle, that

$$\lim_{t\to\infty} q(t) = q^\star(t)\,, \qquad \lim_{t\to\infty} p(t) = p^\star(t)\,.$$

Remark 2. It is worth remarking that this kind of control strategy is very similar to the classical tracking control made by inversion of the model and introducing simple proportional and derivative terms (see, e.g., [20], [21], [17]).

2.2 Problem statement

It is now possible to state the input disturbance suppression problem, introducing a controlled n-dof robot manipulator like (4), (5) the exogenous torque disturbance $v(t)$:

$$\dot{\tilde{q}} = \frac{\partial \tilde{H}}{\partial \tilde{p}}$$
$$\dot{\tilde{p}} = -\frac{\partial \tilde{H}}{\partial \tilde{q}} - k_p \frac{\partial \tilde{H}}{\partial \tilde{p}} + \tau + v(t)\,. \tag{6}$$

In (6), $v(t)$ is a torque disturbance belonging to the signal class generated by the linear, neutrally stable autonomous system (*exosystem*)

$$\begin{cases} \dot{z} & = Sz \\ v(t) & = -\Gamma z\,, \end{cases} \tag{7}$$

with $z \in \mathbb{R}^{2k+1}$, $\Gamma \in \mathbb{R}^{(2k+1)\times m}$ a known matrix and S defined as

$$S = \mathrm{diag}\{S_0, S_1, \ldots, S_k\}\,, \tag{8}$$

with

$$S_0 = 0\,, \qquad S_i = \begin{bmatrix} 0 & \omega_i \\ -\omega_i & 0 \end{bmatrix} \qquad \omega_i > 0 \qquad i = 1,\ldots,k \tag{9}$$

and $z(0) \in \mathcal{Z}$, with $\mathcal{Z} \subseteq \mathbb{R}^{2k+1}$ bounded compact set.

In this discussion, the dimension $2k + 1$ of matrix S is known but all characteristic frequencies ω_i are unknown but range within known compact sets, i.e., $\omega_i^{\min} \leq \omega_i \leq \omega_i^{\max}$. In this set-up the lack of knowledge of the exogenous disturbance means lack of knowledge of the initial state $z(0)$ of the exosystem and of the characteristic frequencies; so any $v(t)$ obtained by a linear combination of a constant term and a finite number of sinusoidal signals with unknown frequencies, amplitudes, and phases will be considered.

All the above assumptions allow us to cast the problem of disturbance suppression as a problem of *output regulation* (see [5], [10], [1]) complicated by the lack of knowledge of the matrix S (see [19], [12]) and suggest that we look for a controller that embeds an *internal model* of the exogenous disturbances augmented by an adaptive part to estimate the characteristic frequencies of the disturbances.

To make the design solution more understandable, in the next section we will solve a simpler problem, relaxing the hypothesis of unknown frequencies and introducing the canonical internal model based regulator. In Section 4 the complete solution will be finally presented augmenting the regulator with an adaptive part.

Remark 3. Note again that the whole design method introduced later in this chapter can be easily applied to general mechanical systems described as pHs (6). Hence it is straightforward to consider this method suitable for a generic mechanical system already regulated to accomplish a certain task with a classical control strategy (see [17] for a survey about passivity based control strategies applied to pHs).

3 Canonical internal model unit design

In this section we design a canonical internal model unit able to overcome external torque disturbances (i.e., exogenous sinusoidal torque ripples). The main hypothesis here (removed in the next section) is that the exogenous matrix S is perfectly known.

As previously mentioned, the regulator to be designed will embed the internal model of the exogenous disturbance: this internal model unit is designed according to the procedure proposed in [16] (*canonical internal model*). Given any Hurwitz matrix F and any matrix G such that (F, G) is controllable, denote by Y the unique matrix solution of the Sylvester equation

$$YS - FY = G\Gamma$$

and define $\Psi := \Gamma Y^{-1}$.

Let us introduce the internal model unit as

$$\dot{\xi} = (F + G\Psi)\xi + N(\tilde{p}, \tilde{q}), \tag{10}$$

and set the control law as

$$\tau = \Psi\xi + \tau_{\text{st}},$$

where $N(\tilde{q}, \tilde{p})$ and τ_{st} are additional terms that will be designed later.

Defining the change of coordinates

$$\chi = \xi - Yz - G\tilde{p}, \tag{11}$$

system (6), (11), becomes

$$
\begin{cases}
\dot{\tilde{q}} = M(q)^{-1}\tilde{p} \\[2mm]
\dot{\tilde{p}} = -\dfrac{\partial \tilde{H}}{\partial \tilde{q}} - k_p \dfrac{\partial \tilde{H}}{\partial \tilde{p}} + \Psi \xi + \tau_{\mathrm{st}} - \Psi Y z \\[2mm]
\dot{\chi} = (F + G\Psi)\xi + N(\tilde{p}, \tilde{q}) - Y S z - G\dot{\tilde{p}}.
\end{cases}
\tag{12}
$$

Choosing $\tau_{\mathrm{st}} = -\Psi G\tilde{p}$, simple computation shows that the $\tilde{p}$-dynamics become

$$
\dot{\tilde{p}} = -\frac{\partial \tilde{H}}{\partial \tilde{q}} - k_p \frac{\partial \tilde{H}}{\partial \tilde{p}} + \Psi \chi .
\tag{13}
$$

Concentrating on the χ-dynamics it is possible to design

$$
N(\tilde{q}, \tilde{p}) = -\Psi^{\mathrm{T}} \tilde{p} - FG\tilde{p} - G\left(\frac{\partial \tilde{H}}{\partial \tilde{q}} + k_p \frac{\partial \tilde{H}}{\partial \tilde{p}} + \Psi G \tilde{p} \right)
$$

and write the last equation of (12) as

$$
\dot{\chi} = F\chi - \Psi^{\mathrm{T}} \tilde{p} .
\tag{14}
$$

Consider now the first equation of (12) with (13) and (14). This new system identifies a port-Hamiltonian system described by

$$
\dot{x} = [J(x) - R(x)]\frac{\partial H_x(x)}{\partial x} ,
\tag{15}
$$

with $x = \left(\tilde{q} \ \tilde{p} \ \chi \right)^{\mathrm{T}}$, the Hamiltonian $H_x(x)$ defined by

$$
H_x(x) = \frac{1}{2}\tilde{p}^{\mathrm{T}} M(q)^{-1}\tilde{p} + \frac{1}{2}\tilde{q}^{\mathrm{T}}\tilde{q} + \frac{1}{2}\chi^{\mathrm{T}}\chi ,
$$

and the skew-symmetric interconnection matrix $J(x)$ and the positive-definite damping matrix R defined by

$$
J(x) = \begin{pmatrix} 0 & I & 0 \\ -I & 0 & \Psi \\ 0 & -\Psi^{\mathrm{T}} & 0 \end{pmatrix} , \quad R = \begin{pmatrix} 0 & 0 & 0 \\ 0 & k_p & 0 \\ 0 & 0 & -F \end{pmatrix} .
$$

Proposition 1. *Consider the controlled n-dof robot manipulator (6) with Hamiltonian (5), affected by the torque disturbances generated by (7), (8), (9).*

The additional control law generated by the internal model unit:

$$
\begin{cases}
\dot{\xi} = (F + G\Psi)\xi - \Psi^{\mathrm{T}}\tilde{p} - FG\tilde{p} + G\tilde{q} + G\tilde{p}^{\mathrm{T}}\dfrac{\partial M^{-1}(q)}{\partial q}\tilde{p}+ \\[2mm]
\quad +Gk_p M^{-1}(q)\tilde{p} - G\Psi G\tilde{p} \\[2mm]
\tau = \Psi \xi - \Psi G\tilde{p}
\end{cases}
\tag{16}
$$

assures asymptotically the input disturbance suppression (fault tolerance with respect to torque ripple, i.e., $(\tilde{q}, \tilde{p}) \to (0,0)$ as time $t \to \infty$) and the convergence of the state of the internal model to the fault signal (fault detection, i.e., $\xi \to Yz$).

Proof. Considering $H_x(x)$ as a Lyapunov function the proof is immediate as (remembering that F is an arbitrary Hurwitz matrix)

$$\dot{H}_x \leq -k_p M^{-1}(q)\|\tilde{p}\|^2 + F\|\chi\|^2 \,,$$

and for the La Salle invariant principle the system will asymptotically converge to $\lim_{t \to \infty}(\tilde{p}, \chi) = (0,0)$. Moreover from the first and second equation of (16) it is possible to state that also $\lim_{t \to \infty} \tilde{q}(t) = 0$ and the proposition is proved.

4 Adaptive internal model unit design

In this section we introduce the main result of the chapter, designing an adaptive internal model unit able to overcome external torque disturbances (i.e., exogenous sinusoidal torque ripples) and asymptotically stabilize the origin of system (6). As previously announced, the perfect knowledge of the exogenous matrix S is not assumed, as only the dimension $2k+1$ of the matrix is known. This means that, for instance, any $v(t)$ obtained by linear combination of constant terms and sinusoidal signals with unknown frequencies, amplitudes, and phases can be modeled.

Let us introduce a new change of coordinates:

$$\bar{q} = \tilde{q}$$
$$\bar{p} = \tilde{p} + M(q)k_q\tilde{q}, \tag{17}$$

where k_q is a symmetric design matrix.

Taking into account system (6) and deriving the new error coordinates, we obtain for the $\bar{q}$-dynamics:

$$\dot{\bar{q}} = M^{-1}(q)\bar{p} - k_q\bar{q} \tag{18}$$

and for the $\bar{p}$-dynamics:

$$\begin{aligned}
\dot{\bar{p}} &= -\frac{1}{2}\tilde{p}^{\mathrm{T}}\frac{\partial M^{-1}(q)}{\partial q}\tilde{p} - \tilde{q} - k_p M^{-1}(q)\tilde{p} + \tau + v(t) + \frac{d}{dt}(M(q)k_q\tilde{q}) = \\
&= -\frac{1}{2}(\bar{p} - M(q)k_q\bar{q})^{\mathrm{T}}\frac{\partial M^{-1}(q)}{\partial q}(\bar{p} - M(q)k_q\bar{q}) - \bar{q} + \\
&\quad -k_p M^{-1}\bar{p} + k_p k_q\bar{q} + \tau + v(t) + \frac{d}{dt}(M(q)k_q\bar{q}) = \\
&= -\frac{1}{2}\bar{p}^{\mathrm{T}}\frac{\partial M^{-1}(q)}{\partial q}\bar{p} - \bar{q} - k_p M^{-1}(q)\bar{p} + \tau + v(t) - \tilde{\Pi}(q, q^\star(t), \dot{q}^\star(t), \ddot{q}^\star(t))
\end{aligned} \tag{19}$$

for a suitable defined vector $\tilde{\Pi}(q, q^\star(t), \dot{q}^\star(t), \ddot{q}^\star(t))$.

Defining a new Hamiltonian function as

$$H' = \frac{1}{2}\bar{p}^{\mathrm{T}} M^{-1}(q)\bar{p} + \frac{1}{2}\bar{q}^{\mathrm{T}}\bar{q} \tag{20}$$

and imposing a preliminary control action

$$\tau = \tilde{\Pi}(q, p, q^\star(t), \dot{q}^\star(t), \ddot{q}^\star(t)) + \tau' ,$$

it is possible to write again (18) and (19) as a port-Hamiltonian system:

$$
\begin{aligned}
\dot{\bar{q}} &= -k_q \frac{\partial H'}{\partial \bar{q}} + \frac{\partial H'}{\partial \bar{p}} \\
\dot{\bar{p}} &= -\frac{\partial H'}{\partial \bar{q}} - k_p \frac{\partial H'}{\partial \bar{p}} + \tau' + v(t) .
\end{aligned}
\tag{21}
$$

Remark 4. It is worth pointing out that the suitably defined change of coordinates (17) makes it possible to deal with torque disturbances characterized by the presence also of a constant term. This was not possible in the previous framework presented in [3] and is due to the presence of the damping action k_q that will be instrumental in the following.

It is now possible to suitably design the internal model unit. As the "classical" internal model control introduced in (10) depends on matrix Ψ (now unknown) in the following, an *adaptive* canonical internal model is introduced:

$$
\begin{cases}
\dot{\xi} = (F + G\hat{\Psi})\xi + N(\bar{p}, \bar{q}) \\
\dot{\hat{\Psi}}_i^{\mathrm{T}} = \varphi_i(\xi, \bar{p}, \bar{q}) ,
\end{cases}
$$

calling $\hat{\Psi}_i^{\mathrm{T}}$ with $i = 1, \cdots, n$ the ith column of the matrix $\hat{\Psi}^{\mathrm{T}} \in \mathbb{R}^{2k \times n}$.

Moreover, let us set the control law as

$$\tau' = \hat{\Psi}\xi + \tau_{\mathrm{st}} ,$$

where $N(\bar{q}, \bar{p})$ and τ_{st} are additional terms that will be designed later. The adaptation law $\varphi(\xi, \bar{p}, \bar{q})$ will be designed to assure that asymptotically the internal model unit will provide a torque able to overcome all disturbances.

Defining the change of coordinates

$$
\begin{aligned}
\chi &= \xi - Yz - G\bar{p} \\
\bar{\Psi}_i^{\mathrm{T}} &= \hat{\Psi}_i^{\mathrm{T}} - \Psi_i^{\mathrm{T}} \qquad i = (1, \cdots, n) ,
\end{aligned}
\tag{22}
$$

where Ψ_i^{T} represents the ith column of Ψ^{T}, system (21), (20) becomes

$$\begin{cases} \dot{\bar{q}} &= -k_q \dfrac{\partial H'}{\partial \bar{q}} + \dfrac{\partial H'}{\partial \bar{p}} \\[2mm] \dot{\bar{p}} &= -\dfrac{\partial H'}{\partial \bar{q}} - k_p \dfrac{\partial H'}{\partial \bar{p}} + \hat{\Psi}\xi + \tau_{\mathrm{st}} - \Psi Y z \\[2mm] \dot{\chi} &= (F + G\hat{\Psi})\xi + N(\bar{p},\bar{q}) - YSz - G\dot{\bar{p}} \\[2mm] \dot{\hat{\Psi}}_i^{\mathrm{T}} &= \varphi_i(\xi,\bar{p},\bar{q}) \qquad i = (1,\cdots,n)\,. \end{cases} \tag{23}$$

Note that

$$\dot{\bar{p}} = -\frac{\partial H'}{\partial \bar{q}} - k_p \frac{\partial H'}{\partial \bar{p}} + \hat{\Psi}(\xi - Yz) + \tau_{\mathrm{st}} - \tilde{\Psi}Yz =$$

$$= -\frac{\partial H'}{\partial \bar{q}} - k_p \frac{\partial H'}{\partial \bar{p}} + \hat{\Psi}(\xi - Yz - G\bar{p}) + \hat{\Psi}G\bar{p}_1 + \tau_{\mathrm{st}} - \tilde{\Psi}(\xi - \chi - G\bar{p})$$

and choosing $\tau_{\mathrm{st}} = -\hat{\Psi}G\bar{p}$ it is possible to write

$$\dot{\bar{p}} = -\frac{\partial H'}{\partial \bar{q}} - k_p \frac{\partial H'}{\partial \bar{p}} + \hat{\Psi}\chi + \tilde{\Psi}\xi - \tilde{\Psi}\chi - \tilde{\Psi}G\bar{p}$$

$$= -\frac{\partial H'}{\partial \bar{q}} - k_p \frac{\partial H'}{\partial \bar{p}} + \Psi\chi + \tilde{\Psi}(\xi - G\bar{p})\,.$$

Considering a single element of vector $\bar{p}$ it is possible to write (from now on apex i means the ith element of the vector considered)

$$\dot{\bar{p}}^i = \left(-\frac{\partial H'}{\partial \bar{q}} - k_p \frac{\partial H'}{\partial \bar{p}} + \Psi\chi \right)^i + (\xi - G\bar{p})^{\mathrm{T}}\tilde{\Psi}_i^{\mathrm{T}} \tag{24}$$

with $i = 1,\cdots,n$.

Concentrate now on the χ-dynamics to suitably design the update term $N(\bar{q},\bar{p})$:

$$\dot{\chi} = (F + G\hat{\Psi})\xi + N(\bar{p},\bar{q}) - YMz - G\Gamma z +$$

$$-G\left(-\frac{\partial H'}{\partial \bar{q}} - k_p \frac{\partial H'}{\partial \bar{p}} + \hat{\Psi}\xi + \tau_{\mathrm{st}} - \Gamma z \right) =$$

$$= F\chi + FG\bar{p} + N(\bar{p},\bar{q}) + G\frac{\partial H'}{\partial \bar{q}} + Gk_p\frac{\partial H'}{\partial \bar{p}} - G\tau_{\mathrm{st}}\,.$$

Choosing

$$N(\bar{p},\bar{q}) = -FG\bar{p} - G\frac{\partial H'}{\partial \bar{q}} - Gk_p\frac{\partial H'}{\partial \bar{p}} + G\tau_{st}\,,$$

we obtain

$$\dot{\chi} = F\chi = F\chi - \Psi^{\mathrm{T}}\bar{p} + \Psi^{\mathrm{T}}\bar{p}\,. \tag{25}$$

As all dynamics of (23) have been investigated, it is now possible to design an adaptation law for $\hat{\Psi}^{\mathrm{T}}$. Assume then

$$\varphi_i(\xi, \bar{p}, \bar{q}) = -(\xi - G\bar{p})\bar{p}^i \qquad i = 1, \cdots, n .$$

With this in mind we can write the $\tilde{\Psi}_i^{\mathrm{T}}$-dynamics as

$$\dot{\tilde{\Psi}}_i^{\mathrm{T}} = \dot{\hat{\Psi}}_i^{\mathrm{T}} - \dot{\Psi}_i^{\mathrm{T}} = -(\xi - G\bar{p})\bar{p}^i \qquad i = 1, \cdots, n . \tag{26}$$

Consider now the first equation of (23) with all (24), (25), and (26). This new system (with a small abuse of notation to obtain a more compact and readable formulation) identifies an interconnection described by

$$\dot{x} = [\tilde{J}(x) - \tilde{R}(x)]\frac{\partial H_x(x)}{\partial x} + \Lambda(x) , \tag{27}$$

with state

$$x = \left(\bar{q}\ \bar{p}\ \chi\ \tilde{\Psi}^{\mathrm{T}} \right)^{\mathrm{T}} ,$$

the Hamiltonian $H_x(x)$ defined by

$$H_x(x) = \frac{1}{2}\bar{p}^{\mathrm{T}} M(q)^{-1}\bar{p} + \frac{1}{2}\bar{q}^{\mathrm{T}}\bar{q} + \frac{1}{2}\chi^{\mathrm{T}}\chi + \sum_{i=1}^{n}\frac{1}{2}\tilde{\Psi}_i\tilde{\Psi}_i^{\mathrm{T}} ,$$

the skew-symmetric interconnection matrix $\tilde{J}(x)$ defined by

$$\tilde{J}(x) = \begin{pmatrix} 0 & I & 0 & 0 \\ -I & 0 & \Psi\ (\xi - G\bar{p})^{\mathrm{T}} \\ 0 & -\Psi^{\mathrm{T}} & 0 & 0 \\ 0 & -(\xi - G\bar{p}) & 0 & 0 \end{pmatrix} ,$$

the positive-definite damping matrix $\tilde{R}$ defined by

$$\tilde{R} = \begin{pmatrix} k_q & 0 & 0 & 0 \\ 0 & k_p & 0 & 0 \\ 0 & 0 & -F & 0 \\ 0 & 0 & 0 & 0 \end{pmatrix} ,$$

and $\Lambda(x)$ defined by

$$\Lambda(x) = \left(0\ 0\ \Psi^{\mathrm{T}}\bar{p}\ 0 \right)^{\mathrm{T}} .$$

Proposition 2. *Consider the controlled n-dof robot manipulator (21) with Hamiltonian (20), affected by the torque disturbances generated by (7), (8), (9).*

The additional control law generated by the adaptive internal model unit:

$$\begin{cases} \dot{\xi} = (F + G\hat{\Psi})\xi - FG\bar{p} + G\frac{1}{2}\bar{p}^{\mathrm{T}}\frac{\partial M(q)^{-1}}{\partial q}\bar{p} + G\bar{q} + \\ \qquad -G\hat{\Psi}G\bar{p} - Gk_p M^{-1}(q)\bar{p} \\ \dot{\hat{\Psi}} = -\bar{p}(\xi - G\bar{p})^{\mathrm{T}} \\ \tau = \hat{\Psi}\xi - \hat{\Psi}G\bar{p} , \end{cases} \tag{28}$$

assures asymptotically the input disturbance suppression (fault tolerance with respect to torque ripple, i.e., $(\bar{q}, \bar{p}) \to (0,0)$ as time $t \to \infty$) and the convergence of the state of the adaptive internal model to the fault signal (fault detection, i.e., $\xi \to Y z$).

Proof. Consider for system (23) (obtained connecting (21) with (28)) the following Lyapunov function:

$$V = H_x(x) \,.$$

Easy computations (remembering the skew-symmetry of interconnection matrix $J(x)$) show that there exist real numbers $\eta_{k_p} \in \mathbb{R}^-$, $\eta_F \in \mathbb{R}^-$, $\eta_{k_q} \in \mathbb{R}^-$ (depending on design matrices k_p, F, k_q) and $\eta_\Psi \in \mathbb{R}$, such that

$$\dot{V} \le \eta_{n_q} \|\bar{q}\|^2 + \eta_{k_p} \|\bar{p}\|^2 + \eta_F \|\chi\|^2 + \Psi^\mathrm{T} \bar{p} \chi$$

$$\le \eta_{n_q} \|\bar{q}\|^2 + \eta_{k_p} \|\bar{p}\|^2 + \eta_F \|\chi\|^2 + \eta_\Psi \|\bar{p}\| \|\chi\| \,.$$

Using a Young's inequality argumentation we can write

$$\dot{V} \le \eta_{k_q} \|\bar{q}\|^2 + \eta_{k_p} \|\bar{p}\|^2 + \eta_F \|\chi\|^2 + \frac{\eta_\Psi}{2} \varepsilon \|\bar{p}\|^2 + \frac{\eta_\Psi}{2\varepsilon} \|\chi\|^2 \,,$$

for a certain value of ε. Now choosing $\varepsilon = -\eta_{k_p}/\eta_\Psi$, we obtain

$$\dot{V} \le \eta_{k_q} \|\bar{q}\|^2 + \frac{\eta_{k_p}}{2} \|\bar{p}\|^2 + \left(\eta_F - \frac{\eta_\Psi^2}{2\eta_{k_p}} \right) \|\chi\|^2 \,;$$

hence choosing matrix F such that

$$\eta_F < \frac{\eta_\Psi^2}{2\eta_{k_p}} \,,$$

we have that $\dot{V} \le 0$.

System asymptotic behavior for $t \to \infty$ and for any bounded torque disturbance generated by (7) with arbitrary bounded initial conditions $z(0) \in \mathcal{Z}$ is then defined by La Salle's invariant principle: $\dot{V} = 0$ implies that asymptotically, as time $t \to \infty$, $\bar{q} \to 0$, $\bar{p} \to 0$ and $\chi \to 0$ (and consequently, by the definition of χ (22), $\lim_{t \to \infty} \xi = Y z$).

The original tracking problem is then asymptotically solved[2] and the exogenous disturbances are perfectly compensated by the adaptive internal model unit.

Remark 5. The FDI phase can be performed by checking the state of the fault compensation unit that automatically offsets the fault effect. In this framework the *detection phase* can be easily carried out by comparing $\|\xi(t)\|$ with a suitably tuned threshold. In fact, as proved in Proposition 2, $\xi(t)$

---

[2]Note again that $\bar{q} = 0$ and $\bar{p} = 0$ means that the tracking is achieved as $q \to q^\star(t)$ and $p \to p^\star(t)$.

asymptotically converge to $Yz(t)$, which is zero in the un-faulty case and different from zero when a fault occurs. Also the *isolation procedure* is possible in this framework. According to the modeling presented, it is clear that faults acting on different actuators are represented by different components of the exogenous signal $\Gamma z(t)$; the fault isolation is then solved if we are able to reconstruct this fault signal.

In view of this, let us point out the behavior of the adaptive dynamics. From (26), simple computations show that, asymptotically, $\dot{\tilde{\Psi}} \to 0$; hence the matrix $\tilde{\Psi}$ tends to a certain constant $\tilde{\Psi}_\infty$. Taking into account the asymptotic behavior of (24) (considered for $i = 1, \cdots, n$), the following holds:

$$0 = \tilde{\Psi}_\infty Y z \qquad \Rightarrow \qquad \hat{\Psi}_\infty Y z = \Psi Y z .$$

This means that, asymptotically, the adaptive algorithm is able to reproduce the disturbance effects, even if the perfect identification of all unknown frequencies is not assured (to this aim an additional persistence of excitation hypothesis should be stated).

Hence it is possible to state that, in steady state,

$$\Gamma z(t) = \Psi Y z(t) = \hat{\Psi}_\infty Y z(t) = \hat{\Psi}_\infty \xi .$$

The estimation of the fault signals is then easily obtained. Let us define some residual signals able to identify the faulty actuator as follows:

$$r_i = \begin{cases} 1 & \text{if} \quad \left\| \hat{\Psi}_\infty \xi \big|_i \right\| \geq T_i \\ 0 & \text{otherwise} \end{cases} \qquad (i = 1 \ldots n) ,$$

where T_i, $(i = 1 \ldots n)$ are n positive thresholds and $\hat{\Psi}_\infty \xi \big|_i$ is the ith component of vector $\hat{\Psi}_\infty \xi$.

5 Simulation results

To check the performances of the regulator designed above, a certain number of tests have been made, simulating the response of a 2-dof fully actuated mechanical manipulator without dissipation and inertia matrix defined by

$$M = \begin{bmatrix} 5 & 2.7 \\ 2.7 & 2.7 \end{bmatrix} ,$$

subject to various sinusoidal torque disturbances acting on both joints. The results of these tests are shown in Figures 2 and 3. In particular the disturbance ripple to be rejected was a sinusoidal signal $\delta(t) = V \sin(\Omega t)$ with amplitude $V = 10$ Nm and (unknown) frequency $\Omega = 0.4$ rad/sec occurring at time $t = 100$ sec and affecting only the first joint torque.

The internal model unit was connected at time $t = 200$ sec, but the frequency adaptation unit was connected only at time $t = 300$ sec. In Figure 2 it is possible to see clearly the action of the adaptation unit, zeroing completely the effect of the disturbance ripple on the tracking error for the angle position of both joints. Moreover it is worth pointing out that the adaptation unit makes it possible to obtain a perfect detection of the fault occurring on the first joint. In Figure 3, while the upper plot represents the disturbance ripple, the last two plots show the behavior of the controlled torques and how the adaptation unit makes it possible to clearly see the fault affecting only the control torque acting on the first joint. The adaptation unit is then disconnected again at time $t = 500$ sec for 200 sec to point out that the effect of the exogenous ripple is always present and that a classical internal model designed for a wrong frequency is not able to completely overcome this disturbance.

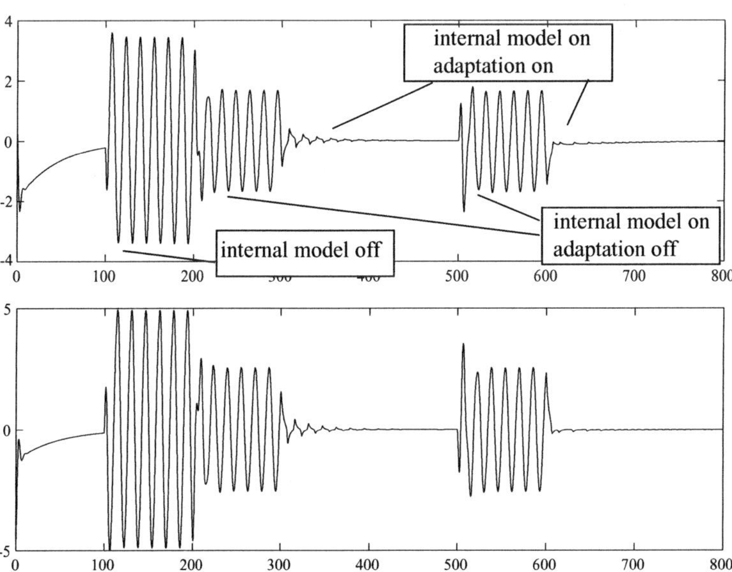

Fig. 2. Tracking error for the first and second joint angle positions (rad) θ_1 (upper plot) and θ_2 (lower plot).

6 Conclusions

The main result of this chapter is contained in Section 4, where an adaptive internal model unit is designed to compensate unknown spurious torque harmonics that degrade performances of an n-dof fully actuated mechanical robot. We have shown how a standard tracking robot control can be "augmented" with an internal model unit to achieve *global implicit fault tolerance*

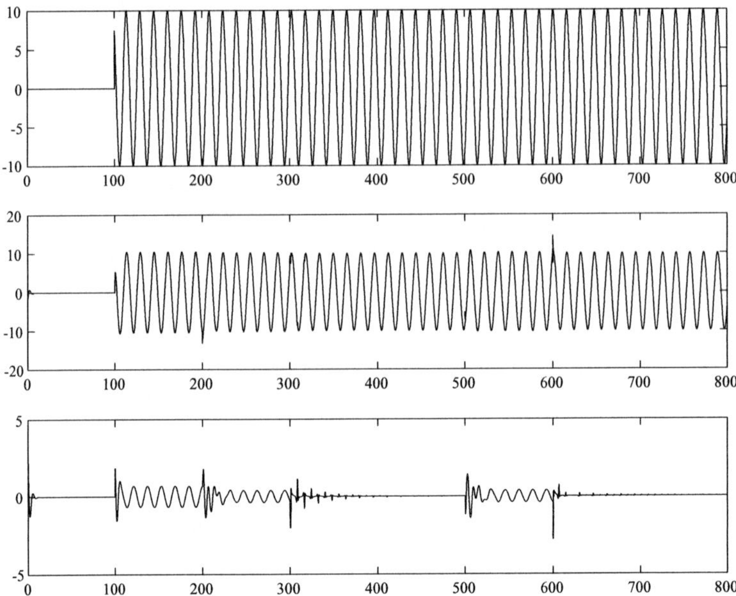

Fig. 3. From upper to lower plot: disturbance torque ripple acting on the first joint (Nm), controlled torque τ_1 on the first joint (Nm), and controlled torque τ_2 on the second joint (Nm).

to all the faults belonging to the model embedded in the regulator. We also have shown how it is possible to perform fault detection and isolation simply, by testing the state of the internal model.

References

1. Astolfi A, Isidori A, Marconi L (2003) A note on disturbance suppression for Hamiltonian systems by state feedback. 2nd IFAC Workshop on Lagrangian and Hamiltonian Methods for Nonlinear Control, Seville, Spain
2. Blanke M, Kinnaert M, Lunze J, Staroswiecki M (2003) Diagnosis and fault-tolerant control. Springer-Verlag, London
3. Bonivento C, Gentili L, Paoli A (2004) Internal model based fault tolerant control of a robot manipulator. Proc. 43rd Conference on Decision and Control, Paradise Island, Bahamas
4. Bonivento C, Isidori A, Marconi L, Paoli A (2004) Implicit fault tolerant control: Application to induction motors. Automatica 40(3):355–371
5. Byrnes C, Delli Priscoli F, Isidori A (1997) Output regulation of uncertain nonlinear systems. Birkhäuser, Boston
6. Byrnes C, Delli Priscoli F, Isidori A, Kang W (1997) Structurally stable output regulation of nonlinear systems. Automatica 33(2):369–385
7. Canudas de Wit C, Praly L (2000) Adaptive eccentricity compensation. IEEE Transactions on Control Systems Technology 8(5):757–766

8. Frank P (1990) Fault diagnosis in dynamic systems using analytical and knowledge based redundancy: a survey and some new results. Automatica 26(3):459–474

9. Fujimoto K, Sakurama K, Sugie T (2003) Trajectory tracking control of port-controlled hamiltonian systems via generalized canonical transformations. Automatica 39(12):2059–2069

10. Gentili L, van der Schaft A (2003) Regulation and input disturbance suppression for port-controlled hamiltonian systems. 2nd IFAC Workshop LHMNLC, Seville, Spain

11. Isidori A (1995) Nonlinear control systems. Springer-Verlag, London

12. Isidori A, Marconi L, Serrani A (2003) Robust autonomous guidance: an internal model-based approach. Limited series: advances in industrial control. Springer-Verlag, London

13. Lee P, Anderson T (1990) Fault tolerance: principles and practice. Springer-Verlag, Vienna, Asutria

14. De Luca A, Mattone R (2003) Actuator failure detection and isolation using generalized momenta. ICRA, Taipei, Taiwan

15. Maschke B, van der Schaft A (1992) Port-controlled hamiltonian system: modelling origins and system theoretic approach. Proc. 2nd IFAC NOLCOS, Bordeaux, 282–288

16. Nikiforov V (1998) Adaptive nonlinear tracking with complete compensation of unknown disturbances. European Journal of Control 4:132–139

17. Ortega R (2003) Some applications and recent results on passivity based control. 2nd IFAC Workshop on Lagrangian and Hamiltonian Methods for Nonlinear Control, Seville, Spain

18. Patton R, Frank P, Clark R (2000) Issues of fault diagnosis for dynamical systems. Springer-Verlag, London

19. Serrani A, Isidori A, Marconi L (2001) Semiglobal output regulation with adaptive internal model. IEEE Transaction on Automatic Control 46(8):1178–1194

20. Takegaki G, Arimoto S (1981) A new feedback method for dynamic control of manipulators. ASME Journal of Dynamic Systems Measurement and Control 103(2):119–125

21. van der Schaft A (1999) L_2-gain and passivity techniques in nonlinear control. Springer-Verlag, London

Set Membership Localization and Map Building for Mobile Robots

Nicola Ceccarelli, Mauro Di Marco, Andrea Garulli, Antonio Giannitrapani, and Antonio Vicino

Dipartimento di Ingegneria dell'Informazione
Università di Siena
Via Roma, 56
53100 Siena, Italia
`{ceccarelli,dimarco,garulli,giannitrapani,vicino}@dii.unisi.it`

Summary. Autonomous navigation of mobile robots requires the continuous estimation of the vehicle position and orientation in a given reference frame (localization problem). When moving in unknown environments, the more challenging problem of building a map, while at the same time localizing within it, must be faced (simultaneous localization and map building, SLAM). By adopting a landmark-based description of the environment, both tasks can be cast as a state estimation problem for an uncertain dynamic system, based on noisy measurements.

Under the assumption that both process disturbances and measurement errors are unknown but bounded, the estimation process can be carried out in terms of feasible sets. This chapter reviews efficient set membership localization and mapping techniques for different kinds of available measurements and different classes of approximating regions. An extension of the SLAM algorithm to the case of a team of cooperating robots is also presented. The proposed techniques are validated through extensive numerical simulations and experimental tests performed in a laboratory setup.

1 Introduction

Self-localization of a mobile robot is a fundamental issue for achieving long-range autonomy. Almost all the tasks an autonomous agent is asked to perform require the knowledge of the vehicle position within a global reference frame. The problem has been deeply studied in the last decades, and several solutions have been proposed, providing the estimates of the robot position once a map of its surroundings is available. However, in real-world applications, an autonomous agent is often called to face more challenging situations, where the operating environment is only partially known (uncertain map) or even completely unknown. In all these cases (e.g., exploration tasks or missions in hostile environments), a mobile robot must build a map of the environment it is navigating and simultaneously localize itself within the map. Consequently, in recent years, great effort has been devoted to the development of efficient solutions to the simultaneous localization and map building (SLAM) problem.

Localization and SLAM problems can be cast as state estimation problems for an uncertain dynamic system. Depending on the assumptions on the uncertainty, the estimation problem can be tackled in different ways. When a statistical description of the disturbances is adopted, standard solutions are provided by the extended Kalman filter (EKF) [26, 9, 31, 16, 34] or other probabilistic techniques [11, 33, 29]. All the methods based on the EKF generally model the uncertainty affecting the vehicle dynamics and the measurement process as zero-mean, white Gaussian noise. Unfortunately, real-world uncertainties seldom satisfy these hypotheses. Moreover, even when these assumptions are fulfilled, the Kalman filter is guaranteed to converge, but the EKF is not. Special care must be taken when neglecting correlated noise or systematic errors. In these cases, the Kalman filter estimates tend to be overoptimistic [22, 7]. These considerations motivated the adoption of alternative data fusion techniques. One of the most popular is set-theoretic estimators whose main feature is that the estimation uncertainty is represented by bounded sets in the state space.

One of the first papers in which the set-theoretic paradigm has been applied to robotic localization is [32], in which fusion of multiple sensor information with bounded uncertainty is used to construct a geometric representation of features in the environment. A set-based localization algorithm using images from a stereo vision system has been proposed in [3]. To cope with nonwhite, non-Gaussian noise, a set-theoretic approach to the problem of tracking a mobile robot, based on angular measurements, has been introduced in [20]. Under the hypothesis of bounded errors, information fusion is obtained via set intersections. This work has been later extended to a mixed stochastic/set-theoretic framework [19], in which a probabilistic uncertainty description is associated to ellipsoidal approximations of the admissible robot poses. In [30] a localization method is proposed, based on goniometrical observations of indistinguishable landmarks, taken by a panoramic camera. The robot evolution field is iteratively subdivided into rectangles containing at least one admissible landmark/measurement matching, until the desired estimation quality is achieved. In [23, 21] the problem of guaranteed robot localization is addressed via interval analysis, the optimal solution being computed via set inversion. A new set-theoretic localization method, based on angular measurements, has been recently proposed in [6]. Recasting the original problem in a higher dimensional space yields accurate implicit polynomial descriptions of the admissible vehicle positions. As far as the SLAM problem is concerned, preliminary ideas on how to deal with bounded errors are given in [10, 15].

This chapter presents a set-theoretic approach to localization and SLAM, which has been recently developed and successfully applied to several problems in mobile robotics [17, 15, 12, 13, 14, 8]. The scenario considered in this chapter consists of a mobile robot navigating in a *two-dimensional (2D) environment*, whose description is given in terms of point-wise features *(land-marks)*. The robot is supposed to be equipped with suitable sensors, providing only metric information related to the surrounding landmarks, which therefore turn out to be *indistinguishable*. The disturbances affecting both the robot dynamic model and the measurement process are supposed to be *unknown but bounded* (UBB), while no statistical assumptions concerning the nature of the errors are made. As a consequence, the solutions of the estimation problems related to localization and SLAM are given in terms of *feasible uncertainty regions*, i.e., sets containing the robot pose and/or the landmark positions. The basic reference theory for the technical development of the chapter relies on the recently developed set membership estimation theory (see, e.g., [28, 27]). Though most of the theory is developed for linear estimation, the specific structure of the nonlinear localization problem can be exploited to get efficient solutions, based on recursive approximations of the uncertainty regions.

This chapter is organized as follows. In Sections 2 and 3, the localization and SLAM problems, respectively, are stated in a set-theoretic framework. In Section 4, the strategy leading to efficient approximate solutions of the above problems is summarized. In Section 5, it is shown how to exploit the set-valued estimates to tackle the matching problem in the presence of indistinguishable features. In Section 6, the extension of the SLAM algorithm to the case of a team of cooperating robots is outlined. In Section 7, a set-theoretic path planning algorithm for designing minimum uncertainty trajectories is presented. In Section 8, results of simulated and real-world experiments are discussed. Finally, some conclusions are drawn in Section 9.

2 Mobile robot localization

Let us consider a robot navigating in a 2D environment, whose *pose* (position and orientation) at time k is denoted by

$$p(k) = [x(k) \; y(k) \; \theta(k)]' \in \mathcal{Q},$$

with $\mathcal{Q} \triangleq \mathbb{R}^2 \times [-\pi \; \pi]$ being the set of all possible robot configurations. The coordinates $(x(k), \; y(k))$ represent the position of the vehicle, while its heading, with respect to the positive x-axis, is given by $\theta(k)$. Under the hypothesis of slow robot dynamics, a simple linear model can be adopted to describe the time evolution of the robot pose

$$p(k + 1) = p(k) + u(k) + w(k), \tag{1}$$

where $u(k) \in \mathbb{R}^3$ denotes the measurements of robot displacements, coming from the odometric sensors, and $w(k) \in \mathbb{R}^3$ models the error affecting such measurements.

As far as the localization problem is concerned, it is assumed that an accurate map of the environment is available, in terms of pointwise static landmarks, whose positions are *known*. Let

$$l_i = [x_{l_i}\ y_{l_i}]',\quad i = 1,\dots,n$$

denote the coordinates of the ith landmark. The robot is supposed to be equipped with exteroceptive sensors, providing measurements related to each landmark. Depending on the sensory system, this information may consist of the distance of the vehicle from a landmark (provided by proximity sensors, such as sonars), or in the *visual angle* under which a landmark is seen (i.e., the angle between robot heading and the direction of the landmark, provided by panoramic cameras), or in both (provided by laser range finders or stereocams). In general, the measurement equations are nonlinear functions of the current robot pose $p(k)$ and of the coordinates of the sensed landmark. Since the latter are known, the sensor readings at time k can be described by equations of the form

$$c_i(k) = \mu_i(p(k)) + v_i(k),\quad i = 1,\dots,n,\tag{2}$$

where $\mu_i(p(k))$ models the ith measurement process and $v_i(k)$ represents the noise affecting that measurement. When a deterministic description of the uncertainty is adopted, no statistical hypothesis is made on the nature of the errors, the only assumption being that the disturbances are UBB:

$$|w_i(k)| \le \epsilon_i^w(k),\ i = 1,2,3\tag{3}$$

$$|v_i(k)| \le \epsilon^{v_i}(k),\ i = 1,\dots,n,\tag{4}$$

where $\epsilon_i^w(k)$ and $\epsilon^{v_i}(k)$ are known positive scalars.[1] The assumptions (3)–(4) naturally lead to a formulation of the localization problem in a set-theoretic framework. As a matter of fact, from (4) it is possible to define, for each measurement $c_i(k)$, a set

$$\mathcal{M}_i(k) = \{p(k):\ |c_i(k) - \mu_i(p(k))| \le \epsilon^{v_i}\}\tag{5}$$

representing all the vehicle poses $p(k)$ compatible with the sensor reading $c_i(k)$ and the UBB hypothesis of bounded error (4). If at time k the robot performs m different measurements, the admissible robot poses will be those belonging to the *measurement set*

$$\mathcal{M}(k) = \bigcap_{i=1}^{m} \mathcal{M}_i(k).\tag{6}$$

The set $\mathcal{M}(k)$ contains all the robot poses compatible with all the measurements gathered at time k and with the bounds (4). If the hypothesis (4) of

---

[1]From now on, for notational convenience, the time dependency of the error bounds will be omitted.

bounded errors is correct, the measurement set $\mathcal{M}(k)$ is not empty. Conversely, an empty intersection in (6) implies that at least one of the constraints (4) is violated.

The shape of each set $\mathcal{M}_i(k)$ depends on the nonlinear functions $\mu_i(p(k))$ modeling each measurement process. Hence, $\mathcal{M}_i(k)$ is, in general, a nonconvex set, bounded by nonlinear curves.

Since process disturbances are also supposed to be UBB, it is possible to introduce the notion of *feasible pose set* $P(k|k)$, defined as the set of all robot poses at time k compatible with all the information collected up to that time. Now, the set membership dynamic localization problem can be stated as follows.

Set Membership (SM) Localization Problem. *Let $P(0) \subset \mathcal{Q}$ be a set containing the initial pose $p(0)$. Given the dynamic model (1) and the measurement equation (2), find at each time $k = 1, 2, \ldots$, the* feasible pose set *$P(k|k) \subset \mathcal{Q}$ containing all vehicle poses $p(k)$ that are compatible with the robot dynamics, the assumptions (3)–(4) on the disturbances and the measurements collected up to time k.*

The exact solution to the SM localization problem is given by the following recursion, which stems out directly from the robot dynamics (1) and the UBB assumptions (3)–(4):

$$P(0|0) = P(0) \tag{7}$$

$$P(k+1|k) = P(k|k) + u(k) + \mathrm{Diag}[\epsilon^w]\mathcal{B}_\infty \tag{8}$$

$$P(k+1|k+1) = P(k+1|k) \cap \mathcal{M}(k+1)\,, \tag{9}$$

where $\epsilon^w = [\epsilon_1^w\ \epsilon_2^w\ \epsilon_3^w]'$, $\mathrm{Diag}[v]$ denotes the diagonal matrix with vector v on the diagonal, and $\mathcal{B}_\infty$ is the unit ball in the ∞-norm.

The update of the feasible pose set is carried out within a prediction-correction scheme, as in the Kalman filter, processing at each time step the current odometric and exteroceptive measurements. Equation (8) is often referred to as *time update*: during this step, information provided by the robot dynamic model and odometric sensors are used to update the feasible set. The set $\mathrm{Diag}[\epsilon^w]\mathcal{B}_\infty$ describes all the admissible unmodeled dynamics and odometric errors satisfying assumption (3); due to these disturbances, the size of the feasible pose set grows during the time update (see Figure 1(a)).

Equation (9) is known as *measurement update* as it exploits all the exteroceptive measurements to reduce the total robot uncertainty. The main property of recursion (7)–(9) is to provide, at each time step, all the pose values that are compatible with the available information. The true state is guaranteed to belong to set $P(k|k)$, and the size of such set gives a measure of the uncertainty associated to the estimate (see Figure 1(b)).

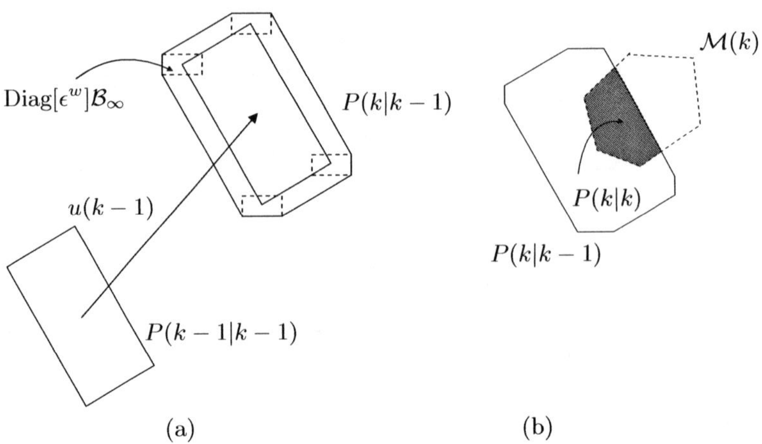

Fig. 1. Updating the feasible pose set: time update (a) and measurement update (b).

3 Simultaneous localization and map building

In several real-world applications a map of the working space is not available. In those cases, the harder SLAM problem has to be faced.

Let us consider the scenario outlined in Section 2. To state the SLAM problem as a state estimation problem, a model describing the time evolution of the landmark positions is needed. As long as only static landmarks are considered, their coordinates $l_i(k)$ satisfy the equations

$$l_i(k+1) = l_i(k). \tag{10}$$

Since both robot pose and landmark positions are unknown, the dimension of the state vector to be estimated can be very large, as it depends on the number of remarkable features present in the environment. Indeed, when n landmarks are considered, the state vector is given by

$$\xi(k) = [p'(k)\, l'_1(k)\ldots l'_n(k)]' \in \mathbb{R}^{(3+2n)}. \tag{11}$$

From (1) and (10), the state update equation is

$$\xi(k+1) = \xi(k) + E_3 u(k) + E_3 w(k), \tag{12}$$

where $E_3 = [I_3\, 0\ldots 0]' \in \mathbb{R}^{(3+2n)\times 3}$. Concerning the exteroceptive measurements, (2) can be properly rewritten as

$$c_i(k) = \mu_i(p(k), l_i(k)) + v_i(k), \quad i = 1,\ldots,n \tag{13}$$

to emphasize the dependence of the measurement process μ_i on the landmark position $l_i(k)$. Notice that in this case each equation (13) provides a nonlinear relation among some of the state variables (namely the robot pose and the ith landmark coordinates).

Under the UBB assumptions (3)–(4), the definition of *measurement set* $\mathcal{M}(k)$ is still given by (6), provided that (5) is replaced by

$$\mathcal{M}_i(k) = \{\xi(k):\ |c_i(k) - \mu_i(p(k), l_i(k))| \leq \epsilon^{v_i}\}. \tag{14}$$

Hence, the SLAM problem can be phrased in a set membership framework as follows.

SM SLAM Problem. *Let $\Xi(0) \subset \mathbb{R}^{(3+2n)}$ be a set containing the initial vehicle pose and landmark positions, $\xi(0)$. Given the dynamic model (12) and the measurement equations (13), find at each time $k = 1, 2, \ldots$ the set $\Xi(k|k)$ of state vectors $\xi(k)$ that are compatible with the robot dynamics, the assumptions (3)–(4) on the disturbances and the measurements collected up to time k.*

The *solution* to the above problem is still provided by the algorithm outlined in (7)–(9), where the robot feasible pose set $P(k)$ is replaced by the extended feasible state set $\Xi(k)$:

$$\Xi(0|0) = \Xi(0) \tag{15}$$
$$\Xi(k+1|k) = \Xi(k|k) + E_3 u(k) + E_3 \mathrm{Diag}[\epsilon^w]\mathcal{B}_\infty \tag{16}$$
$$\Xi(k+1|k+1) = \Xi(k+1|k) \cap \mathcal{M}(k+1). \tag{17}$$

The *initialization* step of the algorithm, i.e., the choice of the set $\Xi(0)$, depends on the specific problem tackled. If there is no available information on the initial position of any element of the problem (i.e., robot and landmarks, as it happens in the SLAM case), the initial estimate set $\Xi(0)$ should be chosen as $\mathbb{R}^{(3+2n)}$. Nonetheless, since all measurements are relative, in this case one is allowed to choose an arbitrary reference frame. Hence, without loss of generality, it is possible to set the origin of the reference frame in the initial position of the robot, choosing as x-axis the robot initial heading. On the other hand, when an uncertain map of the environment is a priori given, $\Xi(0)$ can be chosen accordingly to the available information.

4 Suboptimal solutions

The major drawback of the above localization and SLAM algorithms is their high computational burden. The main reason for this is intimately related to the nature of the measurement sets $\mathcal{M}(k)$, defined in (5), (6), and (14). As already mentioned, such a set is given by the intersection of nonlinear and nonconvex sets, so that its shape can become arbitrarily complex. During the measurement update steps (9) and (17), such complexity affects the feasible state set, making the overall recursion too computationally demanding for real-time applications. Moreover, the exact feasible set is not only expensive to compute, but also its explicit, analytical expression is hard to maintain, since as new measurements are processed, the complexity of its shape increases. Finally, as far as the SLAM problem is concerned, computational issues are worsened by the high dimension of the true state space. For these reasons, suboptimal solutions to the SM localization and SLAM problems, trading off computational complexity and estimation accuracy, are sought.

The guidelines inspiring the suboptimal strategy can be summarized as follows.

- The devised algorithm should provide a *recursive, efficient* solution, suitable for *on-line* implementation.
- The set-valued estimates should be *guaranteed* to contain the actual robot pose and landmark locations.
- Simple, analytical expressions of the feasible sets are desired.

To meet the above requirements, outer approximations of the feasible sets, through simple structure regions, are adopted. Specifically, two approximations are introduced at different stages, leading to suboptimal solutions:

- decomposition of the state vector into subsets of state variables;
- outer approximations of the true feasible subsets through classes of simple regions.

The computation of the feasible sets in the entire state space can be replaced by the evaluation of their projections on suitable subspaces, whose Cartesian product gives an outer approximation of the overall feasible sets. This policy allows us to perform sums and intersections in lower-dimensional subspaces, yielding a significant reduction of the computational burden. This proves to be especially useful in the SLAM context, where a high-dimensional state vector is involved. A natural choice is to estimate separately the *feasible robot position set* Ξ_{xy}, the *feasible robot orientation set* Ξ_θ, and, in the SLAM case, the *feasible landmark* position sets Ξ_{l_i}. With this strategy, sequential updates of only 1D and 2D sets are required.

The exact computation of the intersections required during the measurement update, although performed in low-dimensional spaces, is still computationally demanding, due to the complex shape of the measurement sets. To improve the algorithm efficiency, as well as to come up with tractable analytical expressions for the set-valued estimates, the exact feasible subsets can be approximated through regions belonging to a class $\mathcal{R}$ of fixed and simple structure sets. To minimize the estimation uncertainty, the minimum volume region in the chosen class containing the corresponding feasible set is selected (see Figures 2–3).

Efficient algorithms for localization and SLAM, using boxes or parallelotopes as approximating regions, have been proposed in [17, 12, 14]. The main features of the devised techniques can be summarized as follows.

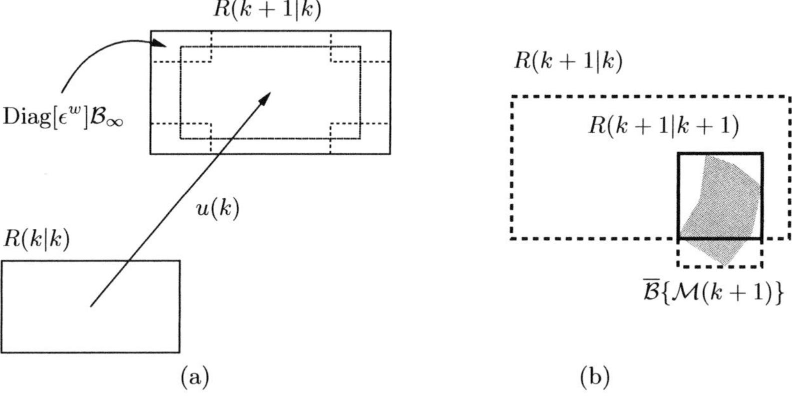

Fig. 2. Time update (a) and measurement update (b) using boxes. $\overline{\mathcal{B}}\{\mathcal{M}(k + 1)\}$ denotes the minimum volume box containing the measurement set $\mathcal{M}(k + 1)$.

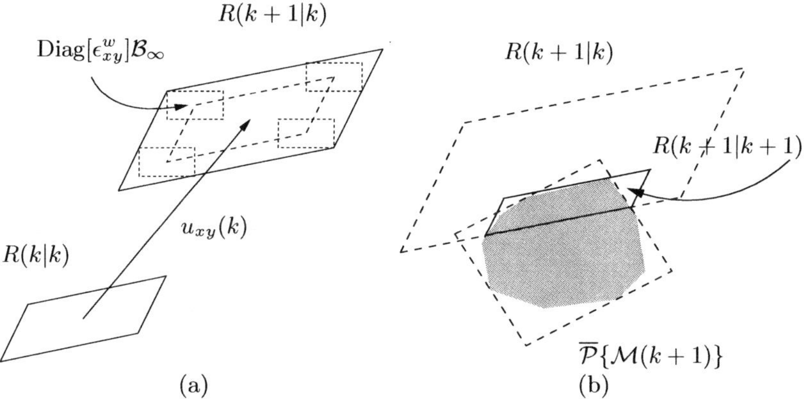

Fig. 3. Time update (a) and measurement update (b) using parallelotopes. $\overline{\mathcal{P}}\{\mathcal{M}(k + 1)\}$ denotes the minimum volume parallelotope containing the measurement set $\mathcal{M}(k + 1)$.

- The estimation algorithms provide guaranteed set-valued estimates, in the sense that the actual state vector is guaranteed to belong to the computed uncertainty regions *(feasibility property)*.
- No statistical assumptions on the nature of the errors are required, only an upper bound is assumed to be available.
- The suboptimal strategy adopted leads to fast algorithms suitable for on line implementation. Low computationally demanding estimates are obtained at the price of some conservativeness. This feature turns out to be especially useful when facing the SLAM problem. In this case, the storage requirements of the proposed set membership techniques grow linearly in

the number of landmarks present in the environment, whereas its computational burden is linear in the number of features sensed at each time instant (as opposed to the quadratic complexity required by EKF-based algorithms).

- The desired trade-off between computational effort and estimation accuracy can be achieved by a suitable choice of the class of approximating regions.

- Set approximations involved in the measurement update can be iteratively repeated, by processing several times the same measurements. This generally leads to more accurate approximations, at the price of a higher computational load. Moreover, in the SLAM problem this allows us to implicitly account for the relationships between different state variables (see [14] for details).

The formulation of the localization and mapping problems presented so far implicitly assumes that the robot is able to correctly associate each measurement to the corresponding sensed feature. For example, this happens if the observed landmarks are distinguishable. However, in real-world applications, especially if natural beacons are used as landmarks and sensors provide only metric information, landmarks are indistinguishable and consequently matching between measurements and landmarks is of paramount importance. In the next section, we will describe a data association algorithm exploiting the feasibility property of the set membership estimates [14].

5 Data association

At each time instant $k+1$, before the measurements are taken, the SM localization algorithm computes a region $R_{xy}(k+1|k)$ containing the prediction of the feasible robot positions according to the dynamic model (1) and the error bounds (3). Since measurement noise is supposed to be bounded, for each measurement c_i it is possible to define a set $\mathcal{M}_{l_i}(k+1)$ containing all the admissible ith landmark locations (in a robot-centered reference frame) compatible with the noise bounds (4). Since the robot position is known up to an uncertainty region $R_{xy}(k+1|k)$, the set in which the ith landmark is guaranteed to lie is

$$R_{xy}(k+1|k) + \mathcal{M}_{l_i}(k+1).$$

As a consequence, any landmark position $l_j, j = 1, \ldots, n$ such that

$$l_j \in R_{xy}(k+1|k) + \mathcal{M}_{l_i}(k+1), \tag{18}$$

can be associated to the ith measurement. An example of such a set is depicted in Figure 4, when boxes are used as approximating regions and range and bearing measurements are available.

By considering each measurement, it is possible to build a matching matrix T, where all the admissible measurement-landmark matchings are listed. For

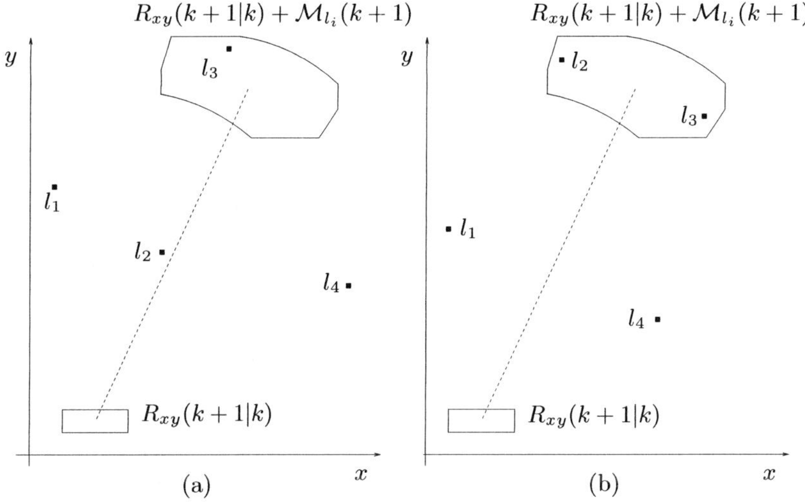

Fig. 4. Uncertainty sets can be employed to perform matching between measurements and features: (a) measurement i is not ambiguous, since only landmark l_3 lies in $R_{xy}(k+1|k) + \mathcal{M}_{l_i}(k+1)$; (b) measurement i is ambiguous, since it can be associated both to landmark l_2 and to landmark l_3.

instance, one can set $t_{ij} = 1$ if the jth landmark lies inside $R_{xy}(k+1|k) + \mathcal{M}_{l_i}(k+1)$, while $t_{ij} = 0$ otherwise. If on the ith row of matrix T there is only one nonzero entry, then the ith measurement is not ambiguous. Should all the entries be null, then there is no landmark compatible with the measurement (consequently, the ith measurement is a spurious one). Finally, a row with more than one nonzero entry corresponds to an ambiguous measurement (see Figure 5a).

Even if some measurements are ambiguous, it is often possible to find a unique admissible solution to the matching problem (see Figure 5b). Indeed, the problem of determining the existence of a perfect matching is widely studied in the operation research field, and efficient algorithms are available to determine the solution of such problems [24].

The above procedure can be applied, with minor changes, to the SLAM problem. As long as a map of the environment is not exactly known, in order to check whether a previously sensed landmark l_j can be accountable for the ith measurement, the uncertainty of its current estimate must be suitably taken into account. Denoting by $R_{l_j}(k+1|k)$ the region approximating the feasible jth landmark set at time $k+1$, this amounts to replace condition (18) by

$$R_{l_j}(k+1|k) \cap (R_{xy}(k+1|k) + \mathcal{M}_{l_i}(k+1)) \neq \emptyset. \qquad (19)$$

Accordingly, each landmark l_j satisfying (19) is eligible for the ith measurement. The most significant difference with respect to the scenario of known landmarks concerns the definition of spurious measurement. In a typ-

meas.\land.	a	b	c	d	e	f
1	0	1	0	0	1	1
2	1	0	1	0	0	0
3	0	0	0	1	0	1
4	1	0	0	0	1	0
5	1	1	0	0	0	0
6	0	0	0	0	1	0

(a)

meas.\land.	a	b	c	d	e	f
1	0	0	0	0	0	1
2	0	0	1	0	0	0
3	0	0	0	1	0	0
4	1	0	0	0	0	0
5	0	1	0	0	0	0
6	0	0	0	0	1	0

(b)

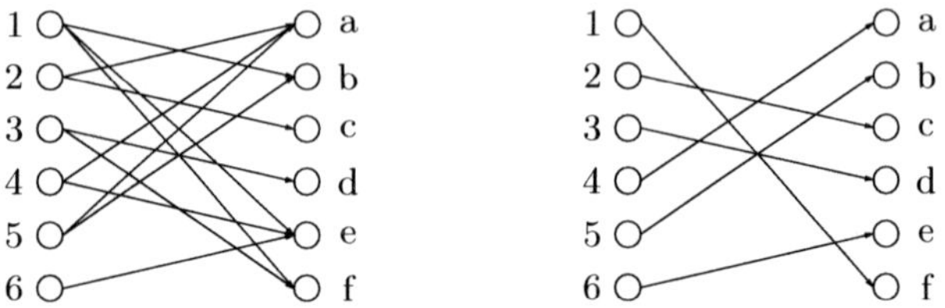

Fig. 5. Example of matching matrix T: (a) matrix with ambiguous measurements; (b) unique admissible matching.

ical exploration task, where neither the locations nor the total number of remarkable features is a priori known, the landmarks to be mapped are usually incrementally discovered. In this case, an empty intersection between $R_{xy}(k+1|k) + \mathcal{M}_{l_i}(k+1)$ and every $R_{l_j}(k+1|k)$, may occur because

- the ith measurement is spurious, or
- the ith measurement is related to a newly discovered landmark.

In the first case, one would simply discard the uninformative measurement. On the contrary, if a new feature is detected, the state vector should be properly augmented and the set-valued estimate of the new landmark should be initialized according to the set $R_{xy}(k+1|k)+\mathcal{M}_{l_i}(k+1)$. Unfortunately, there is no chance to certainly discriminate between these two options. Nonetheless, some heuristic techniques borrowed from statistical approach can be adopted to avoid the introduction of spurious landmarks. The inclusion of a new feature into the state vector may be deferred until sufficient evidence of its presence is gathered, using a tentative list (see, e.g., [16]). A tentative landmark is initialized on receipt of a measurement and is then inserted into the state vector when a sufficiently high number of consecutive hits is reached. A possible alternative is to generate many data association hypotheses whenever a measurement is taken, and later discard all of them but one as more sensor data are collected (see, e.g., [5]).

6 Cooperative SLAM

In recent years, the employment of a team of cooperating agents to carry out complex tasks has received more and more attention by the robotics research community (see, e.g., the special issues [2, 4, 1]). Fusion of information provided by different robots moving in the same area can indeed improve the exploration performance for several reasons. At each time instant the same feature can be perceived by more than one robot. If the robots share mapping information, this can lead to a more accurate, faster converging global map. Moreover, each robot plays the role of a moving landmark for all other agents, thus improving localization accuracy in poorly informative environments. For instance, if only one robot at a time is moving, the others can act as a landmark base in regions where it is difficult to extract reliable features [18].

The SM localization and SLAM algorithms can be naturally extended to the cooperative scenario. Let

$$p_j(k) = [x_j(k) \ y_j(k) \ \theta_j(k)]'$$

denote the pose of the jth robot at time k. The setting hereafter considered consists of M agents, moving in an environment containing n unknown static landmarks $l_1, \ldots, l_n$. Then, the state vector to be estimated becomes

$$X(k) = [p'_1(k) \ \ldots \ p'_M(k) \ l'_1 \ \ldots \ l'_n]' \in \mathbb{R}^{3M+2n}. \tag{20}$$

Following the suboptimal approach illustrated in Section 4, the underlying idea is to decompose the approximation of the overall feasible set into $M + n$ approximations of 2D feasible subsets for the position of each feature in the environment, plus M interval approximations for the feasible orientation of each robot.

The main issue to be addressed is the fusion of the local maps, built by each agent, into a global one, when the relative starting positions of the agents are completely unknown. To this purpose, two additional assumptions are needed: (i) each feature is supposed to have a unique signature (distinguishable landmarks), (ii) absolute orientation measurements (such as the ones provided by compasses) are available. Nonetheless, according to the devised algorithm, map fusion is performed only once, at the beginning of the exploration in order to generate a common map, in a global reference frame. Afterward, each agent is able to consistently update the global map even in the presence of indistinguishable features and only relative bearing information.

In the single robot SLAM problem, since all measurements are relative, the initial position of the exploring agent can be fixed arbitrarily (in the robot-centered reference frame) and all the environment features are then estimated with respect to that initial position. When operating with multiple agents, whose initial positions are not known, the global reference frame shared by all the robots must be chosen carefully. Assuming that absolute orientation measurements are available, it is possible to rotate the initial 2D uncertain maps onto an absolute orientation reference frame. By computing the initial 2D feasible position sets of each feature, the jth agent is able to produce a self-centered initial map of the environment, within an absolute orientation reference system. This means that only relative translation among the different initial maps produced by the robots is unknown. Clearly, all the single-robot maps are a valid, suboptimal representation of the environment. The quality of the maps will generally vary from robot to robot. While it is possible to choose the "most accurate" map as a global one for all the robots, this choice does not use all the available information. Exploiting the feasibility property, it is possible to obtain a global refined map, by finding the minimum uncertainty description that satisfies all the constraints of each map. Moreover, when box-based approximations are adopted, the 2D map fusion can be decomposed into two 1D problems, whose solutions can be efficiently computed via linear programming algorithms. The interested reader is referred to [13], for a detailed description of SM cooperative SLAM.

7 Set membership approach to path planning

Motion planning has been recognized as a hard problem for a long time. Nonetheless, as far as mobile robots (characterized by few degrees of freedom) are concerned, several solutions have been proposed, which proved to be effective in practice (see [25] for a comprehensive review). While uncertainty and disturbances are crucial issues in localization and map building, they have been often neglected in the path planning phase, i.e., the robot pose is usually assumed to be perfectly known.

In this section, we briefly show how the set-theoretic framework is well suited to design minimum uncertainty paths. In particular, the problem of finding a trajectory minimizing the overall position uncertainty along the path is tackled. This is motivated by applications in which it is crucial to localize the robot as precisely as possible along the whole path and not only in the final target, typical examples being the exploration of unknown environments or the accomplishments of several tasks requiring a prescribed precision along the traveled path. The appealing property of SM localization algorithms of delivering guaranteed set estimates provides a simple way of quantifying the estimation uncertainty in terms of the size of the feasible pose sets.

Consider a robot with initial pose p_0 whose target is to reach a pose p_T, after T moves. The objective is to plan a path $\mathcal{P} = [p(0), p(1), \ldots, p(T)]$, such that the average uncertainty associated to the path is minimized. This corresponds to tackling the following problem:

$$\min_{\mathcal{P}} \frac{1}{T} \sum_{k=1}^{T} V\left[\mathcal{M}(k)\right]$$

$$\text{s.t.}$$

$$|x(k+1) - x(k)| \leq b_x \; \forall k = 0 \ldots T - 1$$

$$|y(k+1) - y(k)| \leq b_y \; \forall k = 0 \ldots T - 1$$

$$p(0) = p_0 \; p(T) = p_T \,,$$

$$(21)$$

where $V\left[\mathcal{M}(k)\right]$ is the volume of the feasible set $\mathcal{M}(k)$, and b_x, b_y are bounds on the maximum x- and y-displacement for each move, respectively.

The solution of problem (21) presents several difficulties, since

a) the sets $\mathcal{M}(k)$ depend on the actual realization of measurement noises;
b) the size of $\mathcal{M}(k)$ does not depend on the robot orientation $\theta(k)$ if the visibility range is unlimited and all landmarks can be seen from any robot pose; in the more realistic situation of limited field of view, $V\left[\mathcal{M}(k)\right]$ depends on the subset of features falling in the robot field of view.

Facing problem a) in a worst-case scenario, i.e., by maximizing $V\left[\mathcal{M}(k)\right]$ over all possible noise realizations, would lead to an intractable high-dimensional min-max optimization problem. A more viable approach is that of neglecting the measurement errors in the planning phase, which amounts to considering zero noise realization. Notice that this is generally an unfavorable case in set-theoretic estimation, where the maximum uncertainty reduction is usually achieved when the noise sequence takes values on the boundary as often as possible. In order to cope with problem b), the orientation interval can be suitably partitioned according to the subset of visible landmarks. By choosing the orientation subset for which the size of the feasible set $\mathcal{M}(k)$ is minimum, one can get rid of the dependence on θ in the cost function (21), thus reducing the number of optimization variables from $3T$ to $2T$. Moreover, the minimization problem can be further simplified if set membership localization techniques are used to approximate the feasible sets $\mathcal{M}(k)$. Notice that the proposed method is flexible enough to incorporate a priori knowledge on the robot working space. For instance, obstacles in the environment can be easily taken into account by combining artificial potential functions with the cost function in (21) (see Figure 6). Further details can be found in [8].

8 Experimental validation

The SM localization and SLAM algorithms outlined so far have been extensively tested in both simulated and real experiments.

Monte Carlo simulations have shown that the SM localization algorithms are characterized by a good accuracy, fairly degrading with the increase of the measurement error bounds. Thanks to their flexibility, parallelotopic approximations usually delivers better performance, with respect to box-based ones. Extensive simulations have been performed also in different scenarios, showing

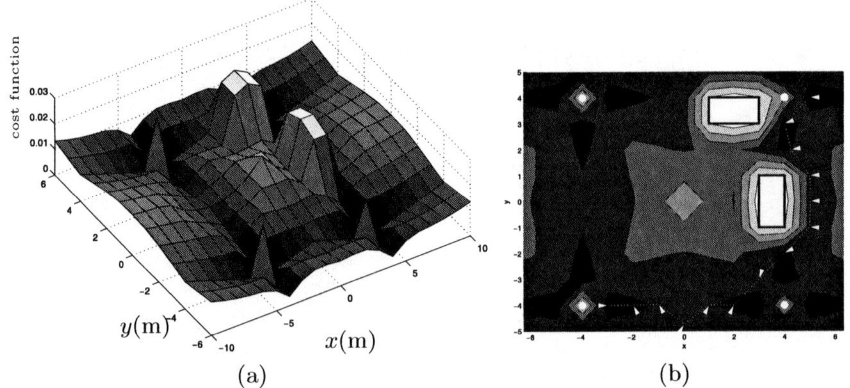

Fig. 6. Cost function (a) and computed path (b), for a scenario including two rectangular obstacles and four landmarks (white circles). Triangles represent the chosen robot poses along the optimal path.

that SM algorithms are able to face the localization problem with different robot motion models and different exteroceptive sensors. Specifically, guaranteed estimates have been computed even in the presence of correlated process disturbances or measurement noise. Moreover, the uncertainty affecting the set estimates turns out to rapidly approach its steady state value, especially when parallelotopic approximations are adopted (see [17, 12, 14]).

Similar simulation results have been obtained for the SM SLAM algorithm (see [15]). SM SLAM techniques have been compared to EKF-based algorithms, taking into account different noise models. Since a key performance index is the quality of the constructed map, the comparison is focused on the nominal error and the associated uncertainty of the final landmark position estimates. When measurement errors are modeled as white Gaussian noise, SM and EKF algorithms yield map estimates with comparable accuracy, whereas in case of non-Gaussian colored noise, the SM approach usually performs better (see [14]). It is worth noticing that the set-theoretic framework requires a limited modeling effort (only upper bounds on the involved errors have to be set), while in the statistical context the noise covariance matrices have to be carefully tuned to preserve EKF consistency. Moreover, a major advantage of the SM SLAM algorithm lies in its low computational burden and memory requirements. At each time step, the SM algorithm performs a number of operations proportional to the number of currently sensed landmark, whereas the EKF algorithm requires the propagation of the whole covariance matrix (a very expensive task when the number of landmarks increases). Such a feature is of paramount importance when exploring large areas, densely populated with landmarks.

Concerning the cooperative SLAM problem, simulations of the map fusion step have been run to evaluate the accuracy improvement of the global map

Fig. 7. Experimental setup: the mobile robots Nomad XR4000 (left) and Pioneer 3AT (right).

with respect to the maps built by each single robot. It turns out that the uncertainty reduction becomes more significant when the number M of robots (and consequently of available maps) increases. This happens because more maps will generally provide overall tighter constraints in the map fusion step. In addition, for a fixed number of robots, improvements get less remarkable as the number n of nonsensing features increases, since adding nonsensing features to the environment increases the map uncertainty without providing additional means for uncertainty reduction. The SLAM algorithm has been simulated in a dynamic setting, in order to evaluate the accuracy improvement of robot localization and map estimation. The adoption of the cooperative scheme results in a faster convergence of landmark estimation accuracy to its steady-state value. In addition, each robot is able to precisely localize itself at large distances from its starting point, thanks to the accurate map of initially faraway areas built by the other agents (see [13]).

The SM localization and SLAM techniques have been validated in a laboratory setup, featuring the mobile robots Nomad XR4000 and Pioneer 3AT (see Figure 7). These vehicles have different kinematics, the latter being a nonholonomic platform, while the former has a fully holonomic drive system. Both robots are equipped with a laser rangefinder and all the experiments involved indistinguishable, artificial landmarks, easily detectable from the laser scans. The proposed techniques are flexible enough to handle different vehicle motion models. The bounded error assumption turns out to be a reasonable description of the disturbances involved. The SM estimation techniques allow the robots to safely localize and navigate in spite of odometric error accumulation, and the data association algorithm proved to be effective when dealing with indistinguishable features. The results of the experimental tests are generally in good agreement with simulations and confirm the effectiveness of SM approach in real-world applications (see [14]).

9 Conclusions

In this chapter, a set-theoretic framework for addressing some relevant problems of autonomous navigation has been presented. Specifically, the attention has been mainly focused on localization and map building problems for a single robot or a team of cooperating agents. Such problems can be cast as a state estimation problem for an uncertain dynamic system, based on nonlinear observations. Under the hypothesis of bounded model disturbances and measurement noise, the estimation process can be naturally stated in terms of feasible sets. Since real-time suitability represents an essential requirement, one of the main results is the design of efficient algorithms, trading off computational complexity and estimation accuracy. The extensive numerical simulations and experimental tests carried out show that the set-theoretic approach represents a valuable alternative to statistical localization and map building methods, whenever the bounded error assumption holds.

A path planning algorithm for holonomic vehicles, which explicitly takes into account the localization uncertainty affecting the robot along its trajectory, has been devised. The proposed design procedure represents an attempt to consider in the planning stage the whole information available along the path. The next goal is the extension of the proposed planner when the more challenging simultaneous localization and map building problem has to be addressed. This should be intended as a first step toward the development of advanced exploration strategies, through the integrated solution of different tasks, like localization, map building, and path planning. It is believed that set membership techniques provide a suitable framework for dealing with estimation uncertainty in such complex scenarios.

References

1. Arai T, Pagello E, Parker L (Eds) (2002) Advances in multirobot systems. IEEE Transactions on Robotics and Automation 18(5):655–864
2. Arkin R C, Bekey G A (Eds) (1997) Special issue on robot colonies. Autonomous Robots 4(1):1–153
3. Atiya S, Hager G D (1993) Real-time vision-based robot localization. IEEE Transactions on Robotics and Automation 9(6):785–800
4. Balch T, Parker L E (2000) Special issue on heterogeneous multi-robot systems. Autonomous Robots 8(3)

5. Bar-Shalom Y, Fortman T (1988) Tracking and data association. Academic Press, Reading, MA
6. Briechle K, Hanebeck U D (2004) Localization of mobile robot using relative bearing measurements. IEEE Transactions on Robotics and Automation 20(1):36–44
7. Castellanos J M, Neira J, Tardos J D (2004) Limits to the consistency of EKF-based SLAM. 1244–1249. In: Proceedings of the 5th IFAC Symposium on Intelligent Autonomous Vehicles, Lisbon, Portugal
8. Ceccarelli N, Di Marco M, Garulli A, Giannitrapani A (2004) A set theoretic approach to path planning for mobile robots. In: IEEE Conference on Decision and Control, Atlantis, Bahamas
9. Cox I J (1991) Blanche–an experiment in guidance and navigation of an autonomous mobile robot. IEEE Transactions on Robotics and Automation 7(3):193–204
10. Csorba M, Uhlmann J K, Durrant-Whyte H F (1996) A new approach to simultaneous localization and dynamic map building. 26–36. In: Proceedings of SPIE–The International Society for Optical Engineering, Orlando, FL
11. Dellaert F, Burgard W, Fox D, Thrun S (1999) Using the condensation algorithm for robust, vision-based mobile robot localization. In: Proceedings of the IEEE International Conference on Computer Vision and Pattern Recognition, Ft. Collins, CO
12. Di Marco M, Garulli A, Giannitrapani A, Vicino A (2001) Set membership pose estimation of mobile robots based on angle measurements. 3734–3739. In: IEEE Conference on Decision and Control, Orlando, FL
13. Di Marco M, Garulli A, Giannitrapani A, Vicino A (2003) Simultaneous localization and map building for a team of cooperating robots: a set membership approach. IEEE Transactions on Robotics and Automation 19(2):238–249
14. Di Marco M, Garulli A, Giannitrapani A, Vicino A (2004) A set theoretic approach to dynamic robot localization and mapping. Autonomous Robots 16:23–47
15. Di Marco M, Garulli A, Lacroix S, Vicino A (2001) Set membership localization and mapping for autonomous navigation. International Journal of Robust and Nonlinear Control 11(7):709–734
16. Dissanayake M, Newman P, Clark S, Durrant-Whyte H F, Csorba M (2001) A solution to the simultaneous localization and map building (SLAM) problem. IEEE Transactions on Robotics and Automation 17(3):229–241
17. Garulli A, Vicino A (2001) Set membership localization of mobile robots via angle measurements. IEEE Transactions on Robotics and Automation 17(4):450–463
18. Grabowski R, Navarro-Serment L E, Paredis C J J, Khosla P K (2000) Heterogeneous teams of modular robots for mapping and exploration. Autonomous Robots 8(3):293–308
19. Hanebeck U D, Horn J (1999) A new estimator for mixed stochastic and set theoretic uncertainty models applied to mobile robot localization. 1335–1340. In: IEEE Int. Conf. Robotics and Automation, Detroit, MI
20. Hanebeck U D, Schmidt G (1996) Set theoretical localization of fast mobile robots using an angle measurement technique. 1387–1394. In: IEEE Int. Conf. Robotics and Automation, Minneapolis, MN

21. Jaulin L, Walter E, Leveque O, Meizel D (2000) Set inversion for chi-algorithms, with application to guaranteed robot localization. Math. Comput. Simulation 52:197–210
22. Julier S J, Uhlmann J K (2001) A counter example to the theory of simultaneous localization and map building. 4238–4243. In: Proceedings of the 2001 IEEE International Conference on Robotics and Automation, Seoul, Korea
23. Kieffler M, Jaulin L, Walter E, Meizel D (1999) Nonlinear identification based on unreliable priors and data, with application to robot localization In: Garulli A, Tesi A, Vicino A (Eds), Robustness in identification and control. Springer-Verlag, London
24. Korte B, Vygen J (2000) Combinatorial optimization: theory and algorithms. Springer, New York
25. Latombe J (1991) Robot motion planning. Kluwer Academic Publishers, Boston, MA
26. Leonard J J, Durrant-Whyte H F (1991) Mobile robot localization by tracking geometric beacons. IEEE Transactions on Robotics and Automation 7(3):376–382
27. Milanese M, Norton J P, Piet-Lahanier H, Walter E (Eds) (1996) Bounding approaches to system identification. Plenum Press, New York
28. Milanese M, Vicino A (1991) Optimal estimation theory for dynamic systems with set membership uncertainty: an overview. Automatica 27(6):997–1009
29. Montemerlo M, Thrun S, Koller D, Wegbreit B (2003) Fastslam 2.0: An improved particle filtering algorithm for simultaneous localization and mapping that provably converges. In: Proceedings of the International Joint Conference on Artificial Intelligence (IJCAI), Acapulco, Mexico
30. Mouaddib E M, Marhic B (2000) Geometrical matching for mobile robot localization. IEEE Transactions on Robotics and Automation 16(5):542–552
31. Neira J, Tardos J D, Horn J, Schmidt G (1999) Fusing range and intensity images for mobile robot localization. IEEE Transactions on Robotics and Automation 15(1):76–84
32. Sabater A, Thomas F (1991) Set membership approach to the propagation of uncertain geometric information. 2718–2723. In: IEEE International Conference on Robotics and Automation, Sacramento, CA
33. Thrun S, Fox D, Burgard W, Dellaert F (2000) Robust Monte Carlo localization for mobile robots. Artificial Intelligence 128(1-2):99–141
34. Thrun S, Liu Y, Koller D, Ng A, Ghahramani Z, Durrant-Whyte H F (2002) Simultaneous localization and mapping with sparse extended information filters. In: Proceedings of the Fifth International Workshop on Algorithmic Foundations of Robotics, Nice, France

Visual Servoing with Central Catadioptric Camera

Gian Luca Mariottini, Eleonora Alunno, Jacopo Piazzi, and Domenico
Prattichizzo

Dipartimento di Ingegneria dell'Informazione
Università di Siena
Via Roma, 56
53100 Siena, Italia
{gmariottini,alunno,piazzi,prattichizzo}@dii.unisi.it

Summary. In this chapter we present an epipolar-based visual servoing for holonomic mobile robots equipped with panoramic camera. The proposed visual servoing
is based on epipolar geometry and exploits the *auto-epipolar property*, a special configuration for the epipoles that occurs when the desired and the current panoramic
views undergo a pure translation. This occurrence is detectable directly in the image
plane simply controlling when the so-called biosculating conics all co-intersect at only
two points. Our visual servoing control law exploits the auto-epipolar property in
order to retrieve the equal orientation between target and current camera. Translation is performed by exploiting the epipoles. Simulation results and Lyapunov-based
stability analysis demonstrate the parametric robustness of the proposed method.
We also provide a short introduction to the Epipolar Geometry Toolbox (EGT), a
free MATLAB software package developed at the University of Siena, with which
all simulation results have been obtained. EGT can be downloaded from the EGT
web site together with a detailed manual and code examples.

1 Introduction

In visual servoing, the control goals and the feedback laws are directly designed
in the image domain. Designing the feedback at the sensor level increases
system performances, especially when uncertainties and disturbances affect
the robot model and the camera calibration [14].

Visual servoing has also been applied to mobile robots [16],[5], however,
the problem of keeping the features in the field of view remains one of the
main issues. Several approaches have been proposed to deal with the image
constraints of pinhole camera perspective projection, such as switching or
hybrid control techniques [3],[18] or via navigation functions based on the
knowledge of a geometrical model of the object [7].

Recently, omnidirectional cameras have been introduced in computer vision [20],[2]. These new sensors are characterized by a wide field of view that allows us to design visual servoing strategies without worrying about image constraints.

In this chapter we propose a visual servoing algorithm for a holonomic mobile robot equipped with a fixed omnidirectional camera. In particular, this proposed servoing law is directly designed in the image domain and does not need the knowledge of three-dimensional (3D) scene geometry or features position with respect to the robot.

While visual servoing with pinhole cameras has been deeply investigated in the literature, fewer results exist for omnidirectional cameras. One of the first works on mobile robots performing visual servoing with panoramic cameras is presented in [9], where the authors propose a method for the visual-based navigation of a mobile robot in indoor environments by tracking visual landmarks. In [6] a vision-based control law is implemented to control a formation of mobile robots equipped with omnidirectional cameras. A switching control law for the pose control of a car-like vehicle with a mounted panoramic camera has been studied in [25].

We propose a visual servoing strategy for a holonomic mobile robot based on epipolar geometry retrieved from current and desired images taken by the on-board omnidirectional camera. Epipolar geometry is the intrinsic projective geometry between two views [10]. It is independent of the observed scene structure and only depends on the cameras internal parameters and the relative pose of the two views [13].

This work builds upon previous contributions [4],[21] where the servoing problem was considered for pinhole cameras. We here exploit, in the case of panoramic views, the so-called *auto-epipolar property* as the main feature of the visual controller. This auto-epipolar condition occurs when the desired and current panoramic views undergo a pure translation; it is detectable directly in the image domain by computing the so-called *biconic*. Our servoing law is divided into two independent steps dealing with the compensation, respectively, of rotation and translation occurring between the actual and desired views. In the first step a set of biosculating conics, or biconic for short, is controlled to gain the same orientation between the two views. In the second step, epipoles and corresponding feature points are employed to execute the translational step necessary to reach the target position. Simulated experiments and Lyapunov-based stability analysis validate the proposed method.

In the Appendix we also provide a short introduction to the Epipolar Geometry Toolbox (EGT), a free MATLAB software package developed at the University of Siena, with which all simulation results have been obtained. EGT can be downloaded from the EGT web site together with a detailed manual and code examples.

2 Background: Epipolar Geometry for Panoramic Cameras

For the reader's convenience we now review basic facts from epipolar geometry relating a pair of central catadioptric cameras [23, 22, 24].

A central catadioptric camera with a hyperbolic mirror consists of a pinhole camera looking into a convex hyperbolic mirror centered at f (Figure 1). Note that the pinhole camera is centered on the focus f' of the opposite sheet of the hyperboloid. By construction, every scene point p projects at x onto the mirror surface through the focus f. The image point m (pixels) is then obtained via perspective projection of x through f'. The fact that all incoming rays intersect at a single viewpoint f is an important property and is a necessary condition for the existence of the epipolar geometry [1].

Consider two fully calibrated central panoramic cameras, referred to as $\{a\}$ *actual* and $\{d\}$ *desired* centered in f_a and f_d, respectively (Figure 2). Without loss of generality, we choose the world reference frame coincident with the desired camera frame; further, we suppose the cameras having the same calibration and mirror parameters.

The segment $\overline{f_a f_d}$, called the *baseline*, intersects the two mirror surfaces at the *epipoles*. Note that the baseline intersects each mirror at the *mirror epipoles*, namely c_{a1}, c_{a2} for the actual camera and c_{d1}, c_{d2} for the desired one (Figure 2). The mirror epipoles are projected to the image planes in the *image epipole* e_{a1}, e_{a2} and e_{d1}, e_{d2}, respectively.

A plane π containing the baseline is called an *epipolar plane*. It intersects each mirror surface at a *mirror epipolar conic C* that is projected to the *image epipolar conic* on the image plane.

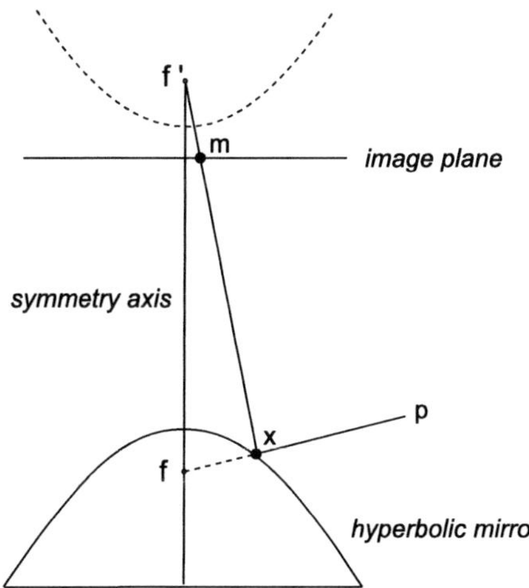

Fig. 1. Imaging model of a panoramic camera with a hyperbolic mirror.

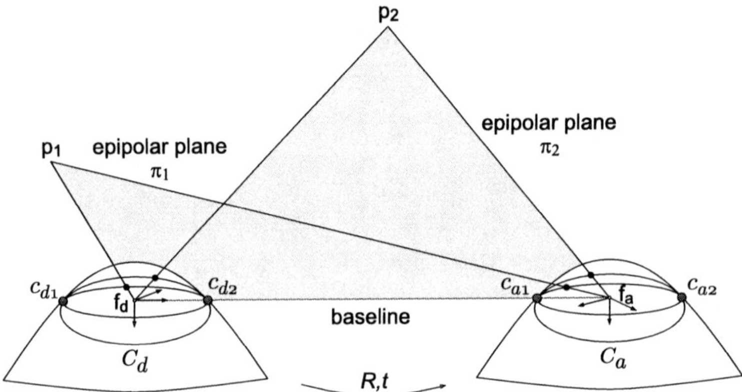

Fig. 2. Basic epipolar geometry setup for panoramic cameras. Every epipolar plane contains the baseline. The intersection between an epipolar plane and the mirror surface characterizes a mirror epipolar conic.

Given one scene point p_i, there exists an associated epipolar plane that generates, for each camera, a mirror epipolar conic C_i containing both the epipoles and the projections of p_i (Figure 2). Note that since the epipoles c_a and c_d belong to every mirror epipolar conics, their projections to the image plane e_a and e_d also belong to every image epipolar conic.

In this two-view setup, each one of the N scene points p_i with $i = 1, ..., N$ projects to the actual and desired mirrors, respectively, in points x_{ai}, x_{di} and then to the image planes in points m_{ai} and m_{di}. In general there exists a function $\eta : \mathbb{R}^3 \to \mathbb{R}^2$, dependent on calibration and mirror parameters, such that, given a point on the mirror x_i, it returns the point projection to the image plane m_i, namely $\eta(x_i) = m_i$ [2]. In the case of fully calibrated cameras, for every image point m_i it is then possible to recover its back-projection on the mirror x_i through the inverse of η.

Given a pair of cameras and their views of a scene, there exists a matrix $E \in \mathbb{R}^{3\times3}$, called the *essential matrix* [8], such that

$$x_{di}^T E x_{ai} = 0 \qquad (1)$$

for all corresponding mirror points x_{di} and x_{ai} $i = 1, ..., N$, obtained as a back-projection of the corresponding image points m_{di} and m_{ai}. In general the essential matrix has rank 2 and is defined up to an arbitrary scale.

For any point x_a in one mirror, $n_d = E x_a$ defines the normal vector, with respect to the desired mirror frame, to the epipolar plane. Likewise, given any point x_d, $n_a = E^T x_d$ defines the epipolar plane normal vector expressed with respect to the actual camera frame. In the case of a pinhole camera, the normal vectors n_a and n_d represent, in homogeneous coordinates, the epipolar lines associated with points x_a and x_d, respectively.

Since the epipoles belong to any epipolar plane, we have thus

$$c_a^T n_{ai} = 0, \quad c_d^T n_{di} = 0 \quad \forall i . \tag{2}$$

Equations (2) hold true for every i, therefore c_a and c_d represent the null spaces of the matrices E^T and E, respectively. The essential matrix depends on the relative position and orientation of the two cameras, namely

$$E = \widehat{t}R , \tag{3}$$

where $\widehat{t}$ is the skew-symmetric matrix obtained from the relative translation vector t and rotation matrix R between the two cameras [13].

3 Auto-Epipolar Configurations

In general, an essential matrix takes the form of (3), but there are certain configurations that cause the essential matrix to be skew-symmetric. For example, under pure translational displacement between two cameras (i.e., $R = I$), the essential matrix becomes

$$E = \widehat{t} = \widehat{c_d} . \tag{4}$$

However, another configuration exists that provides a skew symmetric essential matrix as pointed out by the following theorem.

Theorem 1 (Ma et al. [17]). *Let $t \in \mathbb{R}^3$ and $R \in SO(3)$. If $\widehat{t}R$ is skew symmetric, then $R = I$ or $R = e^{\hat{u}\pi}$ where[1] $u = \frac{t}{\|t\|}$. Further, $\hat{t}e^{\hat{u}\pi} = -\hat{t}$.* □

Theorem 1 describes the two configurations where the essential matrix is skew symmetric: pure translation and relative rotation of $180°$ about the baseline.

Definition 1 (Auto-epipolar configurations). *The relative configurations between two cameras yielding to a skew-symmetric essential matrix are referred to as auto-epipolar configurations.*

Remark 1. The auto-epipolar configuration is fundamental to designing our visual servoing algorithm. In what follows we will introduce the concept of biosculating conics, which will allow us to observe the occurrence of the auto-epipolar configurations for panoramic cameras, only through the image points without estimating the epipolar geometry. The main idea presented in Section 4 is to design a visual servoing that will steer the camera-robot in the auto-epipolar configuration where $R = I$, thus compensating the orientation disparity.

For pinhole cameras, the auto-epipolar property has been addressed describing the bitangent idea [15]. In the case of panoramic cameras we extend the bitangent line concept defining the *biosculating conics* or *biconic*, for short. The choice of this name is motivated by our guess that the tangent concept for lines becomes the more involved concept of osculating conics.

---

[1]The exponential representation of rotation matrices has been used.

Definition 2 (Biosculating conics, or biconic). *Consider two images of the same scene with a pair of corresponding image points m_d and m_a. Overlap the two images. Let x_d and x_a be their back-projections onto one of the known mirror surfaces. Let the mirror biosculating conic be the intersection between the plane ψ of normal vector $n = x_d \times x_a$ and the desired camera mirror.[2] The image biosculating conic is the projection of the mirror biosculating conic to the image plane.*

Proposition 1 (Common intersection line). *Consider a set of N corresponding points m_{di} and m_{ai} for $i = 1, ..., N$ and let x_{di} and x_{ai} be their back-projections to the mirror. Consider the matrix $M \in \mathbb{R}^{N \times 3}$ containing all the normal vectors $n_i = x_{di} \times x_{ai}$ to the planes ψ_i, namely $M = [x_{d1} \times x_{a1}, ..., x_{dN} \times x_{aN}]^T$. If matrix M has rank 2 then all planes ψ_i intersect at a common line $\ell \in \mathbb{R}^{3 \times 1}$ or equivalently*

$$x_{di}^T \hat{\ell} x_{ai} = 0, \ \forall i.$$

In other terms ℓ is perpendicular to n_i, $\forall i$.

Proof. If M has rank 2 then there exists a right null vector ℓ such that $M\ell = 0$, namely $(x_{di} \times x_{ai})^T \ell = 0 \ \forall i$, which means each plane ψ_i contains the vector ℓ and since all planes pass through the origin of the desired mirror frame then it follows that there exists a unique common intersection line ℓ between all planes. From the property of the triple scalar product the expression $(x_{di} \times x_{ai})^T \ell = 0$ is equivalent to $x_{di}^T \hat{\ell} x_{ai} = 0 \ \forall i$, as required. $\square$

It is easy to verify that if N different planes ψ_i all share a common intersection line then all the biosculating conics pass through the same two points (Figure 2).

The relationship between biosculating conics and auto-epipolar configurations is stated in the following theorem for points in general configuration.

Theorem 2. *Consider eight corresponding points m_{di} and m_{ai} in two views and let x_{di} and x_{ai} be their back-projections on the mirror surfaces. Assume that the matrix $A = [a_1, a_2, \ldots, a_8]^T$ has full rank, where $a_i = (x_{ai} \otimes x_{di})^T$.[3] If the matrix M has rank 2 (i.e., from Proposition 1 there exists a common intersection line between all planes ψ_i), then the essential matrix E is skew symmetric.*

Proof. Since M is rank 2 from Proposition 1 there exists a common intersection line for at least eight planes, therefore $x_{di}^T \hat{\ell} x_{ai} = 0$ holds true. Then at least one essential matrix $E = \hat{\ell}$ for the eight corresponding mirror points x_{di} and x_{ai} exists. But since the matrix A has rank 8, then there will exist a unique (up to a scale) essential matrix E [12], thus ending the proof. $\square$

---

[2]Note that the biosculating conic passes through x_a and x_d, since these are generated by overlapping images and then are vectors applied to the origin of the same mirror frame, the desired or actual one.

[3]The symbol $\otimes$ is the Kronecker product and the resulting vector $a_i \in \mathbb{R}^{9 \times 1}$.

Figure 3 (*bottom left*) shows the image plane biconic when the plane ψ_i does not share a common intersection line, i.e., the camera is not in auto-epipolar configuration. On the other hand, Figure 3 (*bottom center*) shows the image plane biconic when all planes ψ_i have a common intersection line and the camera is in auto-epipolar configuration.

The previous theorems provide an important tool to recognize when the cameras are in auto-epipolar configuration (Def. 1) only from visual data. The camera is in auto-epipolar configuration if matrix M has rank 2 (recall that $M \in \mathbb{R}^{N \times 3}$). It is worthwhile to mention that to build matrix M we only need to know back-projection mirror points x_{ai} and x_{di}, which can be retrieved from the image points m_{ai} and m_{di}, and the internal camera calibration matrix.

3.1 Planar Motions

In the presence of points in general configurations (matrix A full rank in Theorem 1), there exist only two auto-epipolar configurations. However, since the robot is constrained to move on a plane, the co-intersection between all the mirror biconics implies the pure translational displacements of the cameras.

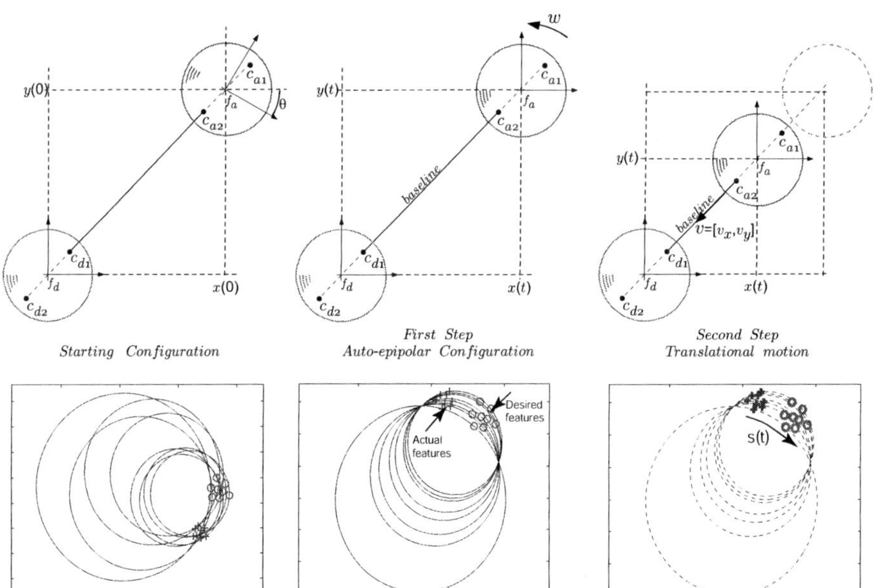

Fig. 3. The visual servoing algorithm steers the holonomic camera-robot from the starting configuration centered at f_a (no common intersection between biconic exists), through an intermediate configuration when both views have the same orientation (common intersection exists), toward the position f_d ($s(t) = 0$).

The other auto-epipolar configuration occurs, in fact, when one camera is rotated 180° around the baseline with respect to the other; this will not occur for planar camera motions. The planar case allows us to design a rotational vision-based controller exploiting the co-intersection of the biconic as objective or, in other words, when the matrix M has rank 2. To test if matrix M is rank deficient we propose to exploit the smallest singular value σ_3 obtained by singular value decomposition (SVD) of M. As the displacement approaches a pure translation, the smallest singular value approaches zero. In practical cases, however, we experimentally verified that better results are obtained if the smallest singular value is squared and normalized with respect to the second singular value, i.e.,

$$\xi = \frac{\sigma_3^2}{\sigma_2}. \tag{5}$$

If the two cameras are purely translated then ξ is exactly zero as shown in Figure 4.

Remark 2. Function $\xi(t)$ will be used as a control variable for the rotational part of the proposed visual servoing. Note that the derivative of $\xi(t)$ is an odd function in a large interval of $\theta = 0°$ (Figure 4).

4 Mobile Robot Controller

In this section we present a new image-based visual servoing strategy for holonomic mobile robots with a fixed catadioptric camera, based on the results of the previous sections.

Let $q = (x, y, \theta)^T$ be the robot configuration vector with respect to the world frame $< Oxyz >_w$ (Figure 5). Let v_x, v_y be the translational velocities and ω the rotational velocity, respectively,

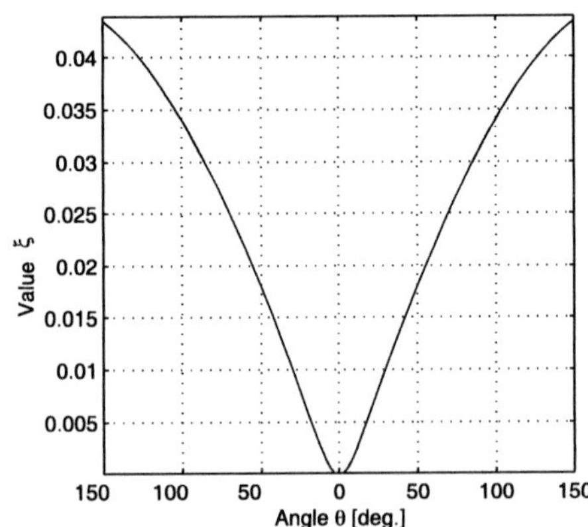

Fig. 4. Singular value ratio $\xi = \frac{\sigma_3^2}{\sigma_2}$ with respect to the actual camera orientation variation.

$$\dot{x} = v_x$$
$$\dot{y} = v_y \qquad (6)$$
$$\dot{\theta} = \omega \, .$$

Suppose that a fixed catadioptric camera is mounted on the mobile robot such that its optical axis is perpendicular to the xy-plane. Without loss of

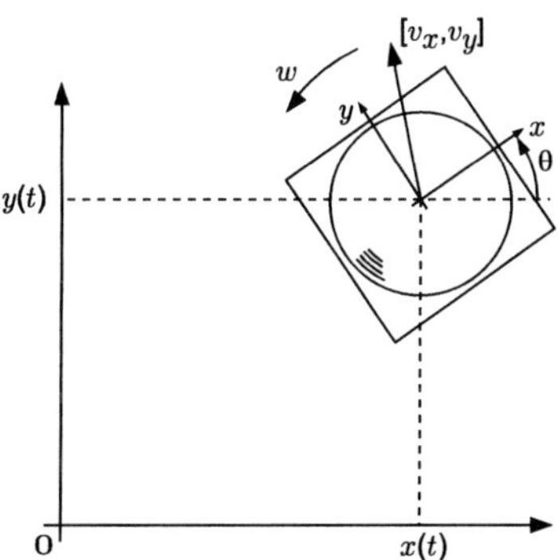

Fig. 5. The holonomic mobile robot with a fixed catadioptric camera moving on a plane.

generality we suppose that a target panoramic image, referred to as *desired*, has been previously acquired in the desired configuration $q_d = [0, 0, 0]^T$. Moreover, another panoramic view, the *actual* one, is available at each time instant from the camera robot in the actual position. As shown in the initial setup of Figure 3 (*upper left*), the mobile robot disparity between the actual and desired poses is characterized by rotation $R \in SO(1)$ and translation $t \in \mathbb{R}^2$.

The control law will be able to drive the robot disparity between the actual and desired configuration to zero only using visual information. In particular, we will regulate separately the rotational disparity (*first step*) and the translational displacement (*second step*). The main feature of our control scheme is that it exploits the auto-epipolar condition (Section 3) to compensate for rotational disparity, thus not involving any estimation procedure.

The goal of the first step is to bring the actual omnidirectional view to the same orientation as the desired one. During the first step, the translational velocity is set to zero, $v_x = v_y = 0$, while the angular velocity is set to

$$\omega = -\alpha_\omega \Psi(t) \qquad \text{where} \qquad \Psi(t) = \sigma_s \xi(t) \, , \qquad (7)$$

where α_ω is a positive scalar, $\sigma_s = sign(\dot{\xi}(t))$, and $\xi(t)$ is the control variable as defined in (5).

The test for local asymptotic convergence can be done considering the definite positive Lyapunov function $V = \frac{1}{2}\theta^2$, whose derivative is $\dot{V} = \theta\dot{\theta} =$

$-\theta\alpha_\omega\Psi(t)$. Due to the properties of $\xi(t)$ (Remark 2), it follows that $\dot{V} < 0$ in a large interval around the equilibrium $\theta = 0°$.

Note that the domain of attraction is related to the singular values behavior with respect to θ that has been experimentally evaluated.

The rotational error of the mobile camera robot is regulated to zero during the first step with the control law proposed in (7). Note that the proposed controller is a simple proportional controller on variable ξ. More complex and high-performance controllers can be tested; however, the main focus of this chapter is to design epipole-based control laws more than stressing controller performances.

Remark 3. When a translation occurs between the cameras, as at the end of the first step, the intersection of biconic coincides with the epipoles of the camera (Figure 3, *bottom center*). This important property aims to retrieve epipoles that will be used in the second step to choose the translational direction.

Let us now address the second step, which will start when $\|\xi(t)\| < \epsilon$, i.e., when only a pure translation is needed to reach the target configuration. In this case, according to Figure 6(a), all biconics intersect at the epipoles and the corresponding feature points m_{ai} and m_{di} both slide on the epipolar conic C_{di} during the translational motion.

Due to this property, if the camera robot is constrained to move along the baseline, then the translation will stop when all features are matching, i.e., when the norm of $s(t)$, the vector containing all the distances $s_i(t)$ between $m_{ai}(t)$ and $m_{di}(t)$ computed along the conic C_{di}, is equal to zero (Figure 6(a)).

The direction of translation, which is coincident with the baseline, can be retrieved from the knowledge of the epipole $\tilde{c}_d$ pointing toward f_a (Fig-

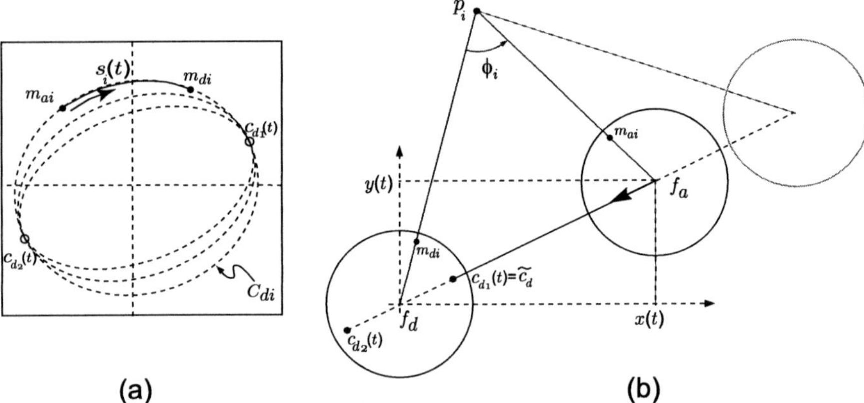

Fig. 6. (a) The distance $s_i(t)$ between corresponding points m_{ai} and m_{di}. After the first step all epipolar conics intersect at the epipoles. (b) If the actual robot moves to the desired position, then ϕ tends to zero.

ure 6(b)), chosen between the two epipoles c_{d1} and c_{d2} (Remark 3). Even if the baseline provides the translation vector, we have an uncertainty on the direction to follow to reach the target. To remove this uncertainty, i.e., to find the epipole $\widetilde{c}_d$, provide an initial guess $\widetilde{c}_{dg}$ and consider the variation $\Delta\phi$ of the angle $\phi = \sum_{i=1}^{N} \phi_i$ (Figure 6(b)), where $\phi_i = \arctan(\frac{m_{aiy}}{m_{aix}}) - \arctan(\frac{m_{diy}}{m_{dix}})$ obtained after a small and finite motion along the direction of the initial guess $\widetilde{c}_{dg}$. It is straightforward to check that

$$ if \quad \Delta\phi < 0 \quad \Rightarrow \quad \widetilde{c}_d = \widetilde{c}_{dg} \,; $$

otherwise the initial guess was wrong and the other epipole is chosen to perform the second step of the visual servoing.

Set the angular velocity to zero and choose the translational velocity as

$$ v = -s(t)[\,\widetilde{c}_{dx}\,,\,\widetilde{c}_{dy}\,]^T. \tag{8} $$

To prove the asymptotic convergence of this control law, consider the Lyapunov function $V = \frac{x^2+y^2}{2}$ whose derivative is negative for all values of x and y. In fact this function ensures that $\dot{V} = -2s(t)\lambda\|\widetilde{c}_d\|^2$ where λ is such that $[x,y]^T = \lambda\widetilde{c}_d$ with $\lambda > 0$ (Figure 6(b)).

Recall that $s(t) = \sum s_i(t)$ is a distance and is zero only when the desired configuration is reached (Figure 6(a)).

5 Simulation Results

To validate the proposed visual servoing, simulations have been performed with the use of the EGT [19], a set of MATLAB functions to build multi-camera simulation setups and to manipulate the multiple-view epipolar geometry. More details about the EGT will be provided in the Appendix.

5.1 First Step: Rotational Controller

Without loss of generality consider the desired camera robot with configuration $q_d = [0,0,0]^T$, while the initial one has been placed at $q_a = [10,7,\pi/4]^T$. We here present simulation results for the first step, which rotates the actual camera robot until the same orientation of the desired camera is gained. In this step the translational velocity is constrained to zero.

Figure 7 gathers some simulation results. The two top windows show the actual and desired camera locations, respectively, together with scene points.

On the left bottom window of Figure 7 we can observe the feature points (crosses) moving from initial position toward the epipolar conics. In the right bottom window the angular velocity ω provided by the controller is shown. Note that when ω approaches zero, all biconics intersect at the same two points (i.e., the epipoles) and also the error variable $\xi(t)$ in (5) approaches zero. In this case the biconic coincides with epipolar conics.

5.2 Second Step: Translational Controller

After the two cameras gain the same orientation, the translational controller starts to steer the robot toward the desired configuration. In this case the control law proposed in (8) is applied, with $\omega = 0$, and simulation results are reported in Figure 8. In the top left window the translational velocities $v_x(t)$ and $v_y(t)$ are reported. In the top right window the distance $s(t)$ between all feature points goes to zero, as reported in the bottom window, where crossed points (actual positions) go to the desired one (circles).

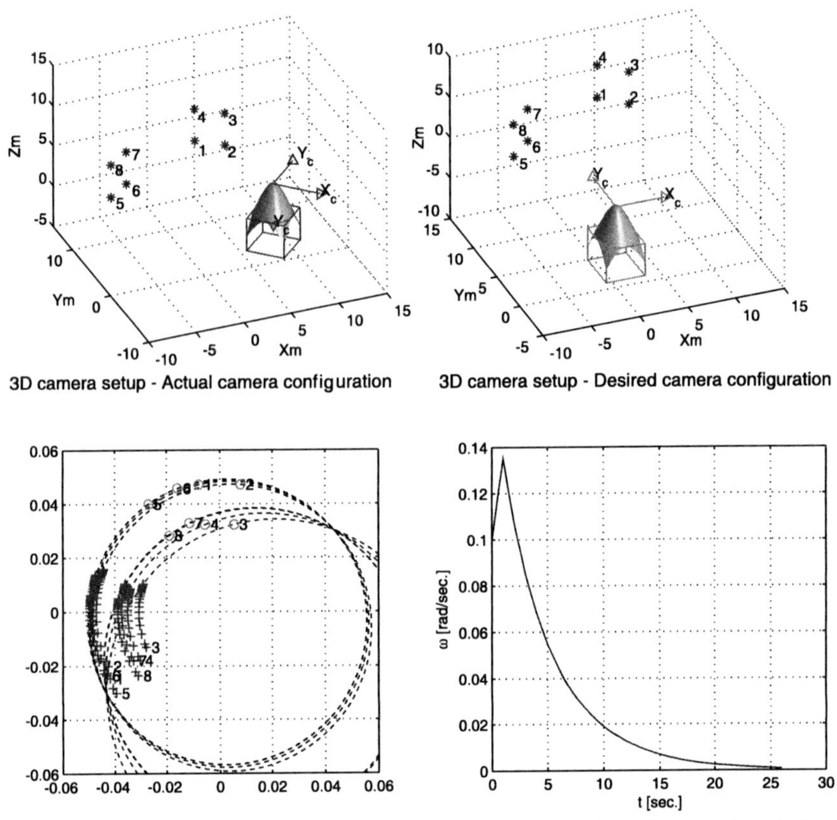

Fig. 7. Simulation results for the first step. The robot, placed in the actual configuration, rotates until the same orientation of the desired configuration is obtained.

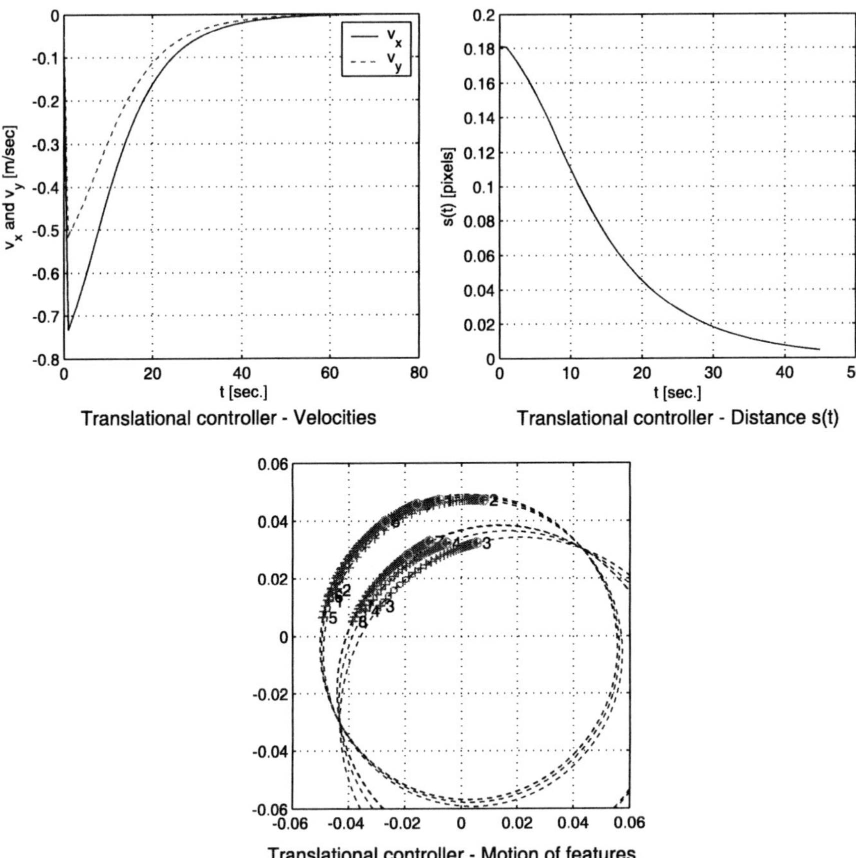

Translational controller - Velocities

Translational controller - Distance s(t)

Translational controller - Motion of features

Fig. 8. Simulation results for the second step. After the first step only a translation occurs to reach the desired configuration. The translational controller guarantees the motion of the robot along the baseline (the feature slide along the biconic). Then it stops when the distance $s(t)$ between corresponding features is equal to zero, i.e., the robot is in the desired pose.

6 Conclusions and Future Work

In this chapter we presented an epipolar-based visual servoing for a mobile robot equipped with a panoramic camera. The main feature of this technique is the use of epipolar geometry to visually control the robot motion. We here present a completely decoupled rotational and translational controller, based on the auto-epipolar property, which makes use of visual information provided by a central catadioptric camera. Lyapunov-based stability analysis and simulation results have confirmed the validity of our new approach.

The calibration parameters represent an open issue. In this work the mirror and camera parameters are a mandatory requirement for computing the biosculating mirror conics. Some uncertainties on the calibration knowledge may actually imperil the control process. Nevertheless, work is in progress to relax the calibration knowledge hypothesis [15].

Our auto-epipolar-based controller splits the rotation and translation in two different tasks with the purpose of obtaining more simple visual servoing tasks. The main idea is to split the entire visual servoing process in several subtasks with the aim to tackle each of them with robust and reliable control laws and with different type of cameras. It is our belief that such controllers can be attached together to design a globally convergent vision-based navigation system for mobile robots.

A The Epipolar Geometry Toolbox (EGT)

Simulation results presented in this work have been realized with the EGT for MATLAB. EGT works for both pinhole and central catadioptric cameras. We present here the basic EGT functions inherent to catadioptric cameras. A more detailed description of EGT, its functions, and some examples can be found in [11].

EGT was created at the University of Siena and is freely available on the web [19]. Its development was mainly due to the growing interest in the robotics community in the use of cameras as the main sensors for robot navigation. Moreover, we observed the need for a software environment that could help researchers to rapidly create a multiple camera setup and extract and use visual data to design new visual servoing algorithms.

With EGT we provide a wide set of easy-to-use functions, which are also completely customizable, to create general multi-camera scenarios and manipulate the visual information between them.

A.1 Camera Placing and Scene Feature Imaging

Camera placement is obtained simply describing the homogeneous transformation matrix between mirror and world frame:

$$\mathbf{H} = \begin{bmatrix} \mathbf{R} & \mathbf{t} \\ \mathbf{0}^T & 1 \end{bmatrix}.$$

Let us place the camera in $\mathtt{t=[-5,-5,0]}$' with orientation $\mathtt{R} \equiv R_{z,\pi/4}$. EGT provides a function to visualize the panoramic camera in the 3D world frame:

```
>> H=[rotoz(pi/4) , [-5,-5,0]';
>>         0 0 0 ,    1    ];
>> f_3Dpanoramic(H); %visualize camera
```

Moreover, for assigned camera calibration matrix $\mathbf{K}$:

```
>>K=[10^3  0    320;
      0   10^3  240;
      0    0     1 ];
```

the projection of a 3D point $Xw=[0,0,4]$' in both the camera (m) and mirror (Xh) frames can be obtained from

```
>> [m,Xh] = f_panproj(Xw,H,K);
>>
m =
    4.1048e+002
    2.4000e+002
    1.0000e+000

Xh =
    6.0317e-001
    3.7881e-017
    3.4120e-001
    1.0000e+000
```

Graphical results are accurately described in the *EGT Reference Manual*. More code examples can be found in the directory **demos** of EGT.

A.2 Epipolar Geometry

As in the pinhole cameras, the epipolar geometry is defined here when a pair of central catadioptric cameras is available. Suppose that their position and orientation is completely described by homogeneous transformation matrices H1 and H2. By using the EGT function:

```
>> [e1c,e1pc,e2c,e2pc] = f_panepipoles(H1,H2);
```

the CCD coordinates of the epipoles are easily computed and plotted.

Now place N feature points $\mathbf{X}_w$ placed in the space. The CCD projections of all epipolar conics $\mathbf{C}_1^i$ and $\mathbf{C}_2^i$ where $i = 1, ..., N$, corresponding to all 3D points $\mathbf{X}_w$, are computed and visualized by

```
>> [C1m,C2m] = f_epipconics(Xw,H1,H2);
```

A.3 Download

EGT can be freely downloaded at

`http://egt.dii.unisi.it`

and requires MATLAB 6.5 or higher. The detailed manual provides a large set of examples, figures, and source code also suitable for beginners. EGT is shareware and the source code is available. The current version of EGT is v1.1 and it will be extended in the near future with other functionalities.

References

1. Baker S, Nayar S (1999) A theory of single-viewpoint catadioptric image formation. International Journal of Computer Vision 35(2):175–196
2. Benosman R, Kang S (2001) Panoramic vision: sensors, theory and applications. Springer-Verlag, New York
3. Chesi G, Hashimoto K, Prattichizzo D, Vicino A (2004) Keeping features in the field of view in eye-in-hand visual servoing: a switching approach. IEEE Trans. on Robotics 20(5):908–914
4. Chesi G, Piazzi J, Prattichizzo D, Vicino A (2002) Epipole-based visual servoing using profiles. In: IFAC World Congress, Barcelona, Spain
5. Conticelli F, Prattichizzo D, Bicchi A, Guidi F (2000) Vision-based dynamic estimation and set-point stabilization of nonholonomic vehicles. 2771–2776. In: Proc. IEEE Int. Conf. on Robotics and Automation, San Francisco, CA
6. Cowan N, Shakernia O, Vidal R, Sastry S (2003) Vision-based follow the leader. 1798–1801. In: Proc. International Conference on Intelligent Robots and Systems, Las Vegas, NV
7. Cowan N, Weingarten J, Koditshek D (2002) Visual servoing via navigation functions. IEEE Trans. on Robotics and Automation 18:521–533
8. Faugeras O (1993) 3-D computer vision, a geometric viewpoint. MIT Press, Cambridge, MA
9. Gaspar J, Winters N, Santos-Victor J (2000) Vision-based navigation and environmental representations with an omnidirectional camera. In: IEEE Trans. on Robotics and Automation 16(6):890–898
10. Geyer C, Daniilidis K (2001) Catadioptric projective geometry. International Journal of Computer Vision 45(3):223–243
11. Mariottini GL, Alunno E, Prattichizzo D (2004) The epipolar geometry toolbox (EGT) for Matlab, Technical Report 07-21-3-DII, University of Siena, Siena, Italy
12. Hartley R (1995) In defence of the 8-point algorithm. 1064–1070. In: Proc. of IEEE Int. Conference on Computer Vision, Cambridge, MA
13. Hartley R, Zisserman A (2000) Multiple view geometry in computer vision. Cambridge University Press, Cambridge, UK
14. Hutchinson SA, Hager GD, Corke PI (1996) A tutorial on visual servo control. IEEE Trans. on Robotics and Automation 12(5):651–670
15. Piazzi J, Prattichizzo D (2003) An auto-epipolar strategy for mobile robot visual servoing. In: IEEE IROS03 Conference on Intelligent Robots and Systems, Las Vegas, NV

16. Hashimoto K, Noritsugu T (1997) Visual servoing of nonholonomic cart. 1719–1724. In: IEEE Int. Conf. Robotics and Automation, Albuquerque, NM

17. Ma Y, Soatto S, Košecká J, Sastry SS (2003) An invitation to 3-D vision, from images to geometric models. Springer-Verlag, New York

18. Malis E, Chaumette F, Boudet S (1998) 2D 1/2 visual servoing stability analysis with respect to camera calibration errors. 691. In: Proceedings of the 1998 IEEE/RSJ International Conference on Intelligent Robots and Systems, Vancouver, BC

19. Mariottini G, Prattichizzo D (2004) Epipolar Geometry Toolbox for Matlab, University of Siena, http://egt.dii.unisi.it

20. Nayar S (1997) Catadioptric omnidirectional camera. 482–488. In: Proc. of International Conference on Computer Vision and Pattern Recognition, Puerto Rico

21. Piazzi J, Cowan N, Prattichizzo D (2004) Auto-epipolar visual servoing. In: IEEE Conf. on Intelligent Robots and Systems (IROS), Sendai, Japan

22. Svoboda T (1999) Central panoramic cameras design, geometry, egomotion, PhD thesis, Center for Machine Perception, Czech Technical University, Prague, Czech Republic

23. Svoboda T, Pajdla T, Hlaváč V (1998) Epipolar geometry for panoramic cameras. 218–232. In: Fifth European Conference on Computer Vision LNCS 1406, Freiburg, Germany

24. Svoboda T, Pajdla T, Hlaváč V (2001) Epipolar geometry for central panoramic catadioptric cameras In: Panoramic vision: sensors, theory and applications. Springer-Verlag, New York

25. Usher K, Ridley P, Corke P (2003) Visual servoing for a car-like vehicle– An application of omnidirectional vision. 4288–4293. In: IEEE International Conference on Robotics and Automation, Taiwan

Motion Control and Coordination in Mechanical and Robotic Systems*

Iliya V. Miroshnik

Laboratory of Cybernetics and Control Systems
State University of Information Technologies, Mechanics, and Optics
14 Sablinskaya, Saint Petersburg, 197101 Russia
miroshnik@yandex.ru

Summary. The chapter focuses on concepts and methodologies of coordinating and motion control aimed at maintaining complex spatial behaviour of nonlinear dynamical systems. The main approach is discussed in connection with problems of control of mechanical systems (rigid bodies, robotic manipulators, and mobile robots) and is naturally extended to coordinating the motions of redundant robots, underactuated mechanisms, and walking machines.

1 Introduction

New theoretical and applied problems of control theory and mechanics often involve a spatial anisotropy of nonlinear systems and attractivity of nontrivial sets (goal submanifolds) [10, 13, 14, 15, 16, 20, 23, 24, 27]. These geometric objects and appropriate problems of spatial motion control of multivariable processes are the subject of the differential geometric approach [9, 13], theory of multi-input/multi-output (MIMO) systems (*coordinating control principle* [10, 16]), and stability theory (concepts of *partial stability* [10, 27, 23, 31]).

Recent research demonstrated the efficiency of the geometric approach for maintaining prespecified trajectories and spatial motion of manipulators, mobile robots, walking machines, and pendular systems. Restricted motion of mechanical systems and robots interacting with complex environment, keeping up the required orientation of their links and well-coordinated behavior of the multilink chain, can be considered as stability of the relevant nonlinear MIMO system with respect to spatial attractors. Then the control problem is directly reduced to spatial motion control.

This chapter reviews the spatial behavior of complex dynamical systems when the required performance of the system is associated with achieving a

---

*This chapter outlines the results of recent research supported by the Russian Foundation for Basic Research and Program 17 of Presidium of RAS (project 1.4).

desired mode of spatial motion and providing attractivity of nontrivial geometric objects. The main problems are considered in view of their connection with automatic control of mechanical systems: rigid bodies, robotic manipulators, and mobile robots. The general approach is naturally extended to the problems of coordinating control of multilink redundant robots, underactuated mechanisms, and walking machines.

2 Spatial motion and coordination

Consider properties of the composite (MIMO) system consisting of m independent or interacting subsystems with the principal (output) variables y_j and the control (input) variables u_j:

$$\dot{x} = f(x) + g(x)u, \tag{1}$$
$$y = h(x), \tag{2}$$

where $x \in R^n$ is the state vector, $y = \{y_j\} \in R^m$ is the output vector, and $u = \{u_j\} \in R^m$ is the input (control) vector produced by a feedback controller of the general form

$$u = \mathcal{U}(x). \tag{3}$$

An ordinary problem of MIMO control referred to as *output stabilization* is connected with a simplest attractivity property of the system with respect to a point $y = y^*$. More complex multivariable problems arise when the conventional task, involving the controlled evolution of the system as a whole, is complemented by rules of the subsystem interaction. The latter is usually given in the form of holonomic relationships between the system outputs, or the so-called *coordination conditions*

$$\varphi_{yj}(y) = 0, \tag{4}$$

where φ_{yj} $(j = 1, 2, ..., m - 1)$ are smooth functions, and predetermines the need for coordination of the controlling actions u_j [10, 14, 16].

Motion of system (1)–(3) is called *coordinated* when the outputs $y_j(t)$ obey the conditions (4). The coordination conditions correspond to the implicit form of the curve

$$S^* = \{y \in \mathcal{Y} : \varphi_{yj}(y) = 0\},$$

and, for coordinated motion, it is a one-dimensional invariant set of system (1)–(3) (Figure 1).

The composite system satisfying condition (4) is associated with a unit, whose evolution is characterized by a certain generalized output $s(t)$. The variable s is chosen as one of the system outputs, the arithmetical mean, or in the general form

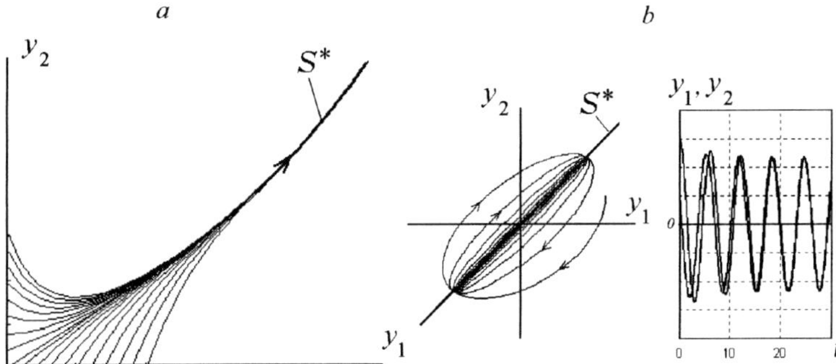

Fig. 1. (a) Coordinated motion and attractor of R^2; (b) synchronization and linear attractor of R^2.

$$s = \psi_y(y), \tag{5}$$

where ψ_y is the smooth scalar function. This variable corresponds to a local coordinate of the curve and *longitudinal dynamics* of the system.

Introduce the vector

$$\varepsilon = \varphi_y(y), \tag{6}$$

where $\varphi_y = \{\varphi_{yj}\}$ $(j = 1, 2, ..., m - 1)$. For coordinated motion and $y(t) \in S^*$, it holds that $\varepsilon = 0$, while in a neighborhood of the set S^* the vector ε characterizes violation of condition (4) and the system's *transversal dynamics*. In the general case, when $y(0)$ fails to belong to S^*, condition (4) can be satisfied only with time (see Figure 1).

The coordinated motion of system (1)–(3) is called *asymptotically coordinated* when

$$\lim_{t \to \infty} \varepsilon(t) = 0. \tag{7}$$

The behavior of the system, corresponding to asymptotically coordinated motion, is often associated with attractivity of the invariant set S^* (Section 3).

The general problem of *coordinating control* is posed as finding control actions, or a closed-loop control law (3), which provides asymptotically coordinated motion and a desired longitudinal dynamics $s(t)$.

The issue of coordination and coordinating control arises in many practical cases such as identical behavior of similar plants of composite systems [14, 16], synchronization of several oscillating or nonperiodic processes [1, 19, 29] (Figure 1(b)), and orbital stabilization and trajectory (curve-following) control [5, 10] (Figure 2(a)).

The problem of trajectory control is of the most transparent geometric nature. It is directly defined through the description (4) of the curve S^* and represents a special case of the more general problems of *spatial motion control*. The latter arises when condition (4) includes $m - \mu$ equations and corresponds to a smooth μ-dimensional hypersurface with local coordinates s_j $(j = 1, 2, ..., \mu < m)$ (Figure 2(b)). The problems consist of maintaining the

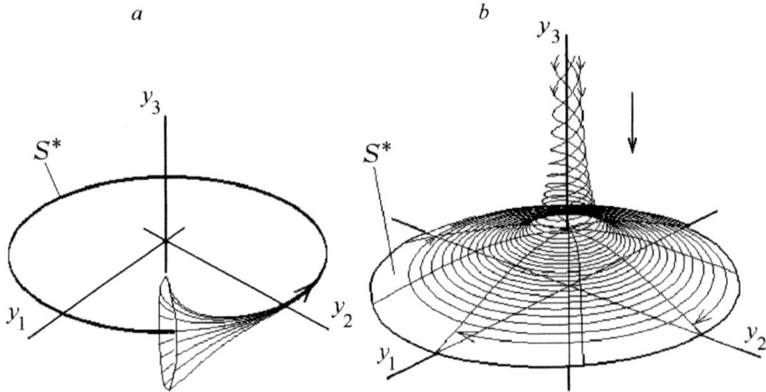

Fig. 2. (a) Orbital stability and one-dimensional attractor, (b) surface-following and two-dimensional attractor.

system motion along prespecified curves, surfaces, and other multidimensional objects of the output space R^m and a desired mode of the internal (longitudinal) behavior $s(t)$. In practice, such a problem concerns a variety of mechanical/robotic systems and their restricted motion in the physical (Cartesian) space.

3 Attractivity and partial stability

In the general case, to analyze dynamics of system (1)–(3) with respect to some sets of the output space R^m and find a correct solution of the relevant control problem, it is necessary to go over to the state space R^n, define a nonlinear equation of the form

$$\varphi(x) \;=\; 0, \tag{8}$$

and consider the system behavior with respect to a state space goal set $\mathcal{Z}^*$ (ν-dimensional hypersurface)

$$\mathcal{Z}^* \;=\; \{x \in \mathcal{X} \subset r^n : \;\; \varphi(x) = 0\},$$

where $\varphi = \{\varphi_i\}$ $(i = 1, 2, \ldots, n - \nu)$ is the smooth $n - \nu$-dimensional vector function, $\nu < n$.

In output stabilization problems [10, 12], a zero (or constant) output vector y is produced by a collection of the system trajectories $x(t)$ in the state space R^n that belong to an invariant hypersurface (8) (the so-called zero dynamics submanifold), where the components of the vector-function φ are usually found as

$$\varphi_1(x) = h(x), \quad \varphi_i(x) = \mathcal{L}_f^{i-1} h(x).$$

In coordination and curve/surface-following problems, (8) is induced by coordination conditions (4),

$$\varphi_1(x) = \varphi_y \circ h(x), \quad \varphi_i(x) = \mathcal{L}_f^{i-1}\varphi_1(x),$$

and the given curve $\mathcal{S}^*$ of the output space R^m defined by (4) is a projection of the hypersurface $\mathcal{Z}^* \in R^n$.

These problems imply the achievement, by using appropriate control actions, of special properties of the system to be designed such as invariance and attractivity with respect to the state space hypersurfaces $\mathcal{Z}^*$.

Analyze the behavior of smooth closed loop system (1)–(3) or that of the autonomous nonlinear system

$$\dot{x} = f_c(x), \tag{9}$$

with respect to a connected geometric object $\mathcal{Z}^*$ defined by (8). Assume that $f_c = f + gU$ is smooth and complete in $\mathcal{X}$, and the function φ satisfies the local *regularity condition* rank $\varphi\,(x) = n - \nu$. Under the latter assumption, the ν-dimensional set $\mathcal{Z}^*$ is a regular hypersurface that can be endowed by the structure of a smooth manifold or an embedded submanifold of $\mathcal{X}$ with the vector *of local coordinates* $z \in \mathcal{Z} \subset R^\nu$. To analyze the motion in the vicinity of $\mathcal{Z}^*$, define a neighborhood $\mathcal{E}(\mathcal{Z}^*) \subset \mathcal{X}$.

The set $\mathcal{Z}^*$ is called an *invariant submanifold* of system (9) when, for all $x_0 \in \mathcal{Z}^*$, the solutions $x = x(t, x_0)$ for all $t \geq 0$ belong to $\mathcal{Z}^*$. The invariant set $\mathcal{Z}^*$ is called an *attracting submanifold* of system (9) (or simply an *attractor*) when it is uniformly (in $x_0 \in \mathcal{E}(\mathcal{Z}^*)$) attractive, i.e.,

$$\lim_{t \to \infty} \operatorname{dist}(x(t, x_0), \mathcal{Z}^*) = 0. \tag{10}$$

The notion of attractivity is evidently relative to concepts of stability theory. For the points $x \in \mathcal{E}(\mathcal{Z}^*)$, introduce a vector $\xi \in \Xi \subset R^{n-\nu}$ as

$$\xi = \varphi(x) \tag{11}$$

and consider the solutions of system (9)–(11) $\xi(t) = \xi(t, x_0) = \varphi(x(t, x_0))$.

The point $\xi = 0$ is called a *partial equilibrium* of system (9)–(11) when, for all $x_0 \in \mathcal{Z}^*$ and $t \geq 0$, the solutions $\xi(t)$ satisfy the identity

$$\xi(t, x_0) = 0. \tag{12}$$

System (9)–(11) at the equilibrium point $\xi = 0$ is called *partially (uniformly in $x_0 \in \mathcal{E}(\mathcal{Z}^*)$) asymptotically stable* when

$$\lim_{t \to \infty} \xi(t, x_0) = 0. \tag{13}$$

The concept of partial equilibrium is associated with system invariance; however, a partially stable nonlinear system sometimes fails to demonstrate attractivity to an invariant set and the converse also holds. Relations between the concepts are considered in [10, 27], where their identity is connected with an additional restriction of *metric regularity.*

Local conditions under which the phenomena of partial stability or set attractivity exist are obtained making use of approaches developed both in stability theory and geometric theory of control [10, 20, 23, 27].

4 Spatial motion control

Let us restrict our attention to the problem of trajectory (curve) or surface motion control of the m-input/m-output nonlinear system, each channel of which is represented by a two-order model, or

$$\ddot{y} + a(y, \dot{y}) = B(y, \dot{y})u, \tag{14}$$

where $y \in \mathcal{Y} \subset R^m$, $u \in R^m$, the system state is represented by $2m$-dimensional vector $x = \begin{vmatrix} y \\ \dot{y} \end{vmatrix} \in \mathcal{X} \subset R^n$, and the matrix B is supposed to be invertible in $\mathcal{X}$. This model is inherent in a variety of mechanical systems, consisting of several multiconnected similar dynamical components (see Sections 5–6).

4.1 General case

Let a μ-dimensional geometric object $\mathcal{S}^* \subset \mathcal{Y}$ be given in the form

$$\varphi_y(y) = 0, \tag{15}$$

where φ_y is an $(m - \mu)$-dimensional smooth function, and the longitudinal motion is characterized by μ-dimensional vector

$$s = \psi_y(y). \tag{16}$$

Consider the coordinate transformation

$$\begin{vmatrix} s \\ \varepsilon \end{vmatrix} = \begin{vmatrix} \psi_y(y) \\ \varphi_y(y) \end{vmatrix}. \tag{17}$$

Supposing that φ_y satisfies the regularity condition, we choose the function ψ_y such that in a small enough neighborhood of $\mathcal{S}_y^*$ the Jacobian matrix

$$\phi_y = \begin{vmatrix} \partial\varphi_y/\partial y \\ \partial\psi_y/\partial y \end{vmatrix}$$

is invertible, and there exists an inverse mapping $y = \gamma(s, \varepsilon)$.

Prespecify the required longitudinal motion of the system $s^*(t)$ by using the reference model

$$\ddot{s}^* + a_s(s^*, \dot{s}^*) = 0 . \tag{18}$$

Then the problem of *spatial motion control* is formulated as follows.

Problem 1. Find the feedback control law (3) that provides
a) desired longitudinal dynamics, or

$$\lim_{t \to \infty} \Delta s(t) = 0, \tag{19}$$

where $\Delta s = s^* - s$ is the longitudinal error;
b) desired transversal dynamics, or

$$\lim_{t \to \infty} \varepsilon(t) = 0. \tag{20}$$

Note that the latter is connected with asymptotic coordination of the system outputs and local attractivity of the state space 2μ-dimensional hypersurface

$$\mathcal{Z}^* = \left\{ (y, \dot{y}) \in \mathcal{X} \subset R^{2m} : \quad \varphi_y(y) = 0, \frac{\partial \varphi_y}{\partial y} \dot{y} = 0 \right\}.$$

The control design procedure includes transformation of the original model of the system, its partial linearization, and the design of local controllers of transversal and longitudinal motions. Transformation of system (14) using the coordinate change (17) and partial linear approximation for small enough $(\varepsilon, \dot{\varepsilon})$ leads to the so-called *task-oriented model*

$$\left|\begin{matrix} \ddot{s} \\ \ddot{\varepsilon} \end{matrix}\right| + A(s, \dot{s}) \left|\begin{matrix} \dot{s} \\ \dot{\varepsilon} \end{matrix}\right| = \phi(y)\Big(- a(y, \dot{y}) + B(y, \dot{y})u\Big). \tag{21}$$

For solving the control problem, we obtain partial exact linearization, choosing the basic (linearizing) control law

$$u = B(y, \dot{y})^{-1}(x_1, x_2)(a(y, \dot{y}) + \tilde{u}), \tag{22}$$

where $\tilde{u} \in R^m$ is the *spatial motion control vector*, and decouple the model with respect to the inputs by using the linear control variables transformation

$$\tilde{u} = \phi_y^{-1} \left|\begin{matrix} u_s \\ u_e \end{matrix}\right|, \tag{23}$$

where u_s is the μ-dimensional vector of *longitudinal control* and u_e is the $(m - \mu)$-dimensional vector of *transversal control*. This leads to the expression

$$\left|\begin{matrix} \ddot{s} \\ \ddot{\varepsilon} \end{matrix}\right| + A(s, \dot{s}) \left|\begin{matrix} \dot{s} \\ \dot{\varepsilon} \end{matrix}\right| = \left|\begin{matrix} u_s \\ u_e \end{matrix}\right|, \tag{24}$$

corresponding to weakly connected models of the longitudinal and transversal dynamics, which enables us to consider Problems 1a and 1b separately.

Problem 1a is solved for the trajectories $(y(t), \dot{y}(t)) \in \mathcal{Z}^*$ by using the control

$$C^s: \quad u_s = a_s(s^*, \dot{s}^*) - A_{11}(s^*, \dot{s}^*)\dot{s}^* - K_{s1}\Delta s - K_{s2}\Delta \dot{s}, \qquad (25)$$

where the feedback gain matrices $K_{si}(s)$ provide the required asymptotic convergence of the longitudinal transient processes (condition (19)).

To solve Problem 1b, consider the motion in the vicinity of $\mathcal{Z}^*$ and find the transversal control law as

$$C^e: \quad u_e = A_{21}(s^*, \dot{s}^*)\dot{s}^* + K_{e1}e + K_{e2}\dot{e}, \qquad (26)$$

where the relevant choice of the feedback gain matrices K_{ei} provides the required stability of the model (24) with respect to $\varepsilon = 0$ (condition (19)) and local attractivity of the hypersurface $\mathcal{Z}^*$.

The resulting control system includes the basic linearizing algorithm (22), the control variable transformation (23), the transformation of the output variables (17), and the local control laws (25)–(26), which arrange feedforward and feedback loops, necessary for solving Problem 1. The solution can be simplified in the cases when condition (15) corresponds to the ortho-normalized description of the goal set $\mathcal{S}^*$ [10, 20].

4.2 Trajectory control

Consider a partial case, when $\mu = 1$, that corresponds to widespread trajectory control problems. Here the function φ_y has the dimension $m - 1$ and the longitudinal motion is characterized by scalar function s.

Let the functions φ_y and ψ_y be smooth and such that, for $y \in \mathcal{S}^*$, the Jacobian matrix $\phi(y)$ is orthogonal, or

$$\phi_y = \begin{vmatrix} \partial\varphi_y/\partial y \\ \partial\psi_y/\partial y \end{vmatrix} = T^*.$$

The matrix $T^* \in SO(m)$ defines an orthogonal frame fixed to the curve $\mathcal{S}^*$ (*Frenet frame*, Figure 3) and satisfies the following Frenet-like differential equation:

$$\dot{T}^* = \dot{s}S(\xi)T^*, \qquad (27)$$

where $S(\xi)$ is a skew-symmetrical matrix, being a function on the vector $\xi(s) \in R^m$ of geometric invariants of the curve.

Note that, in the case considered, the function s is associated with the path length and the variables ε_j are orthogonal deviations from the curve.

We can pose the following simplified spatial motion control problem.

Problem 2. Find the feedback control law (3) that provides

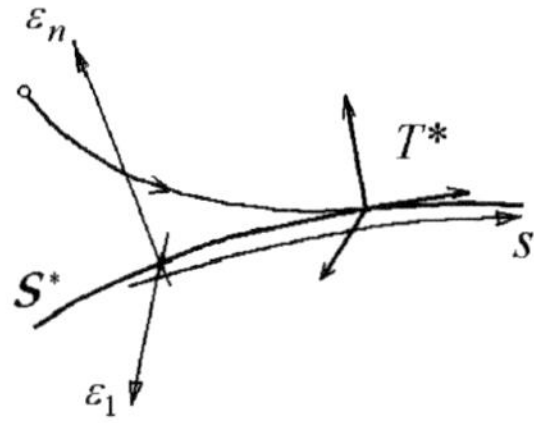

Fig. 3. Trajectory motion and Frenet frame.

a) desired longitudinal dynamics of the system $s(t)$ established as proportional motion at the reference rate $\dot{s}^* = const$, or

$$\lim_{t \to \infty} \Delta \dot{s}(t) = 0, \tag{28}$$

where $\Delta \dot{s} = \dot{s}^* - \dot{s}$;

b) desired transversal dynamics of the system, or asymptotic rejection of the deviation vector $e(t)$ according to (20).

Note that the latter is associated with local attractivity of the state space two-dimensional hypersurface $\mathscr{Z}^*$.

In the case under consideration the task-oriented model of system (14) takes the form

$$\left| \begin{matrix} \ddot{s} \\ \ddot{e} \end{matrix} \right| + \dot{s} S^T(\xi) \left| \begin{matrix} \dot{s} \\ \dot{e} \end{matrix} \right| = T^* \left(-a(y, \dot{y}) + B(y, \dot{y})u \right), \tag{29}$$

and the control variable transformation is

$$\tilde{u} = (T^*)^T \left| \begin{matrix} u_s \\ u_e \end{matrix} \right|. \tag{30}$$

The exact linearization (22) and transformation (30) lead to the model

$$\left| \begin{matrix} \ddot{s} \\ \ddot{e} \end{matrix} \right| + \dot{s} S^T(\xi) \left| \begin{matrix} \dot{s} \\ \dot{e} \end{matrix} \right| = \left| \begin{matrix} u_s \\ u_e \end{matrix} \right|, \tag{31}$$

or,

$$\ddot{s} - S_{12}\dot{e}\dot{s} = u_s, \tag{32}$$
$$\ddot{e} + S_{22}\dot{s}\dot{e} + S_{12}\dot{s}^2 = u_e. \tag{33}$$

Then the controller of longitudinal motion, solving Problem 2a, takes the form

$$C^s: \quad u_s = -K_{s2}\Delta\dot{s}, \tag{34}$$

and the controller of transversal motion, solving Problem 2b, is

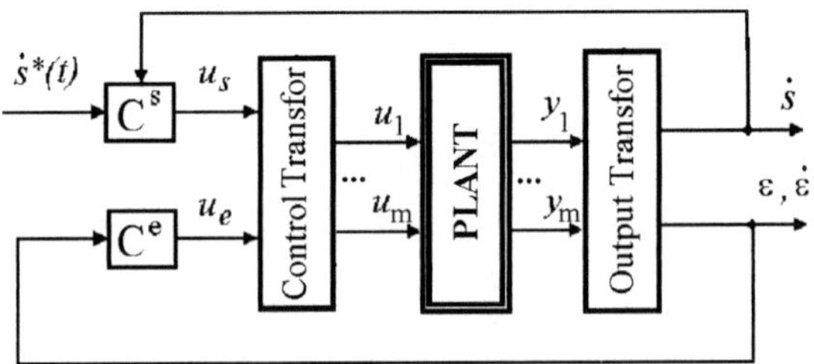

Fig. 4. Trajectory control system.

$$C^e: \quad u_e = S_{12}\dot{s}^{*^2} + K_{e1}\varepsilon + K_{e2}\dot{\varepsilon}. \tag{35}$$

The resulting control system involves the general control variable transformation (22), (30), the transformation of the output variables (17), and the local control laws (34)–(35) (Figure 4). The simplest structure of the control system is obtained for the cases $m = 2$ and 3, which is typical for spatial motion of mechanical and robotic systems (see Sections 5–6).

5 Spatial motion control of rigid body

In the next sections we discuss applied problems of nonlinear control and pre-specified spatial motion of mechanical systems. The rigid body establishes a good benchmark for control of a variety of multilink mechanisms such as manipulation robots and wheeled and walking mechanisms (see Section 6). On the other hand, it imposes an additional requirement for spatial motion control which consists of maintaining a desired orientation with respect to the trajectory.

Consider the motion of a rigid body in the Cartesian space R^3 where the position of the body is characterized by the pair

$$y, T(\alpha), \tag{36}$$

and $y \in R^3$ is the vector of Cartesian coordinates of the body center point, T is an orthogonal matrix: $T \in SO(3)$. The latter is associated with a frame fixed to the body (Figure 5), can be expressed through the vector of *Euler angles* $\alpha = \mathrm{col}(\alpha_1, \alpha_2, \alpha_3)$, and possesses the so-called *kinematic equation* (see (38)) [10].

The model of a rigid body is represented by equations of translational and rotational motion, or

$$m\,\ddot{y} = T^T(\alpha)u_y \tag{37}$$

and

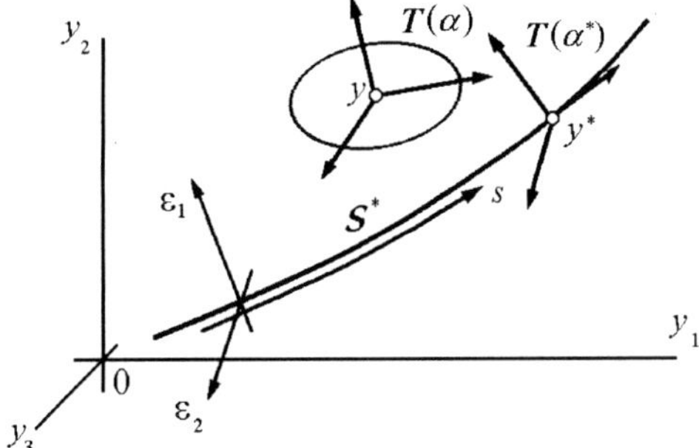

Fig. 5. Trajectory motion of a rigid body.

$$\dot{T}(\alpha) = S(\Omega)\,T(\alpha), \qquad (38)$$

$$J\,\dot{\Omega} + S^T(\Omega)J\Omega = T^T(\alpha)u_\alpha, \qquad (39)$$

where $\Omega \in R^3$ is the vector of instant angular velocities, given in the body-fixed frame, $S(\Omega)$ is a skew-symmetric matrix: $S(\Omega) \in so(3)$, $u_y \in R^3$ is the vector of internal acting forces, $u_y \in R^3$ is the vector of internal torques, and m and J are mass-inertia parameters.

Equations (37)–(39) describe a 6-channel 12-order system with the state defined by $y, \dot{y}, \alpha, \Omega$, the outputs y, α and the inputs (controls) u_y, u_α.

The motion of the rigid body in Cartesian space is considered with respect to a prescribed smooth curve $\mathcal{S}^*$ given by (15), where $\varphi = (\varphi_1, \varphi_2)$, while the path length s is defined in the form (16). Using the orthonormalized description of the curve (see Subsection 4.2), we set that the functions φ_y and ψ_y are assumed to be such that, on the curve $\mathcal{S}^*$, the Jacobian matrix $\phi(y)$ is orthogonal, or

$$\phi(y) = \begin{vmatrix} \partial\psi/\partial y \\ \partial\varphi/\partial y \end{vmatrix} = T(\alpha^*),$$

where $\alpha^* \in R^3$ is the vector of Euler angles. The matrix $T(\alpha^*) \in SO(3)$ corresponds to the Frenet frame fixed to the curve $\mathcal{S}^*$ and satisfies the Frenet-like differential equation:

$$\dot{T}(\alpha^*) = \dot{s}S(\xi)T(\alpha^*), \qquad (40)$$

where $S(\xi) \in so(2)$ and $\xi \in R^3$.

By using the frames $T(\alpha)$ and $T(\alpha^*)$, we can prespecify the body angular orientation with respect to $\mathcal{S}^*$ in the form of angle relations

$$T(\alpha) = T(\triangle\alpha)\,T(\alpha^*), \qquad (41)$$

where $\triangle\alpha$ is the vector of the desired relative rotation and $T(\triangle\alpha) \in SO(3)$.

Thus, rigid body motion control is associated with two groups of coordination conditions: equation of the curve (15) introduces relationships between

Cartesian coordinates y_i, and (41) represents relations of the absolute angular coordinates α_i. They are complemented by a description of the desired mode of the mass point longitudinal motion $s(t)$ usually given through the reference variable $s^*(t)$ or the trajectory velocity $\dot{s}^*(t)$.

The violation of condition (15), associated with orthogonal deviations from the curve $\mathcal{S}^*$, is specified by the error vector

$$\varepsilon = \varphi(y), \tag{42}$$

where $\varepsilon = \mathrm{col}(\varepsilon_1, \varepsilon_2)$. Current violation of the angle relations (41) is specified by the vector of angular errors $\delta \in R^3$ or the angular deviation matrix

$$T(\delta) = T(\alpha)\, T^T(\alpha^*)\, T^T(\triangle\alpha). \tag{43}$$

Then the problem of *rigid body trajectory control* can be formulated as follows.

Problem 3. Find a feedback control law that provides
a) desired longitudinal motion of the system at the rate $\dot{s}^* = const$, or asymptotic rejection of the rate error $\triangle\dot{s}$;
b) asymptotic rejection of the error $e(t)$;
c) asymptotic rejection of the error δ.

Problems 3a and 3b concern the motion of a mass-point described by (37) and are solved by means of the controlling action u_y. The control design procedure is accomplished as in Section 4.2, which leads to the control variables transformation

$$u_y = mT(\triangle\alpha)\begin{vmatrix} u_s \\ u_e \end{vmatrix} \tag{44}$$

and the local controllers (34)–(35) (Figure 6).

To ensure the desired body orientation and solve Problem 3c, define the vector of instant angular velocity errors $\triangle\Omega = \Omega - \dot{s}\,\xi^\triangle$, where $\xi^\triangle = T(\triangle\alpha)\xi$. Then the task-oriented model of the body rotation, for small enough δ and $\triangle\Omega$, takes the form

$$\dot{\delta} + \dot{s}\, S^T(\xi^\triangle)\, \delta = \triangle\Omega, \tag{45}$$
$$\triangle\dot{\Omega} + \dot{s}A(s)\triangle\Omega + \ddot{s}\,\xi^\triangle + \dot{s}\,\dot{\xi}^\triangle = J^{-1}u_\alpha. \tag{46}$$

Introduce the control of the body rotation u_δ by the expression

$$u_\alpha = Ju_\delta \tag{47}$$

and find the controller (Figure 6)

$$C^\delta: \quad u_\delta = J(\ddot{s}\,\xi^\triangle + \dot{s}\,\dot{\xi}^\triangle - K_{\delta 1}\,\dot{\delta} - K_{\delta 2}\,\delta), \tag{48}$$

where the feedback matrices $K_{\delta 1}$ and $K_{\delta 2}$ provide the desired asymptotic stabilization of the model (45)–(46).

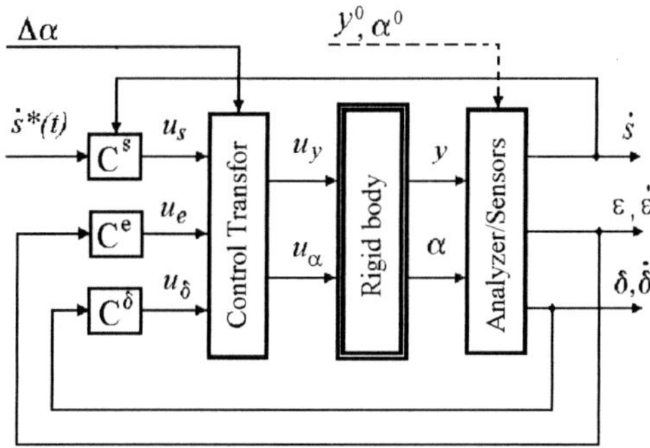

Fig. 6. Rigid body control system.

The solution is easily generalized for *surface-following* (see Section 6.2 and [10]) and *time-varying trajectory problems* [3, 22]. The latter is established by new technological tasks accomplished by mechanical systems in a dynamic environment represented by mobile external objects (obstacles, working pieces, and external mechanisms).

6 Coordination and motion control of robots

Very promising task-based solutions for a variety of robot control problems, including those of wheeled, manipulation, and walking robotic systems, can be obtained on the basis of the unified techniques of output coordination and spatial motion control represented in Sections 4–5. The approach enables us to overcome known difficulties caused by a special configuration of the robot chain, the kinematic structure redundancy, and complexity of prespecified motions.

6.1 Wheeled robot control

The robot platform is considered as a rigid body whose position in the absolute coordinate system R^2 is characterized by the pair $(y,\ T(\alpha))$, where $y \in R^2$ is the vector of the mass-center position, $T(\alpha) = \begin{vmatrix} \cos\alpha & \sin\alpha \\ -\sin\alpha & \cos\alpha \end{vmatrix} \in SO(2)$, and α is the angle of the platform orientation (Figure 7). The platform dynamics are described by Newton-Euler equations with force-torque actions u_y and u_α produced by a wheeled system. The latter consists of several controlled wheeled modules that provide the required motion of the wheels due to appropriate steering and driving actions (controls) u and u_β [3, 10].

The basic problem of wheeled robot trajectory control is posed through coordination conditions, represented in Section 5, and is reduced to the design

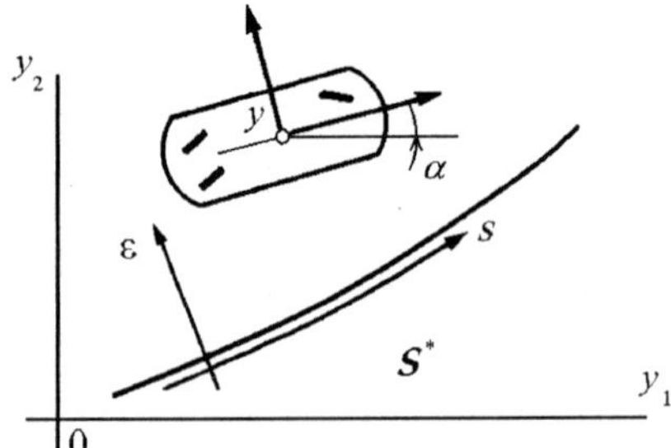

Fig. 7. Wheeled robot trajectory motion.

of the controlling inputs u and u_β that solve Problem 3. However, taking into the account peculiarities of the robot model, the design procedure is divided into two steps [3, 10].

The first step (*forth-torque control*) is accomplished by analogy with the procedure represented in Section 5 and leads to finding the required force-torque actions u_y and u_α. At the second step we must solve the *inverse kinematic-dynamic problem* and obtain the relevant wheel control actions u and u_β.

Note that, for simplest vehicles with two driving (steering) control loops, variables u_y, u_α are interconnected, and the design procedure has to be modified. On the contrary, the advanced multidrive samples of mobile robots are kinematically redundant, which increases the mobility of the robot but leads to a nonunique solution of the inverse problem. To specify the cooperative behaviour of the wheeled modules, additional conditions are introduced in the form

$$\varphi_\beta(\beta, \dot{\alpha}) = 0, \tag{49}$$

where β is the vector of current angles of the wheel rotations and φ_β is a smooth function, which predetermines the requirement for the coordination of the controls.

6.2 Manipulation robot control

Multilink robots represented by an m-degree-of-freedom (DOF) spatial kinematic mechanism are described by Lagrangian equations that establish the connection of the vector of joint (generalized) coordinates $q \in R^m$ and the vector of generalized controlling torques $u \in R^m$. The position of the jth link in the Cartesian space R^3 is characterized by the pair

$$y^j, T^j(\alpha^j), \quad j = 1, 2, \ldots, m, \tag{50}$$

where $y^j \in \mathcal{Y} \subset R^3$ is the vector of Cartesian coordinates of a principal point of the link and $T^j \in SO(3)$ is the matrix associated with the frame fixed to the link depending on the vector of Euler angles $\alpha^j \in R^3$. A description of the relations of the pairs $(y^j,\, T^j)$ with the generalized coordinates is given in the form of *direct* kinematic equations.

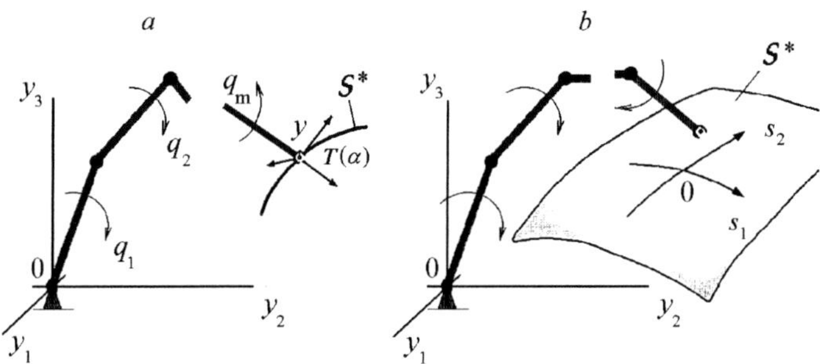

Fig. 8. (a) Robot trajectory motion and (b) surface-following.

The trajectory motion of the robot end-point in the space R^3 (see Figure 8a) characterized by the pair

$$(y, T(\alpha)) \;=\; (y^m, T^m(\alpha^m))$$

is prespecified a desired spatial curve $\mathcal{S}^*$ given in the form (15), and the required orientation with respect to $\mathcal{S}^*$ is given in the form of angle relationship (41). Then the basic problem of robot trajectory control is posed as Problem 3. The problem is solved by using the methodology considered in Section 5. However, taking into account that in this case we face the two-step change of system coordinates: $q \mapsto (y, T(\alpha)) \mapsto (s, e, \delta)$, a two-step model transformation is also needed. The first step implies deriving the robot model in Cartesian space, and the second leads to a task-oriented model.

The general robot control includes terms corresponding to the robot exact linearization (or a so-called *computed torque method*) and the control variable transformation, being typical for the coordinating control. The local control laws associated with partial problems posed in the task-oriented space complete the scheme of robot trajectoy control (see [10, 17, 18]).

For surface-following problems (Figure 8(b)), where $s = (s_1, s_2)$ and ε is a scalar, the procedure of the design has to be modified taking into account the results represented in Section 4.1 and [10].

6.3 Coordinating control of redundant manipulators

A redundant number of DOF is a necessary condition for the dexterity and versatility of robotic manipulators. Dexterous multilink robots are able to perform nontrivial locomotion tasks such as penetrating into hard-to-come domains of the operational zone, perfect obstacle avoidance, suitable approaching external objects, and motion along complex curvilinear trajectories [4, 7, 21, 26, 30]. Execution of these tasks and the necessity for fruitful utilization of the extra DOF in the course of the robot motion induce certain difficulties of control referred to as *redundancy problem.*

A natural way to overcome the uncertainty of control actions caused by robot redundancy is to impose additional constrains [6, 17, 21, 26, 30] written in the form of coordination conditions (4). The most evident conditions are given by analytical descriptions of the trajectories of certain principal points y^j of the robot kinematic chain. Like this, equations of the form

$$\varphi_y^j(y^j) \;=\; 0 \tag{51}$$

describe the trajectory in R^3, motion along which corresponds to a desired behaviour of the jth link. The required jth link angular orientation in the Cartesian space R^3 is given by the equation

$$T(\alpha^j) \;=\; T(\Delta\alpha^j), \tag{52}$$

where $\Delta\alpha^j = \text{const}$ is the vector specifying the desired relative orientation of the jth link. At last, the restrictions on the joint coordinates in the space R^3 are given by relations of the form

$$\varphi_q^j(q^j) \;=\; 0. \tag{53}$$

The collection of control problems posed in the form of coordination conditions (51)–(53) is reduced to that of stabilization of the appropriate deviations, the number of which is assumed to be equal to m. This ensures the existence of a solution of Problem 1 (see Section 4.1) and well-coordinated behaviour of the redundant robot.

The most difficult control problems are connected with spatial motion of hyperredundant robotic systems (articulated and snake-like robots, variable

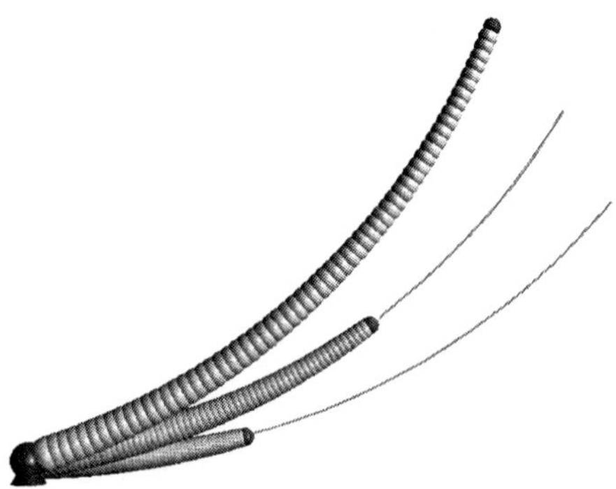

Fig. 9. Motion along moving circle.

geometry truss manipulators, etc. [7, 26]) and time variations of the given trajectory. Here we face the necessity to maintain compactness and desired time-invariant configuration of the kinematic chain in the course of the end-point displacement along a given trajectory. A direct solution for these problems, based on coordinating control, implies introducing the necessary number of coordination conditions of the form

$$\varphi_y(y^j, t) \;=\; 0, \tag{54}$$

which defines a desired variable shape of the robot. After introducing the deviations $\varepsilon^j = \varphi_y(y^j, t)$ and the required longitudinal velocity $\dot{s}^*$, this problem is also solved by using the methodology represented in Section 4. Curvilinear motion of the a 26-link snake-like robot is shown in Figure 9.

6.4 Trajectory motion of underactuated mechanisms

Underactuated mechanisms, for which the number of actuators is less than DOF, are widely used to perform a variety of specific technological tasks where the lack of controlling inputs is admissible but a small number of actuators is desirable taking into account their reduced cost and weight performances [5, 8, 11]. Known peculiarities of the underactuated mechanical systems prevent the realization of the usual modes of the controlled motion and the use of standard solutions of stabilization and regulation problems, as well those of trajectory control [25].

According to the methodology of the analysis of the trajectory motion discussed in Section 4.2, we can define task-oriented coordinates ε, s and pose the problem of the asymptotic rejection of $\varepsilon(t)$ (Problem 2b). It is worth noting that the system under consideration is not completely controllable, and therefore the longitudinal motion $s(t)$ of the end-point cannot be prespecified. Moreover, the required mode of the trajectory motion is reached only in a bounded domain of the Cartesian space.

Figure 10 illustrates the potentialities of two underactuated mechanisms under the control considered.

7 Coordinating control of biped robot

The most challenging directions of biped robot control can be associated with geometric control strategies and the relevant techniques of coordinating control [6, 11]. The approaches mentioned point out a natural way to overcome the difficulties connected with complex behaviour of these multilink mechanisms by imposing additional constrains on the robot motion in order to keep the center of gravity well supported.

Consider an mth-link planar robot (Figure 11) and its motion in $\mathcal{Y} \in R^2$ along a horizontal supporting surface with a simplest walking scheme that

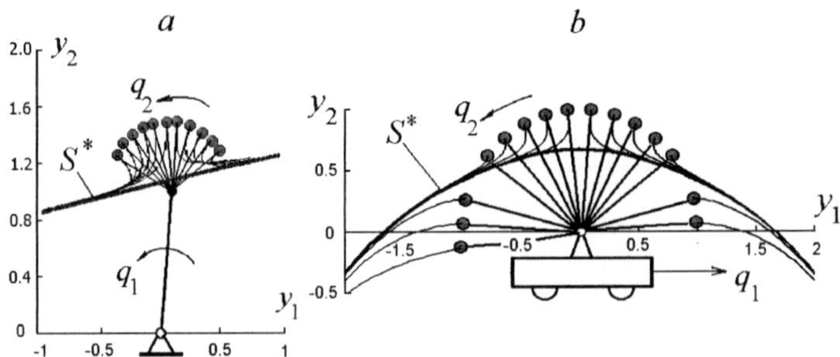

Fig. 10. Trajectory motion of (a) 2-link pendulum and (b) pendulum on a cart.

consists of two phases: body displacement and step. The displacement is a double support phase when the legs provide motion of the gravity center point C along the straight line (or curve, if necessary) S^j at the desired rate $V = \dot{s}^*(t)$. Step one of the leg provides single support and the other is moved from the backward to forward position.

To prespecify the biped mechanism locomotion in each of the phases, we introduce $m-1$ coordination conditions of the form (15) and the required mode of the motion of the gravity center. Some of these restrictions are represented as the relations of joint Cartesian coordinates

$$\varphi_y(y^j) = 0 \tag{55}$$

and describe the trajectories S^j of the motion of certain principal points including the point C. The others are given in the form of the relations of the angels of link orientations

$$\varphi_\alpha(\alpha^j, \alpha^k, \ldots) = 0 \tag{56}$$

and those of the generalized coordinates

$$\varphi_q(q) = 0. \tag{57}$$

Defining task-oriented coordinates ε and s, we pose the the control problem by analogy with Problem 1. Its solution leads to the robot locomotion represented in Figure 11.

Acknowledgements. The content of the chapter was initiated by the lectures that the author gave in Italian universities in 2002, as well as the research accomplished in collaboration with Dipartimento di Informatica, Sistemi e Produzione of Rome University "Tor Vergata," with the kind assistance of Professor Salvatore Nicosia.

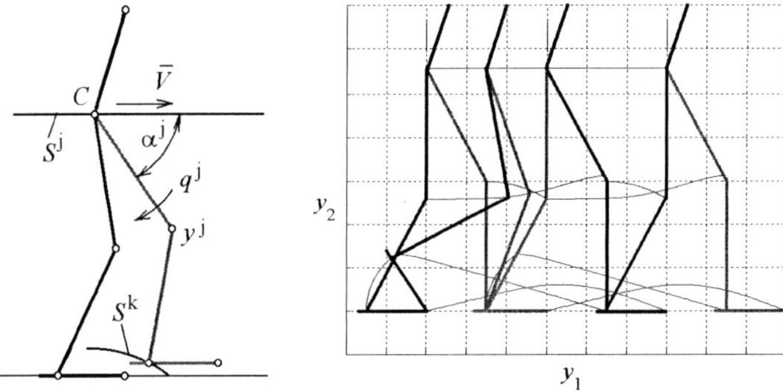

Fig. 11. Biped robot locomotion.

References

1. Blekhman II, Fradkov AL, Nijmeijer H, Pogromsky AY (1997) On self-synchronization and controlled synchronization. Systems & Control Letters 31: 299-305
2. Bloch AM, Krishnaprasad PS, Marsden JE, Sanchez De Alvarez G (1992) Stabilization of rigid body dynamics by internal and external torques. Automatica 28:745-756
3. Burdakov SF, Miroshnik IV, Stelmakov RE (2001) Motion control of wheeled robot. Nauka, Saint Petersburg (in Russian)
4. Caccavale F, Siciliano B (2001). Kinematic control of redundant free-robotic systems. Advanced Robotics 15:429-448
5. Canudas de Wit C, Espiau B, Urrea C (2002). Orbital stabilization of underactuated mechanical systems. 15th triennial World Congress of IFAC, Barcelona, Spain
6. Chevallereau C, Abba G, Aoustin Y, Plestan F, Westervelt ER, Canudas de Wit C, Grizzle JW (2003) RABBIT: A testbed for advanced control theory. IEEE Control Systems Magazine 23(5):57-79
7. Chirikjian GS (1995) Hyperredundant manipulator dynamics: a continuum approximation. Advanced Robotics 9:217-243
8. De Luca A, Lantini S et al. (2001). Control problems in underactuated manipulators. Proc. of Int. Conf. on Advanced Intel. Mechatronics, Como, Italy
9. Elkin VI (1998) Reduction of nonlinear controlled systems. A deferentially geometric approach. Kluwer Academic Publishers, Dordrecht
10. Fradkov AL, Miroshnik IV, Nikiforov VO (1999) Nonlinear and adaptive control of complex systems. Kluwer Academic Publishers, Dordrecht
11. Furuta K (2002) Super mechano-systems: fusion on control and mechanisms. 15th triennial World Congress of IFAC, Barcelona, Plenary Papers, survey papers and milestones: 35-44
12. Isidori A (1995) Nonlinear control systems, 3rd edition. Springer-Verlag, New York
13. Kolesnikov AA (2000) Scientific fundamentals of synergetic control, Ispo-Service, Moscow (in Russian)

14. Miroshnik IV, Usakov AV (1977) The synthesis of an algorithm for synchronous control of quasi-similar systems. Autom. and Rem. Control 11: 22-29

15. Miroshnik IV (1986) On stabilization of the motion about a manifold, Avtomatyca, no. 4 (in Russian). English translation: Soviet J. of Autom. and Inf. Science, Scripta Technica Publ. Company, Washington, DC

16. Miroshnik IV (1990) Coordinating control of multivariable systems. Leningrad, Energoatomizdat (in Russian)

17. Miroshnik IV, Nikiforov VO (1995a) Redundant manipulator motion control in the task oriented space. IFAC Workshop on Motion Control, Munich, 648–655

18. Miroshnik IV, Nikiforov VO (1995b) Coordinating control of robotic manipulator. Int. J. Robotics and Automation 10(3):101–105

19. Miroshnik IV, Bobtzov A (2000) Stabilization of motions of multipendulum systems. In: 2nd Int. Conf. on Control of Oscillation and Chaos. St. Petersburg 1:22–25

20. Miroshnik IV, Nikiforov VO, Fradkov AL (2000) Nonlinear control of dynamical systems. Nauka, Saint Petersburg (in Russian)

21. Miroshnik IV, Nikiforov VO, Shiegin VV (2001). Motion control for kinematically redundant manipulator robots. J. Computer and System Sciences Int. (Izvestia Rossiiskoi Akademii Nauk. Teoriya i Systemy Upravleniya), 1(40)

22. Miroshnik IV, Sergeev KA (2001) Nonlinear control of robot spatial motion in dynamic environments. Proc. Int. IEEE Conf. on Advanced Intel. Mechatronics, Como, Italy 2:1303–1306

23. Miroshnik IV (2002a) Partial stabilization and geometric problem of nonlinear control. 15th Triennial World Congress of IFAC, Barcelona, Spain

24. Miroshnik IV (2002b) Attractors and geometric problems of nonlinear dynamics. In: Attractors, Signals, and Synergetics. Pabst Science Pub., Lengerich, 114–141

25. Miroshnik IV, Chepinsky SA (2003) Trajectory motion control of underactuated manipulators. IFAC Symposium on Robot Control, Wrozlaw

26. Miroshnik IV, Boltunov GI, Gorelov DP (2003) Control of configuration and trajectory motion of redundant robots. IFAC Symposium on Robot Control, Wrozlaw, 259–264

27. Miroshnik IV (2004) Attractors and partial stability of nonlinear dynamical systems. Automatica 40(3):473–480

28. Murray RM, Zexiang IL, Sastry SS (1993). A mathematical introduction to robotic manipulation. CRC Press, Boca Raton, FL

29. Pecora LM, Carroll TL (1999) Master stability functions for synchronized coupled systems. Int. J. of Bifurcation and Chaos 9(12): 2315-2320

30. Seraji H (1989) Configuration control of redundant manipulators: theory and implementation. IEEE Trans. on Robotics and Automation 5:472–490

31. Vorotnikov VI (1998) Partial stability and control. Birkhauser, New York

Coordination of Robot Teams:
A Decentralized Approach*

Rafael Fierro[1] and Peng Song[2]

[1]Oklahoma State University, School of Electrical and Computer Engineering, 202
Engineering South, Stillwater, OK 74078, USA. `rfierro@okstate.edu`
[2]Rutgers, The State University of New Jersey, Department of Mechanical and
Aerospace Engineering, 98 Brett Road, Engineering B242, Piscataway, NJ 08854,
USA. `pengsong@jove.rutgers.edu`

Summary. In this chapter, we present two main contributions: (1) a leader-follower
formation controller based on dynamic feedback linearization, and (2) a framework
for coordinating teams of mobile robots (i.e., swarms). We derive coordination algorithms that allow robot swarms having *independent goals* but sharing a common
environment to reach their target destinations. Derived from simple potential fields
and the hierarchical composition of potential fields, our framework leads to a decentralized approach to coordinate complex group interactions. Because the framework
is decentralized, it can potentially scale to teams of tens and hundreds of robots.
Simulation results verify the scalability and feasibility of the proposed coordination
scheme.

1 Introduction

New developments in complex networks of interacting systems, like multitar-
geting/multiplatform groups, unmanned air vehicle systems, and intelligent
highway/vehicle systems, place severe demands on the design of decision-
making supervisors, cooperative control schemes, and communication strategies in the presence of a changing environment. The large-scale interconnected
nature of such systems requires coordination strategies with increased capabilities for reconfiguration.

Suppose that a search-and-rescue team of unmanned ground vehicles
(UGVs) are deployed within an unknown, possibly unfriendly, environment.
Some robots are able to localize and navigate using global positioning system
(GPS), others may have biological or chemical sensors, some might have a
vision system, and so on. Each team is responsible for performing a particular

---

*The work of the first author is partially supported by NSF grants #0311460
and #0348637 (CAREER), and by the U.S. Army Research Office under grant
DAAD19-03-1-0142 (through the University of Oklahoma).

task and cooperating with other teams. A mission may be a parallel or sequential composition of subtasks based on robot capabilities and resources. Also, some of the functions can be done individually or cooperatively. For instance, exploration of a large area can be done more efficiently using a team of robots instead of a single robot. The robot teams may have to compete periodically for limited resources and may come into conflict with one another throughout the course of their activities. The success of robot teams in accomplishing a mission ultimately depends on their ability to manage these interrelationships through appropriate coordination strategies.

In this chapter we consider situations in which there may be no access to any GPS and the main sensing modality is vision. Our platform of interest is a nonholonomic car-like robot shown in Figure 1. Details of the platform design are given in [9]. Each robot is capable of autonomous operation or following

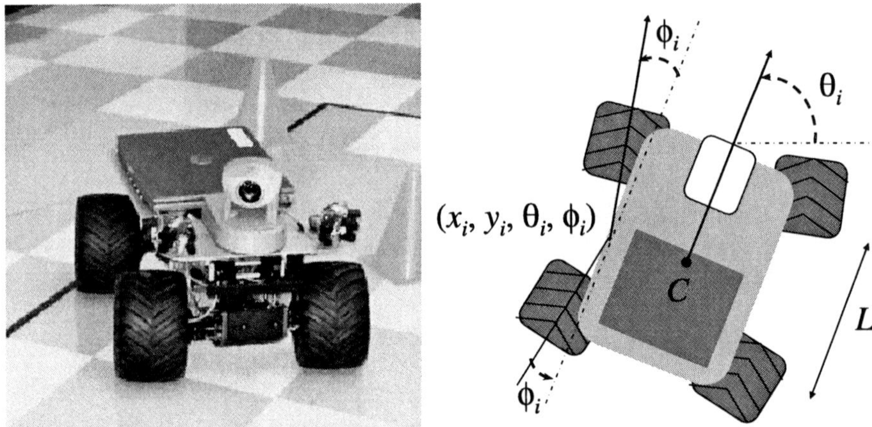

Fig. 1. The MARHES car-like nonholonomic mobile robot.

one or two robots. In this chapter, we present a trajectory tracking control algorithm for a car-like robot based on dynamic feedback linearization. This algorithm is applied to the leader-follower formation control problem. Additionally, we describe a framework for decentralized cooperative control of multirobot systems that emphasizes simplicity in planning, coordination, and control. Specifically, each robot plans its own trajectory based on the sensory information of its surroundings and neighboring robots. The proposed framework is also suitable for solving cooperative manipulation problems, where a relatively *rigid* formation may be necessary to transport a grasped object to a prescribed location [24].

The rest of the chapter is organized as follows. The motivation, problem definition, and some preliminaries are given in Section 2. Section 3 presents a leader-follower formation control algorithm based on dynamic feedback linearization. Then, a decentralized coordination scheme is described in Section 4. Finally, some concluding remarks and future work directions are given in Section 5.

2 Multiple Robot Teams with Independent Goals

Most current multirobot literature focuses on coordination and cooperative control of teams of robots having a common goal. There are several research groups making important theoretical and practical contributions in this area, see, for instance, [6, 12]. The research challenges encountered in cooperative multirobot systems require the integration of different disciplines including control systems, artificial intelligence (AI), biology, and optimization. There-fore, it is not surprising that the related literature enjoys the flavor of a broad spectrum of approaches that have been utilized in attempt to come up with a solution for cooperative control problems [17, 5].

A natural extension to the problem of coordinating a group of robots hav-ing a common goal is coordinating teams of robots having different goals. Our work is motivated by [18] and [1]. In the former, the authors address the op-timal motion planning problem for multiple robots having independent goals. The task considered in this paper is to simultaneously drive the robots from an initial configuration to a target configuration by minimizing independent performance measures for each robot. In the latter, authors develop a decen-tralized approach based on potential functions to coordinate a group of planar vehicles within a dynamic unfriendly environment.

We are interested in designing a flexible, feasible, and decentralized frame-work that allows teams of robots with different mission objectives to interact with each other while accomplishing their goals. More formally, let $\mathcal{T}$, M, and N_i denote a team of robots, the number of teams, and the number of robots in $\mathcal{T}_i$, respectively. Also, the motion of an individual robot $\mathcal{R}_{ij}$ with $i \in \{1, 2, \ldots, M\}$ and $j \in \{1, 2, \ldots, N_i\}$ is given by

$$\dot{q}_{ij}(t) = f(q_{ij}(t), u_{ij}(t)), \qquad \forall i, j, \tag{1}$$

where $u_{ij} \in \mathcal{U}_{ij}$ is the control input for robot $\mathcal{R}_{ij}$ with $\mathcal{U}_{ij}$ the set of allowable control inputs, and $q_{ij} = [x_{ij} \; y_{ij}]^T \in \Re^2$ is the Cartesian position of $\mathcal{R}_{ij}$. For a nonholonomic robot, we have $q_{ij} = [x_{ij} \; y_{ij} \; \theta_{ij}]^T$.

As in [18] we define a state space $\mathcal{X}_i$ that describes the configurations of all the robots in team i. The coordination task is to drive each team from some initial state $X_i^o \in \mathcal{X}_i$ to its goal configuration $X_i^g \in \mathcal{X}_i$. As teams navigate toward their goals they might interact with other teams (e.g., groups of robots manipulating objects within a common workspace). Moreover, it would be convenient to define a reference point $\bar{X}_i \in \mathcal{X}_i$ (e.g., the position of center of mass of the group of robots) describing the motion of the entire team as follows (cf., multirobot abstractions are proposed in [3]):

$$\dot{\bar{X}}_i(t) = F_i(\bar{X}_i(t), \bar{U}_i(t)), \qquad i = 1, 2, \ldots, M. \tag{2}$$

Note that conventional planning approaches would add some collision avoidance conditions to the above coordination task. We, on the other hand, use *rigid body contact dynamics* models to allow collisions between the robots and their surroundings instead of avoiding them. Another important component of our decentralized framework is the use of artificial potential fields. The main idea behind potential field approaches is to define a scalar field $\phi(q)$ (called *potential function*) over the robot's free space. This artificial field produces a force $-\nabla\phi$ acting on the robot. Obstacles and goals produce repulsive and attractive potentials, respectively. The resultant forces are mapped to controller/actuator commands. Thus the robot, at least in theory, would navigate toward its goal destination while avoiding collisions. Several researchers have extended potential field methods to make them suitable for multirobot systems [20, 1]. However, the main drawback of potential field methods is that the robot might get stuck in a local minimum before reaching the goal. Several variants have been proposed to overcome this limitation.

In this chapter, each team $\mathcal{T}_i$ has an attractive potential function $\phi_i(\mathcal{T}_i)$ associated to it while obstacles are handled by using contact dynamics models.

3 Formation Control

Formation control is a fundamental group behavior required by robot teams engaged in tasks within spatial domains. Simply speaking, formation control refers to the problem of controlling the robots such that they maintain a geometric shape. The *shape* of the formation is defined by relative positions and orientations of the vehicles with respect to a reference frame. Hence, formation control can be seen as an extension of control laws developed for a single mobile robot. A common method to implement trajectory tracking is defining a *virtual* reference robot with state q_r, then designing a control law that makes $q \to q_r$ as $t \to \infty$. If instead of using a virtual robot we use a *real* robot, then the trajectory tracking controller becomes a *leader-follower* controller.

Many systems in nature exhibit stable formation behaviors e.g., swarms, schools, and flocks [21]. In these highly robust systems, individuals follow distant leaders without colliding with neighbors. Thus, a coordinated grouping behavior emerges by composing individual control actions and sensor information in a distributed control architecture. One possibility to realize such a grouping behavior is using artificial potential functions as a coordination mechanism [19].

We consider nonholonomic robots (e.g., differential drive, car-like platforms), since they are quite common in real-world applications. A formation controller based on dynamic feedback linearization [14, 22] is presented in the next section.

3.1 Dynamic Feedback Linearization

In our previous work [7, 11] we developed formation control algorithms for the unicycle model based on *static* input-output (I/O) feedback linearization. It is well known that analyzing the stability of the internal dynamics when using static I/O linearization is not straightforward [8]. To overcome this limitation, we develop a dynamic full input-state-output linearizing controller.

Consider the car-like (double-steering) robot shown in Figure 1 (right) where the pair x, y denotes the Cartesian position of the center of mass C, θ is the orientation of the robot, and ϕ is the steering angle. The kinematic model is given by

$$\begin{bmatrix} \dot{x} \\ \dot{y} \\ \dot{\theta} \\ \dot{\phi} \end{bmatrix} = \begin{bmatrix} \cos\theta \\ \sin\theta \\ \frac{2\tan\phi}{L} \\ 0 \end{bmatrix} v + \begin{bmatrix} 0 \\ 0 \\ 0 \\ 1 \end{bmatrix} w, \tag{3}$$

where v and w are the linear velocity and steering rate, respectively. Using

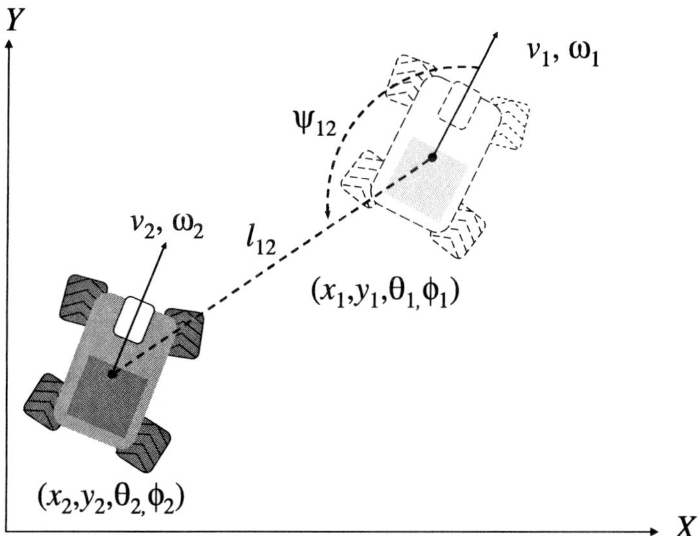

Fig. 2. Trajectory tracking based leader-follower control.

the reference (sometime called virtual) vehicle approach, we can rewrite the kinematic equations in a more suitable form for trajectory tracking or leader-follower control design purposes. We choose the separation l and relative bearing ψ (see Figure 2) as output variables, since they can easily be measured with a vision sensor [25]. The relative distance and bearing become

$$l_{12} = \sqrt{(x_1 - x_2)^2 + (y_1 - y_2)^2} \tag{4}$$

$$\psi_{12} = \pi - \arctan 2\left(y_2 - y_1, x_1 - x_2\right) - \theta_1 \,, \tag{5}$$

where (x_1, y_1, θ_1) and (x_2, y_2, θ_2) are the state vectors of the reference (*leader*) and actual (*follower*) robots, respectively. It can be seen that

$$\dot{l}_{12} = -v_1 \cos\psi_{12} + v_2 \cos\gamma \tag{6}$$

$$\dot{\psi}_{12} = \frac{v_1}{l_{12}}\sin\psi_{12} - \frac{v_2}{l_{12}}\sin\gamma - \frac{2v_1}{L}\tan\phi_1 \tag{7}$$

$$\dot{\theta}_{12} = \omega_1 - \omega_2 \tag{8}$$

$$\dot{\phi}_{12} = w_1 - w_2 \,, \tag{9}$$

where $\gamma = \psi_{12} + \theta_1 - \theta_2$, w_1 and w_2 represent the steering rate, the reference, and follower robots, respectively. Similarly, ω_1 and ω_2 are the angular velocities of the leader and follower robots.

The output of interest is $z = [l_{12} \quad \psi_{12}]^T$; then we have

$$\dot{z} = \begin{bmatrix} \cos\gamma & 0 \\ -\frac{\sin\gamma}{l_{12}} & 0 \end{bmatrix}\begin{bmatrix} v_2 \\ w_2 \end{bmatrix} + \begin{bmatrix} -v_1\cos\psi_{12} \\ \frac{v_1\sin\psi_{12}}{l_{12}} - \frac{2v_1\tan\phi_1}{L} \end{bmatrix}. \tag{10}$$

Since the decoupling matrix is singular, we apply the dynamic extension procedure by adding an integrator (whose state is given by ζ_1) on the linear velocity input. Thus, we have

$$\begin{aligned} v_2 &= \zeta_1 \\ \dot{\zeta}_1 &= a_2 \,, \end{aligned} \tag{11}$$

where a_2 is the linear acceleration of the car-like robot, and

$$\dot{z} = \begin{bmatrix} \cos\gamma & 0 \\ -\frac{\sin\gamma}{l_{12}} & 0 \end{bmatrix}\begin{bmatrix} \zeta_1 \\ w_2 \end{bmatrix} + \begin{bmatrix} -v_1\cos\psi_{12} \\ \frac{v_1\sin\psi_{12}}{l_{12}} - \frac{2v_1\tan\phi_1}{L} \end{bmatrix}. \tag{12}$$

Taking the derivative of (12) with respect to time, we get

$$\ddot{z} = \begin{bmatrix} \cos\gamma & 0 \\ -\frac{\sin\gamma}{l_{12}} & 0 \end{bmatrix}\begin{bmatrix} a_2 \\ w_2 \end{bmatrix} + \begin{bmatrix} s_1 \\ s_2 \end{bmatrix}, \tag{13}$$

where

$$s_1 = v_1\dot{\psi}_{12}\sin\psi_{12} - \zeta_1\dot{\gamma}\sin\gamma - a_1\cos\psi_{12}$$

$$s_2 = \frac{1}{l_{12}^2}(\zeta_1\dot{l}_{12}\sin\gamma - \zeta_1 l_{12}\dot{\gamma}\cos\gamma + a_1 l_{12}\sin\psi_{12} - v_1\dot{l}_{12}\sin\psi_{12}$$

$$+ v_1 l_{12}\dot{\psi}_{12}\cos\psi_{12}) - \frac{2}{L}(a_1\tan\phi_1 + v_1 w_1\sec^2\phi_1)\,,$$

where a_1 is the linear acceleration of the reference robot. The decoupling matrix in (13) is still singular; hence, we add an extra integrator on the first input. That is,

$$a_2 = \zeta_2$$
$$\dot{\zeta}_2 = j_2,\tag{14}$$

where j_2 is the *jerk* of the robot. Using (14) and taking the derivative of (13), we obtain

$$\ddot{z} = A(l_{12}, \gamma, \zeta_1, \phi_2)\begin{bmatrix} j_2 \\ w_2 \end{bmatrix} + B = Au + B.\tag{15}$$

The actual expressions for $A(\cdot)$ and $B(\cdot)$ are too long to be included here. It can be verified that the dynamic controller has a singularity at $\zeta_1 = v_2 = 0$. Finally, the auxiliary control input $u = \begin{bmatrix} j_2 & w_2 \end{bmatrix}^T$, computed as

$$u = A^{-1}(R - B),$$

with $R = \begin{bmatrix} r_1 & r_2 \end{bmatrix}^T$, transforms the original leader-follower system into a linear decoupled system:

$$\ddot{z} = \begin{bmatrix} r_1 \\ r_2 \end{bmatrix}.\tag{16}$$

The original system has six states (with two controller states included) and the output differentiation order is also six. Therefore the system is fully linearized in the new coordinates. Thus, the closed loop system becomes

$$r_1 = \ddot{l}^{d}_{12} + k_1(\ddot{l}^{d}_{12} - \ddot{l}_{12}) + k_2(\dot{l}^{d}_{12} - \dot{l}_{12}) + k_3(l^{d}_{12} - l_{12})$$
$$r_2 = \ddot{\psi}^{d}_{12} + k_4(\ddot{\psi}^{d}_{12} - \ddot{\psi}_{12}) + k_5(\dot{\psi}^{d}_{12} - \dot{\psi}_{12}) + k_6(\psi^{d}_{12} - \psi_{12}),\tag{17}$$

where the desired variables l^{d}_{12}, ψ^{d}_{12} define the formation geometry, and k_i, $i = 1, \ldots, 6$ are positive feedback gains whose values can be found using well-known linear control techniques.

3.2 Simulation Results

The above algorithm was extensively tested and verified in simulation. We present two examples. First, a follower robot (dashed line) is to maintain a formation shape defined by $l^{d}_{12} = 5m$ and $\psi^{d}_{12} = 4\pi/3$ (see Figure 3). In the second case (shown in Figure 4), the desired relative bearing is changed from $\psi^{d}_{12} = \pi/2$ to $4\pi/3$. In both simulation studies, the steering servo of the reference robot is modeled as a first-order system

$$\dot{\phi}_1 = \lambda(\phi^{d}_1 - \phi_1),$$

where ϕ^{d}_1 is the desired steering angle, and λ is a constant that depends on the characteristics of the servo (assumed equal to one for simplicity).

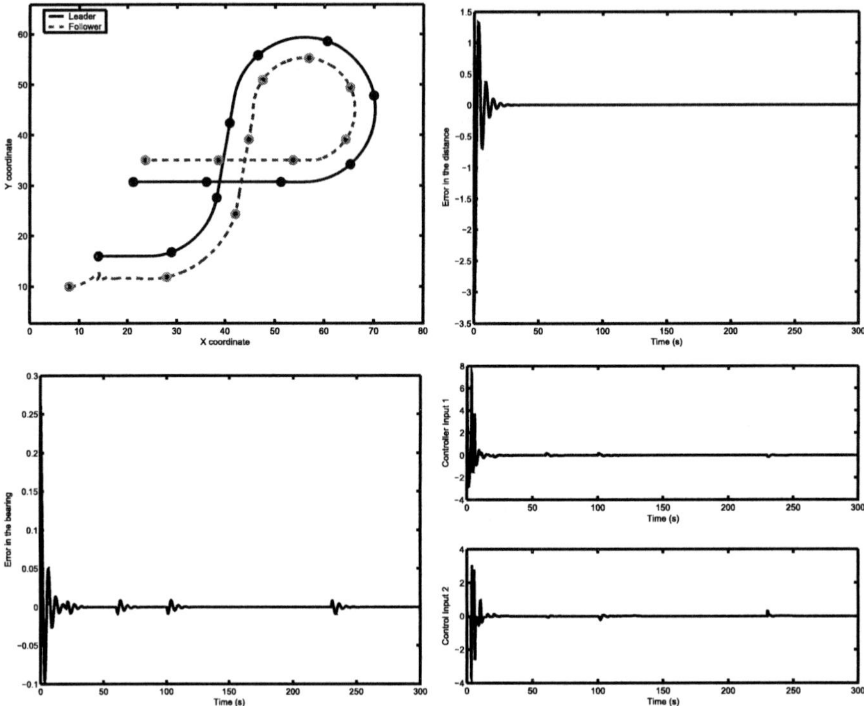

Fig. 3. Example 1. Top left: leader-follower trajectory; top right: separation error; bottom left: relative bearing error; bottom right: control inputs.

4 Trajectory Generation for Decentralized Coordination

In this section, we describe a scheme for sensor-based trajectory generation for teams of robots having independent goals. The key idea that distinguishes our approach from previous work is the use of rigid body contact dynamics models to allow virtual collisions among the robots and the surroundings instead of avoiding them. Consider a group of mobile robots moving in an environment with the presence of obstacles. We first characterize the surrounding spatial division of each mobile robot with three zones as depicted in Figure 5.[1] Use robot R_1 as an example; the sensing zone denotes the region within which a robot can detect obstacles and other robots. The contact zone is a collision warning zone. The robot starts estimating the relative positions and velocities of any objects that may appear inside its contact zone. The innermost circle is the protected zone, which is modeled as a rigid core during a possible contact to provide a collision-free environment for the actual robot. The ellipse within the protected zone represents the *reachable* region of the nonholonomic robot

---

[1] Compare with [13]. Our definition of *contact zone* is derived from the principles of contact dynamics.

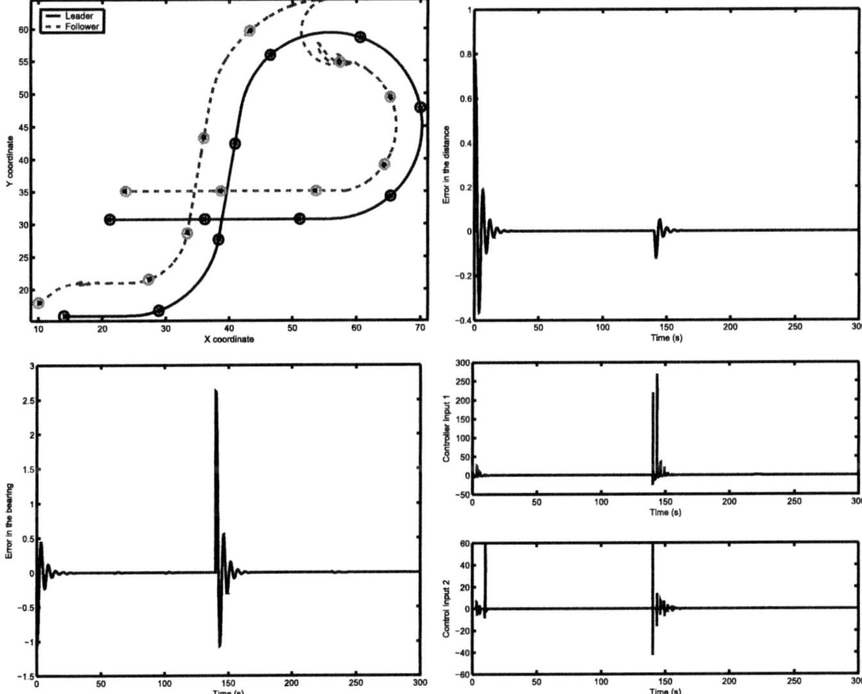

Fig. 4. Example 2. Top left: leader-follower trajectory; top right: separation error; bottom left: relative bearing error; bottom right: control inputs.

for a given time buffer. During the planning process, we will use the protected zone as an abstraction of the agent itself.

For the planar case, the dynamics equations of motion for the ith agent in a n-robot group are given by

$$M_i(q_i)\ddot{q}_i + h_i(q_i, \dot{q}_i) = u_i + \sum_{j=1}^{k} W_{ij} F_{ij}, \qquad i = 1 \ldots n, \qquad (18)$$

where $q_i \in \Re^2 \times \mathcal{S}$ is the vector of generalized coordinates for the ith robot, M_i is an 3×3 positive definite symmetric inertia matrix, $h_i(q_i, \dot{q}_i)$ is a 3×1 vector of nonlinear inertial forces, and u_i is the 3×1 vector of applied (external) forces and torques that can be provided through the local controller. k is the number of the contacts between the ith agent and all other objects that could be either obstacles or other robots. $F_{ij} = (F_{N,ij} \ F_{T,ij})^T$ is a 2×1 vector of contact forces corresponding to the jth contact, and $W_{ij} \in \Re^{3 \times 2}$ is the Jacobian matrix that relates the velocity at the jth contact point to the time derivatives of the generalized coordinates of the agent. For the time being, we will assume that nonholonomic constraints are not present.

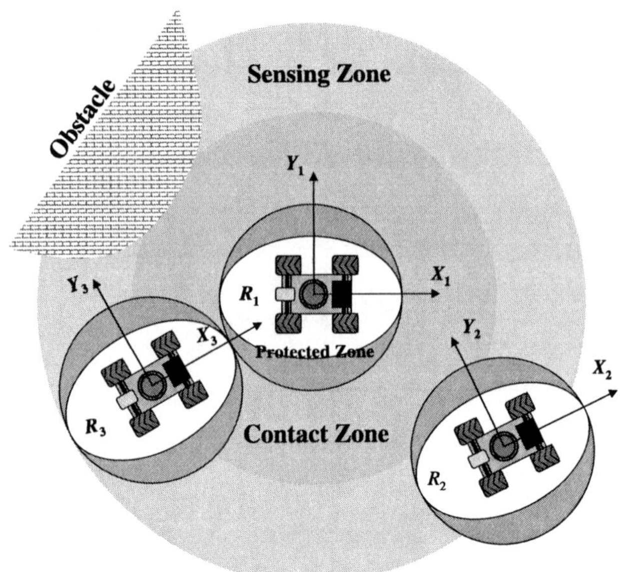

Fig. 5. Zones for the computation of contact response.

We adopt a state-variable-based compliant contact model described in [23] to compute the contact forces. At the jth contact of the agent i, the normal and tangential contact forces $F_{N,ij}$ and $F_{T,ij}$ are given by

$$F_{N,ij} = f_N(\delta_{N,ij}) + g_N(\delta_{N,ij}, \dot{\delta}_{N,ij}), \quad j = 1,\ldots,k, \tag{19}$$

$$F_{T,ij} = f_T(\delta_{T,ij}) + g_T(\delta_{T,ij}, \dot{\delta}_{T,ij}), \quad j = 1,\ldots,k, \tag{20}$$

where the functions f_N and f_T are the elastic stiffness terms and g_N and g_T are the damping terms in the normal and tangential directions, respectively. Similar to handling rigid body contact, these functions can be designed to adjust the response of the robot. $\delta_{N,ij}(q)$ and $\delta_{T,ij}(q)$ are the local normal and tangential deformations, which can be uniquely determined by the generalized coordinates of the system. The details and variations on the compliant contact model are discussed in [23]. A key feature of this model is that it allows us to resolve the ambiguous situations when more than three objects came into contact with one robot.

Figures 6–7 illustrate our approach by an example in which two teams of robots try to arrange themselves around their goals. The grouping is done dynamically using a decentralized decision-making process. A quadratic well type of potential function [15, 16] is constructed to drive the robots toward the goal. The expression of the potential function is given by

$$\phi(q) = \frac{k_p}{2}(q - q_g)^T(q - q_g), \tag{21}$$

where q_g is the coordinates of the goal. The input u_i for the ith agent can be obtained by the gradient of the potential function

$$u_i = -\nabla\phi(q_i) = -k_p(q - q_g),\qquad(22)$$

which is a proportional control law. Asymptotic stabilization can be achieved by adding dissipative forces [15].

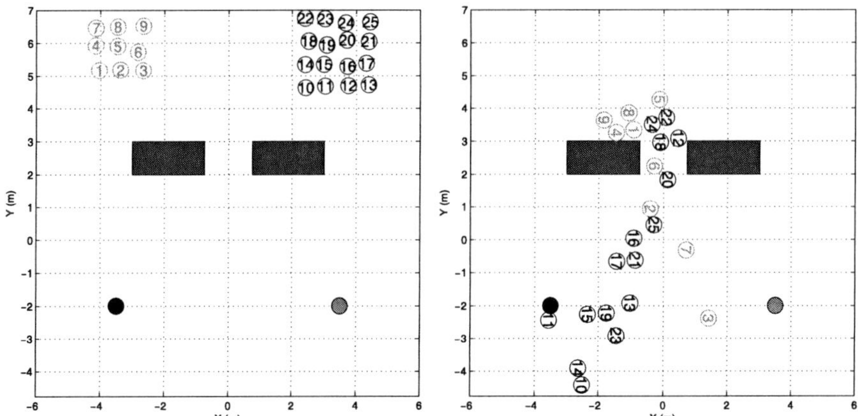

Fig. 6. Two teams of robots. Left: initial configuration; right: intermediate configuration.

The nonholonomic nature of most autonomous robots requires substantial care when developing the local level controllers [2, 4]. The dynamic model for a car-like nonholonomic agent can be expressed as

$$M_i(q_i)\ddot{q}_i + h_i(q_i,\dot{q}_i) = B(q_i)u_i - A_i(q_i)^T\lambda_i + \sum_{j=1}^{k} W_{ij}F_{ij},\quad i = 1\ldots n,\quad(23)$$

where $q_i \in SE(2)$, and B is an input transformation matrix. λ is the constraint force due to the following nonholonomic constraints:

$$A_i(q_i)\dot{q}_i = 0,\ \text{ or }\ \dot{q}_i = S_i(q_i)v_i.\qquad(24)$$

We can project the contact forces $\sum_{j=1}^{k} W_{ij}F_{ij}$ onto the reduced space while eliminating the constraint forces $A_i(q_i)^T\lambda_i$ in (23). The complete dynamics of the reduced system are given by

$$\begin{cases} \dot{q} = Sv \\ \dot{v} = \bar{M}^{-1}(MS^TBu + S^T\sum_{j=1}^{k} W_jF_j - SM\dot{S}v - S^Th), \end{cases}\qquad(25)$$

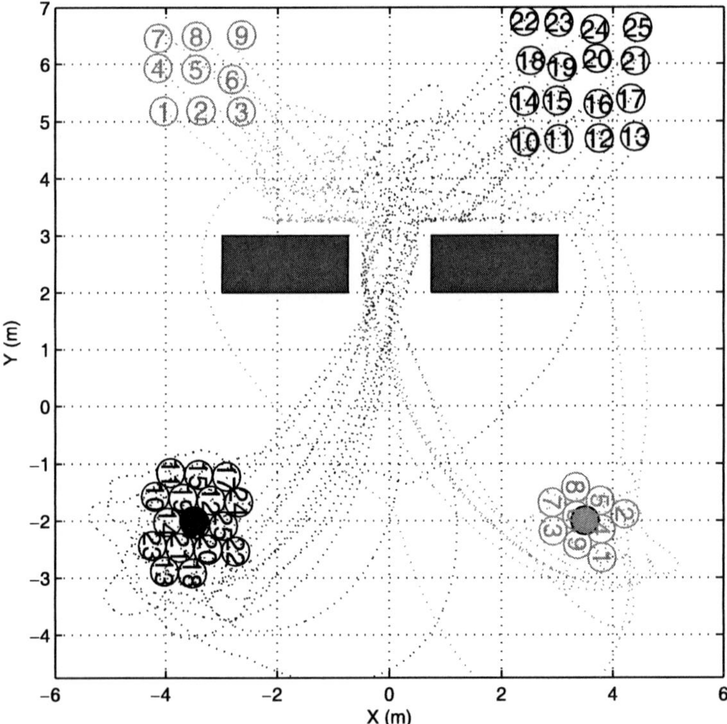

Fig. 7. Two teams of robots arrange themselves around the assigned destinations using the decentralized trajectory generation scheme.

where $\bar{M} = S^T M S \in \Re^{2\times2}$ is a symmetric, positive definite matrix. Note that the index i in the above equations is omitted for the sake of simplicity. We use the dynamic feedback linearization scheme presented in Section 3 and the *backstepping* technique developed in [10] to generate a control law u that gives exponentially convergent solutions for the state variables $(q,\ v)$. The projected contact forces are treated as external disturbances during this process. We illustrate this approach by an example in which four nonholonomic mobile robots are commanded to a goal position in an unknown environment. In this experiment, each robot runs its own trajectory generator and controller. As can be seen in Figure 9, the robots are able to follow the generated trajectories and navigate and eventually group themselves at the goal position.

In the next scenario, we use same robot model described to coordinate cooperative manipulation tasks as seen in Figure 8, in which a team of robots tries to reach and surround a target and eventually transport the target to a desired destination. In this case, the robots are running on three control modes—*approach*, *organization*, and *transportation*. In the *approach* mode, the robots swarm to the object by following an attractive potential field cen-

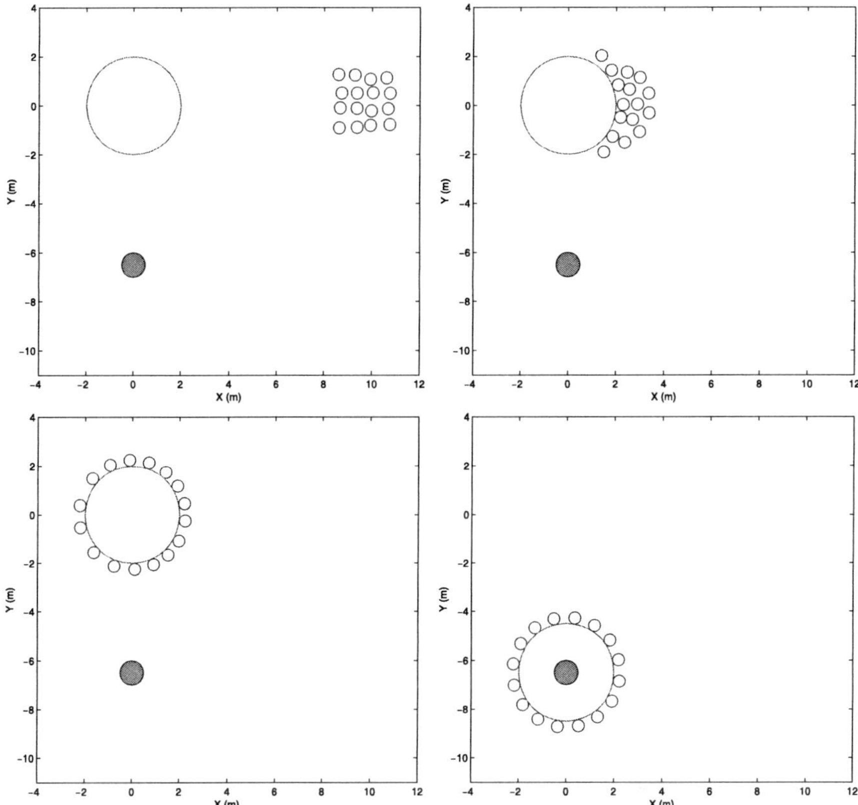

Fig. 8. Decentralized controllers are used to get nonholonomic robots to surround a target. Each robot uses one of the three different controllers (modes) depending on information about the neighbor's state.

tered at object location. After reaching the object, each robot (independently) enters into an *organization* mode where it moves away from its neighbors while staying near the object. This is done by stacking a repulsive potential onto the existing approach potential to redistribute the robots. The repulsive potential is designed to organize the robots into the desired formation, trapping the object in the process. Each robot autonomously transitions into the final *transportation* mode after it senses a quorum. In this phase, an added transportation potential similar to the one used in the approach phase, but with a much lower intensity and centered at the destination location, attracts the robots and the object to the goal position. See [24] for the details of the potential fields and the three control modes. In each control mode, the gradient of the corresponding potential fields and an appropriately designed dissipative function provide the driving force to the robots and the trajectory is calculated by simulating the dynamics of the system.

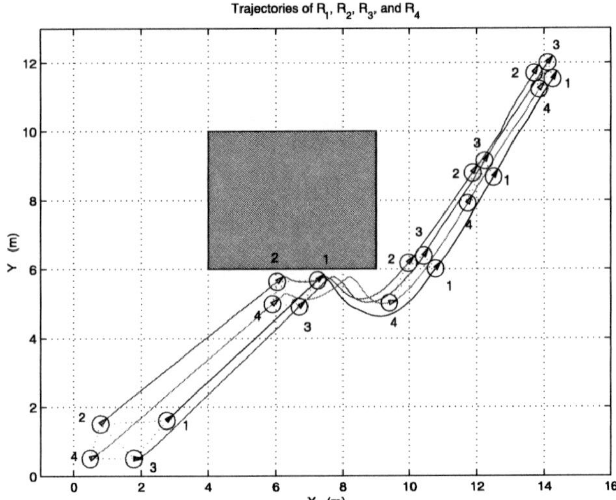

Fig. 9. Grouping of nonholonomic robots.

We illustrate our approach by examples in which teams with different number of robots are commanded to transport an object to a goal location within an unknown environment as shown in Figure 10.

In all these scenarios, the computation of the trajectory for each robot is based on information that is available to it through its sensors or through the communication network. The relevant information is the relative state information of other robots and obstacles in the contact zone. Collisions can occur only if this information is not available, either because of faulty sensors or failed communication channels. Each robot runs a simulator of the world and the same algorithm for computing trajectories. Thus, penetration of two contact zones, for example, will result in both robots being bounced away from each other with equal and opposite contact forces. Obstacles do not have viscoelastic shells. However, obstacles are stationary. Thus the contact zone must be sized and the properties of the viscoelastic shell must be carefully selected. Even if there is a head-on confrontation with an obstacle, it should be possible to completely dissipate the energy of the robot and allow it come to a complete standstill without causing its protected zone to touch the obstacle.

5 Conclusions

After solving some of the basic problems in single-team-robot control, the natural next step is to address problems in the coordination of multiple teams of cooperative robotic systems that are critical to applications requiring concurrent operation of robots, distributed sensing, parallel execution, and team

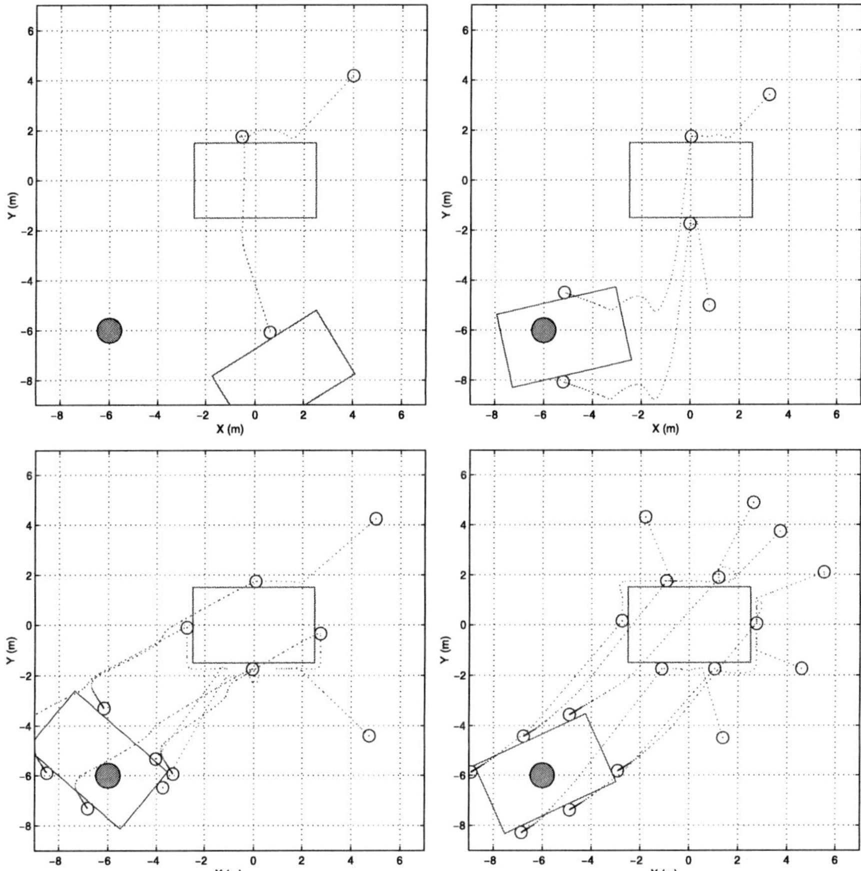

Fig. 10. Sample trials by using decentralized controllers to get teams of one, two, four, or six robot(s) to surround an object and transport the object to a desired destination.

interaction. We show that a contact dynamics-based model can lead to decentralized trajectory generation schemes and we demonstrate through simulated examples that the decentralized framework is applicable for controlling and coordinating teams of intelligent robots with independent goals. We assume that each robot has approximate information about the object position, its goal position, and the number of team members, and is equipped with an omnidirectional range sensor. The sensor has a limited range, but the robots can see the neighbors in this range. Because the framework is decentralized at both trajectory generation level and the estimation and control agent level, our framework can easily scale to any number (tens and hundreds) of vehicles and is flexible enough to support many formation shapes and mission requirements.

Currently, we are implemented the algorithms presented herein on the MARHES experimental testbed.[2]

Acknowledgements. We thank Ghulam M. Hassan for his assistance in simulating the dynamic feedback linearization controller.

References

1. Baras JS, Tan X, Hovareshti P (2003) Decentralized control of autonomous vehicles 1532–1537. In: Proc. IEEE Conf. on Decision and Control, Maui, HI
2. Barraquand J, Latombe J (1993) Non-holonomic multibody mobile robots: controllability and motion planning in the presence of obstacles. Algorithmica 10:121–155
3. Belta C, Kumar V (2004) Abstraction and control for groups of robots. IEEE Trans. on Robotics and Automation 20(5):865–875
4. Bemporad A, De Luca A, Oriolo G (1996) Local incremental planning for a car-like robot navigating among obstacles 1205–1211, In: Proc. IEEE Int. Conf. on Robotics and Automation, Minneapolis, MN
5. Butenko S, Murphey R, Pardalos P (eds) (2003) Cooperative control: models, applications and algorithms, vol. 1 of Applied optimization. Kluwer Academic Publishers, Dordrecht
6. Chaimowicz L, Kumar V, Campos M (2004) A paradigm for dynamic coordination of multiple robots. Autonomous Robots 17(1):7–21
7. Das AK, Fierro R, Kumar V, Ostrowski JP, Spletzer J, Taylor CJ (2002) A vision-based formation control framework. IEEE Trans. on Robotics and Automation 18(5):813–825
8. De Luca A, Oriolo G, Samson C (1998) Feedback control of a nonholonomic car-like robot In: Laumond J.-P (ed), Robot motion planning and control, 171–253. Springer-Verlag, London
9. Fierro R, Clark J, Hougen D, Commuri S (2005) A multi-robot testbed for bilogically inspired cooperative control In: Schultz A, Parker L, Schneider F (eds), Multi-robot systems: from swarms to intelligent automata, Naval Research Laboratory, Washington, DC
10. Fierro R, Lewis FL (1997) Control of a nonholonomic mobile robot: backstepping kinematics into dynamics. J. Robotic Systems 14(3):149–163
11. Fierro R, Song P, Das AK, Kumar V (2002) Cooperative control of robot formations In: Murphey R, Pardalos P (eds), Cooperative control and optimization, vol. 66 of Applied optimization, chapter 5, 73–93. Kluwer Academic Publishers, Dordrecht
12. Gerkey BP, Matarić MJ (2004) A formal analysis and taxonomy of task allocation in multi-robot systems. Int. J. Robot. Research 23(9):939–954
13. Ghosh R, Tomlin C (2000) Maneuver design for multiple aircraft conflict resolution 672–676. In: Proc. American Control Conference, Chicago, IL
14. Isidori A (1995) Nonlinear control systems, Springer-Verlag, London

---

[2]http://marhes.okstate.edu.

15. Khatib O (1986) Real-time obstacle avoidance for manipulators and mobile robots. International Journal of Robotics Research 5:90–98
16. Koditschek D (1987) Exact robot navigation by means of potential functions: some topological considerations 1–6. In: Proc. IEEE Int. Conf. Robot. Automat.
17. Kumar V, Leonard N, Morse A (eds) (2004) A post-workshop volume, 2003 Block Island Workshop on Cooperative Control Series, vol. 309 of LNCIS. Springer-Verlag, London
18. LaValle S, Hutchinson S (1998) Optimal motion planning for multiple robots having independent goals. IEEE Trans. on Robotics and Automation 14(6):912–925
19. Leonard NE, Fiorelli E (2001) Virtual leaders, artificial potentials and coordinated control of groups 2968–2973. In: Proc. IEEE Conf. on Decision and Control, Orlando, FL
20. Ögren P, Fiorelli E, Leonard N (2004) Cooperative control of mobile sensor networks: adaptive gradient climbing in a distributed environment. IEEE Trans. on Automatic Control 49(8):1292–1302
21. Okubo A (1985) Dynamical aspects of animal grouping: swarms, schools, flocks and herds. Advances in Biophysics 22:1–94
22. Oriolo G, De Luca A, Vendittelli M (2002) WMR control via dynamic feedback linearization. IEEE Trans. on Robotics and Automation 10(6):835–852
23. Song P, Kraus P, Kumar V, Dupont P (2001) Analysis of rigid-body dynamic models for simulation of systems with frictional contacts. ASME Journal of Applied Mechanics 68:118–128
24. Song P, Kumar V (2002) A potential field based approach to multi-robot manipulation 1217–1222. In: Proc. IEEE Int. Conf. Robot. Automat., Washington, DC
25. Spletzer J, Das A, Fierro R, Taylor CJ, Kumar V, Ostrowski JP (2001) Cooperative localization and control for multi-robot manipulation 631–636. In: IEEE/RSJ Int. Conf. on Intelligent Robots and Systems, Maui, HI

Part IV

Control of Electromechanical Systems

Transient Stabilization of Multimachine Power Systems

Martha Galaz,[1] Romeo Ortega,[1] Alessandro Astolfi,[2] Yuanzhang Sun,[3] and Tielong Shen[4]

[1] Lab. des Signaux et Systémes, Supelec, Plateau du Moulon, 91192 Gif-sur-Yvette, France. `{Martha.Galaz,Romeo.Ortega}@lss.supelec.fr`
[2] Electrical Engineering Department, Imperial College London, Exhibition Road, London SW7 2AZ, UK. `a.astolfi@imperial.ac.uk`
[3] Department of Electrical Engineering, Tsinghua University, Beijing 10084, China. `yzsun@mail.tsinghua.edu.cn`
[4] Department of Mechanical Engineering, Sophia University, Kioicho 7-1, Chiyoda-ku, Tokyo 102-8554, Japan. `tetu-sin@sophia.ac.jp`

Summary. In this chapter we provide a solution to the long-standing problem of transient stabilization of multimachine power systems with nonnegligible transfer conductances. More specifically, we consider the full $3n$-dimensional model of the n-generator system with lossy transmission lines and loads and prove the existence of a nonlinear static state feedback law for the generator excitation field that ensures asymptotic stability of the operating point with a well-defined estimate of the domain of attraction provided by a bona fide Lyapunov function. To design the control law we apply the recently introduced interconnection and damping assignment passivity-based control methodology that endows the closed-loop system with a port-controlled Hamiltonian structure with desired total energy function. The latter consists of terms akin to kinetic and potential energies, thus has a clear physical interpretation. Our derivations underscore the deleterious effects of resistive elements that, as is well known, hamper the assignment of simple "gradient" energy functions and compel us to include nonstandard cross terms. A key step in the construction is the modification of the energy transfer between the electrical and the mechanical parts of the system, which is obtained via the introduction of state-modulated interconnections.

1 Introduction

Traditional analysis and control techniques for power systems have undergone a major reassessment in recent years—see [10] for an excellent tutorial. This worldwide trend is driven by multiple factors including the adoption of new technologies, like flexible AC transmission systems, which offer improvements in power angle and voltage stability but give rise to many unresolved modeling and control issues. Also, the ever-increasing utilization of power electronic converters is drastically modifying the energy consumption profile, as well as the underlying distributed generation. The new deregulated market, on the other hand, has seen the emergence of separate entities for generation that impose more stringent requirements on the dynamic behavior of voltage regulated units and the task of coordinating a large number of (small and large) active and reactive control units in the face of significant load uncertainty. It is, by now, widely recognized that the existing methods and tools to approach power systems should be revisited to ensure reliable and secure planning—the recent dramatic blackouts in North America and Italy provide compelling evidence of this fact.

In this chapter we study the fundamental problem of *transient stability* of power systems whose reliable assessment has become a major operating constraint, particularly in regions that rely on long distance transfers of bulk power. Transient stability is concerned with a power system's ability to reach an acceptable steady state following a fault, e.g., a short circuit or a generator outage, that is later cleared by the protective system operation, see [1, 11, 12, 24] and the tutorial paper [3] for more details. The fault modifies the circuit topology—driving the system away from the stable operating point—and the question is whether the trajectory will remain in the basin of attraction of this (or other) equilibrium after the fault is cleared. The key analysis issue is then the evaluation of the domain of attraction of the system's operating equilibrium, while the *control objective* is the enlargement of the latter.

Transient stability *analysis* dates back to the beginning of the electric age [4] with the problem originally studied via numerical integration and, starting in the 1947 seminal paper [14], with Lyapunov-like methods. A major open problem in this area is the derivation of Lyapunov functions for transmission systems that are *not lossless*, i.e., with transfer conductances between busses.[1] While the transmission system itself can be modeled as being lossless without loss of accuracy, the classical network reduction of the load busses induces transfer conductances between the rest of the system busses, rendering the negligible transfer conductances assumption highly unsatisfactory [1, 12]. Although considerable efforts have been made to find Lyapunov functions for lossy line systems, to the best of the authors' knowledge this research has unfortunately been in vain. In [17] it is claimed that, even for the simple swing equation model,[2] the standard energy function of a lossless system cannot be extended (in general) to a lossy system. (See also [27] for some surprising results obtained via local stability analysis.)

Our interest is in the design of *excitation controllers* to enhance transient stability. These controllers are proposed to replace the traditional automatic

---

[1]More precisely, the conductances represent partial losses caused by the line and the loads in the nodes. For the sake of simplicity, we say that the line is lossless or lossy if the conductances are neglected or not, respectively.

[2]The situation for the more realistic flux-decay model [1, 12] which we consider in this chapter, is of course much more complicated.

voltage regulator (AVR) plus power system stabilizer (PSS) control structure. Questions about the benefits of this replacement have not yet been answered. Given the highly nonlinear nature of power system models, the applicability of linear controller design techniques for transient stability enhancement is severely restricted. On the other hand, the application of more promising nonlinear control methods has attracted much attention in the literature, with feedback linearization being one of the early strategies to be explored [9, 32, 15, 12]. The well-known robustness problems, both against parameter uncertainties and unmodeled dynamics, of these nonlinearity cancellation schemes have motivated the more recent works on *passivity-based* techniques [19, 30]. Most results along these lines are based on the application of damping injection (also called $L_g V$) controllers (see [13, 16, 8].) See also [21] for an alternative passivation approach. In [2] a dynamic damping injection controller is proposed which is proven to enlarge the estimate of the region of attraction and is shown (in simulation studies) to increase critical clearing times for a single machine infinite bus (SMIB) system with lossless transmission lines.

Attention has been given also to passivity-based methods that rely on *port-controlled Hamiltonian* (PCH) descriptions of the system [30, 22, 23, 6]. As explained in Section 3, these techniques go beyond $L_g V$ schemes endowing the closed-loop system with a PCH structure, with stability of the desired equilibrium established assigning an energy function that qualifies as a Lyapunov function. In [29], we exploit the fact that the lossless SMIB open-loop system is in PCH form, to give conditions for a *constant* field control action to shape the energy function. In [26], again for this class of systems, we add an adaptive $\mathcal{L}_2$-disturbance attenuation controller (which belongs to the class of $L_g V$ controllers). An energy function for the *multimachine* case was first suggested in [28], where a domination design is used to cope with the effect of the losses. In the recent interesting paper [31] the existence of a static state feedback that assigns a PCH structure—using the same energy function and still retaining the lossless assumption—is established. This result is important because it paves the way to additionally apply, in the spirit of [26], an $\mathcal{L}_2$-disturbance attenuation controller to this "PCH-ized" system. Unfortunately, neither one of these papers proves that the energy function qualifies as a Lyapunov function, hence the stability of the desired equilibrium, which is the issue of main concern in transient stability studies, is left unclear.[3]

To the best of our knowledge, even in the lossless case, the problem of designing a state-feedback controller that ensures asymptotic stability of the desired equilibrium point for multimachine systems remains open. The main contribution of this chapter is to provide an affirmative answer to this problem for the lossy case. Our work is the natural extension of [7] where, for the lossless

---

[3]In [31] this claim is established if we make the assumption that the load angles remain within $(-\pi, \pi)$. Even though this is true for the open-loop system, which lives in the torus, this structure is destroyed by the control, rendering the assumption a priori unverifiable.

SMIB system, we propose a state-feedback controller that effectively shapes the total energy function and enlarges the domain of attraction. As in [7] the control law is derived applying the recently introduced interconnection and damping assignment passivity-based control (IDA-PBC) methodology [23, 18]. (See also the recent tutorials [22, 18] and the closely related work [6].) The parameterization of the energy function, motivated by our previous works on mechanical [20] and electromechanical systems [25], consists of terms akin to kinetic and potential energies and thus has a clear physical interpretation. Our derivations underscore the deleterious effects of *transfer conductances* which, as is well known, hamper the assignment of simple "gradient" energy functions [17, 3] and compel us to include nonstandard cross terms. As usual in IDA-PBC designs, a key step in the construction is the modification of the energy transfer between the electrical and the mechanical parts of the system, which is obtained via the introduction of state-modulated interconnections, which play the role of multipliers in classical passivity theory [5]. As a byproduct of our derivations we also present the first "globally"[4] asymptotically stabilizing law for the lossy SMIB system.

The remainder of the chapter is organized as follows. The problem is formulated in Section 2. In Section 3 we briefly review the IDA-PBC technique. We give the slightly modified version presented in [18] where the open-loop system is not in PCH form, as is the case for the problem at hand. In Section 4, we apply the method to the lossy SMIB system where we show that a separable Lyapunov function, which ensures "global" stability, can be assigned. In Section 5 we show that, for the two-machine case, the presence of transfer conductances hampers the assignment of a separable Lyapunov function, therefore, cross-term must be included. The general n-machine case is treated in Section 6. A simulation study is presented in Section 7 showing the enlargement of the domains of attraction and its effect on the enhancement of critical clearing times for a two-machine system. We conclude with some final remarks in Section 8.

2 Model and Problem Formulation

We consider the problem of transient stabilization of a large-scale power system consisting of n generators interconnected through a transmission network that we assume is lossy; that is, we explicitly take into account the presence of *transfer conductances*. The dynamics of the ith machine with excitation are represented by the classical three-dimensional flux decay model ((9) of [3], or (6.47) of [12]):

---

[4]The qualifier "global" is used here in the sense that we provide an estimate of the domain of attraction that contains all the operating region of the system.

$$\dot{\delta}_i \;\; = \omega_{i0}\omega_{Mi}$$

$$M_i\dot{\omega}_{Mi} = -D_{Mi}\omega_{Mi} + P_{mi} - G_{Mii}E_{qi}^{'2}$$

$$-E_{qi}' \sum_{j=1,j\neq i}^{n} E_{qj}' \{G_{Mij}\cos(\delta_i - \delta_j) + B_{Mij}\sin(\delta_i - \delta_j)\}$$

$$T_{di}\dot{E}_{qi}' = -[1 - B_{Mii}(x_{di} - x_{di}')]E_{qi}' + E_{fsi} + u_{fi} -$$

$$(x_{di} - x_{di}') \sum_{j=1,j\neq i}^{n} E_{qj}' \{G_{Mij}\sin(\delta_i - \delta_j) - B_{Mij}\cos(\delta_i - \delta_j)\}\,.$$

The state variables of this subsystem are the rotor angle δ_i, the rotor speed ω_{Mi}, and the quadrature axis internal voltage E_{qi}'; hence the overall system is of dimension $3n$.[5] The control input is the field excitation signal u_{fi}. The parameters $G_{Mij} = G_{Mji}$, $B_{Mij} = B_{Mji}$, and G_{Mii} are, respectively, the conductance, susceptance, and self-conductance of the generator i. E_{fsi} represents the constant component of the field voltage and P_{mi} the mechanical power, which is assumed to be constant. The parameters $x_{di}, x_{di}', \omega_{i0}$ and D_{Mi} represent the direct-axis—synchronous and transient—reactances, the synchronous speed and damping coefficient, respectively. We note that all parameters are positive and $x_{di} > x_{di}'$. See [24, 12] for further details on the model.

Using the following identities:

$$G\cos\delta + B\sin\delta = Y\sin(\delta + \alpha) \qquad G\sin\delta - B\cos\delta = -Y\cos(\delta + \alpha) \quad (1)$$

with $Y^2 = G^2 + B^2$ and $\alpha = \arctan(G/B)$, and introducing the parameters
$$E_{fi} \triangleq \tfrac{E_{fsi}}{T_{di}}; \;\; u_i \triangleq \tfrac{u_{fi}}{T_{di}}; \;\; D_i \triangleq \tfrac{D_{Mi}}{M_i}; \;\; P_i \triangleq \tfrac{P_{mi}\omega_{i0}}{M_i}; \;\; G_{ij} \triangleq \tfrac{G_{Mij}\omega_{i0}}{M_i}; \;\; B_{ij} \triangleq$$
$$\tfrac{B_{Mij}\omega_{i0}}{M_i}, \omega_i \triangleq \omega_{i0}\omega_{Mi},$$ we can rewrite the system in the more compact form:

$$\dot{\delta}_i = \omega_i$$

$$\dot{\omega}_i = -D_i\omega_i + P_i - G_{ii}E_i^2 - E_i \sum_{j=1,j\neq i}^{n} E_j Y_{ij}\sin(\delta_i - \delta_j + \alpha_{ij})$$

$$\dot{E}_i = -a_iE_i + b_i \sum_{j=1,j\neq i}^{n} E_j\cos(\delta_i - \delta_j + \alpha_{ij}) + E_{fi} + u_i\,, \qquad (2)$$

where, in order to simplify the notation, we use E_i instead of E_{qi}', and we have defined

$$Y_{ij} \triangleq \sqrt{G_{ij}^2 + B_{ij}^2} \qquad\qquad \alpha_{ij} \triangleq \arctan\frac{G_{ij}}{B_{ij}}$$
$$a_i \triangleq \frac{1}{T_{di}}[1 - B_{Mii}(x_{di} - x_{di}')] \qquad b_i \triangleq \frac{x_{di} - x_{di}'}{T_{di}}Y_{ij}\,. \qquad (3)$$

Observe that $a_i, b_i > 0$, $\alpha_{ij} = \alpha_{ji}$ and that, *if* $M_i = M_j$, $Y_{ij} = Y_{ji}$. This assumption will be made throughout to simplify the derivations, see Remark

---

[5]In the following we will consider that the full system consists of the interconnection of n subsystems of dimension 3, and talk about the ith subsystem only.

5. We underscore that if $G_{ij} = 0$ then $\alpha_{ij} = 0$. As shown below, this ubiquitous assumption considerably simplifies the stabilization problem. (It is convenient at this point to make the following clarification: even in the case when the line is lossless, the classical reduction of transmission lines and *load buses* will lead to a reduced model with $G_{ij} \neq 0$. To simplify the notation we will refer to the case $G_{ij} = 0 \ (\neq 0)$ as lossless (resp., lossy) network cases.)

Problem Formulation Assume the model (2), with $u_i = 0$, has a stable equilibrium point at $[\delta_{i*}, 0, E_{i*}]$, with $E_{i*} > 0$.[6] Find a control law u_i such that in closed loop

- an operating equilibrium is preserved,
- we have a Lyapunov function for this equilibrium and,
- it is asymptotically stable with a well-defined domain of attraction.

Two additional requirements are that the domain of attraction of the equilibrium is enlarged by the controller and that the Lyapunov function has an energy-like interpretation.

A detailed analysis of the equilibria of (2) is clearly extremely involved—even in the two-machine case. To formulate our claims we will make some assumptions on these equilibria that will essentially restrict $|\delta_{i*} - \delta_{j*}|$ to be small, a scenario that is reasonable in practical situations. We will also sometimes assume that the line conductances are sufficiently small.

3 Interconnection and Damping Assignment Control

To solve this problem we will use the IDA-PBC methodology proposed in [23], see also [22]. IDA-PBC is a procedure that allows us to design a static state feedback that stabilizes the equilibria of nonlinear systems of the form

$$\dot{x} = f(x) + g(x)u, \tag{4}$$

where $x \in \mathbb{R}^n$ is the state vector and $u \in \mathbb{R}^m, m < n$, is the control action, endowing the closed loop with a port-controlled Hamiltonian (PCH) structure[7]

$$\dot{x} = [J_d(x) - \mathcal{R}_d(x)]\nabla H_d, \tag{5}$$

where the matrices $J_d(x) = -J_d^\top(x)$ and $\mathcal{R}_d(x) = \mathcal{R}_d^\top(x) \geq 0$, which represent the desired interconnection structure and dissipation, respectively, are selected by the designer—hence the name IDA—and $H_d : \mathbb{R}^n \to \mathbb{R}$ is the desired total stored energy. If the latter has an isolated minimum at the desired equilibrium $x_\star \in \mathbb{R}^n$, that is, if

---

[6] Because of physical constraints E_i is restricted to be positive.

[7] We note that all vectors are *column* vectors, even the gradient of a scalar function. We use the notation $\nabla_x = \frac{\partial}{\partial x}$, when it is clear from the context the subindex in ∇ will be omitted.

$$x_\star = \arg\min H_d(x)\,, \tag{6}$$

then $x_\star$ is stable with Lyapunov function $H_d(x)$. As stated in the simple proposition below [18], the admissible energy functions are characterized by a parameterized partial differential equation (PDE).

Proposition 1. *Consider the system (4). Assume there exists matrices* $J_d(x) = -J_d^\top(x), \mathcal{R}_d(x) = \mathcal{R}_d^\top(x) \geq 0$ *and a function* $H_d(x)$, *which satisfies (6), such that the PDE*

$$g^\perp(x)f(x) = g^\perp(x)[J_d(x) - \mathcal{R}_d(x)]\nabla H_d \tag{7}$$

is solved, where $g^\perp(x)$ *is a left annihilator of* $g(x)$, *i.e.,* $g^\perp(x)g(x) = 0$. *Then, the closed-loop system (4) with*

$$u = [g^\top(x)g(x)]^{-1}g^\top(x)\{[J_d(x) - \mathcal{R}_d(x)]\nabla H_d - f(x)\},$$

will be a PCH system with dissipation of the form (5) with $x_\star$ *a (locally) stable equilibrium. It will be asymptotically stable if, in addition, the largest invariant set under the closed-loop dynamics (5) contained in*

$$\left\{x \in \mathbb{R}^n \mid (\nabla H_d)^\top \mathcal{R}_d(x)\nabla H_d = 0\right\} \tag{8}$$

equals $\{x_\star\}$. *An estimate of its domain of attraction is given by the largest bounded level set* $\{x \in \mathbb{R}^n \mid H_d(x) \leq c\}$.

A large list of applications of this method may be found in the recent tutorial paper [18]. In particular, it has been applied in [7] for transient stabilization of SMIB systems fixing $J_d(x)$ and $\mathcal{R}_d(x)$ to be constant and solving the PDE (7) for $H_d(x)$. Inspired by [6], where the energy function is fixed and then a set of algebraic equations for $J_d(x)$ and $\mathcal{R}_d(x)$ are solved, we applied this variant of the method for a class of electromechanical systems in [25] with a quadratic in increments desired energy function. In [20] these two extremes were combined for application in mechanical systems. Namely, we fixed $H_d(x)$ to be of the form

$$H_d(q,p) = \frac{1}{2}p^\top M_d^{-1}(q)p + V_d(q),$$

where the state $x = (q,p)$ consists of the generalized position and momenta, and $M_d(q) = M_d^\top(q) > 0$ and $V_d(q)$ represent the (to be defined) closed-loop inertia matrix and potential energy function, respectively. PDEs for $M_d(q)$ and $V_d(q)$ were then established and solved in several examples selecting suitable functions for $J_d(q,p)$.

We also proceed as in [20] for the system (2), fix the kinetic energy of the total energy function as $\frac{1}{2}\sum_{i=1}^n \omega_i^2$, and choose a quadratic function in the electrical coordinates E_i, which similarly to the matrix $M_d(q)$ above, is parameterized by a function of the angular positions δ_i. The proposed energy function is completed adding a potential-energy-like function of δ_i.

For the sake of clarity of presentation we will illustrate the procedure first with the simplest SMIB case.

Remark 1. We have concentrated here on the design of the energy shaping term of the controller. As usual in IDA-PBC, it is possible to add a damping injection term of the form $-K_{di}g^{\top}(x)\nabla H_d$, where $K_{di} = K_{di}^{\top} > 0$, that enforces the seminegativity of $\dot{H}_d$.

4 Single Machine System with Lossy Network

In the case $n = 1$ the model (2) reduces to the well-known SMIB system

$$\dot{\delta} = \omega$$
$$\dot{\omega} = -D\omega + P_m - GE^2 - EY\sin(\delta + \alpha)$$
$$\dot{E} = -aE + b\cos(\delta + \alpha) + E_f + u\,, \tag{9}$$

where we have introduced some obvious simplifying notation. We underscore the fact that, due to the presence of losses in the line, there appears a quadratic term GE^2 and a phase α in the trigonometric functions. Compare with (1) of [7]. We are interested in the behavior of the system in the set $\mathcal{D} = \{(\delta, \omega, E) \mid \delta \in [-\frac{\pi}{2}, \frac{\pi}{2}], E > 0\}$. Mimicking the derivations in [7] it is possible to show that inside this set there is a stable equilibrium that we denote $(\delta_\star, 0, E_\star)$, and the corresponding Lyapunov function provides estimates of its domain of attraction which are strictly contained in $\mathcal{D}$. The objective is to design a control law that assigns a Lyapunov function which provides larger estimates of the domain of attraction of this equilibrium.

As will be shown below, in this particular case we will be able to assign a separable energy function, more precisely a function of the form

$$H_d(\delta, \omega, E) = \frac{1}{2}\omega^2 + \psi(\delta) + \frac{\gamma}{2}(E - E_\star)^2\,, \tag{10}$$

where $\gamma > 0$ is some weighting coefficient and $\psi(\delta)$ is a potential-energy-like function that should satisfy $\delta_\star = \arg\min\psi(\delta)$. Our choice of energy function candidate (10), together with the first equation of (9), fixes the first row and consequently the first column, and the $(2,2)$ term of the matrix $J_d(\delta, E) - \mathcal{R}_d$. Hence, we propose

$$J_d(\delta, E) = \begin{bmatrix} 0 & 1 & 0 \\ -1 & 0 & J_{23}(\delta, E) \\ 0 & -J_{23}(\delta, E) & 0 \end{bmatrix}, \quad \mathcal{R}_d = \begin{bmatrix} 0 & 0 & 0 \\ 0 & D & 0 \\ 0 & 0 & r \end{bmatrix},$$

where $J_{23}(\delta, E)$ is a function to be determined and $r > 0$ is a constant damping injection gain. According to Proposition 1, the PDE to be solved is

$$-\nabla_\delta H_d - D\nabla_\omega H_d + J_{23}(\delta, E)\nabla_E H_d = -D\omega + P_m - GE^2 - EY\sin(\delta + \alpha),$$

which, upon replacement of (10), yields the ordinary differential equation

$$-\nabla\psi + \gamma J_{23}(\delta, E)(E - E_\star) = P_m - GE^2 - EY\sin(\delta + \alpha). \tag{11}$$

Evaluating (11) at $E = E_\star$, we obtain

$$\nabla\psi = -P_m + GE_\star^2 + E_\star Y\sin(\delta + \alpha).$$

Replacing this expression back in (11) we can compute $J_{23}(\delta, E)$ as

$$J_{23}(\delta, E) = -\frac{1}{\gamma}[(E + E_\star)G + Y\sin(\delta + \alpha)]. \tag{12}$$

From the construction above it is clear that $\nabla\psi(\delta_\star) = 0$, therefore to ensure the minimum condition (6) and to estimate the domain of attraction, it only remains to study the second derivative of $\psi(\delta)$, which is given by

$$\nabla^2\psi = E_\star Y\cos(\delta + \alpha) = E_\star(B\cos\delta - G\sin\delta),$$

where we have used the trigonometric identities (1) and the definitions (3) to obtain the last equation. Some simple calculations prove that the function $\psi(\delta)$ is strictly convex in the interval $(-\frac{\pi}{2}, \arctan\frac{B}{G})$, where we remark that the right-hand-term of the interval approaches $\frac{\pi}{2}$ as the line resistance G tends to 0.[8] We have established the following proposition.

Proposition 2. *The SMIB system with losses (9) in closed loop with the control law*

$$u = -b[\cos(\delta + \alpha) - \cos(\delta_\star + \alpha)] - J_{23}(\delta, E)\omega,$$

where $J_{23}(\delta, E)$ is given in (12) and γ is an arbitrary positive constant, ensures asymptotic stability of the desired equilibrium $(\delta_\star, 0, E_\star)$ with Lyapunov function

$$H_d(\delta, \omega, E) = \frac{1}{2}\omega^2 + \frac{\gamma}{2}(E - E_\star)^2 + (GE_\star^2 - P_m)\delta - E_\star Y\cos(\delta + \alpha),$$

and a domain of attraction containing the largest connected component inside the set

$$\left\{(\delta, \omega, E) \mid H_d(\delta, \omega, E) \le \max_{\delta \in (-\frac{\pi}{2},\, \arctan\frac{B}{G})} H_d(\delta, 0, E_\star)\right\},$$

which contains the equilibrium $(\delta_\star, 0, E_\star)$. In particular, if the system is lossless, then the control takes the simpler proportional-plus-derivative form

$$u = -\left(b + \frac{1}{\tilde{\gamma}}p\right)[\cos(\delta) - \cos(\delta_\star)],$$

where $p \triangleq \frac{d}{dt}$, $\tilde{\gamma} > 0$ is a free parameter, and the estimate of the domain of attraction above coincides with $\mathcal{D}$.

---

[8] Actually, in the case $n = 1$, we have that $B \gg G$ in most practical scenarios. Therefore, the interval is typically of the form $(-\frac{\pi}{2}, \frac{\pi}{2} - \epsilon]$, with $\epsilon > 0$ a small number.

Proof. The proof of stability is completed by computing the control law as suggested in Proposition 1, selecting the free coefficients to satisfy $r\gamma = a$, and using the equilibrium equations to simplify the controller expression. The Lyapunov function is obtained evaluating the integral of $\nabla\psi$ and replacing in (10). Asymptotic stability follows also from Proposition 1 noting, first, that the closed-loop system lives in the set[9] $[-\pi, \pi] \times \mathbb{R} \times \mathbb{R}$ and the energy function $H_d(\delta, \omega, E)$ is positive definite and proper throughout this set. Second, since $\dot{H}_d \leq 0$, we have that all solutions are bounded. Finally, we have that $\dot{H}_d = 0 \Rightarrow E = E_\star, \omega = 0$, and this in its turn implies $\delta = \delta_\star$.

The claim for the lossless transmission line is established by noticing that, in this case, $G = \alpha = 0$, reducing the control law to the expression given in the proposition (with $\tilde{\gamma} \triangleq \frac{\gamma}{B}$) and enlarging the estimate of the domain of attraction.

Remark 2. The construction proposed above should be contrasted with the one given in [7]. In the latter, $J_{23}(\delta, E)$ is fixed to be a constant and no particular structure is assumed a priori for $H_d(\delta, \omega, E)$. It turns out that, in this case, the energy function obtained from the solution of the PDE contains quadratic terms in the full state plus trigonometric functions of δ. A nice feature of this approach is that the resulting controller is linear. On the other hand, the estimate of the domain of attraction does not cover the whole operating region as proven here.

Remark 3. It has been mentioned before that, because of physical constraints, $E > 0$. The controller of Proposition 2 does not guarantee that this bound is satisfied; however, it can easily be modified to do so. To this aim, we propose instead of (10) a function that grows unbounded as E approaches zero, for instance,

$$H_d(\delta, \omega, E) = \frac{1}{2}\omega^2 + \psi(\delta) + \gamma[E - E_\star \log(E)].$$

Mimicking the derivations above we obtain the same function $\psi(\delta)$, but with a new interconnection term

$$J_{23}(\delta, E) = -\frac{E}{\gamma}[G(E + E_\star) + Y\sin(\delta + \alpha)],$$

that accordingly modifies the control.

5 Necessity of Nonseparable Energy Functions

In the previous section we have shown that for the single machine case it is possible to assign a separable Lyapunov function of the form (10). Before

---

[9]Notice that, in contrast to [32], the proposed controller is periodic in δ, hence leaves the system living in the torus.

considering the n-machine case we will show now that, already for the two-machine case, separable Lyapunov functions are not assignable via IDA-PBC, thus it is necessary to include cross-terms in the energy function. We then propose a procedure to add this cross-term for the two-machine system.

Caveat In the two-machine case the dynamics can be considerably simplified defining a new coordinate $\delta_1 - \delta_2$. This simplification is, however, of little interest in the general case. Since the derivations in this section will help to set up the notation that will be used for the general n-machine problem we avoid this simplification and consider all the equations of the system.

5.1 Separable Lyapunov Function for the Lossless Case

We consider a power system represented by two machines connected via a lossy transmission line. The dynamics of this system are obtained from (2) resulting in the sixth-order model[10]

$$
\begin{aligned}
\dot{\delta}_1 &= \omega_1 \\
\dot{\omega}_1 &= -D_1\omega_1 + P_1 - G_{11}E_1^2 - YE_1E_2\sin(\delta_1 - \delta_2 + \alpha) \\
\dot{E}_1 &= -a_1E_1 + b_1E_2\cos(\delta_1 - \delta_2 + \alpha) + E_{f1} + u_1 \\
\dot{\delta}_2 &= \omega_2 \\
\dot{\omega}_2 &= -D_2\omega_2 + P_2 - G_{22}E_2^2 + YE_1E_2\sin(\delta_1 - \delta_2 - \alpha) \\
\dot{E}_2 &= -a_2E_2 + b_2E_1\cos(\delta_2 - \delta_1 + \alpha) + E_{f2} + u_2 \, .
\end{aligned}
\tag{13}
$$

We want to investigate the possibility of assigning, via IDA-PBC, an energy function of the form

$$
H_d(\delta, \omega, E) = \psi(\delta) + \frac{1}{2}|\omega|^2 + \frac{1}{2}(E - E_*)^\top \Gamma(E - E_*),
$$

with $\delta = [\delta_1, \delta_2]^\top$, $\omega = [\omega_1, \omega_2]^\top$, $E = [E_1, E_2]^\top$, $E_\star = [E_{1*}, E_{2*}]^\top$, $|\cdot|$ the Euclidean norm and $\Gamma = \text{diag}\{\gamma_1, \gamma_2\} > 0$. Reasoning as in the single-machine case we propose

$$
J_d(\delta, E) - \mathcal{R}_d =
\begin{bmatrix}
0 & 1 & 0 & 0 & 0 & 0 \\
-1 & -D_1 & J_{23}(\delta, E) & 0 & 0 & J_{26}(\delta, E) \\
0 & -J_{23}(\delta, E) & -r_1 & 0 & J_{35}(\delta, E) & 0 \\
0 & 0 & 0 & 0 & 1 & 0 \\
0 & 0 & -J_{35}(\delta, E) & -1 & -D_2 & J_{56}(\delta, E) \\
0 & -J_{26}(\delta, E) & 0 & 0 & -J_{56}(\delta, E) & -r_2
\end{bmatrix}, \tag{14}
$$

---

[10]Using the property $\sin(x) = -\sin(-x)$ we have inverted the sign of the last term in the fifth equation to underscore the nice antisymmetry property that appears if $\alpha = 0$, which is lost in the lossy case. As will be proven below, this implies a "loss of integrability," which constitutes the main stumbling block for the assignment of energy functions.

where the $J_{ij}(\delta, E)$ are functions to be defined and $r_i > 0$.

The system of PDEs to be satisfied is

$$-\nabla_{\delta_1}\psi + J_{23}(\delta, E)\gamma_1(E_1 - E_{1*}) + J_{26}(\delta, E)\gamma_2(E_2 - E_{2*}) = F_1(\delta, E)$$
$$-\nabla_{\delta_2}\psi - J_{35}(\delta, E)\gamma_1(E_1 - E_{1*}) + J_{56}(\delta, E)\gamma_2(E_2 - E_{2*}) = F_2(\delta, E), \quad (15)$$

where we introduced the functions

$$F(\delta, E) = \begin{bmatrix} F_1(\delta, E) \\ F_2(\delta, E) \end{bmatrix} \stackrel{\triangle}{=} \begin{bmatrix} P_1 - G_{11}E_1^2 - YE_1E_2\sin(\delta_1 - \delta_2 + \alpha) \\ P_2 - G_{22}E_2^2 - YE_1E_2\sin(\delta_2 - \delta_1 + \alpha) \end{bmatrix}. \quad (16)$$

Now, evaluating (15) at $E = E_*$ we get $\nabla\psi = -\tilde{F}^{E_*}(\delta)$, where we have defined $\tilde{F}^{E_*}(\delta) \stackrel{\triangle}{=} F(\delta, E_*)$. Recalling Poincare's Lemma,[11] we see that there exists a scalar function $\psi : \mathbb{R}^2 \to \mathbb{R}$ such that the equation above holds if and only if

$$\nabla\tilde{F}^{E_*} = (\nabla\tilde{F}^{E_*})^\top. \quad (17)$$

Now,

$$\nabla_{\delta_1}\tilde{F}_2^{E_*} = YE_{1*}E_{2*}\cos(\delta_1 - \delta_2 - \alpha), \quad \nabla_{\delta_2}\tilde{F}_1^{E_*} = YE_{1*}E_{2*}\cos(\delta_1 - \delta_2 + \alpha),$$

where $\cos(x) = \cos(-x)$ was used now to invert the arguments in the first equation. These two functions are equal if and only if $\alpha = 0$—that is, if the line is lossless—concluding then that

- a separable Lyapunov function is assignable via IDA-PBC *if and only if* the line is lossless.

In this case, we can integrate $\nabla\psi$ to get

$$\psi(\delta) = -YE_{1*}E_{2*}[\sin(\delta_{1*} - \delta_{2*})(\delta_1 - \delta_2) + \cos(\delta_1 - \delta_2)],$$

where we invoked again the equilibrium equations to simplify the expression. The Hessian of $\psi(\delta)$ is positive semidefinite in the interval $|\delta_1 - \delta_2| \le \frac{\pi}{2}$. Hence, the proposed energy function will have a minimum at the desired equilibrium provided $|\delta_{1*} - \delta_{2*}| \le \frac{\pi}{2}$, and the IDA-PBC will ensure its asymptotic stability.

5.2 Lossy Line

If the line is not lossless we have to introduce a cross-term in the energy function and, reasoning like in [20, 25], propose to include a function $\lambda : \mathbb{R}^2 \to \mathbb{R}$ as

$$H_d(\delta, \omega, E) = \psi(\delta) + \frac{1}{2}|\omega|^2 + \frac{1}{2}[E - \lambda(\delta)E_*]^\top \Gamma[E - \lambda(\delta)E_*], \quad (18)$$

where $\lambda(\delta_*) = 1$. The PDEs now take the form

---

[11]Poincare's Lemma: given $f : \mathbb{R}^n \to \mathbb{R}^n$, $f \in C^1$ in $S \subset \mathbb{R}^n$. There exists $\psi : \mathbb{R}^n \to \mathbb{R}$ such that $\nabla\psi = f$ if and only if $\nabla f = (\nabla f)^\top$.

$$F_1(\delta, E) = -\nabla_{\delta_1}\psi + \gamma_1[E_1 - E_{1*}\lambda(\delta)]E_{1*}\nabla_{\delta_1}\lambda + \gamma_2[E_2 - E_{2*}\lambda(\delta)]E_{2*}\nabla_{\delta_1}\lambda +$$
$$+ J_{23}(\delta, E)\gamma_1[E_1 - E_{1*}\lambda(\delta)] + J_{26}(\delta, E)\gamma_2[E_2 - E_{2*}\lambda(\delta)]$$
$$F_2(\delta, E) = -\nabla_{\delta_2}\psi + \gamma_1[E_1 - E_{1*}\lambda(\delta)]E_{1*}\nabla_{\delta_2}\lambda + \gamma_2[E_2 - E_{2*}\lambda(\delta)]E_{2*}\nabla_{\delta_2}\lambda -$$
$$- J_{35}(\delta, E)\gamma_1[E_1 - E_{1*}\lambda(\delta)] + J_{56}(\delta, E)\gamma_2[E_2 - E_{2*}\lambda(\delta)], \qquad (19)$$

which evaluated at $E = \lambda(\delta)E_*$ yield

$$\nabla\psi = -F^{E_*}(\delta), \qquad (20)$$

with[12]

$$F^{E_*}(\delta) \triangleq F(\delta, \lambda(\delta)E_*). \qquad (21)$$

The problem is now to prove the existence of a function $\lambda(\delta)$ such that the integrability conditions of Poincare's Lemma, i.e., $\nabla_{\delta_1}F_2^{E_*} = \nabla_{\delta_2}F_1^{E_*}$, are satisfied. The identity above defines a PDE for $\lambda(\delta)$ that we could try to solve. With an eye on the general case, $n > 2$, when we have to deal with a set of PDEs that becomes extremely involved, we will proceed in an alternative way and directly construct this function as follows. First, we postulate that $\psi(\delta)$ is a function of $\delta_1 - \delta_2$, in which case $\nabla_{\delta_1}\psi + \nabla_{\delta_2}\psi = 0$, and consequently, invoking (20), we also have that $F_1^{E_*}(\delta) + F_2^{E_*}(\delta) = 0$. Evaluating the latter (recall (16) and (21)) we get $P_1 + P_2 - \lambda^2(\delta)R(\delta) = 0$ with

$$R(\delta) = \left\{ G_{11}E_{1*}^2 + G_{22}E_{2*}^2 + YE_{1*}E_{2*}[\sin(\delta_1 - \delta_2 + \alpha) + \sin(\delta_2 - \delta_1 + \alpha)] \right\}.$$

$R(\delta)$ evaluated at δ_* equals $P_1 + P_2$. This implies, on the one hand, that $\lambda(\delta_*) = 1$ as desired while, on the other hand, it ensures the existence of a neighborhood of $\delta_{1*} - \delta_{2*}$ where this expression is bounded away from zero thus we can define

$$\lambda(\delta) = +\sqrt{\frac{P_1 + P_2}{P_1 + P_2 + 2G_{12}E_{1*}E_{2*}[\cos(\delta_1 - \delta_2) - \cos(\delta_{1*} - \delta_{2*})]}}, \qquad (22)$$

which is obtained using the equilibrium equations and (3). It is clear from (22) that the neighborhood increases as the line conductance G_{12} (and α) approaches zero, and in the limit of a lossless line it becomes the whole real axis and we recover the previous derivations with $\lambda(\delta) = 1$. Some lengthy but straightforward calculations prove that $F(\delta, \lambda(\delta)E_*)$—with $\lambda(\delta)$ given by (22)—satisfies the integrability conditions of Poincare's Lemma, ensuring the existence of the function $\psi(\delta)$ that solves (20). The design of the IDA-PBC is completed evaluating the functions $J_{ij}(\delta, E)$ from (19) and computing the control law[13] according to Proposition 1. The result is summarized in the proposition below.

---

[12] Due to the presence of $\lambda(\delta)$, the functions $\tilde{F}^{E_*}(\delta)$ and $F^{E_*}(\delta)$ are not equal.

[13] We should underscore that the controller depends on $\nabla\psi$; hence we do not require the knowledge of the function $\psi(\delta)$ itself.

Proposition 3. *Consider the two-machine system with losses (13) and stable operating equilibrium satisfying*
Assumption A.1

$$|\delta_{1*} - \delta_{2*}| \le \frac{\pi}{2}; \tag{23}$$

Assumption A.2

$$P_1 + P_2 > 4G_{12}E_{1*}E_{2*}; \tag{24}$$

Assumption A.3

$$(P_1 + P_2)B_{12}\cos(\delta_{1*} - \delta_{2*}) + (P_1 - P_2)G_{12}\sin(\delta_{1*} - \delta_{2*}) > 0. \tag{25}$$

Then the system in closed loop with the control law

$$u_1 = a_1 E_1 - b_1 E_2 \cos(\delta_1 - \delta_2 + \alpha) - E_{f1} - J_{23}(\delta, E)\omega_1 - \\ -r_1\gamma_1[E_1 - E_{1*}\lambda(\delta)] + J_{35}(\delta, E)\omega_2$$

$$u_2 = a_2 E_2 - b_2 E_1 \cos(\delta_2 - \delta_1 + \alpha) - E_{f2} - J_{26}(\delta, E)\omega_1 - \\ -r_2\gamma_2[E_2 - E_{2*}\lambda(\delta)] - J_{56}(\delta, E)\omega_2,$$

where $\lambda(\delta)$ is given in (22) and

$$J_{23}(\delta, E) = -\frac{YE_{2*}}{\gamma_1}\lambda(\delta)\sin(\delta_1 - \delta_2 + \alpha) - \frac{G_1}{\gamma_1}[E_1 + E_{1*}\lambda(\delta)] - E_{1*}\nabla_{\delta_1}\lambda$$

$$J_{26}(\delta, E) = -\frac{Y}{\gamma_2}E_1\sin(\delta_1 - \delta_2 + \alpha) + E_{2*}\nabla_{\delta_1}\lambda$$

$$J_{35}(\delta, E) = \frac{YE_{2*}}{\gamma_1}\lambda(\delta)\sin(\delta_2 - \delta_1 + \alpha) + E_{1*}\nabla_{\delta_1}\lambda$$

$$J_{56}(\delta, E) = -\frac{Y}{\gamma_2}E_1\sin(\delta_2 - \delta_1 + \alpha) - \frac{G_2}{\gamma_2}[E_2 + E_{2*}\lambda(\delta)] + E_{2*}\nabla_{\delta_1}\lambda$$

$$\nabla_{\delta_1}\lambda = \frac{\sqrt{P_1 + P_2}G_{12}E_{1*}E_{2*}\sin(\delta_1 - \delta_2)}{2\left\{P_1 + P_2 + 2G_{12}E_{1*}E_{2*}[\cos(\delta_1 - \delta_2) - \cos(\delta_{1*} - \delta_{2*})]\right\}^{3/2}}$$

has an asymptotically stable equilibrium at $(\delta_\star, 0, E_\star)$. Furthermore, a Lyapunov function for this equilibrium is given by (18) with (22) and

$$\psi(\delta) = \int_0^{\delta_1 - \delta_2}\left[\frac{(P_1 + P_2)[G_{11}E_{1*}^2 + YE_{1*}E_{2*}\sin(\tau + \alpha)]}{P_1 + P_2 + 2G_{12}E_{1*}E_{2*}[\cos(\tau) - \cos(\delta_{1*} - \delta_{2*})]}\right]d\tau - P_1(\delta_1 - \delta_2),$$

and an estimate of its domain of attraction *is the largest* bounded *level set*

$$\{(\delta, \omega, E) \in \mathbb{R}^6 \mid H_d(\delta, \omega, E) \le c\}.$$

Proof. First note that (23) and (24) assure that $\lambda(\delta)$, given by (22), is well defined (in some neighborhood of the desired equilibrium). Now, we will compute from (19) the functions $J_{ij}(\delta, E)$. To this aim, with some obvious abuse of notation, define the vector function

$$F^{E_1}(\delta) \overset{\triangle}{=} F(\delta, E_1, \lambda(\delta)E_{2*}),$$

that is, the value of the function $F(\delta, E)$, given in (16), at $E_2 = \lambda(\delta)E_{2*}$. Evaluating the first equation of (19) at this point and using the notation just defined we get

$$F_1^{E_1}(\delta) = -\nabla_{\delta_1}\psi + \gamma_1[E_1 - E_{1*}\lambda(\delta)]E_{1*}\nabla_{\delta_1}\lambda + J_{23}(\delta, E)\gamma_1[E_1 - E_{1*}\lambda(\delta)].$$

Now, from (20) we have that $\nabla_{\delta_1}\psi = -F_1^{E_*}(\delta)$, hence

$$F_1^{E_1}(\delta) - F_1^{E_*}(\delta) = -[E_1 - E_{1*}\lambda(\delta)]\{G_{11}[E_1 + E_{1*}\lambda(\delta)] + YE_{2*}\lambda[\sin(\delta_1 - \delta_2 + \alpha)]\}.$$

Plugging this expression back in the first of equations (19) and eliminating the common factor $[E_1 - E_{1*}\lambda(\delta)]$ yields, after the calculation of $\nabla_{\delta_1}\lambda$ from (22), the expression of $J_{23}(\delta, E)$ in (26). With this definition of $J_{23}(\delta, E)$ substituted in (19) we immediately obtain $J_{26}(\delta, E)$.

Proceeding in exactly the same way, with the second of equations (19) we calculate $J_{35}(\delta, E)$ and $J_{56}(\delta, E)$.

It remains only to prove that the proposed energy function has indeed a minimum at the desired equilibrium point. Introduce a partial change of coordinates

$$z = E - \lambda(\delta)E_*, \tag{26}$$

and—recalling that $\lambda(\delta_\star) = 1$, with $\delta_\star = (\delta_{1\star}, \delta_{2\star})$—we look at the minima of the function

$$\tilde{H}_d(\delta, \omega, z) = \psi(\delta) + \frac{1}{2}|\omega|^2 + \frac{1}{2}z^\top \Gamma z,$$

which are obviously determined solely by $\psi(\delta)$. From (16) and (20) we have that $\nabla\psi(\delta_\star) = 0$. Some lengthy calculations establish that $\nabla^2\psi(\delta_\star) \geq 0$ if and only if

$$P_2\cos(\delta_1 - \delta_2 + \alpha) + P_1\cos(\delta_1 - \delta_2 - \alpha) > 0,$$

which, using (1), can be shown to be equivalent to (25), completing the proof.

Remark 4. Assumption A.1 captures the practically reasonable constraint that the normal operating regime of the system should not be "overly stressed." (See also [27].) Notice that, as indicated in the proof, the assumption provides a sufficient condition ensuring that $\lambda(\delta)$, given by (22), is well defined, but it is far from being necessary. The last two assumptions are obviated in the lossless case, but may be satisfied even for large values of the conductance G_{12}. In particular, Assumption A.2—which is given in this way for ease of presentation—can clearly be relaxed restricting our analysis to the set

$$\{(\delta, \omega, E) \in \mathbb{R}^6 \mid |P_1 + P_2 + 2G_{12}E_{1*}E_{2*}[\cos(\delta_1 - \delta_2) - \cos(\delta_{1*} - \delta_{2*})]| > 0 \},$$

which covers the whole $\mathbb{R}^6$ as G_{12} tends to zero.

Remark 5. If the rotor inertias M_i of the two machines are different we have that $Y_{12} \neq Y_{21}$. However, retracing the derivations above we can easily derive the new control law taking into account this fact.

6 The n-Machine Case

For the n-machine case with losses, the obstacle of integrability discussed in the previous section cannot be overcome with a scalar function $\lambda(\delta)$ and we need to consider a vector function. Therefore, we propose a total-energy function of the form

$$H_d(\delta, \omega, E) = \psi(\delta) + \frac{1}{2}|\omega|^2 + \frac{1}{2}[E - \mathrm{diag}\{\lambda_i(\delta)\}E_*]^\top \Gamma [E - \mathrm{diag}\{\lambda_i(\delta)\}E_*],$$
(27)

where $\delta = [\delta_1, ..., \delta_n]^\top$, $\psi(\delta) = [\psi_1(\delta), ..., \psi_n(\delta)]^\top$, $\omega = [\omega_1, ..., \omega_n]^\top$, $E = [E_1, ..., E_n]^\top$, with $\lambda_i, \psi : \mathbb{R}^n \to \mathbb{R}$, functions to be defined.

Omitting details for brevity, we state the following result.

Proposition 4. *Consider the n-machine system with losses (2) and assume the line conductances are sufficiently small. More precisely, for all $i \neq j$, $G_{ij} \leq \epsilon$ for some sufficiently small $\epsilon > 0$.*

Then, for any arbitrary C^1 function $\psi(\delta)$ such that $\delta_\star = \arg \min \psi$, there exists a (locally-defined) IDA-PBC that ensures asymptotic stability of the equilibrium $(\delta_\star, 0, E_\star)$ with a Lyapunov function of the form (27) and estimate of its domain of attraction *the largest* bounded *level set*

$$\{(\delta, \omega, E) \in \mathbb{R}^{3n} \mid H_d(\delta, \omega, E) \leq c\}.$$

Remark 6. Unlike the two-machine case considered before, Proposition 4 does not explicitly impose an assumption on the operating equilibrium of the form $|\delta_{i*} - \delta_{j*}| \leq \frac{\pi}{2}$. In view of the results of [27], it is expected that a similar condition will be needed to ensure stability of the open-loop equilibrium. Also, as in the two-machine case it will appear in the definition of the admissible domain of operation. Furthermore, Proposition 4 allows us to assign *arbitrary* potential energy functions $\psi(\delta)$, of course, at the price of only proving the existence of the controller.

7 Simulations: Two-Machine System

This section presents simulations for the two-machine system depicted in Figure 1. In this case the disturbance is a three-phase fault in the transmission line that connects buses 3 and 5, cleared by isolating the faulted circuit simultaneously at both ends, which modifies the topology of the network and consequently induces a change in the equilibrium point. The parameters of the model (13) are given in Table 1. The equilibrium point after the fault is $(\delta_{1\star}, \omega_{1\star}, E_{1\star}, \delta_{2\star}, \omega_{2\star}, E_{2\star}) = (0.6481, 0, 1.0363, 0.8033, 0, 1.0559)$, which verifies the conditions of Proposition 3. We remark that the rotor inertias of the two machines, M_i, are different thus we have that $Y_{12} \neq Y_{21}$ and, consequently, the control law of Proposition 3 has to be slightly modified as indicated in Remark 5.

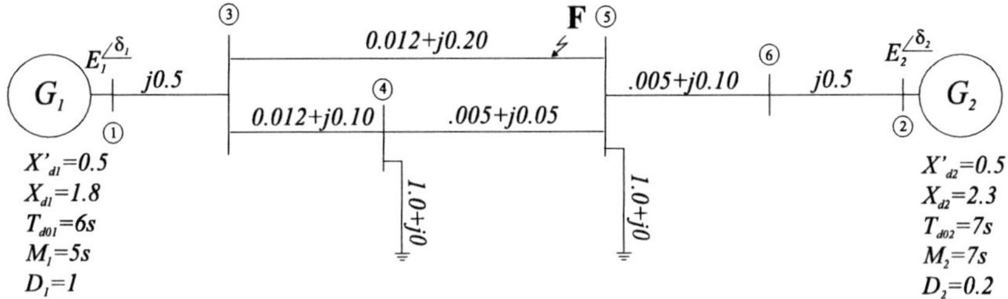

Fig. 1. Two-machine system.

Parameter	Gen 1	Gen 2
a	16.7255	14.2937
b	11.1059	9.4147
Y	51.2579	36.6127
G_{ii}	28.9008	20.3936
α	0.5430	0.5430
E_f	5.8103	7.9279
P	52.2556	48.4902

Table 1: Parameters of the post-fault system.

The simulation scenario corresponds to an overly stressed system. Without control the system is highly sensitive to the fault and the critical clearing time is almost zero. Although this scenario may be practically unrealistic we decided to present it to show that, even in this extreme case, the proposed controller enhances performance. Figure 2 presents the system's response to a short-circuit with clearing time $t_{cl} = 80\,m\sec$. The control inputs are depicted in Figure 3, which were clipped at ± 10.

8 Conclusions

We have presented a static state feedback controller that ensures asymptotic stability of the operating equilibrium for multimachine power systems with lossy transmission lines. The controller is derived using the recently developed IDA-PBC methodology, hence endows the closed-loop system with a PCH structure with a Hamiltonian function akin to a true total energy of an electromechanical system. A key step in the procedure is the inclusion of an interconnection between the electrical and the mechanical dynamics that may be interpreted as multipliers of the classical passivity theory.

Unfortunately, in the general n-machine case only *existence* of the IDA-PBC is ensured and we need to rely on a "sufficiently small" transfer conductances assumption. On the other hand, for the single and two-machine problems we give a complete constructive solution. Some preliminary calculations for the three-machine system suggest that, with a suitable selection of

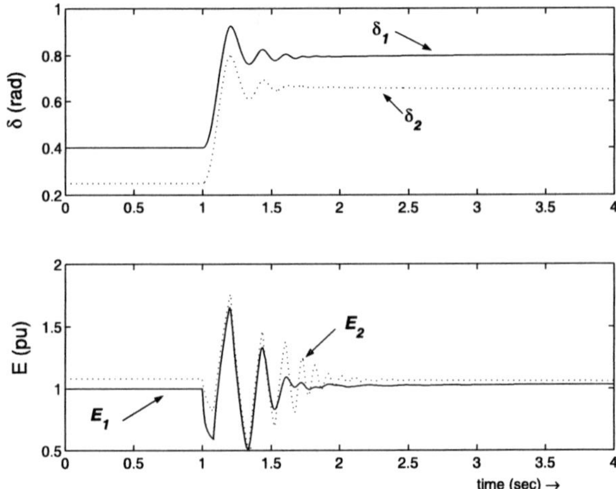

Fig. 2. Two-machine system in closed loop with the proposed IDA-PBC and $t_{cl} = 80\,m\,\mathrm{sec}$. Behavior of load angles and internal voltages.

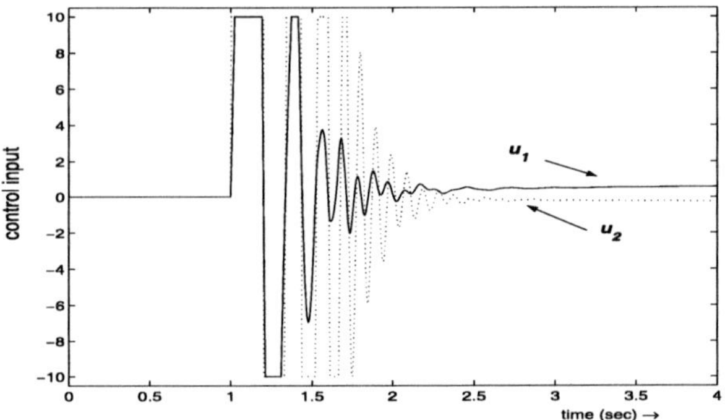

Fig. 3. Two-machine system in closed loop with the proposed IDA-PBC and $t_{cl} = 80\,m\,\mathrm{sec}$. Behavior of the control inputs.

the "potential energy " term, it is possible to obtain an explicit expression for the controller in the general case. In any case, the complexity of the resulting control law certainly stymies its practical application, and the result must be understood only as a proof of assignability of a suitable energy function. Alternative routes must be explored to come out with a practically feasible design—probably trying other parameterizations for the energy function. In this respect it is interesting to note that the proposed function (27) differs from the ones used in mechanical and electromechanical systems ((2.3) of [20] and (5) of [25], respectively). It is easy to see that neither one of these forms is suitable for the power systems problem at hand. Simulations were carried out to evaluate, in an academic example, the performance of the proposed scheme.

References

1. Anderson P, Fouad A (1977) Power systems control and stability. Iowa State University Press, Ames, IA
2. Bazanella A, Kototovic P, de Silva A (1999) A dynamic extension for $L_g V$ controllers. IEEE Transactions on Automatic Control 44:588–592
3. Chiang H, Chu C, Cauley G (1995) Direct stability analysis of electric power systems using energy functions: theory, applications, and perspective. Proceedings of the IEEE 83(11):1497–1529
4. Dahl O (1938) Electric power circuits: theory and applications. McGraw-Hill, New York
5. Desoer C, Vidyasagar M (1975) Feedback systems: input-output properties. Academic Press, New York
6. Fujimoto K, Sugie T (2001) Canonical transformations and stabilization of generalized Hamiltonian systems. Systems and Control Letters 42(3):217–227
7. Galaz M, Ortega R, Bazanella A, Stankovic A (2003) An energy-shaping approach to excitation control of synchronous generators. Automatica 39(1):111–119
8. Ghandhari M, Andersson G, Pavella M, Ernst D (2001) A control strategy for controllable series capacitor in electric power systems. Automatica 37:1575–1583
9. King C, Chapman J, Ilic M (1994) Feedback linearizing excitation control on a full-scale power system model. IEEE Transactions on Power Systems 9:1102–1109
10. Kirschen D, Bacher R, Heydt G (2000) Scanning the special issue on the technology of power system competition. Proceedings of the IEEE 88(2):123–127
11. Kundur P (1994) Power system stability and control. McGraw-Hill, New York
12. Lu Q, Sun Y, Mei S (2001) Nonlinear excitation control of large synchronous generators. Kluwer Academic Publishers, Boston, MA
13. Machowski J, Bialek J, Bumby J (1997) Power system dynamics and stability. John Wiley and Sons, New York
14. Magnusson P (1947) The transient-energy method of calculating stability. IEEE Transactions on Automatic Control 66:747–755
15. Mielczarsky W, Zajaczkowski A (1994) Nonlinear field voltage control of a synchronous generator using feedback linearization. Automatica 30:1625–1630
16. Moon Y, Choi B, Roh T (2000) Estimating the domain of attraction for power systems via a group of damping-reflected energy functions. Automatica 36:419–425
17. Narasimhamurthi N (1984) On the existence of energy function for power systems with transmission losses. IEEE Transactions on Circuits and Systems 31:199–203
18. Ortega R (2003) Some applications and extensions of interconnection and damping assignment passivity-based control. In: IFAC Workshop on Lagrangian and Hamiltonian Methods in Nonlinear Systems, Seville, Spain

19. Ortega R, Loria A, Nicklasson P, Ramirez HS (1998) Passivity-based control of Euler-Lagrange systems in Communications and Control Engineering. Springer-Verlag, Berlin

20. Ortega R, Spong M, Gomez F, Blankenstein G (2002) Stabilization of underactuated mechanical systems via interconnection and damping assignment. IEEE Transactions on Automatic Control 47(8):1218–1233

21. Ortega R, Stankovic A, Stefanov P (1998) A passivation approach to power systems stabilization. In: IFAC Symp. Nonlinear Control Systems Design, Enschede

22. Ortega R, van der Schaft A, Mareels I, Maschke B (2001) Putting energy back in control. IEEE Control Systems Magazine 21(2):18–33

23. Ortega R, van der Schaft A, Maschke B, Escobar G (2002) Interconnection and damping assignment passivity-based control of port-controlled Hamiltonian systems. Automatica 38(4):585–96

24. Pai M (1989) Energy function analysis for power system stability. Kluwer Academic Publishers, Boston, MA

25. Rodriguez H, Ortega R (2003) Interconnection and damping assignment control of electromechanical systems. International Journal on Robust and Nonlinear Control 13(12):1095–1111

26. Shen T, Ortega R, Lu Q, Mei S, Tamura K (2003) Adaptive disturbance attenuation of Hamiltonian systems with parametric perturbations and application to power systems. Asian Journal of Control 5(1):143–52

27. Skar S (1980) Stability of multimachine power system with nontrivial transfer conductances. SIAM Journal on Applied Mathematics 39(3):475–491

28. Sun Y, Shen T, Ortega R, Liu Q (2001) Decentralized controller design for multimachine power systems based on the Hamiltonian structure. In: Proceedings of the 40th CDC, Orlando, FL

29. Sun Y, Song Y, Li X (2000) Novel energy-based lyapunov function for controlled power systems. IEEE Power Engineering Review 20(5):55–57

30. van der Schaft A (2000) L_2-gain and passivity techniques in nonlinear control, 2nd edition LNCIS, Springer-Verlag, Berlin

31. Wang Y, Daizhan C, Chunwen L, You G (2003) Dissipative Hamiltonian realization and energy-based l_2 disturbance attenuation control of multimachine power systems. IEEE Transactions on Automatic Control 48(8):1428–1433

32. Wang Y, Hill D, Middleton R, Gao L (1993) Transient stability enhancement and voltage regulation of power system. IEEE Transactions on Power Systems 8:620–627

Robust Controllers for Large-Scale Interconnected Systems: Applications to Web Processing Machines*

Prabhakar R. Pagilla and Nilesh B. Siraskar

School of Mechanical and Aerospace Engineering, Oklahoma State University, Stillwater, OK 74078, USA
`pagilla@ceat.okstate.edu,nilesh.siraskar@okstate.edu`

Summary. Decentralized control of large-scale systems with uncertain, unmatched interconnections is considered. A stable decentralized adaptive controller is proposed for both linear and nonlinear interconnections. Sufficient conditions under which the overall large-scale system is stable are developed; results from algebraic Riccati equations are used to develop these conditions. The choice of the class of large-scale systems considered is directly motivated by control of web processing machines used in processing of materials. A large experimental web platform is considered for experimental study of the proposed decentralized adaptive controller and its comparison to a decentralized PI controller, which is widely used in the web processing industry. Extensive experiments were conducted; a representative sample of the experiments is presented and discussed.

1 Introduction

Large-scale interconnected systems can be found in a variety of engineering applications such as power systems, large structures, communication systems, transportation, and manufacturing processes. Decentralized control of large-scale interconnected systems has been a topic of interest for several decades. Decentralized control schemes present a practical and efficient means for designing control algorithms that utilize only the state of individual subsystems without any information from other subsystems. A large body of research in this area can be found in [20] and references therein, where classical application areas such as power systems were discussed. Recent applications of decentralized controllers include control of wafer temperature for multizone rapid thermal processing systems [17], platoons of underwater vehicles [21], and cooperative robotic vehicles [6].

---

*This work was supported by the US National Science Foundation under the CAREER grant No. CMS 9982071.

The decentralized adaptive control problem for large-scale systems has received considerable attention in the last two decades [9, 7, 20, 13, 14, 12]. Much of the existing research assumed that the interconnection function of each subsystem of the large-scale system satisfies the matching condition. In this research, we consider a class of large-scale systems with unmatched interconnections. The web processing application under consideration directly falls into this class of large-scale systems. Both linear and nonlinear interconnections are considered; further, the nonlinear interconnections are assumed to be polynomially bounded. We design a decentralized adaptive controller and develop sufficient conditions under which the overall large-scale system is stable. In the case of linear interconnections, global stability of the overall system is achieved, whereas for nonlinear interconnections, semiglobal stability is achieved. The proposed decentralized adaptive controller is applied to control of web processing machines, which is a large-scale system. In the following, we give a description of the web processing application together with related research.

A web is any material that is manufactured and processed in continuous, flexible strip form. Examples include paper, plastics, textiles, strip metals, and composites. Web processing pervades almost every industry today. It allows us to mass produce a rich variety of products from a continuous strip material. Products that utilize web processing somewhere in their manufacturing include aircraft, appliances, automobiles, bags, books, diapers, boxes, and newspapers. Web tension and velocity are two key variables that influence the quality of the finished web and hence the products manufactured from it.

Web handling refers to the physical mechanics related to the transport and control of web materials through processing machinery. The primary goal of research in web handling is to define and analyze the underlying sciences that govern unwinding, web guiding, web transport, and rewinding in an effort to minimize the defects and losses that may be associated with handling of the web. Web handling systems facilitate transport of the web during its processing, which is typically an operation specific to a product. For example, in the case of an aluminum web, the web is brought to a required thickness, cleaned, heat-treated, and coated; and in the case of some consumer products, the web is laminated and printed.

Early development of mathematical models for longitudinal dynamics of a web can be found in [5, 8, 10, 2]. In [5], a mathematical model for longitudinal dynamics of a web span between two pairs of pinch rolls, which are driven by two motors, was developed. This model does not predict tension transfer and does not consider tension in the entering span. A modified model that considers tension in the entering span was developed in [10]. In [2, 3], the moving web was considered as a moving continuum and general methods of continuum mechanics were used in the development of a mathematical model. In [18], equations describing web tension dynamics are derived based on the fundamentals of web behavior and the dynamics of the drives used for web transport. Nonideal effects such as temperature and moisture change on web tension were studied in [19]. An overview of lateral and longitudinal dynamic behavior and control of moving webs was presented in [23]. A review of the problems in tension control of webs can be found in [22]. Dynamics and control of accumulators in continuous strip processing lines were considered in [16]. A robust centralized H_∞ controller for a web winding system consisting of an intermediate driven roller and unwind/rewind rolls was proposed in [11]. The role of active dancers in attenuation of periodic tension disturbances was considered in [15].

A large experimental web platform, which mimics most of the features of an industrial web processing machine, is considered as an example for model development and experimentation with the proposed decentralized adaptive controllers. The experimental web platform consists of a number of intermediate rollers, some of which are driven, to provide web transport from unwind to rewind roll. First, a model of the unwind (rewind) roll is developed by explicitly considering the variation of radius and inertia resulting from release (accumulation) of material to (from) the process. A dynamic model of the entire experimental web platform is presented by considering each section of the web between two driven rollers as a tension zone; a natural subsystem formulation is proposed based on this strategy. This model can be easily extended to other web processing machines by the addition of more subsystems. Extensive experiments were conducted with the proposed adaptive decentralized controller and results are compared with those of a classical decentralized PI controller, which is mostly used in the web processing industry.

The rest of the chapter is organized as follows. Section 2 gives the problem description together with some existing results required for subsequent developments. The decentralized adaptive controller is given in Section 3, and sufficient conditions for stability of the overall large-scale system for both linear and nonlinear interconnections are derived. Section 4 gives a description of the experimental platform, dynamic model development, control inputs, and experimental results. Conclusions are given in Section 5.

2 Description of the Problem

We consider a large-scale system, S, consisting of $(N + 1)$ subsystems; each subsystem S_i is described by

$$\dot{x}_i(t) = A_i x_i(t) + b_i u_i(t) + g_i(t, x),\tag{1}$$

where $x_i(t) \in \mathbb{R}^{n_i}$ and $u_i(t) \in \mathbb{R}$ are the state vector and control input, respectively, of the ith subsystem $i \in I = \{0, 1, \ldots, N\}$, $g_i(t, x)$ represents interconnection of the ith subsystem with other subsystems, and $x^T = \left[x_0^T, x_1^T, \cdots, x_N^T \right]$ is the overall state of the large-scale system. Two types of interconnection functions $g_i(t, x)$ are considered, namely linear and nonlinear. Each nonlinear interconnection function $g_i(t, x)$ is assumed to be bounded by a polynomial function of states of all the subsystems. The two cases considered are given by the following:

$$g_i(t, x) = \sum_{j=0, j \neq i}^{N} A_{ij} x_j(t) \tag{2}$$

$$g_i^T(t, x) g_i(t, x) \leq \sum_{j=2}^{p_i} \sum_{k=0}^{N} \delta_{i,jk} ||x_k||^j , \tag{3}$$

where $A_{ij} \in \mathbb{R}^{n_i \times n_j}$ are linear interconnection matrices, p_i is the order of the bounding polynomial for the ith subsystem, and $\delta_{i,jk}$ are positive constants for i, j, and k.

The following assumptions are made

1. The constant vectors $b_i \in \mathbb{R}^{n_i}$ are known.
2. The matrices $A_i \in \mathbb{R}^{n_i \times n_i}$ are unknown, but it is assumed that there exist constant vectors $k_i \in \mathbb{R}^{n_i}$ such that

$$\bar{A}_i = A_i - b_i k_i^T , \tag{4}$$

   where $\bar{A}_i$ is an asymptotically stable matrix.
3. The matrices $A_{ij} \in \mathbb{R}^{n_i \times n_j}$ are unknown, but it is assumed that there exist positive numbers η_{ij} such that

$$\eta_{ij}^2 \geq \lambda_{max}(A_{ij}^T A_{ij}) , \tag{5}$$

   where $\lambda_{max}(M)$ denotes the largest eigenvalue of the matrix M.
4. The constant bounds $\delta_{i,jk}$ for all i, j, and k are known.

For the $N + 1$ subsystems given by (1) and the interconnections given by (2) or (3), the goal is to generate bounded decentralized control inputs $u_i(x_i(t))$, such that $x_i(t)$ is bounded and $\lim_{t \to \infty} x_i(t) = 0$ for all $i \in I$.

2.1 Preliminaries

The following results are required to establish sufficient conditions that ensure stability of the overall large-scale system for the proposed controllers and result in asymptotic convergence of $x_i(t)$, $i \in I$, to zero.

Lemma 1. *Consider the algebraic Riccati equation (ARE)*

$$A^\top P + PA + PRP + Q = 0. \tag{6}$$

If $R = R^\top \geq 0$, $Q = Q^\top > 0$, A is Hurwitz, and the associated Hamiltonian matrix $\mathcal{H} = \begin{bmatrix} A & R \\ -Q & -A^\top \end{bmatrix}$ is hyperbolic, i.e., $\mathcal{H}$ has no eigenvalues on the imaginary axis, then there exists a unique $P = P^\top > 0$, which is the solution of the ARE (6).

In the ARE (6), if $A = \bar{A}_i$, $Q = \xi_i^2 I$, where $\xi_i > 0$, and $R = NI$, then the following lemma gives a computable condition under which the Hamiltonian matrix $\mathcal{H}_i = \begin{bmatrix} \bar{A}_i & NI \\ -\xi_i^2 I & -\bar{A}_i^\top \end{bmatrix}$ is hyperbolic.

Lemma 2. $\mathcal{H}_i$ *is hyperbolic if and only if*

$$\min_{\omega \in R} \sigma_{min}(\bar{A}_i - j\omega I) > \xi_i \sqrt{N}, \tag{7}$$

where $\sigma_{min}(M)$ *denotes the minimum singular value of* M.

Lemma 2 is a special case of Theorem 2 in [1] and is obtained by setting $C = 0$ in that theorem.

Remark 1. An efficient numerical algorithm for the computation of $\min_{\omega \in R} \sigma_{min}(\bar{A}_i - j\omega I)$ using the bisection method can be found in [4].

3 Decentralized Adaptive Controllers

For each $i \in I$, consider the control input (8) where the estimate $\widehat{k}_i(t)$ of k_i is obtained by the adaptation law (9):

$$u_i(t) = -\widehat{k}_i^T(t) x_i(t) \tag{8}$$

$$\dot{\widehat{k}}_i(t) = -(x_i^T(t) P_i b_i) x_i(t), \tag{9}$$

where P_i is a positive definite gain matrix. Using (4) and defining the gain estimation error as $\widetilde{k}_i(t) = k_i - \widehat{k}_i(t)$, the state dynamics for each subsystem upon simplification become

$$\dot{x}_i = \bar{A}_i x_i + b_i \widetilde{k}_i^T x_i + g_i(t, x). \tag{10}$$

The stability of the overall large-scale system can be shown by choosing the quadratic Lyapunov function candidate

$$V(x_i, \widetilde{k}_i) = \sum_{i=0}^{N} \left(x_i^T P_i x_i + \widetilde{k}_i^T \widetilde{k}_i \right). \tag{11}$$

In the following, we will derive sufficient conditions under which stability of the overall large-scale system can be achieved for both linear and nonlinear interconnections $g_i(t, x)$, as given by (2) and (3), respectively.

3.1 Linear Interconnections

For the linear interconnection case, the derivative of the Lyapunov function candidate along the trajectories of (9) and (10) can be obtained as

$$\dot{V}(x_i, \widetilde{k}_i) = \sum_{i=0}^{N} x_i^T (\bar{A}_i^T P_i + P_i \bar{A}_i) x_i$$

$$+ x_i^T P_i \sum_{j=0, j \neq i}^{N} A_{ij} x_j + \left(\sum_{j=0, j \neq i}^{N} A_{ij} x_j \right)^T P_i x_i. \qquad (12)$$

Using the inequality

$$(A_{ij} x_j)^T P_i x_i + x_i^T P_i (A_{ij} x_j) \leq (A_{ij} x_j)^T (A_{ij} x_j) + (P_i x_i)^T (P_i x_i)$$

$$= x_j^T (A_{ij}^T A_{ij}) x_j + x_i^T (P_i P_i) x_i \qquad (13)$$

for $i \in I$ in the last two terms of (12), we obtain

$$\dot{V}(x_i, \widetilde{k}_i) \leq \sum_{i=0}^{N} x_i^T (\bar{A}_i^T P_i + P_i \bar{A}_i) x_i + x_i^T (N P_i P_i) x_i + \sum_{j \neq i}^{N} x_j^T (A_{ij}^T A_{ij}) x_j.$$

$$(14)$$

Also, we can obtain the following bound for the last term of $\dot{V}$:

$$\sum_i \sum_{j \neq i} x_j^T A_{ij}^T A_{ij} x_j \leq \sum_i \sum_{j \neq i} \eta_{ij}^2 x_j^T x_j = \sum_i \left(\sum_{j \neq i} \eta_{ji}^2 \right) x_i^T x_i.$$

As a result, $\dot{V}$ satisfies

$$\dot{V}(x_i, \widetilde{k}_i) \leq \sum_{i=0}^{N} x_i^T \left(\bar{A}_i^T P_i + P_i \bar{A}_i + N P_i P_i + \xi_i^2 I \right) x_i, \qquad (15)$$

where $\xi_i^2 = \sum_{j=0, j \neq i}^{N} \eta_{ji}^2$. Therefore, we have the following. If there exist positive definite solutions P_i to the AREs

$$\bar{A}_i^T P_i + P_i \bar{A}_i + N P_i P_i + (\xi_i^2 + \epsilon_i) I = 0, \qquad (16)$$

then

$$\dot{V}(x_i, \widetilde{k}_i) \leq - \sum_{i=0}^{N} \epsilon_i x_i^T x_i. \qquad (17)$$

Thus, $V(x_i, \widetilde{k}_i)$ is a Lyapunov function. As a result $x_i(t) \in \mathcal{L}_2 \cap \mathcal{L}_\infty$ for all $i \in I$. Further from the closed loop dynamics (10), $\dot{x}_i(t) \in \mathcal{L}_\infty$ for all $i \in I$. This implies that $\lim_{t \to \infty} x_i(t) = 0$ for all $i \in I$.

Using Lemma 1 and Lemma 2, it can be seen that there exist positive definite solutions to the AREs

$$\bar{A}_i^T P_i + P_i \bar{A}_i + N P_i P_i + \xi_i^2 I = 0, \tag{18}$$

if

$$\min_{\omega \in R} \sigma_{min}(\bar{A}_i - j\omega I) > \xi_i \sqrt{N} \tag{19}$$

holds for each $i \in I$.

Finally, the existence of $\epsilon_i, i \in I$, in the AREs (16) follows from the continuity of the functions $f_i(\xi_i^2) := \min_{\omega \in R} \sigma_{min}(\bar{A}_i - j\omega I) - \sqrt{N\xi_i^2}$, that is, if there exists a ξ_i such that $f_i(\xi_i^2) > 0$, then there exists an $\epsilon_i > 0$ such that $f_i(\xi_i^2 + \epsilon_i) > 0$.

3.2 Nonlinear Interconnections

Differentiating V along the trajectories of (10) with nonlinear interconnections, and using the adaptation law (9), we obtain

$$\dot{V} = \sum_{i=0}^{N} x_i^T [\bar{A}_i^T P_i + P_i \bar{A}_i] x_i + g_i^T(t,x) P_i x_i + x_i^T P_i g_i(t,x). \tag{20}$$

Using the inequality $X^T Y + Y^T X \leq X^T X + Y^T Y$ for two matrices X and Y, the derivative of the Lyapunov function candidate satisfies

$$\dot{V} \leq \sum_{i=0}^{N} x_i^T [\bar{A}_i^T P_i + P_i \bar{A}_i] x_i + x_i^T (P_i P_i) x_i + g_i^T(t,x) g_i(t,x). \tag{21}$$

Now, using the bounds on the $g_i(t,x)$ as given by (3), we obtain

$$\dot{V} = \sum_{i=0}^{N} x_i^T [\bar{A}_i^T P_i + P_i \bar{A}_i + P_i P_i] x_i + \sum_{j=2}^{p_i} \sum_{k=0}^{N} \delta_{i,jk} ||x_k||^j. \tag{22}$$

Therefore, if there exist positive definite solutions to the AREs

$$\bar{A}_i^T P_i + P_i \bar{A}_i + P_i P_i + Q_i = 0, \tag{23}$$

then

$$\dot{V} = \sum_{i=0}^{N} -\lambda_{min}(Q_i) ||x_i||^2 + \sum_{j=2}^{p_i} \sum_{k=0}^{N} \delta_{i,jk} ||x_k||^j \tag{24}$$

$$\leq -\sum_{i=0}^{N} ||x_i||^2 \left(\lambda_{min}(Q_i) - \sum_{j=2}^{p_i} \sum_{k=0}^{N} \delta_{k,ji} ||x_i||^{j-2} \right). \tag{25}$$

Now, if it is assumed that the initial state at time t_0 of each subsystem satisfies

$$||x_i(t_0)|| \le R_i ,\tag{26}$$

where R_i is a positive number, then the choice of Q_i such that

$$\lambda_{min}(Q_i) = \sum_{j=2}^{p_i}\sum_{k=0}^{N} \delta_{k,ji}(R_i)^{j-2} + \gamma_i ,\tag{27}$$

where γ_i is any positive constant and $\lambda_{min}(M)$ denotes the minimum eigenvalue of M, will lead to

$$\dot{V} \le -\sum_{i=0}^{N} \gamma_i ||x_i||^2.\tag{28}$$

Hence, for each $i \in I$, if the initial condition $x_i(t_0)$ lies in the region defined by $\Omega_i = \{x_i(t) : ||x_i(t)|| \le R_i\}$, and if there exist positive definite solutions to the AREs (23) with Q_i satisfying (27), then $\lim_{t\to\infty} x_i(t) = 0$.

Using Lemma 1 and Lemma 2, the condition for the existence of positive definite solutions to the AREs (23) is given by

$$\min_{\omega \in R} \sigma_{min}(\bar{A}_i - j\omega I) > \sqrt{\lambda_{min}(Q_i)}.\tag{29}$$

Remark 2. If the condition given by (19) or (29) is not satisfied, then we must choose another asymptotically stable matrix $\bar{A}_i$. Since each $\bar{A}_i$ is stable, $\min_{\omega \in R} \sigma_{min}(\bar{A}_i - j\omega I)$ is the distance to the set of unstable matrices [4].

4 Web Processing Application

Figure 1 shows an experimental web processing line. A number of configurations can be used to thread the web through the platform to transport it from unwind and rewind while processing it. Figure 2 shows the experimental platform and the web path used for conducting experiments with the proposed controllers. The line mimics most of the features of an industrial web processing machine and is developed with the aim of open-architecture design that allows for the modification of the web path and hardware to test specific research experimentation. The machine contains a number of different stations, as shown in Figure 1, and a number of driven rollers. For the experimentation, the web is threaded through four driven rollers M_0 to M_3 as shown, and through many other idle rollers throughout the line to facilitate transport of the web from the unwind to rewind. The nip rollers (denoted by NR), which are pneumatically driven, are used to maintain web contact with the driven rollers. Two controlled lateral guides (guides are denoted by DG and the web edge sensors by E), near unwind and rewind sections, respectively, are used to maintain the lateral position of the web on the rollers during web transport.

Three-phase induction motors with 30 HP capacity, from Rockwell Automation, are used to drive the unwind and rewind rolls, whereas the master speed and process section rollers are driven by 15 HP induction motors. The motor drive system, the real-time architecture that includes micro-processors, I/O cards, and real-time software (AUTOMAX), are from Rockwell Automation. In the experimental platform, each motor is driven by a dedicated vector controller. Reference torque and flux signals for vector controllers are gen-

Fig. 1. The experimental platform.

erated by microprocessors, which are a part of the AUTOMAX distributed control system. To implement the desired control algorithms, programs in AUTOMAX can be modified using an off-line personal computer and then uploaded to the dedicated microprocessors. Similar to a typical industrial distributed control system, microprocessors used in the experimental platform are located in two racks termed A00 and A01. Rack A00 has microprocessors and vector control drives for the rewind roll (M_3) and process section roller (M_2). Rack A01 has microprocessors and vector control drives for the unwind roll (M_0) and master speed roller (M_1). Depending upon the number of process sections, actual industrial setup may have a large number of such racks. Decentralized controllers are widely used in the web handling industry due to ease in tuning controllers for individual stations without information exchange from other subsystems, and also due to flexibility in reconfiguring the large-scale system. Further, decentralized strategies are known to provide reliable operation of the process line in the event of occasional actuator and sensor malfunctions.

In the following, we will systematically partition the process line shown in Figure 2 into subsystems, develop dynamic models for each subsystem, derive

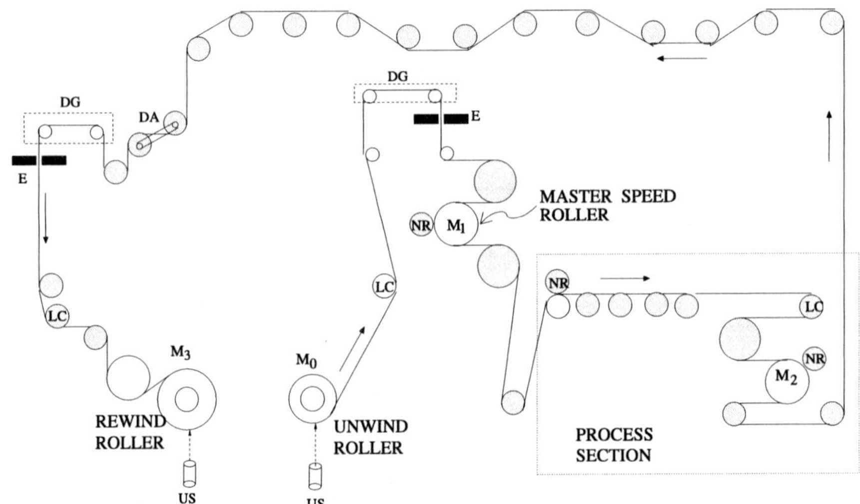

Fig. 2. Sketch of the experimental platform showing web path.

control inputs and reference velocities that will keep the web at the forced equilibrium, and finally obtain the dynamic model for each subsystem in the form given by (1).

4.1 Dynamic Models

It is common in the web handling industry to divide the process line into several tension zones by calling the span between two successive driven rollers a tension zone, thus ignoring the effect of the free rollers that lie between two driven rollers. Since the free roller dynamics have an effect on the web tension during the transients due to acceleration/deceleration of the web line and negligible effect during steady state operation, the assumption that the free rollers do not contribute to web dynamics during steady state operation is reasonable and is extensively used in the industry. This assumption will be used in developing the dynamic model in this section. Further, it is also assumed that the web is elastic and there is no web slip on the rollers.

Figure 3 shows a line sketch of the web line with three tension zones; the line consists of the unwind/rewind rolls and two intermediate driven rollers. In this figure, LC denotes the loadcell roller, which is mounted on a pair of loadcells on either side for measuring web tension. The driving motors are represented by M_i for $i = 0, 1, 2, 3$, τ_i represents input torque from the ith motor, v_i represents the transport velocity of the web on the ith roller, and t_i represents web tension in the span between $(i-1)$th and ith driven rollers. There are four sections in the web line shown in Figure 3: unwind section, master speed roller, process section, and rewind section. The name master speed roller is given to a driven roller that sets the reference web transport

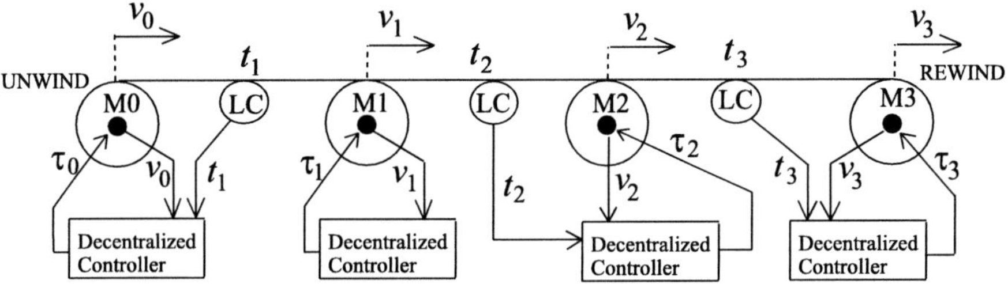

Fig. 3. Web line sketch showing tension zones.

speed for the entire web line and is generally the first driven roller upstream of the unwind roll in almost all web process lines. The master speed roller regulates web line speed and is not used to regulate tension in the spans adjacent to it. The unwind (rewind) rolls release (accumulate) material to (from) the processing section of the web line. Thus, their radii and inertia are time-varying. The dynamics of each of the four sections are presented in the following.

Unwind section: The local state variables for the unwind section are web velocity v_0 and tension t_1. At any instant of time t, the effective inertia $J_0(t)$ of the unwind section is given by

$$J_0(t) = n_0^2 J_{m0} + J_{c0} + J_{w0}(t), \tag{30}$$

where n_0 is the ratio of motor speed to unwind roll shaft speed and J_{m0} is the inertia of all the rotating elements on the motor side, which includes inertia of motor armature, driving pulley (or gear), and driving shaft. The inertia of the driven shaft and the core mounted on it is given as J_{c0}, and $J_{w0}(t)$ is the inertia of the cylindrically wound web material on the core. Both J_{m0} and J_{c0} are constant, but $J_{w0}(t)$ is not constant because the web is continuously released into the process. The inertia $J_{w0}(t)$ is given by

$$J_{w0}(t) = \frac{\pi}{2} b_w \rho_w (R_0^4(t) - R_{c0}^4), \tag{31}$$

where b_w is the web width, ρ_w is the density of the web material, R_{c0} is the radius of the empty core mounted on the unwind roll-shaft, and $R_0(t)$ is the radius of the material roll.

The velocity dynamics of the unwind roll can be written as

$$\frac{d}{dt}(J_0\omega_0) = t_1 R_0 - n_0\tau_0 - b_{f0}\omega_0$$

$$\dot{J}_0\omega_0 + J_0\dot{\omega}_0 = t_1 R_0 - n_0\tau_0 - b_{f0}\omega_0, \tag{32}$$

where ω_0 is the angular velocity of the unwind roll and b_{f0} is the coefficient of friction in the unwind roll shaft. The rate of change in $J_0(t)$ only depends on the change in $J_{w0}(t)$, and from (31), the rate of change of $J_0(t)$ is given by

$$\dot{J}_0(t) = \dot{J}_{w0}(t) = 2\pi b_w \rho_w R_0^3 \dot{R}_0. \tag{33}$$

The transport velocity of the web coming off the unwind roll is related to the angular velocity of the unwind roll by $v_0 = R_0 \omega_0$, and hence we can obtain $\dot{\omega}_0$ in terms of v_0 as

$$\dot{\omega}_0 = (R_0 \dot{v}_0 - \dot{R}_0 v_0)/R_0^2. \tag{34}$$

Substituting (33) and (34) into the velocity dynamics given by (32) and simplifying, we obtain

$$(J_0/R_0)\dot{v}_0 = t_1 R_0 - n_0 \tau_0 - (b_{f0}/R_0)v_0 + (\dot{R}_0 v_0 J_0)/R_0^2 - 2\pi \rho_w b_w R_0^2 \dot{R}_0 v_0. \tag{35}$$

But the rate of change of radius $\dot{R}_0$ is a function of the transport velocity v_0 and the web thickness t_w and is approximately given by

$$\dot{R}_0 \approx -(t_w v_0(t))/(2\pi R_0(t)). \tag{36}$$

Notice that (36) is approximate because the thickness affects the rate of change of radius of the roll after each revolution of the roll. The continuous approximation is valid since the thickness is generally very small. Also notice that the last term in the velocity dynamics (35) is often ignored in the literature under the assumption that the roll radius is slowly time-varying. But in practice, since the web transport velocity is kept constant, the last two terms in (35) are significant as the roll radius becomes smaller. Hence, (35) can be simplified to

$$\frac{J_0}{R_0}\dot{v}_0 = t_1 R_0 - n_0 \tau_0 - \frac{b_{f0}}{R_0}v_0 - \frac{t_w}{2\pi R_0}\left(\frac{J_0}{R_0^2} - 2\pi \rho_w b_w R_0^2\right)v_0^2. \tag{37}$$

Dynamic behavior of the web tension t_1 in the span immediately downstream of the unwind roll is given by

$$L_1 \dot{t}_1 = AE[v_1 - v_0] + t_0 v_0 - t_1 v_1, \tag{38}$$

where L_1 is the length of the web span between unwind roller (M_0) and master speed roller (M_1), A is the area of cross-section of the web, E is the modulus of elasticity of the web material, and t_0 represents the wound-in tension of the web in the unwind roll.

Master speed roller: The dynamics of the master speed roller are given by

$$(J_1/R_1)\dot{v}_1 = (t_2 - t_1)R_1 + n_1 \tau_1 - (b_{f1}/R_1)v_1. \tag{39}$$

Process section: The web tension and web velocity dynamics in the process section are given by

$$L_2 \dot{t}_2 = AE[v_2 - v_1] + t_1 v_1 - t_2 v_2 \tag{40}$$

$$\frac{J_2}{R_2}\dot{v}_2 = (t_3 - t_2)R_2 + n_2 \tau_2 - \frac{b_{f2}}{R_2}v_2. \tag{41}$$

Rewind section: The web velocity dynamics entering the rewind roll can be determined along similar lines as those presented for the unwind roll. The web tension and velocity dynamics in the rewind section are

$$L_3 \dot{t}_3 = AE[v_3 - v_2] + t_2 v_2 - t_3 v_3 \tag{42}$$

$$\frac{J_3}{R_3} \dot{v}_3 = -t_3 R_3 + n_3 \tau_3 - \frac{b_{f3}}{R_3} v_3 + \frac{t_w}{2\pi R_3} \left(\frac{J_3}{R_3^2} - 2\pi \rho_w b_w R_3^2 \right) v_3^2. \tag{43}$$

Equations (37) through (43) represent the dynamics of the web and rollers for the web line configuration shown in Figure 3. Extension to other web lines can be easily made based on this model. For web process lines that have a series of process sections between the master speed roller and the rewind roll, (40) and (41) can be written for each process section.

4.2 Equilibrium Inputs and Reference Velocities

The control goal is to regulate web tension in each of the tension zones while maintaining the prescribed web transport velocity. To achieve this, first, we systematically have to calculate the control input required to keep the web line at the forced equilibrium of the reference web tension and web velocity in each of the zones. In the following, equilibrium control inputs and reference velocities are determined for each driven roll/roller based on the reference velocity of the master speed roller and reference tension in each tension zone.

Define the following variables: $T_i(t) = t_i(t) - t_{ri}$ and $V_i(t) = v_i(t) - v_{ri}$, where t_{ri} and v_{ri} are tension and velocity references, respectively, T_i and V_i are the variations in tension and velocity, respectively, around their reference values, τ_{ieq} is the control input that maintains the forced equilibrium at the reference values, and $u_i = \tau_i - \tau_{ieq}$ is the variation of the control input. Define the state vector for the unwind section as $x_0^T = [T_1, V_0]$ and the state for the master speed roller as $x_1 = V_1$. After master speed section, define the state vector for the jth subsystem as $x_j^T = [T_j, V_j]$ for $j = 2, 3$.

The velocity dynamics in the unwind section can be written as

$$(J_0/R_0)\dot{V}_0 = (T_1 + t_{r1})R_0 - n_0(u_0 + \tau_{0eq}) - (b_{f0}/R_0)(V_0 + v_{r0})$$

$$- (J_0/R_0)\dot{v}_{r0} - \frac{t_w}{2\pi R_0} \left(\frac{J_0}{R_0^2} - 2\pi \rho_w b_w R_0^2 \right) (V_0 + v_{r0})^2. \tag{44}$$

Assuming the variations T_i and V_i and their derivatives are zero, the input that maintains forced equilibrium is given by

$$\tau_{0eq} = -\frac{b_{f0}}{n_0 R_0} v_{r0} + \frac{R_0}{n_0} t_{r1} - \frac{J_0}{n_0 R_0} \dot{v}_{r0} - \frac{t_w}{2\pi n_0 R_0} \left(\frac{J_0}{R_0^2} - 2\pi \rho_w b_w R_0^2 \right) v_{r0}^2. \tag{45}$$

The web tension dynamics in the unwind section can be written as

$$L_1 \dot{T}_1 = AE[(V_1 + v_{r1}) - (V_0 + v_{r0})] + t_0(V_0 + v_{r0}) - (T_1 + t_{r1})(V_1 + v_{r1}). \tag{46}$$

From (46), assuming $\dot{T}_1$, T_1, and V_i ares zero at the forced equilibrium, the relationship between the reference velocities v_{r0} and v_{r1} is given by

$$v_{r0} = \frac{AE - t_{r1}}{AE - t_0} v_{r1}. \tag{47}$$

Assuming the product of variations $T_1 V_1$ is negligible, the variational dynamics in the unwind section can be written as

$$\dot{x}_0 = \begin{bmatrix} \dot{T}_1 \\ \dot{V}_0 \end{bmatrix} = A_0 x_0 - b_0 u_0 - b_0 f_0(V_0) + \sum_{j=1}^{3} A_{0j} x_j , \tag{48}$$

where A_{02} and A_{03} are null matrices, and

$$A_0 = \begin{bmatrix} -v_{r1}/L_1 & (t_0 - AE)/L_1 \\ R_0^2/J_0 & -b_{f0}/J_0 \end{bmatrix}, \quad b_0 = \begin{bmatrix} 0 \\ \frac{n_0 R_0}{J_0} \end{bmatrix}$$

$$f_0 = \frac{t_w}{2\pi n_0 R_0} \left(\frac{J_0}{R_0^2} - 2\pi \rho_w b_w R_0^2 \right) (V_0^2 + 2v_{r0} V_0)$$

$$A_{01} = \begin{bmatrix} \frac{AE - t_{r1}}{L_1}, & 0 \end{bmatrix}^T .$$

For the master speed roller, letting $x_1 = V_1$, similar equilibrium analysis gives the following velocity variation dynamics:

$$\dot{x}_1 = \dot{V}_1 = A_1 x_1 + b_1 u_1 + \sum_{j=0, j \neq 1}^{3} A_{1j} x_j , \tag{49}$$

where $A_1 = -b_{f1}/J_1$, $b_1 = n_1 R_1/J_1$, and

$$A_{10} = \begin{bmatrix} \frac{-R_1^2}{J_1}, 0 \end{bmatrix}, \quad A_{12} = \begin{bmatrix} \frac{R_1^2}{J_1}, 0 \end{bmatrix}, \quad A_{13} = [0, 0] ,$$

with the equilibrium input given by

$$\tau_{1eq} = \frac{b_{f1}}{n_1 R_1} v_{r1} - \frac{R_1}{n_1} (t_{r2} - t_{r1}) . \tag{50}$$

For the process section driven roller, the reference velocity and the equilibrium input are given by

$$v_{r2} = \left(\frac{AE - t_{r1}}{AE - t_{r2}} \right) v_{r1} \tag{51}$$

$$\tau_{2eq} = \frac{b_{f2}}{n_2 R_2} v_{r2} - \frac{R_2}{n_2} (t_{r3} - t_{r2}) . \tag{52}$$

Therefore, the variational dynamics in the process section are given by

$$\dot{x}_2 = \begin{bmatrix} \dot{T}_2 \\ \dot{V}_2 \end{bmatrix} = A_2 x_2 + b_2 u_2 + \sum_{j=0, j\neq 2}^{3} A_{2j} x_j , \tag{53}$$

where

$$A_2 = \begin{bmatrix} -v_{r2}/L_2 & (AE - t_{r2})/L_2 \\ -R_2^2/J_2 & -b_{f2}/J_2 \end{bmatrix}, b_2 = \begin{bmatrix} 0 \\ \frac{n_2 R_2}{J_2} \end{bmatrix}$$

$$A_{20} = \begin{bmatrix} \frac{v_{r1}}{L_2} & 0 \\ 0 & 0 \end{bmatrix}, A_{21} = \begin{bmatrix} \frac{t_{r1} - AE}{L_2} \\ 0 \end{bmatrix}, A_{23} = \begin{bmatrix} 0 & 0 \\ \frac{R_2^2}{J_2} & 0 \end{bmatrix} .$$

The dynamics of the rewind section are given by

$$\dot{x}_3 = \begin{bmatrix} \dot{T}_3 \\ \dot{V}_3 \end{bmatrix} = A_3 x_3 + b_3 u_3 + b_3 f_3(V_3) + \sum_{j=0, j\neq 3}^{3} A_{3j} x_j , \tag{54}$$

where $u_3 = \tau_3 - \tau_{3eq}$ and

$$v_{r3} = \left(\frac{AE - t_{r1}}{AE - t_{r3}} \right) v_{r1} \tag{55}$$

$$\tau_{3eq} = \frac{b_{f3}}{n_3 R_3} v_{r3} + \frac{R_3}{n_3} t_{r3} + \frac{J_3}{n_3 R_3} \dot{v}_{r3}$$
$$- \frac{t_w}{2\pi n_3 R_3} \left(\frac{J_3}{R_3^2} - 2\pi \rho_w b_w R_3^2 \right) v_{r3}^2 \tag{56}$$

$$A_3 = \begin{bmatrix} -v_{r3}/L_3 & (AE - t_{r3})/L_3 \\ -R_3^2/J_3 & -b_{f3}/J_3 \end{bmatrix}, \quad b_3 = \begin{bmatrix} 0 \\ \frac{n_3 R_3}{J_3} \end{bmatrix}$$

$$A_{32} = \begin{bmatrix} \frac{v_{r2}}{L_3} & \frac{t_{r2} - AE}{L_3} \\ 0 & 0 \end{bmatrix}$$

$$f_3 = \frac{t_w}{2\pi n_3 R_3} \left(\frac{J_3}{R_3^2} - 2\pi \rho_w b_w R_3^2 \right) (V_3^2 + 2v_{r3} V_3) .$$

The dynamics of the entire system are given by (48), (49), (53), and (54). Notice that subsystems 0 and 1 have the additional matched terms $b_0 f_0(V_0)$ and $b_3 f_3(V_3)$, respectively, which can be compensated using their respective control inputs to bring the subsystem equations to the form given in Section 2.

Therefore, based on the adaptive design given in Section 3, we have the following control inputs:

$$\tau_0 = \tau_{0eq} + \hat{k}_0(t) x_0(t) - f_0(V_0) \tag{57}$$

$$\tau_1 = \tau_{1eq} - \hat{k}_1(t) x_1(t) \tag{58}$$

$$\tau_2 = \tau_{2eq} - \hat{k}_2(t) x_2(t) \tag{59}$$

$$\tau_3 = \tau_{3eq} - \hat{k}_3(t) x_3(t) - f_3(V_3) . \tag{60}$$

The adaptation laws for obtaining $\hat{k}_i$ are given by (9).

4.3 Experimental Results

Three decentralized controllers were compared via extensive experimentation:

1. Industrial decentralized PI controller with inertia compensation
2. Decentralized nonadaptive state feedback controller with equilibrium inputs and inertia compensation
3. Decentralized adaptive controller with equilibrium inputs and inertia compensation

A control block diagram for the often used industrial PI controller is shown in Figure 4. Notice that the output of the tension loop becomes reference velocity error correction for the velocity loop. Decentralized nonadaptive controller is

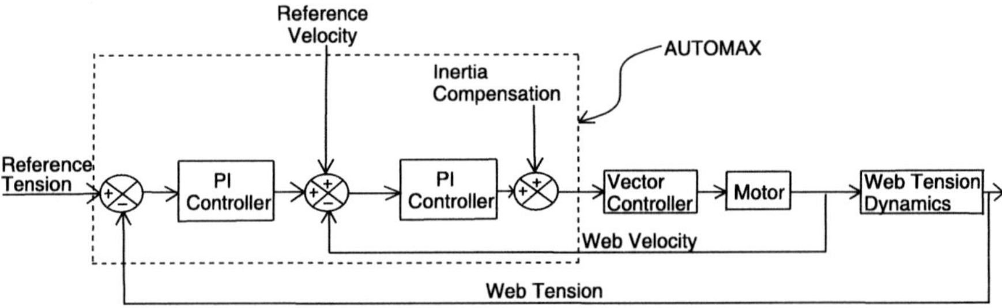

Fig. 4. Control block diagram of the industrial decentralized PI controller.

implemented with the assumption that all the parameters in the dynamic model are exactly known. For the nonadaptive controller, in the control input (8), since k_i is assumed to be known, $\hat{k}_i$ is replaced by k_i. Implementation control block diagram for the proposed decentralized controllers is shown in Figure 5. For inertia compensation and equilibrium control of unwind and

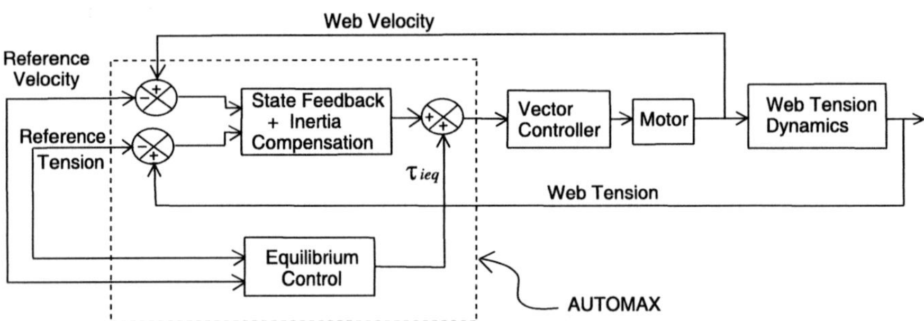

Fig. 5. Control block diagram for the proposed decentralized controller.

rewind rolls, instantaneous radius of each roll is calculated using an encoder signal from the corresponding motor. Effective coefficient of friction b_{fi} for all $i = 0, 1, 2, 3$ is taken as 0.5 lbf-ft-sec/rad. The web material used is Tyvec®, which is a product made by Dupont. The product of the elasticity of the web material and its cross-sectional area AE is equal to 2090 lbf. For both the master speed and process section driven rollers, the values of the radii and inertia are 0.3 ft and 2 lb-ft^2, respectively. The lengths of the three different web spans are $L_1 = 20$ ft, $L_2 = 33$ ft, and $L_3 = 67$ ft.

Extensive experiments with different reference web transport speeds were conducted with the currently used industrial decentralized PI controller and the proposed decentralized controller. The sampling period for control scanning loop and data acquisition was chosen to be 5 milliseconds. Experimental results for the following case are presented: reference velocity $v_{r1} = 1500$ ft/min and reference tension $t_{ri} = 14.35$ lbf, $i = 1, 2, 3$.

Figures 6, 7, and 8 show experimental results using the decentralized well-tuned PI controller, nonadaptive decentralized controller, and adaptive decentralized controller, respectively. Variation of the web line speed at the master speed roller and tension variations in each tension zone are shown. In each figure, the top plot shows the variation in master speed velocity from its reference (V_1) and the remaining three plots show the tension variations in the three tension zones (T_1, T_2, T_3). Results using the proposed decentralized controllers show much improved regulation of web transport speed and tension in each of the zones over the industrial decentralized PI controller. A 20-second 2-norm of the variations for different controllers is given in Table 1.

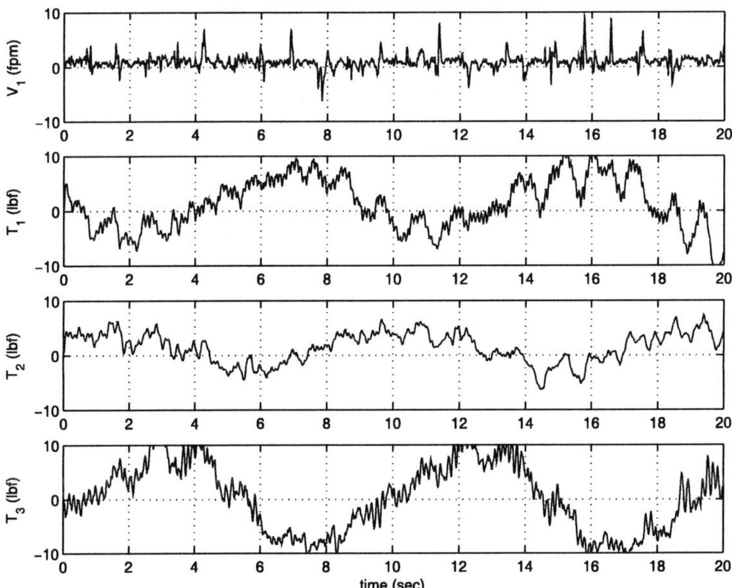

Fig. 6. Decentralized PI controller.

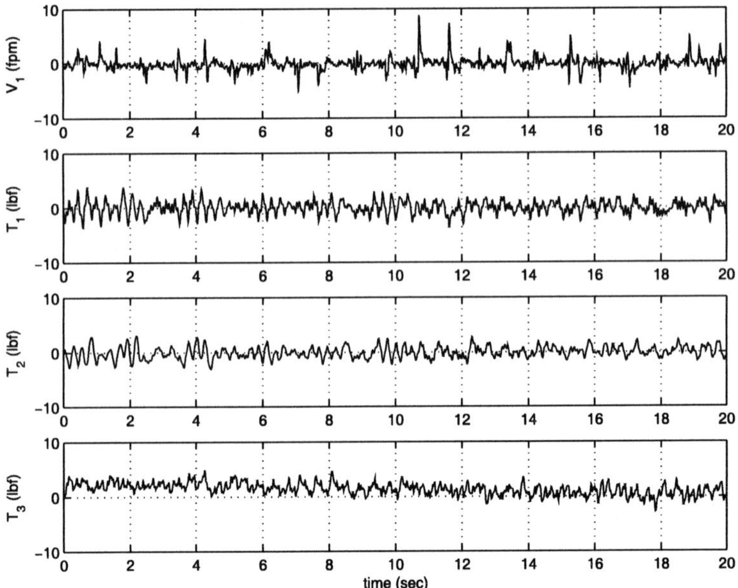

Fig. 7. Decentralized nonadaptive controller.

	PI	Nonadaptive	Adaptive
$\|V_1\|$	217.4	156.7	184.7
$\|T_1\|$	74.8	18.4	19.2
$\|T_2\|$	49.2	16.3	20.3
$\|T_3\|$	97.9	27.3	20.8

Table 1. Comparison of controllers.

5 Conclusions

A stable decentralized adaptive controller is developed for regulation of a class of large-scale systems. Both linear and nonlinear interconnections between subsystems of the large-scale system are considered. Further, these interconnections were assumed to be unmatched and uncertain. The proposed decentralized controller was implemented on a large experimental web platform and results were compared with those of an industrial decentralized PI controller. Experimental results show that, during web transport, web tension variations are much smaller with the proposed decentralized controller than the results obtained from the use of the decentralized industrial PI controller.

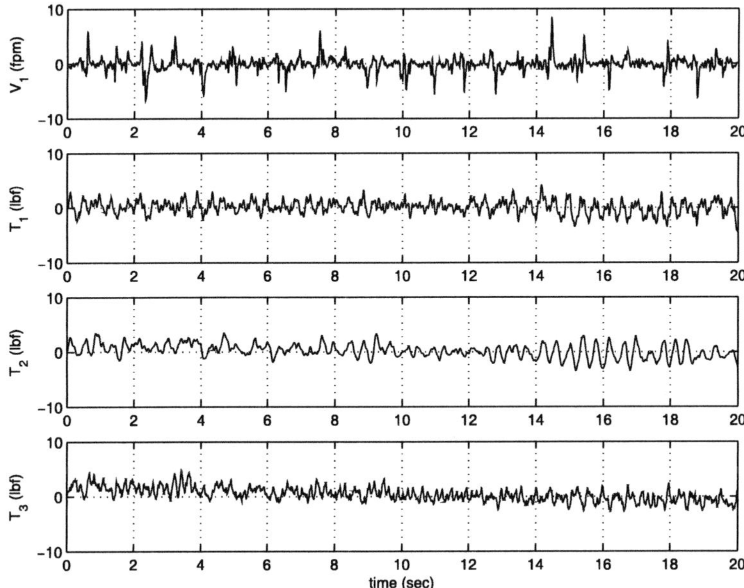

Fig. 8. Decentralized adaptive controller.

Acknowledgements. We would like to thank Ram Dwivedula for his helpful comments and suggestions.

References

1. Aboky C, Sallet G, Vivalda JC (2002) Observers for lipschitz nonlinear systems. International Journal of Control 75(3):204–212
2. Brandenburg G (1972) The dynamics of elastic webs threading a system of rollers. Newspaper techniques; the monthly publication of the INCA-FIEJ Research Association, 12–25
3. Brandenburg G (1977) New mathematical models for web tension and register error. International IFAC Conference on Instrumentation and Automation in the Paper Rubber and Plastics Industry 1:411–438
4. Byers R (1988) A bisection method for measuring the distance of a stable matrix to the unstable matrices. SIAM Journal on Scientific and Statistical Computing 9(3):875–881
5. Campbell DP (1958) John Wiley and Sons, New York
6. Feddema JT, Lewis C, Schoenwald DA (2002) Decentralized control of cooperative robotic vehicles: theory and application. IEEE Transactions on Robotics and Automation 18(5):852–864
7. Gavel DT, Siljak DD (1989) Decentralized adaptive control: structural conditions for stability. IEEE Transactions on Automatic Control 34(4):413–426

8. Grenfell KP (1963) Tension control on paper-making and converting machinery. IEEE 9th Annual Conference on Electrical Engineering in the Pulp and Paper Industry, 20–21

9. Ioannou PA (1986) Decentralized adaptive control of interconnected systems. IEEE Transactions on Automatic Control 31(4):291–298

10. King D (1969) The mathematical model of a newspaper press. Newspaper Techniques, 3–7

11. Koc H, Knittel D, de Mathelin M, Abba G (2002) Modeling and robust control of winding systems for elastic webs. IEEE Transactions on Control Systems Technology 10(2):197–208

12. Mirkin BM, Gutman PO (2003) Decentralized output-feedback MRAC of linear state delay systems. IEEE Transactions on Automatic Control 48(9):1613–1619

13. Narendra KS, Oleng' NO (2002a) Exact output tracking in decentralized adaptive control systems. IEEE Transactions on Automatic Control 47(2):390–395

14. Narendra KS, Oleng' NO (2002b) Decentralized adaptive control. Proceedings of American Control Conference, 3407–3412, Anchorage, AK

15. Pagilla PR, Dwivedula RV, Zhu Y, Perera LP (2003) Periodic tension disturbance attenuation in web process lines using active dancers. ASME Journal of Dynamic Systems, Measurement, and Control 125:361–371

16. Pagilla PR, Garimella SS, Dreinhoefer LH, King EO (2000) Dynamics and control of accumulators in continuous strip processing lines. IEEE Transactions on Industry Applications 36(3):865–870

17. Schaper CD, Kailath T, Lee YJ (1999) Decentralized control of wafer temperature for multizone rapid thermal processing systems. IEEE Transactions on Semiconductor Manufacturing 12(2):193–199

18. Shelton JJ (1986) Dynamics of web tension control with velocity or torque control. Proceedings of the American Control Conference 1423–1427, Seattle, WA

19. Shin KH (1991) Distributed control of tension in multi-span web transport systems. Ph.D. thesis, Oklahoma State University, Stillwater, OK

20. Siljak DD (1991) Decentralized control of complex systems. Academic Press, New York

21. Stilwell DJ, Bishop BE (2000) Platoons of underwater vehicles. IEEE Control Systems Magazine 20(6):45–52

22. Wolfermann W (1995) Tension control of webs, a review of the problems and solutions in the present and future. Proceedings of the Third International Conference on Web Handling, 198–229

23. Young G, Reid K (1993) Lateral and longitudinal dynamic behavior and control of moving webs. ASME Journal of Dynamic Systems, Measurement, and Control 115:309–317

Control Strategy Using Vision for the Stabilization of an Experimental PVTOL Aircraft Setup

Isabelle Fantoni,[1] Amparo Palomino,[1] Pedro Castillo,[1] Rogelio Lozano,[1] and Claude Pégard[2]

[1] Heudiasyc, UMR CNRS 6599,
Université de Technologie de Compiègne,
BP 20529, 60205 Compiègne, France
{ifantoni,apalomino,castillo,rlozano}@hds.utc.fr
[2] CREA - EA 3299, 7 rue du Moulin Neuf, 80000 Amiens, France
Claude.Pegard@u-picardie.fr

Summary. In this chapter, we stabilize the planar vertical takeoff and landing (PVTOL) aircraft using a camera. The camera is used to measure the position and orientation of the PVTOL moving on an inclined plane. We have developed a simple control strategy to stabilize the system in order to facilitate the real experiments. The proposed control law ensures convergence of the state to the origin.

1 Introduction

The planar vertical takeoff and landing (PVTOL) aircraft system is based on a simplified aircraft model with a minimal number of states and inputs. In the last few years, numerous control designs for the stabilization and trajectory tracking of the PVTOL aircraft model have been proposed. The proposed control techniques include the approximate I-O linearization procedure by [2], stabilization algorithm for nonlinear systems in so-called feedforward form by [12], output tracking of nonminimum phase flat systems in [6], linear high gain approximation of backstepping proposed by [10], robust hovering control of the PVTOL using nonlinear state feedback based on optimal control presented by [3]. Furthermore, a paper on an internal model based approach for the autonomous vertical landing on an oscillating platform has been proposed by Marconi et al. [5]. They presented an error-feedback dynamic regulator that is robust with respect to uncertainties of the model parameters and they provided global convergence to the zero-error manifold. Olfati-Saber [7] proposed a global configuration stabilization for the VTOL aircraft with a strong input coupling using a smooth static state feedback. Recently, control methodologies using embedded saturation functions have been proposed for the stabilization of the PVTOL aircraft. Indeed, Zavala et al. in [13] developed a new control strategy that coped with (arbitrarily) bounded inputs and provided global convergence to the origin. Lozano et al. [4] presented a simple control algorithm for stabilizing the PVTOL aircraft using Lyapunov convergence analysis. Experimental results have been provided using a four-rotor mini-helicopter.

The PVTOL system dynamics commonly used are quite simple and constitute a challenging nonlinear control problem. Moreover, the PVTOL problem is important because it retains the main features that must be considered when designing control laws for a real aircraft. It represents a good test bed for researchers, teachers and students working on flying vehicles.

Due to the difficulties in building an experimental platform of the PVTOL, there are very few experimental tests published in the literature. Note that, as far as we are aware, only M. Saeki et al. [9] carried out a real experiment of the PVTOL aircraft. Indeed, they offered a new design method, making use of the center of oscillation and a two-step linearization, and they provided some experimental results for a twin rotor helicopter model.

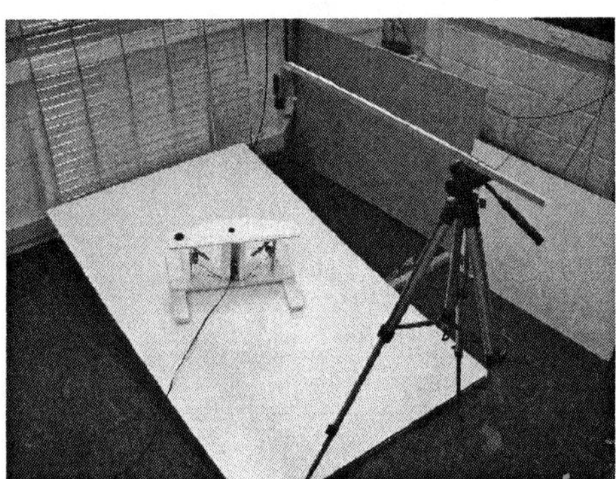

Fig. 1. Experimental setup.

In this chapter, we present both a simple control strategy and experimental results on the stabilization of the PVTOL aircraft, by using a camera for measuring the position and the orientation of the aircraft. We have developed an experimental setup for the PVTOL system. The platform is composed of a two-rotor radio-controlled object moving on an inclined plane (see Figure 1). The control strategy that has been used comes from [1] and [8]. The methodology is relatively simple and gives a satisfactory behavior. The PVTOL aircraft dynamics depicted in Figure 2 are given by the following equations:

$$\ddot{x} = -u_1 \sin\theta + \varepsilon u_2 \cos\theta$$
$$\ddot{y} = u_1 \cos\theta + \varepsilon u_2 \sin\theta - 1 \tag{1}$$
$$\ddot{\theta} = u_2 \, ,$$

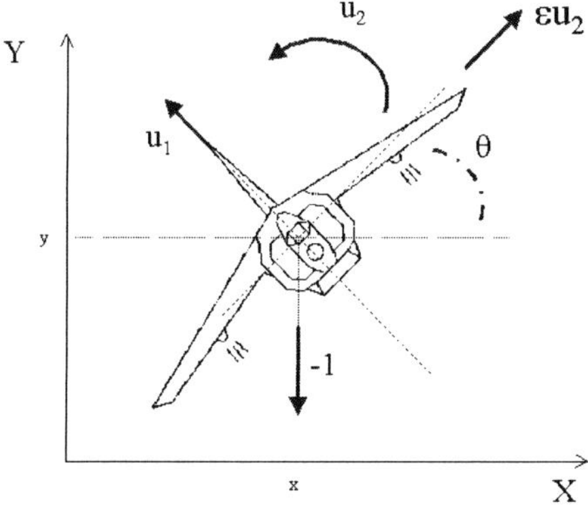

Fig. 2. The PVTOL aircraft (front view).

where x, y denote the center of mass horizontal and vertical positions and θ is the roll angle that the aircraft makes with the horizon. The control inputs u_1 and u_2 are, respectively, the thrust (directed out the bottom of the aircraft) and the angular acceleration (rolling moment). The constant -1 is the normalized gravitational acceleration. The parameter ε is a coefficient characterizing the coupling between the rolling moment and the lateral acceleration of the aircraft. Its value is in general so small that $\varepsilon = 0$ can be supposed in (1) (see for instance [2, §2.4]). Furthermore, several authors have shown that by an appropriate coordinate transformation, we can obtain a representation of the system without the term due to ($\varepsilon \neq 0$) [7, 9, 11]. Consequently, in this study we choose to consider the PVTOL aircraft dynamics with $\varepsilon = 0$, i.e.,

$$\ddot{x} = -u_1 \sin\theta$$
$$\ddot{y} = u_1 \cos\theta - 1 \qquad (2)$$
$$\ddot{\theta} = u_2 ,$$

which means that ε has been neglected.

The main contribution of this chapter is the development of a simple control strategy to stabilize the system and validation of the stabilizing control algorithm in real experiments using a camera as a position/orientation measuring device.

The chapter is organized as follows. In Section 2, we describe the environment used for our experiments. Section 3 describes the methodology used in the vision program to detect the position and the orientation of the PVTOL aircraft. In Section 4, the control approach is presented, and the stability analysis is developed in Section 5. Experimental results are shown in Section 6 and conclusions are finally given in Section 7.

2 Description of the experimental setup

The PVTOL aircraft prototype is shown in Figure 3. The rotors are driven separately by two electric Speed 400 motors, with a gear reduction of 1.85:1. One motor rotates clockwise while the second one rotates counterclockwise. The main thrust is the sum of the thrusts of each motor. The rolling moment is obtained by increasing (decreasing) the speed of one motor while decreasing (increasing) the speed of the second motor. Each motor is linked to a speed variator, which is itself linked to a gyroscope. The two gyroscopes that improve the maneuverability and stability of the object are connected to the receiver of the radio. The radio sends the signals through the transmitter to the receiver located on the PVTOL aircraft. The size of each propeller is 10 cm long. The mass of the PVTOL is 0.7 kg, while the inertia has been neglected. To provide additional information on the experimental setup, Table 1 describes correspondences between voltage of one motor, rpm of one propeller, speed of the wind, and thrust measured in the middle of the motors.

Table 1. Additional information for the PVTOL prototype.

Volts (V)	RPM (x10)	Anemometer (km/h)	Thrust (kg)
5.24	140	29.0	0.20
6.37	190	32.4	0.22
6.74	250	33.0	0.25
7.40	290	35.0	0.31
7.98	300	36.7	0.34

Fig. 3. The PVTOL prototype.

Since the maximal voltage of each motor is 10.8 V and the weight of the PVTOL aircraft is 0.7 kg, we can foresee that it would be difficult for our prototype to take off vertically. The PVTOL prototype is rather designed to move on an inclined plane. The general view of our experimental setup is depicted in Figure 1. The PVTOL moves on an inclined plane, which defines our two-dimensional workspace. The size of the PVTOL prototype is 60 cm (L) x 20 cm (W) x 32 cm (H), while the size of the inclined plane is 200 cm (L) x 122 cm (W) and the size of the camera field of vision on the plane is 128 cm (L) x 106 cm (W). The inclination of the plane is 15 deg.

The PVTOL platform in Figure 1 is an experimental setup designed to study the problems currently found in the navigation at low altitude of small flying objects. At low altitude, global positioning systems (GPS) and even inertial navigation systems are not enough to stabilize the mini-flying objects. Vision using cameras should provide additional information to make autonomous flights near the ground possible. We have therefore chosen to use a camera for measuring position and orientation of the mini-helicopter. For simplicity, at a first stage, we have placed the camera outside the aircraft. In the future, the camera will be located at the base of the mini-helicopter pointing downward or upward. Note that even when the camera is located outside the flying object, we still have to deal with the problems of object localization computation using cameras and delays in the closed loop system.

In the platform, a CCD camera Pulnix is located perpendicular to the plane at a fixed altitude and provides an image of the whole workspace. We have used an acquisition card PCI-1409 by the National Instruments Company. The camera is linked to the personal computer (PC) dedicated to the vision part (called Vision PC). From the image provided by the camera, the program calculates the position (x, y) and the orientation θ of the PVTOL with respect to a given origin. Then the Vision PC sends this information to another PC dedicated to the control part (called Control PC), via a RS232 connection, transmitting at 115200 bauds. The two control inputs are therefore calculated according to the proposed strategy and sent to the PVTOL via the radio. To simplify the implementation of the control law, we have designed the platform so that each of the two control inputs can independently work either in automatic or manual mode. The Vision PC calculates the position and orientation every 40 ms, while the Control PC requires this information every 2 ms. Therefore, the minimum sampling period we are able to obtain in the experimental platform is 40 ms. This includes the computation of the control law, image processing, localization computation, and A/D and D/A conversion in the radio-PC interface. Figure 4 shows a diagram of the radio-PC interface.

3 Position and orientation of the PVTOL aircraft using vision

We have placed two black points on the white PVTOL experiment to obtain a contrasted image (see Figure 1). One of these black points is larger than

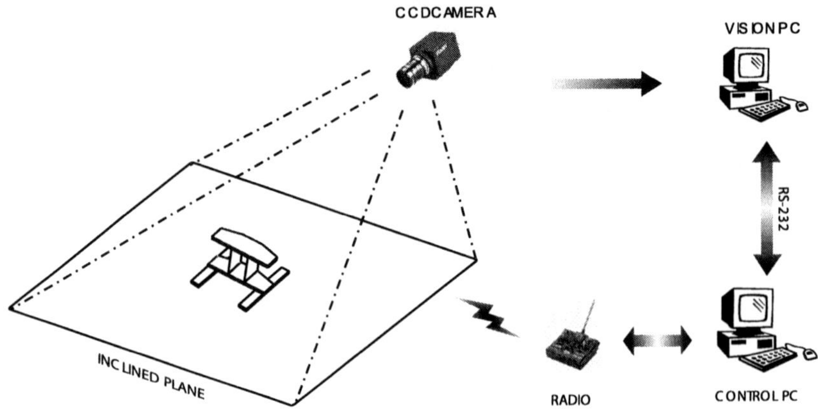

Fig. 4. The system interface.

the other one. The smaller one corresponds to the position of the PVTOL system. Its orientation is determined by the angle between the line linked up the two points and the horizontal axis in the image plane. From the scene, we obtain a two-dimensional image by using the camera as described in Section 2. This image is saved in a computer via an acquisition card PCI-1409 by the National Instruments Company.

The acquired image is black and white (given the real conditions of the scene) and does not need a binarisation process. We detect the black points on the white background in the following way. Starting at the top of the image, we skim through all the pixels, line by line, considering the gray level of each pixel. We save the position of all the pixels having a gray level whose value is 255 or having a "black" gray level, ignoring those having a gray level whose value is 0 or having a "white" gray level. The program classifies all the pixels having a high gray level and being in the same neighborhood. Then it calculates the barycenter of these pixels, which gives the position of each black point in the image plane. The orientation angle of the PVTOL aircraft is then obtained with the help of the straight line linked up the two black points and the horizontal of the reference system in the image plane. The program computing the capture and the image processing was written in the programming language C.

4 Stabilizing control law

In this section, we propose a control law that will be applied to the experimental setup. The control strategy follows the controller synthesis approach developed in [1] and [8].

The controller is obtained by defining the following desired linear behavior for the position x and the altitude y. Let us, therefore, define the following functions r_1 and r_2 as

$$\ddot{x} \triangleq r_1(x, \dot{x}) = -2\dot{x} - x \tag{3}$$

$$\ddot{y} \triangleq r_2(y, \dot{y}) = -2\dot{y} - y. \tag{4}$$

Other choices are possible but the above have been chosen for simplicity. From (2) and (4) it follows that

$$u_1 = \frac{1}{\cos\theta}(1 + r_2), \tag{5}$$

which will not have any singularity provided $\tan\theta$ is bounded. In the stability analysis (Section 5), we will prove that this is indeed the case. Introducing (5) into the system (2) gives

$$\ddot{x} = -\tan\theta(1 + r_2) \tag{6}$$

$$\ddot{y} = r_2. \tag{7}$$

From (7), it follows that $y^{(i)} \to 0$ for $i = 0, 1, \dots$. It means that the altitude is stabilized around the origin: $y^{(i)} \in L_2$ and $r_2 \in L_2$. Independently of the value of $\cos\theta$, (7) holds. Note that if $\cos\theta \to 0$, then from (5), $u_1 \to \infty$. We will prove later that this is not the case.

Let us rewrite (6) as follows:

$$\ddot{x} = -\tan\theta(1 + r_2) \pm r_1(1 + r_2) \tag{8}$$

$$= r_1(1 + r_2) - (\tan\theta + r_1)(1 + r_2). \tag{9}$$

Since we will prove, in the stability analysis, that r_1 will tend to zero, we also would like $(\tan\theta + r_1)$ to converge to zero. Therefore, we introduce the error variable:

$$\nu_1 \triangleq \tan\theta + r_1, \tag{10}$$

then

$$\dot{\nu}_1 = (1 + \tan^2\theta)\dot{\theta} + \dot{r}_1 \tag{11}$$

$$\ddot{\nu}_1 = (1 + \tan^2\theta)(u_2 + 2\tan\theta\dot{\theta}^2) + \ddot{r}_1. \tag{12}$$

We choose a control input u_2, so that the closed loop system above is given by

$$\ddot{\nu}_1 = -2\dot{\nu}_1 - \nu_1, \tag{13}$$

where $s^2 + 2s + 1$ is a stable polynomial. Therefore, $\nu_1 \to 0$. The controller u_2 is then given by

$$u_2 = \frac{1}{1 + \tan^2 \theta} \left(-2\dot{\theta}^2 \tan\theta (1 + \tan^2 \theta) - \ddot{r}_1 \right.$$
$$\left. - \tan\theta - r_1 - 2(1 + \tan^2\theta)\dot{\theta} - 2\dot{r}_1 \right) . \tag{14}$$

u_2 is a function of $\{\theta, \dot{\theta}, r_1, \dot{r}_1, \ddot{r}_1\}$ and all these variables can be expressed as a function of $\{x, y, \theta\}$ and their derivatives.

Indeed,

$$r_1 = -2\dot{x} - x \tag{15}$$

$$\dot{r}_1 = \frac{2}{\cos\theta}(1 - 2\dot{y} - y) - \dot{x} \tag{16}$$

$$\ddot{r}_1 = -\frac{2\sin\theta\dot{\theta}}{\cos^2\theta}(1 - 2\dot{y} - y) + \frac{2}{\cos\theta}(3\dot{y} + 2y)$$
$$+2\dot{x} + x . \tag{17}$$

From (10) through (14) it follows that the closed loop system can be rewritten as

$$\dot{\nu} = A\nu , \tag{18}$$

where A is an exponentially stable matrix and $\nu^T = [\nu_1, \dot{\nu}_1]$. ν_1 converges exponentially to zero and $\nu_1 \in L_2 \cap L_\infty$. The main result is summarized in the following theorem.

Theorem 1. *Consider the PVTOL aircraft model (2) and the control law in (5) and (14). Then the solution of the closed loop system converges asymptotically to the origin, provided that $|\theta(0)| < \frac{\pi}{2}$.*

We present the stability analysis of the above result in the following section.

5 Stability analysis

Let us rewrite the $(x, \dot{x})$ subsystem (9)–(10) as a linear system. Define $z^T = [x, \dot{x}]$, then

$$\dot{z} = Az + Bu \tag{19}$$

$$x = Cz , \tag{20}$$

where $A = \begin{bmatrix} 0 & 1 \\ -1 & -2 \end{bmatrix}$, $C = [1, 0]$, $B = [0, 1]^T$, and $u = r_1 r_2 - \nu_1(1 + r_2)$. Let us define $r_3 = \nu_1(1 + r_2)$. In view of (13), ν_1 converges exponentially to zero. Furthermore, in view of (2), (4), and (5), r_2 converges exponentially to zero.

Since ν_1 and r_2 both converge exponentially to zero, $r_3 \in L_2$, i.e., $\int_0^\infty r_3^2 dt$ is bounded. Therefore, $\int_0^t r_3^2 dt + \int_t^\infty r_3^2 dt = \int_0^\infty r_3^2 dt = $ constant and $\frac{d}{dt} \int_t^\infty r_3^2 dt = -\frac{d}{dt} \int_0^t r_3^2 dt = -r_3^2$. We propose the following Lyapunov function candidate

$$V = z^T P z + 2 \int_t^\infty r_3^2 dt \,, \tag{21}$$

where P is a positive definite matrix, satisfying $A^T P + PA = -2I$. Note that $P = \begin{bmatrix} 3 & 1 \\ 1 & 1 \end{bmatrix}$. Differentiating V, we obtain

$$\begin{aligned}
\dot{V} &= \dot{z}^T P z + z^T P \dot{z} - 2r_3^2 \\
&= \left[z^T A^T + u^T B^T \right] P z + z^T P \left[Az + Bu \right] - 2r_3^2 \\
&= z^T \left[A^T P + PA \right] z + 2z^T PBu - 2r_3^2 \\
&= -2z^T z + 2z^T PBu - 2r_3^2 \\
&= -2(\dot{x}^2 + x^2) + 2(\dot{x} + x) \left[(-2\dot{x} - x)r_2 - r_3 \right] \\
&\quad - 2r_3^2 \,.
\end{aligned} \tag{22}$$

Since

$$\begin{aligned}
|2(\dot{x} + x)r_3| &= |2\dot{x}r_3 + 2xr_3| \\
&\le 2|\dot{x}r_3| + 2|xr_3| \\
&\le \dot{x}^2 + r_3^2 + x^2 + r_3^2 \,,
\end{aligned} \tag{23}$$

$\dot{V}$ becomes

$$\begin{aligned}
\dot{V} &\le -2(1 + 2r_2)\dot{x}^2 - 2(1 + r_2)x^2 - 6x\dot{x}r_2 \\
&\quad + x^2 + \dot{x}^2 + 2r_3^2 - 2r_3^2 \\
&\le -\dot{x}^2 - x^2 - 4r_2\dot{x}^2 - 2r_2 x^2 - 6x\dot{x}r_2 \,.
\end{aligned} \tag{24}$$

Since r_2 is decreasing exponentially to zero, it follows that for any $\bar{k} > 0$ there exists a T large enough such that $r_2 < \bar{k}$, $\forall t > T$. In the sequel the results will hold for $t \ge T$. Since

$$|6r_2 x\dot{x}| \le 6|\bar{k}x\dot{x}| \le 3\bar{k}(x^2 + \dot{x}^2) \,, \tag{25}$$

we therefore obtain

$$\dot{V} \le -(1 - 7\bar{k})\dot{x}^2 - (1 - 5\bar{k})x^2 \quad \forall t > T \,. \tag{26}$$

Choosing $\bar{k} = \frac{1}{10}$, it follows that $\dot{V} < 0$.

Using (3) and (10), (6) becomes

$$\begin{aligned}
\ddot{x} &= -\tan\theta(1 + r_2) \tag{27} \\
&= -(\nu_1 - r_1)(1 + r_2) \tag{28} \\
&= -(\nu_1 + 2\dot{x} + x)(1 + r_2) \,. \tag{29}
\end{aligned}$$

Since both ν_1 and r_2 are exponentially decreasing, then $\ddot{x}$ is linear with respect to x and $\dot{x}$. Therefore x grows at most exponentially and does not exhibit finite escape time. From (21) and (26), we then have x and $\dot{x} \in L_2 \cap L_\infty$. Thus x and $\dot{x} \to 0$. Since

$$\tan \theta = \nu_1 - r_1 \tag{30}$$
$$= \nu_1 + 2\dot{x} + x\,, \tag{31}$$

it follows that $\tan \theta \in L_2 \cap L_\infty$. We have thus proved that $\tan \theta$ is bounded and $\tan \theta \to 0$. Therefore, $\cos \theta \not\to 0$ and then the control law (5) is free from singularities. Finally, the solution of the closed loop system converges asymptotically to zero for any initial condition such that $|\theta(0)| < \frac{\pi}{2}$.

6 Experimental results

In this section, we present the experimental results when the control law given in Section 4 is applied to the PVTOL platform described in Section 2, where the x, y position and the orientation θ measurements are obtained from the image given by the camera. In the computation of the control law, we also require the time derivatives of x, y and θ. They will be obtained using the following approximation:

$$\dot{q}_t \approx \frac{q_t - q_{t-T}}{T}\,, \tag{32}$$

where q represents either x, y or θ and T is the sampling period. The measurement of x, y and θ are expressed in pixels in the image frame, which means that the servoing is done on the basis of image features directly. For the real experiment, we have introduced in the model and in the control law the mass of the PVTOL. In Figure 5, the results of the image acquisition program are shown. We clearly see the detection of the two points located on the PVTOL prototype. From the measurement of these two points, we compute the position x, y and the angle θ of the system as explained in Section 3.

In particular, we started the PVTOL aircraft at the origin, and the objective is to stabilize it at the position $(x, \dot{x}, y, \dot{y}, \theta, \dot{\theta}) = (0, 0, 60, 0, 0, 0)$ in pixels during $t \in [30s, 40s]$, then to bring back the aircraft at the origin. The results are shown on Figures 6, 7, and 8.

In Figure 6, we can see the difference between the real horizontal position of the PVTOL (along x) and the desired horizontal position. Along the horizontal axis x, 1 cm corresponds to 5 pixels, which means that the position error is approximately 2 or 3 cm. Figure 7 describes the difference between the real altitude of the PVTOL (along y) and the desired altitude. We can notice that it satisfactorily follows the desired reference. Along the vertical axis, 1 cm corresponds to 2.5 pixels, which also means that the position error along y is approximately 2 or 3 cm. Figure 8 shows the evolution of the angle θ. In this figure, we visualize the effect of the control law that brings back the angle to zero as the PVTOL goes up toward the altitude of 60 pixels. This also explains the variations of the PVTOL along the horizontal axis x. Moreover,

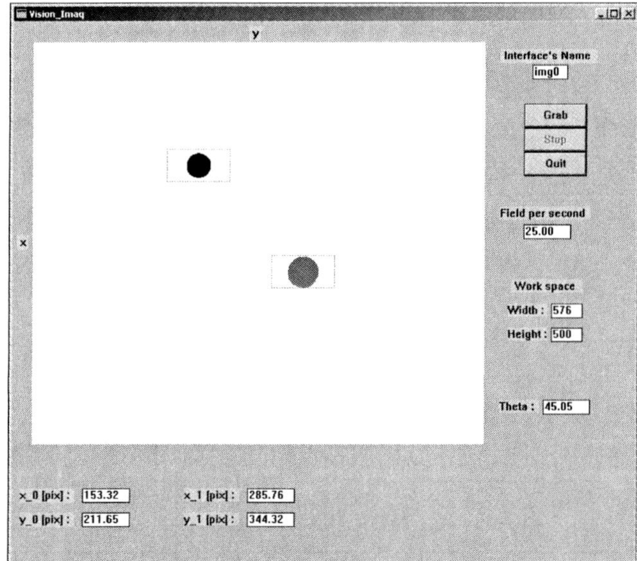

Fig. 5. The vision interface.

the differences between real and desired trajectories and the "staircase" traces appearing in Figure 7 are also due to small frictions when the object moves on the plane and that we deliberately have not considered. The results are nevertheless satisfactory.

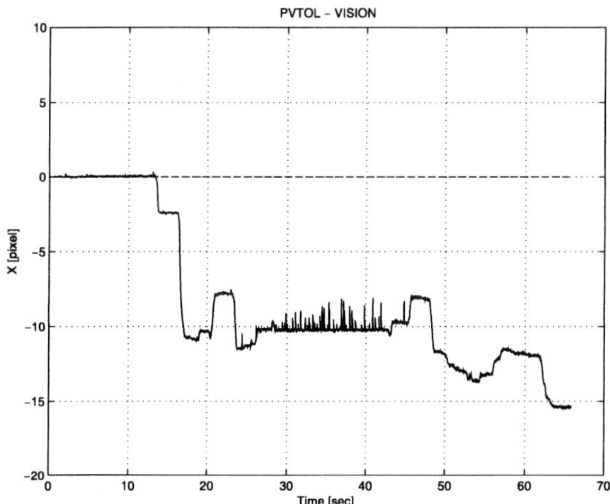

Fig. 6. Position x of the PVTOL system (- - - desired position, — real position).

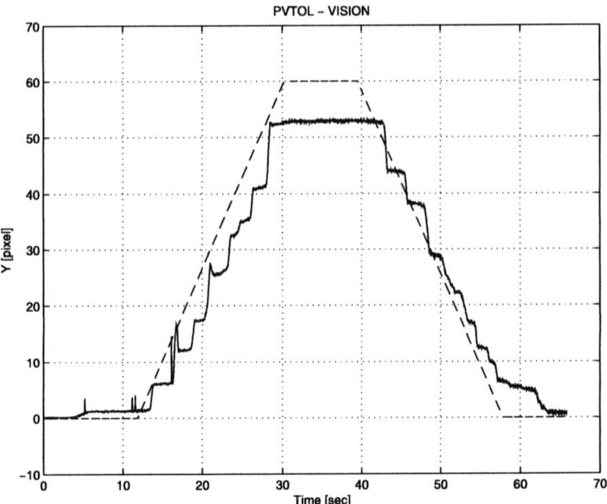

Fig. 7. Position y of the PVTOL system (- - - desired position, — real position).

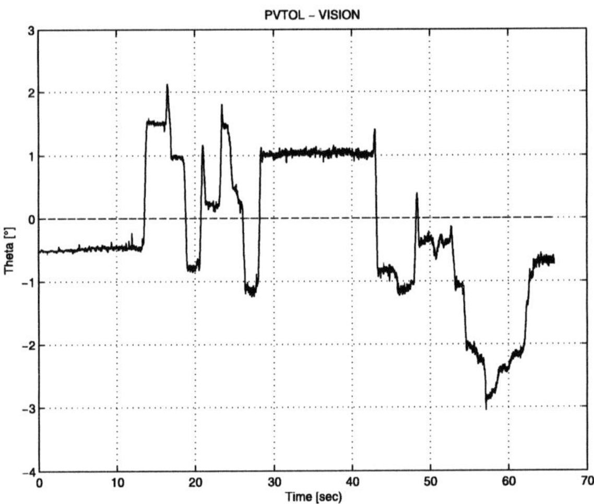

Fig. 8. Angle θ of the PVTOL system (- - - desired position, — real position).

7 Conclusions

We have presented a stabilizing control strategy for the PVTOL and its application in an experimental platform. This platform exhibits the same difficulties found in autonomous flight close to the ground and can be used as a benchmark for developing controllers for unmanned flying vehicles. The position and orientation of the PVTOL are computed using the image provided by a camera. We have developed a real-time environment to be able to validate the proposed control law. The experimental results showed satisfactory behavior of the closed loop system. Future works include visual servoing when the camera is on-board and pointing downward or upward to estimate the flying object position and orientation. We also believe that this experimental platform is a good test bed for educational purposes in the domain of small flying vehicles.

References

1. Fantoni I, Lozano R, Castillo P (2002) A simple stabilization algorithm for the PVTOL aircraft. In: 15th IFAC World Congress, Barcelona, Spain
2. Hauser J, Sastry S, Meyer G (1992) Nonlinear control design for slightly nonminimum phase systems: application to V/STOL aircraft. Automatica, 28(4):665–679
3. Lin F, Zhang W, Brandt RD (1999) Robust hovering control of a PVTOL aircraft. IEEE Transactions on Control Systems Technology, 7(3):343–351
4. Lozano R, Castillo P, Dzul A (2004) Global stabilization of the PVTOL: real-time application to a mini-aircraft. International Journal of Control, 77(18):735–740
5. Marconi L, Isidori A, Serrani A (2002) Autonomous vertical landing on an oscillating platform: an internal-model based approach. Automatica, 38:21–32
6. Martin P, Devasia S, Paden B (1996) A different look at output tracking: control of a VTOL aircraft. Automatica, 32(1):101–107
7. Olfati-Saber R (2002) Global configuration stabilization for the VTOL aircraft with strong input coupling. IEEE Transactions on Automatic Control, 47(11):1949–1952
8. Palomino A, Castillo P, Fantoni I, Lozano R, Pégard C (2003) Control strategy using vision for the stabilization of an experimental PVTOL aircraft setup. In: 42th IEEE Conference on Decision and Control, Maui, HI
9. Saeki M, Sakaue Y (2001) Flight control design for a nonlinear nonminimum phase VTOL aircraft via two-step linearization. In: 40th IEEE Conference on Decision and Control, CDC 01, Orlando, FL
10. Sepulchre R, Janković M, Kokotović P (1997) Constructive nonlinear control. Springer-Verlag, London
11. Setlur P, Dawson YF, Costic B (2001) Nonlinear tracking control of the VTOL aircraft. In: 40th IEEE Conference on Decision and Control, CDC 01, Orlando, FL
12. Teel AR (1996) A nonlinear small gain theorem for the analysis of control systems with saturation. IEEE Transactions on Automatic Control, 41(9):1256–1270
13. Zavala A, Fantoni I, Lozano R (2003) Global stabilization of a PVTOL aircraft with bounded inputs. International Journal of Control, 76(18):1833–1844

Neural Network Model Reference Adaptive Control of Marine Vehicles

Alexander Leonessa, Tannen VanZwieten, and Yannick Morel

University of Central Florida
Department of Mechanical, Materials & Aerospace Engineering
P.O. Box 162450, Orlando, FL 32816, USA
`aleo@mail.ucf.edu,{tannen.vanzwieten,ymorel}@gmail.com`

Summary. A neural network model reference adaptive controller for trajectory tracking of nonlinear systems is developed. The proposed control algorithm uses a single layer neural network that bypasses the need for information about the system's dynamic structure and characteristics and provides portability. Numerical simulations are performed using nonlinear dynamic models of marine vehicles. Results are presented for two separate vehicle models, an autonomous surface vehicle and an autonomous underwater vehicle, to demonstrate the controller performance in terms of tuning, robustness, and tracking.

1 Introduction

Autonomous marine vehicles are used for a wide range of assignments, including oceanographic surveys, coastal patrols, pipeline maintenance, and mine hunting. Such missions require a high degree of agility and maneuverability, which can only be provided by a high-performance motion control system. We distinguish two generic types of marine vehicles: autonomous underwater vehicles (AUVs) and autonomous surface vehicles (ASVs). AUVs and ASVs are complementary, with each vehicle having distinct, mission specific advantages and disadvantages. This is enhanced by the prospect of collaborative control, where the ASVs can be used to transmit real time positioning and communication information to the AUVs through the air-sea interface.

The development of control algorithms for marine vehicles is the focus of several research groups around the world. For example, three different stabilization algorithms are developed in [6] for an underactuated hovercraft. However, the mathematical model for a hovercraft is simple when compared to that of a standard marine vehicle, as it does not contain many of the nonlinear, coupled drag terms that generally characterize marine vehicles. Furthermore, the tracking performance of the algorithm is not assessed, but instead the velocities are regulated to zero, which is a more fundamental problem.

Many tracking controllers are available in the literature (see, for example, [2, 3, 4, 7, 9, 17, 18]). In [18], the case of a surface ship equipped with a pair of propellers is considered. This controller yields interesting results. However, the desired trajectory is limited to straight lines and circles. The same type of underactuation is considered in [17], where the authors derive a controller that uses a state estimator to handle state measurement uncertainties.

In [2] a global controller is designed that accounts for control amplitude saturation effects. Then, in [3], a velocity observer is added to the control algorithm, enabling the controller to work without velocity measurements.

A controller for an underactuated AUV equipped with a propeller and a side thruster is designed in [7]. The controller handles constant and slow varying external perturbations.

Typically, the design of a motion controller relies on the system's mathematical model. However, in the case of marine vehicles it is extremely challenging to obtain a model that will satisfactorily capture the dynamic behavior of the system. To compensate for this, one can use an adaptive approach to handle uncertainties in the geometric and and hydrodynamic constant parameters.

The design of adaptive controllers for marine vehicles has been widely studied. In [5], a nonlinear model-based adaptive output feedback controller was developed for a surface vessel. Global asymptotic position tracking was achieved assuming the structure of the mathematical model was of a particular form, with constant inertia and damping matrices. This structure was also extended to include a bias term representing drift, currents, and wave load. Simulations were presented, but did not demonstrate the controller robustness to unmodeled dynamics.

The dynamical behavior of a marine vehicle can only be partially depicted using current modeling techniques. These dynamics are especially challenging when the vehicle's velocity is not constant, as when it is following a search pattern or when its desired path is constantly being modified. Additionally, the ocean environment is characterized by large unknown perturbations. These features make it desirable to have a control system that is robust to model parameter and structure uncertainty.

This chapter introduces a neural network model reference adaptive controller (NN-MRAC) that has valuable self-tuning capabilities that allows it to adapt to the operating conditions in order to optimize the tracking performance of the closed loop system. The parameter update mechanism is derived using Lyapunov stability theory and guarantees that the tracking error is ultimately bounded when subject to some generalized constraints. The addition of a single layer neural network bypasses the need for information about the system's dynamic structure and characteristics. The control algorithm is applied to the surface vessel's nonlinear model presented in [19] and the AUV's nonlinear model from [14]. Numerical simulation results are presented for both cases.

2 Mathematical Preliminaries

In this section we establish definitions, notation, and a key result used later in the chapter. Let $\mathbb{R}$ denote the set of real numbers, let $\mathbb{R}^n$ denote the set of $n \times 1$ real column vectors, let $\mathbb{R}^{n \times m}$ denote the set of real $n \times m$ matrices, and let $(\cdot)^{\mathrm{T}}$ denote transpose. Furthermore, $\|\cdot\|$ represents the Euclidean vector norm, and $A > 0$ denotes the fact that the Hermitian matrix A is positive definite. For a subset $\mathcal{S} \subset \mathbb{R}^n$, we write $\partial \mathcal{S}$, $\overset{\circ}{\mathcal{S}}$ for the boundary and the interior of $\mathcal{S}$, respectively.

In this chapter we consider nonlinear controlled dynamical systems of the form

$$\dot{x}(t) = f(x(t), u(t)), \qquad x(0) = x_0, \qquad t \geq 0, \tag{1}$$

where $x(t) \in \mathcal{D} \subseteq \mathbb{R}^n$, $t \geq 0$, is the system state vector, $\mathcal{D}$ is an open set, $0 \in \mathcal{D}$, $u(t) \in \mathcal{U} \subseteq \mathbb{R}^m$, $t \geq 0$, is the control input, $\mathcal{U}$ is the set of all admissible controls such that $u(\cdot)$ is a measurable function, and $f(\cdot, \cdot)$ is Lipschitz on $\mathcal{D} \times \mathcal{U}$. The closed-loop dynamical system corresponding to a feedback control $u(x(t), t)$, $t \geq 0$, is given by

$$\dot{x}(t) = \tilde{f}(x(t), t) \triangleq f(x(t), u(x(t), t)), \qquad x(0) = x_0, \qquad t \geq 0. \tag{2}$$

Following [12], the system (2) is said to be *ultimately bounded* if there is a compact set $M_{\mathrm{c}} \subset \mathcal{D}$, $0 \in M_{\mathrm{c}}$, such that corresponding to each solution $x(t)$, $t \geq 0$, of (2) there is a $T > 0$ with the property that $x(t) \in M_{\mathrm{c}}$ for all $t > T$. The following theorem is introduced to establish a sufficient condition for ultimate boundedness.

Theorem 1. *[12] Consider the closed loop nonlinear dynamical system (2) and assume that the forward solution $x(t)$, $t \geq 0$, corresponding to an initial condition $x(0) = x_0 \in \mathcal{D}$ exists. Assume that there exists a continuously differentiable function $V : \mathcal{D} \to \mathbb{R}$ and a compact set $M \subset \mathcal{D}$, $0 \in \overset{\circ}{M}$, such that*

$$V(0) = 0, \qquad V(x) > 0, \qquad x \in \mathcal{D} \backslash \{0\}, \tag{3}$$

$$\frac{\mathrm{d}V(x)}{\mathrm{d}x} \tilde{f}(x, t) \leq -\epsilon < 0, \qquad x \in \mathcal{D} \backslash M, \qquad t \geq 0, \tag{4}$$

where $\epsilon > 0$. Then (2) is ultimately bounded, that is, there exists a time $T > 0$ and a compact set $M_{\mathrm{c}} \supseteq M$ such that $x(t) \in M_{\mathrm{c}}$ for all $t > T$.

Remark 1. From the proof of Theorem 1, it follows that the domain of convergence M_{c} is defined as the smallest level set associated with the Lyapunov function candidate $V(\cdot)$ that entirely contains the compact set M.

3 Modeling

In this section we present a generic mathematical model of a marine vehicle that will be used for the development and testing of the controllers. In particular, our model assumes that pitch, roll and heave motions are negligible and feature only the three degrees of freedom corresponding to surge, sway, and yaw motions. Because of this choice we will assume that the vehicle's state space $\mathcal{D}$ coincides with $\mathbb{R}^6$, although the control algorithm can be easily extended to higher dimensions.

3.1 Equations of Motion

The notation used for the vehicle's generalized equations of motion follows [8], but is reduced to motion in the horizontal plane. The earth fixed frame (EFF), denoted by x_e, y_e and z_e, is chosen so that the vehicle's center of gravity is at the origin at time $t = 0$. The x_e and y_e axes are directed toward the north and the east, respectively, while the z_e axis points downward in accordance with the right-hand rule. This frame is assumed to be inertial, the acceleration due to the earth's rotation being considered negligible. The vehicle's configuration in the EFF is

$$\eta(t) \triangleq [x_N(t), y_E(t), \psi(t)]^T, \qquad t \geq 0, \tag{5}$$

where $x_N(t) \in \mathbb{R}$ and $y_E(t) \in \mathbb{R}$ describe the distance traveled along the x_e and y_e directions, respectively, and $\psi(t) \in \mathbb{R}$ describes the rotation about the z_e axis.

The body fixed frame (BFF) has its origin fixed at the vehicle's center of gravity, the x_b axis points forward, the y_b axis starboard, and the z_b axis downward. The vehicle's velocity is defined in the BFF as

$$\nu(t) \triangleq [u(t), v(t), r(t)]^T, \qquad t \geq 0, \tag{6}$$

where $u(t) \in \mathbb{R}$ and $v(t) \in \mathbb{R}$ are the components of the absolute velocity in the x_b and y_b directions, respectively, and $r(t) \in \mathbb{R}$ describes the angular velocity about the z_b axis. The vectors $\eta(t)$ and $\nu(t)$ are related by the kinematic equation [8],

$$\dot{\eta}(t) = J(\eta(t))\nu(t), \qquad t \geq 0, \tag{7}$$

where

$$J(\eta) \triangleq \begin{bmatrix} \cos\psi & -\sin\psi & 0 \\ \sin\psi & \cos\psi & 0 \\ 0 & 0 & 1 \end{bmatrix}, \tag{8}$$

is the rotation matrix from the BFF to the EFF.

Using the form introduced in [8] and the previous notation, the marine vehicle's equation of motion is given by

$$M\dot{\nu}(t) + C(\nu(t))\nu(t) + D(\nu(t))\nu(t) + g(\eta(t)) = \hat{B}\tau(t), \qquad t \geq 0, \tag{9}$$

where $M \in \mathbb{R}^{3\times 3}$ is the mass matrix (including added mass, see [19] for more details), $C(\nu(t)) \in \mathbb{R}^{3\times 3}$ contains Coriolis, centripetal, and added-mass terms, $D(\nu(t)) \in \mathbb{R}^{3\times 3}$ is the damping matrix, $g(\eta(t)) \in \mathbb{R}^3$ is the vector of restoring forces and moments, $\tau(t) \in \mathbb{R}^m$ is the input vector, and $\hat{B} \in \mathbb{R}^{3\times m}$ characterizes how the control inputs affect the dynamics of the vehicle.

While the rigid body inertia, Coriolis, centripetal, and gravitational terms are described in [8], the hydrodynamic terms are much more challenging to model and depend on the particular geometry of the considered vehicle. In general, even very thorough hydrodynamic modeling efforts are only able to partially describe the dynamic behavior of a marine vehicle, as the assumptions made always considerably affect the final result. In light of these considerations we are going to write the vehicle dynamics as

$$\dot{\nu}(t) = f(x(t)) + B\tau(t), \qquad t \geq 0, \tag{10}$$

where $x \triangleq [\,\eta^{\mathrm{T}}\ \nu^{\mathrm{T}}\,]^{\mathrm{T}}$ is the state vector and

$$B \triangleq M^{-1}\hat{B},$$
$$f(x) \triangleq -M^{-1}[C(\nu) + D(\nu)]\nu - M^{-1}g(\eta),$$

are assumed to be unknown.

4 Adaptive Controller Design

The nonlinearities, unmodeled dynamics, and strong dynamic coupling inherent to marine vehicle models make it desirable to have a multi-input/multi-output control system that is capable of self-tuning. We chose to solve this problem using an adaptive control approach. The stability analysis and the parameter update mechanism are derived using Lyapunov stability theory.

To simplify the presentation, the explicit dependence of the state variables upon time, when obvious, will be omitted.

4.1 Reference System

When using model reference adaptive control, a control algorithm is developed so that the system mimics the behavior of a reference system that provides smooth convergence to the desired trajectory. Choosing the appropriate reference system allows the vehicle to exhibit less overshoot and oscillatory behavior as well as better tracking performance. Furthermore, the control inputs become more realistic, even when the vehicle is far away from the desired trajectory, diminishing the need to implement input amplitude and rate saturation algorithms.

We consider a linear reference system that can be written as

$$\dot{x}_{\mathrm{r}}(t) = A_{\mathrm{r}}x_{\mathrm{r}}(t) + B_{\mathrm{r}}\hat{r}(t), \qquad t \geq 0, \tag{11}$$

where $x_r(t) \in \mathbb{R}^{2m}$ is the reference state, $A_r \in \mathbb{R}^{2m \times 2m}$, $B_r \in \mathbb{R}^{2m \times m}$ are constant matrices, and $\hat{r}(t) \in \mathbb{R}^m$ is the reference input.

The reference system considered here is composed of three uncoupled second-order oscillators. Each oscillator is characterized by a damping coefficient $\zeta_i > 0$, $i = 1, ..., m$ and a natural frequency $w_{0i} > 0$, $i = 1, ..., m$. This choice was mostly motivated by the simplicity of the corresponding reference dynamics. The dynamics of the ith oscillator are given by

$$\ddot{x}_{ri}(t) + 2\zeta_i \omega_{0i} \dot{x}_{ri}(t) + \omega_{0i}^2 x_{ri}(t) = \omega_{0i}^2 \hat{r}_i(t), \qquad t \geq 0, \qquad i = 1, ..., m. \quad (12)$$

The reference system can thus be rewritten as

$$\begin{bmatrix} \dot{x}_{1r}(t) \\ \dot{x}_{2r}(t) \end{bmatrix} = \begin{bmatrix} 0_m & I_m \\ -\omega_0^2 & -A_{rm} \end{bmatrix} \begin{bmatrix} x_{1r}(t) \\ x_{2r}(t) \end{bmatrix} + \begin{bmatrix} 0_m \\ \omega_0^2 \end{bmatrix} \hat{r}(t), \qquad t \geq 0, \qquad (13)$$

where

$$x_{1r}(t) \triangleq \begin{bmatrix} x_{r1}(t) & ... & x_{rm}(t) \end{bmatrix}^{\text{T}}, \qquad t \geq 0, \qquad (14)$$

$$x_{2r}(t) \triangleq \begin{bmatrix} \dot{x}_{r1}(t) & ... & \dot{x}_{rm}(t) \end{bmatrix}^{\text{T}}, \qquad t \geq 0, \qquad (15)$$

and

$$A_{rm} \triangleq \text{diag}(2\zeta_1 \omega_{01}, ..., 2\zeta_p \omega_{0m}), \qquad (16)$$

$$\omega_0 \triangleq \text{diag}(\omega_{01}, ..., \omega_{0m}). \qquad (17)$$

Finally, the desired trajectory for the marine vehicle may be written as

$$\tilde{x}_d(t) = \begin{bmatrix} x_{d1}(t) & ... & x_{dm}(t) \end{bmatrix}^{\text{T}}, \qquad t \geq 0. \qquad (18)$$

By choosing

$$\hat{r}(t) = \omega_0^{-2}(\ddot{\tilde{x}}_d(t) + A_{rm}\dot{\tilde{x}}_d(t) + \omega_0^2 \tilde{x}_d(t)), \qquad t \geq 0, \qquad (19)$$

we find that

$$\begin{bmatrix} \dot{x}_{1r}(t) - \dot{\tilde{x}}_d(t) \\ \dot{x}_{2r}(t) - \ddot{\tilde{x}}_d(t) \end{bmatrix} = A_r \begin{bmatrix} x_{1r}(t) - \tilde{x}_d(t) \\ x_{2r}(t) - \dot{\tilde{x}}_d(t) \end{bmatrix}, \qquad t \geq 0. \qquad (20)$$

Since

$$A_r \triangleq \begin{bmatrix} 0_m & I_m \\ -\omega_0^2 & -A_{rm} \end{bmatrix}, \qquad (21)$$

is Hurwitz, it follows that $x_{1r}(t) - \tilde{x}_d(t) \to 0$ and $x_{2r}(t) - \dot{\tilde{x}}_d(t) \to 0$ as $t \to \infty$, i.e., the reference state converges to the desired trajectory. The remaining problem is to design a control command, $\tau(t) \in \mathbb{R}^m$, such that the tracking error converges to a fixed neighborhood around the origin. Considering that the mass, Coriolis/centrifugal, and damping matrices of the real system contain unknown parameters and unknown terms, a control signal that accounts for these uncertainties needs to be considered.

4.2 Control Command

The next goal is to define a control signal, $\tau \in \mathbb{R}^m$, that guarantees that the tracking error is ultimately bounded.

Theorem 2. *Consider the vehicle dynamics (7), (10) and the reference dynamics (13). Introduce a tracking error $e_1(\eta, x_{1\mathrm{r}}) \in \mathbb{R}^m$, where $\eta \in \mathbb{R}^n$ and $x_{1\mathrm{r}} \in \mathbb{R}^m$ represent the vehicle and reference configuration, respectively, such that $e_1(\eta(t), x_{1\mathrm{r}}(t)) \equiv 0$, $t \geq 0$, if and only if perfect tracking is achieved. Assume that the error dynamics can be written in the form*

$$\dot{e}_1(t) = Q_1(\eta(t), x_{1\mathrm{r}}(t))q_2(x(t), x_{\mathrm{r}}(t), \chi(t)), \qquad t \geq 0, \tag{22}$$

where $\chi(t) \in \mathbb{R}^m$ is an exogenous signal, $x(t) \in \mathcal{D}$ is the state of the system, $x_{\mathrm{r}}(t) \in \mathbb{R}^{2m}$ is the reference state, $Q_1 : \mathbb{R}^n \times \mathbb{R}^m \to \mathbb{R}^{m \times m}$, and $q_2 : \mathbb{R}^{2n} \times \mathbb{R}^{2m} \times \mathbb{R}^m \to \mathbb{R}^m$. Assume also that there exists a Lyapunov function candidate $V_s(e_1)$ such that $\frac{\mathrm{d}V_s(e_1)}{\mathrm{d}e_1} = 0$ if and only if $e_1 = 0$. Next, consider a control command

$$\tau^*(x, x_{\mathrm{r}}, \chi, \hat{r}) = -\Lambda_1 \left[\Theta^* w(x, x_{\mathrm{r}}, \chi, \hat{r}) + \delta^*(x)\right], \tag{23}$$

where $\tau^*(x, x_{\mathrm{r}}, \chi, \hat{r}) \in \mathbb{R}^m$, $\Lambda_1 \in \mathbb{R}^{m \times n}$ is such that $B\Lambda_1$ is nonsingular, $\Theta^* \in \mathbb{R}^{n \times m}$, $w(x, x_{\mathrm{r}}, \chi, \hat{r}) \in \mathbb{R}^m$, and $\delta^*(x) \in \mathbb{R}^n$. Furthermore, let

$$\Theta^* = (B\Lambda_1)^{-1}\Lambda_2, \tag{24}$$

$$w(x, x_r, \chi, \hat{r}) = \left(\frac{\partial q_2(x, x_{\mathrm{r}}, \chi)}{\partial \nu}\Lambda_2\right)^{-1}\left(\frac{\partial q_2(x, x_{\mathrm{r}}, \chi)}{\partial \eta}J(\eta)\nu + \frac{\partial q_2(x, x_{\mathrm{r}}, \chi)}{\partial x_{\mathrm{r}}}\dot{x}_{\mathrm{r}}\right.$$

$$+\frac{\partial q_2(x, x_{\mathrm{r}}, \chi)}{\partial \chi}\dot{\chi} - \dot{q}_{2\mathrm{des}}(x, x_{\mathrm{r}}) + e_2(x, x_{\mathrm{r}}, \chi)$$

$$\left.+G_2^{-1}Q_1^{\mathrm{T}}(\eta, x_{1\mathrm{r}})\frac{\mathrm{d}V_s(e_1)}{\mathrm{d}e_1}^{\mathrm{T}}\right), \tag{25}$$

$$\delta^*(x) = (B\Lambda_1)^{-1}f(x), \tag{26}$$

with

$$q_{2\mathrm{des}}(\eta, x_{1\mathrm{r}}) = -\alpha(\eta, x_{1\mathrm{r}})G_1Q_1^{\mathrm{T}}(\eta, x_{1\mathrm{r}})\frac{\mathrm{d}V_s(e_1)}{\mathrm{d}e_1}^{\mathrm{T}}, \tag{27}$$

$$e_2(x, x_{\mathrm{r}}, \chi) = q_2(x, x_{\mathrm{r}}, \chi) - q_{2\mathrm{des}}(\eta, x_{1\mathrm{r}}), \tag{28}$$

where $\alpha : \mathbb{R}^n \times \mathbb{R}^m \to \mathbb{R}^+$, $G_1 \in \mathbb{R}^{m \times m}$ is positive definite, and $\Lambda_2 \in \mathbb{R}^{n \times m}$ is such that $\frac{\partial q_2(x, x_{\mathrm{r}}, \chi)}{\partial \nu}\Lambda_2$ is nonsingular. Then the error dynamics associated with the closed loop given by (7), (10), (13), and (23) are ultimately bounded.

Proof. Using (22), the derivative of the Lyapunov function candidate $V_s(e_1)$ is given by

$$\dot{V}_s(x, x_{\mathrm{r}}, \chi) = \frac{\mathrm{d}V_s(e_1)}{\mathrm{d}e_1} Q_1(\eta, x_{1\mathrm{r}}) q_2(x, x_{\mathrm{r}}, \chi). \tag{29}$$

Using a backstepping approach derived from [11], we will use $q_2(x, x_{\mathrm{r}}, \chi)$ as a virtual control command. Ideally $q_2(x, x_{\mathrm{r}}, \chi)$ would be equal to $q_{2\mathrm{des}}(\eta, x_{1\mathrm{r}})$ defined by (27), such that

$$\dot{V}_s(x, x_{\mathrm{r}}, \chi)\big|_{q_2 = q_{2\mathrm{des}}} = -\alpha(\eta, x_{1\mathrm{r}}) \frac{\mathrm{d}V_s(e_1)}{\mathrm{d}e_1} Q_1(\eta, x_{1\mathrm{r}}) G_1 Q_1^{\mathrm{T}}(\eta, x_{1\mathrm{r}}) \frac{\mathrm{d}V_s(e_1)}{\mathrm{d}e_1}^{\mathrm{T}}, \tag{30}$$

which is negative definite. Next, consider a new Lyapunov function candidate,

$$V^*(e_1, e_2) = V_s(e_1) + \frac{1}{2} e_2^{\mathrm{T}} G_2 e_2, \tag{31}$$

where $G_2 \in \mathbb{R}^{m \times m}$ is positive definite and e_2 is defined by (28). The time derivative of (31) is of the form

$$\dot{V}^*(x, x_{\mathrm{r}}, \chi, \hat{r}, \tau^*) = \dot{V}_s(e_1) + e_2^{\mathrm{T}}(x, x_{\mathrm{r}}, \chi) G_2 \dot{e}_2(x, x_{\mathrm{r}}, \chi, \hat{r}, \tau^*). \tag{32}$$

Next, taking the derivative of (28) and substituting the kinematic and dynamic equations (7) and (10), we find the error dynamics to be

$$\dot{e}_2(x, x_{\mathrm{r}}, \chi, \hat{r}, \tau^*) = \frac{\partial q_2(x, x_{\mathrm{r}}, \chi)}{\partial \eta} J(\eta)\nu + \frac{\partial q_2(x, x_{\mathrm{r}}, \chi)}{\partial \nu}(f(x) + B\tau^*)$$
$$+ \frac{\partial q_2(x, x_{\mathrm{r}}, \chi)}{\partial x_{\mathrm{r}}} \dot{x}_{\mathrm{r}} + \frac{\partial q_2(x, x_{\mathrm{r}}, \chi)}{\partial \chi} \dot{\chi} - \dot{q}_{2\mathrm{des}}(x, x_{\mathrm{r}}), \tag{33}$$

which, after substituting the control input (23), provides the following closed-loop error dynamics

$$\dot{e}_1(x, x_{\mathrm{r}}, \chi) = Q_1(\eta, x_{1\mathrm{r}}) q_2(x, x_{\mathrm{r}}, \chi), \tag{34}$$

$$\dot{e}_2(x, x_{\mathrm{r}}, \chi) = -e_2(x, x_{\mathrm{r}}, \chi) - G_2^{-1} Q_1^{\mathrm{T}}(\eta, x_{1\mathrm{r}}) \frac{\mathrm{d}V_s(e_1)}{\mathrm{d}e_1}^{\mathrm{T}}. \tag{35}$$

Therefore, (32) becomes

$$\dot{V}^*(x, x_{\mathrm{r}}, \chi) = -\alpha(\eta, x_{1\mathrm{r}}) \frac{\mathrm{d}V_s(e_1)}{\mathrm{d}e_1} Q_1(\eta, x_{1\mathrm{r}}) G_1 Q_1^{\mathrm{T}}(\eta, x_{1\mathrm{r}}) \frac{\mathrm{d}V_s(e_1)}{\mathrm{d}e_1}^{\mathrm{T}}$$
$$- e_2^{\mathrm{T}}(x, x_{\mathrm{r}}, \chi) G_2 e_2(x, x_{\mathrm{r}}, \chi), \tag{36}$$

which is negative definite, proving asymptotic stability of the closed-loop error dynamics. This concludes our proof.

Remark 2. The matrix Θ^* and the function $\delta^*(x)$ in (23) are unknown, while $w(x, x_{\mathrm{r}}, \chi, \hat{r})$ is a known function of the states and the reference input.

Remark 3. The particular choice for the error dynamics (22) is motivated by the individual application of Theorem 2 to marine vehicles. Such a statement is not limiting in any way since we can always choose $Q_1(\eta, x_{1\mathrm{r}}) = I_m$ and $\chi(t) \equiv 0$ for all $t \geq 0$.

Theorem 3. *Consider the system, tracking error, and virtual command described in Theorem 2, with the additional condition that $\frac{\partial q_2(x, x_{\mathrm{r}}, \chi)}{\partial \nu}$ is bounded on $\mathbb{R}^{2n} \times \mathbb{R}^{2m} \times \mathbb{R}^m$ and $B\Lambda_1$ is positive definite. Let the controller (23) be replaced with the following:*

$$\tau(t) = -\Lambda_1 \left[\Theta(t) w(x(t), x_{\mathrm{r}}(t), \chi(t), \hat{r}(t)) + W(t)\sigma(x(t)) \right], \qquad t \geq 0, \quad (37)$$

where $\Theta(t) \in \mathbb{R}^{n \times m}$ and $W(t) \in \mathbb{R}^{n \times q}$, $t \geq 0$, are parameter estimates, and $\sigma(x) \in \mathbb{R}^q$, $x \in \mathbb{R}^n$, is a vector composed of basis functions that we use to approximate the system's dynamics with a maximum approximation error $\varepsilon^* > 0$. Furthermore, let the parameter update laws be

$$\dot{\Theta} = \frac{\partial q_2(x, x_{\mathrm{r}}, \chi)}{\partial \nu}^{\mathrm{T}} G_2^{\mathrm{T}} e_2(x, x_{\mathrm{r}}, \chi) w^{\mathrm{T}}(x, x_{\mathrm{r}}, \chi, \hat{r}) \Gamma_1 - \sigma_1 \Theta, \quad (38)$$

$$\dot{W} = \frac{\partial q_2(x, x_{\mathrm{r}}, \chi)}{\partial \nu}^{\mathrm{T}} G_2^{\mathrm{T}} e_2(x, x_{\mathrm{r}}, \chi) \sigma^{\mathrm{T}}(x) \Gamma_2 - \sigma_2 W, \quad (39)$$

where $\Gamma_1 \in \mathbb{R}^{m \times m}$ and $\Gamma_2 \in \mathbb{R}^{q \times q}$ are positive definite $\sigma_1 \geq 0$, and $\sigma_2 \geq 0$. Then the tracking error and the parameter estimates are ultimately bounded with a domain of convergence defined as

$$M_{\mathrm{c}} \triangleq \{(e_1, e_2, \tilde{\Theta}, \tilde{W}) : V(e_1, e_2, \tilde{\Theta}, \tilde{W}) \leq \alpha\}, \quad (40)$$

where

$$V(e_1, e_2, \tilde{\Theta}, \tilde{W}) = V_s(e_1) + \frac{1}{2} e_2^{\mathrm{T}} G_2 e_2 + \frac{1}{2} \mathrm{tr}\left(B\Lambda_1 \tilde{\Theta} \Gamma_1^{-1} \tilde{\Theta}^{\mathrm{T}}\right) + \frac{1}{2} \mathrm{tr}\left(B\Lambda_1 \tilde{W} \Gamma_2^{-1} \tilde{W}^{\mathrm{T}}\right),$$

$$(41)$$

and

$$\alpha \triangleq \max_{(e_1, e_2, \tilde{\Theta}, \tilde{W}) \in M} V(e_1, e_2, \tilde{\Theta}, \tilde{W}),$$

$$M \triangleq \left\{ (e_1, e_2, \tilde{\Theta}, \tilde{W}) : \left\| G_2^{\frac{1}{2}} e_2(x, x_{\mathrm{r}}, \chi) \right\| \leq \left\| G_2^{\frac{1}{2}} \frac{\partial q_2(x, x_{\mathrm{r}}, \chi)}{\partial \nu} B\Lambda_1 \right\| \varepsilon^*, \right.$$

$$\mathrm{tr}(B\Lambda_1 \tilde{\Theta} \Gamma_1^{-1} \tilde{\Theta}^{\mathrm{T}}) \leq \mathrm{tr}(B\Lambda_1 \Theta^* \Gamma_1^{-1} \Theta^{*\mathrm{T}}),$$

$$\left. \mathrm{tr}(B\Lambda_1 \tilde{W} \Gamma_2^{-1} \tilde{W}^{\mathrm{T}}) \leq \mathrm{tr}(B\Lambda_1 W^* \Gamma_2^{-1} W^{*\mathrm{T}}) \right\}. \quad (43)$$

Proof. Since Θ^* and $\delta^*(x)$ from Theorem 2 are unknown, their estimates need to be introduced. In particular, Θ^* in (23) will be replaced with its estimate

$\Theta(t)$, such that $\Theta(t) = \Theta^* + \tilde{\Theta}(t)$, where $\tilde{\Theta}(t)$ represents the estimation error. Following the approach described in [10], it will be assumed that the vector function $\delta^*(x)$ can be approximated by a linear parameterized neural network with a maximum approximation error given by $\varepsilon^* > 0$. Hence, there exists $\varepsilon(x)$ such that $||\varepsilon(x)|| < \varepsilon^*$ for all $x \in \mathbb{R}^{2n}$, and

$$\delta^*(x) \triangleq W^*\sigma(x) + \varepsilon(x), \qquad x \in \mathbb{R}^{2n}, \tag{44}$$

where $W^* \in \mathbb{R}^{m \times q}$ is the matrix of *unknown* control gain weights (constant) that minimize the approximation error, $\sigma : \mathbb{R}^{2n} \to \mathbb{R}^q$ is a vector of basis functions such that each component of $\sigma(\cdot)$ takes values in $[0, 1]$, $\varepsilon(\cdot)$ is the vector of approximation errors, and $||W^*|| \leq w^*$, where w^* is a bound for the *unknown* control gain optimal weight matrix. Next, following the procedure described in [13], $\delta^*(x(t))$ is replaced in (23) with $W(t)\sigma(x(t))$ where $W(t) \in \mathbb{R}^{m \times q}$ is the estimate of the weights such that $W(t) = W^* + \tilde{W}(t)$, with $\tilde{W}(t)$ representing the estimation error.

When replacing the control command (23) with (37), the corresponding closed-loop error dynamics are given by

$$\dot{e}_1(x, x_r, \chi) = Q_1(\eta, x_{1r})q_2(x, x_r, \chi), \tag{45}$$

$$\dot{e}_2(x, x_r, \chi, \tilde{\Theta}, \tilde{W}, \hat{r}) = -e_2(x, x_r, \chi) - G_2^{-1}Q_1^{\mathrm{T}}(\eta, x_{1r})\frac{\mathrm{d}V_s(e_1)}{\mathrm{d}e_1}^{\mathrm{T}}$$

$$+\frac{\partial q_2(x, x_r, \chi)}{\partial \nu}\gamma(x, x_r, \chi, \tilde{\Theta}, \tilde{W}, \hat{r}), \tag{46}$$

where

$$\gamma(x, x_r, \chi, \tilde{\Theta}, \tilde{W}, \hat{r}) \triangleq -B\Lambda_1\left[\tilde{\Theta}w(x, x_r, \chi, \hat{r}) + \tilde{W}\sigma(x) - \varepsilon(x)\right]. \tag{47}$$

To show ultimate boundedness of the closed-loop error dynamics given by (45) and (46), the following Lyapunov function candidate is considered:

$$V(e_1, e_2, \tilde{\Theta}, \tilde{W}) = V^*(e_1, e_2) + \frac{1}{2}\mathrm{tr}\left(B\Lambda_1\tilde{\Theta}\Gamma_1^{-1}\tilde{\Theta}^{\mathrm{T}}\right) + \frac{1}{2}\mathrm{tr}\left(B\Lambda_1\tilde{W}\Gamma_2^{-1}\tilde{W}^{\mathrm{T}}\right). \tag{48}$$

Note that $V(e_1, e_2, \tilde{\Theta}, \tilde{W})$ is a positive definite scalar function with continuous partial derivatives, in accordance with the hypothesis set by Theorem 1. The corresponding Lyapunov derivative is given by

$$\dot{V}(x, x_r, \chi, \tilde{\Theta}, \tilde{W}, \hat{r}) = \frac{\partial V^*(e_1, e_2)}{\partial e_1}\dot{e}_1(x, x_r, \chi) + \frac{\partial V^*(e_1, e_2)}{\partial e_2}\dot{e}_2(x, x_r, \chi, \tilde{\Theta}, \tilde{W}, \hat{r})$$

$$+\mathrm{tr}\left(B\Lambda_1\tilde{\Theta}\Gamma_1^{-1}\dot{\Theta}^{\mathrm{T}}\right) + \mathrm{tr}\left(B\Lambda_1\tilde{W}\Gamma_2^{-1}\dot{W}^{\mathrm{T}}\right),$$

$$= -\alpha(\eta, x_{1r})\frac{\mathrm{d}V_s(e_1)}{\mathrm{d}e_1}Q_1(\eta, x_{1r})G_1Q_1^{\mathrm{T}}(\eta, x_{1r})\frac{\mathrm{d}V_s(e_1)}{\mathrm{d}e_1}^{\mathrm{T}}$$

$$-e_2^{\mathrm{T}}(x, x_r, \chi)G_2e_2(x, x_r, \chi) + e_2^{\mathrm{T}}(x, x_r, \chi)G_2\frac{\partial q_2(x, x_r, \chi)}{\partial \nu}B\Lambda_1\varepsilon(x)$$

$$-\text{tr}\left(B\Lambda_1\tilde{\Theta}w(x,x_\text{r},\chi,\hat{r})e_2^\text{T}(x,x_\text{r},\chi)G_2\frac{\partial q_2(x,x_\text{r},\chi)}{\partial \nu}-B\Lambda_1\tilde{\Theta}\Gamma_1^{-1}\dot{\Theta}^\text{T}\right)$$

$$-\text{tr}\left(B\Lambda_1\tilde{W}\sigma(x)e_2^\text{T}(x,x_\text{r},\chi)G_2\frac{\partial q_2(x,x_\text{r},\chi)}{\partial \nu}-B\Lambda_1\tilde{W}\Gamma_2^{-1}\dot{W}^\text{T}\right).$$

Finally, the update laws (38) and (39) provide the following bound for the Lyapunov derivative:

$$\dot{V}(x,x_\text{r},\chi,\tilde{\Theta},\tilde{W},\hat{r}) \leq -\frac{\sigma_1}{2}\text{tr}(B\Lambda_1\tilde{\Theta}\Gamma_1^{-1}\tilde{\Theta}^\text{T}) + \frac{\sigma_1}{2}\text{tr}(B\Lambda_1\Theta^*\Gamma_1^{-1}\Theta^{*\text{T}})$$

$$-\frac{\sigma_2}{2}\text{tr}(B\Lambda_1\tilde{W}\Gamma_2^{-1}\tilde{W}^\text{T}) + \frac{\sigma_2}{2}\text{tr}(B\Lambda_1 W^*\Gamma_2^{-1}W^{*\text{T}})$$

$$-\left\|G_2^{\frac{1}{2}}e_2(x,x_\text{r},\chi)\right\|\left(\left\|G_2^{\frac{1}{2}}e_2(x,x_\text{r},\chi)\right\| - \left\|G_2^{\frac{1}{2}}\frac{\partial q_2(x,x_\text{r},\chi)}{\partial \nu}B\Lambda_1\right\|\varepsilon^*\right).$$
$$(49)$$

It follows from Theorem 1 that the solutions of (38), (39), (45), and (46) are ultimately bounded with convergence to the compact set M_c, defined in (40).

5 Applications

Application of the NN-MRAC algorithm will be presented in this section for two different marine vehicles. First, the algorithm is applied to a fully actuated ASV, then to a nonminimum phase AUV. Both models have the same general form as presented in Section 2, and their differences will be pointed out. The theorems from the previous section will be applied to each case to explicitly define the control command and show the stability of the error dynamics. Numerical simulation results will be presented for each case.

5.1 ASV

An example of how to compute the hydrodynamic terms that appear in (9) for an ASV is provided in [19]. Since the modeling trends and dynamical behavior are similar for different surface vessels, using the model developed in [19] helps to quantify the dynamics and performance capabilities of surface vessels and facilitates the testing of the controller.

We assume that the vessel is equipped with two motors in the rear that can be rotated independently. This provides a control input of

$$\tau \triangleq [X,Y,N]^\text{T},\tag{50}$$

consisting of two forces $X,Y \in \mathbb{R}$ along the x_b and y_b axes, respectively, as well as a yawing moment $N \in \mathbb{R}$. This results in a fully actuated vehicle for motion along the horizontal plane. Hence, $\hat{B} = B = M^{-1}$, which is positive definite because of the nature of the mass matrix.

Control Command

The position tracking error is defined as $e_1(\eta, x_{1\mathrm{r}}) \triangleq \eta - x_{1\mathrm{r}}$. The corresponding error dynamics are given by

$$\dot{e}_1(x(t), x_{2\mathrm{r}}(t)) = \dot{\eta}(t) - \dot{x}_{1\mathrm{r}}(t) = J(\eta(t))\nu(t) - x_{2\mathrm{r}}(t), \qquad t \geq 0. \tag{51}$$

Following the notation and procedure introduced in Section 4, let $\chi(t) \equiv 0$ and

$$Q_1(\eta, x_{1\mathrm{r}}) = I_3, \tag{52}$$
$$q_2(x, x_\mathrm{r}) = J(\eta)\nu - x_{2\mathrm{r}}. \tag{53}$$

The quadratic Lyapunov function candidate

$$V_s(e_1) = \frac{1}{2}e_1^{\mathrm{T}}Pe_1 \tag{54}$$

is chosen, where $P > 0$. Hence, (27) and (28) with (52), (53), and $\alpha(\eta, x_{1\mathrm{r}}) = 1$, provide

$$e_2(x, x_\mathrm{r}) = J(\eta)\nu - x_{2\mathrm{r}} + G_1 P(\eta - x_{1\mathrm{r}}), \tag{55}$$

and the new Lyapunov function candidate becomes

$$V^*(e_1, e_2) = \frac{1}{2}e_1^{\mathrm{T}}Pe_1 + \frac{1}{2}e_2^{\mathrm{T}}G_2 e_2. \tag{56}$$

Let $\Lambda_1 = \Lambda_2 = I_3$, so that the control command (23) is given by

$$\tau^*(x, x_r, \hat{r}) = -\left[\Theta^* w(x, x_r, \hat{r}) + \delta^*(x)\right],$$

where the parameters defined in (24), (25), and (26) become

$$\Theta^* = M,$$
$$w(x, x_r, \hat{r}) = J^{\mathrm{T}}(\eta)\left(\frac{\partial q_2(x, x_\mathrm{r})}{\partial \eta}J(\eta)\nu - \dot{x}_{2\mathrm{r}} + G_1 P\left(J(\eta)\nu - x_{2\mathrm{r}}\right) + e_2(x, x_\mathrm{r})\right.$$
$$\left. + G_2^{-1}e_1^{\mathrm{T}}(\eta, x_{1\mathrm{r}})P\right),$$
$$\delta^*(x) = Mf(x). \tag{58}$$

With this control command, we find the derivative of (56) to be

$$\dot{V}^*(e_1, e_2) = -e_1^{\mathrm{T}}G_1 e_1 - e_2^{\mathrm{T}}G_2 e_2, \tag{59}$$

which is negative definite, and the tracking error converges to zero in accordance with Theorem 2.

Introducing the parameter estimates, the above control command is modified according to (37) in Theorem 3:

$$\tau(t) = -\left[\Theta(t)w(x(t), x_{\mathrm{r}}(t), \hat{r}(t)) + W(t)\sigma(x(t))\right], \qquad t \geq 0.$$

The parameter estimates (38) and (39) become

$$\dot{\Theta} = J^{\mathrm{T}}(\eta)G_2^{\mathrm{T}}e_2(x, x_{\mathrm{r}})w^{\mathrm{T}}(x, x_{\mathrm{r}}, \hat{r})\Gamma_1 - \sigma_1\Theta, \tag{61}$$

$$\dot{W} = J^{\mathrm{T}}(\eta)G_2^{\mathrm{T}}e_2(x, x_{\mathrm{r}})\sigma^{\mathrm{T}}(x)\Gamma_2 - \sigma_2 W, \tag{62}$$

which, according to Theorem 3, guarantees that the dynamics of the error and parameter estimates are ultimately bounded.

Numerical Simulations

In this section maneuvers will be performed to test the controller performance capabilities on the ASV.

The reference system, as described previously, consists of three uncoupled second-order differential equations. The constants describing the reference system dynamics are $\omega_0 = 0.2I_3$ and $\zeta = \mathrm{diag}(0.7, 0.7, 0.45)$. Other constants chosen for the controller include $\Gamma_1 = 10\,I_3$, $\Gamma_2 = 10\,I_{12}$, $G_1 = G_2 = I_3$, $\sigma_1 = \sigma_2 = 0.01$, and $P = I_3$.

Circular Trajectory Results

The first maneuver performed on the ASV is a circular trajectory about the origin of the EFF, defined by

$$x_{\mathrm{d}} = A\sin(\omega t + \phi),$$
$$y_{\mathrm{d}} = A\cos(\omega t + \phi),$$
$$\psi_{\mathrm{d}} = -(\omega t + \phi), \qquad t \geq 0.$$

The simulation was performed for a radius $A = 10\,\mathrm{m}$, an angular velocity $\omega = \frac{2\pi}{75}$ rad/s, and a phase angle of $\phi = \frac{\pi}{4}$. The behavior of the vehicle is simulated for 75 s, or one complete cycle. The corresponding trajectory is shown in Figure 1. The vehicle converges quickly to the desired trajectory while staying relatively close to the reference system. However, undesirable fluctuations in the control input can be seen for the first 10 s of the simulation. These are caused by the coupling between sway and yaw, which the reference system fails to account for.

Octomorphic Trajectory Results

The octomorphic trajectory is given by the following parametric equations:

$$x_{\mathrm{d}}(t) = 2A\sin\left(\frac{\omega t}{2}\right), \tag{63}$$

$$y_{\mathrm{d}}(t) = A\sin(\omega t), \qquad t \geq 0. \tag{64}$$

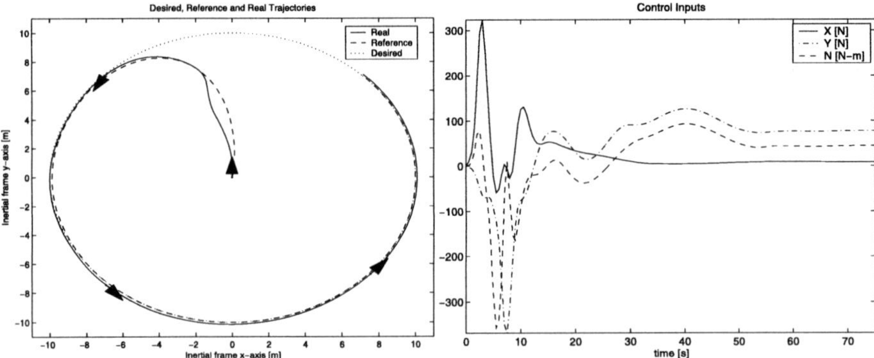

Fig. 1. Circular trajectory and corresponding control command, $\tau(t)$.

The tangential angle for this curve, corresponding to the desired orientation, is given by

$$\psi_{\mathrm{d}}(t) = \tan^{-1}\left(\frac{\dot{y}_{\mathrm{d}}(t)}{\dot{x}_{\mathrm{d}}(t)}\right) = \tan^{-1}\left(\frac{\cos(\omega t)}{\cos\left(\frac{\omega t}{2}\right)}\right), \qquad t \geq 0.$$

This maneuver is more complex than the circular trajectory that was tracked in the previous section. Tracking a circular trajectory involves a single turn with a constant turning radius. The octomorphic trajectory involves both left and right turns as well as straight lines.

Once again undesirable fluctuations in the control input are observed (see Figure 2), caused by the coupling between sway and yaw, which the reference system fails to capture.

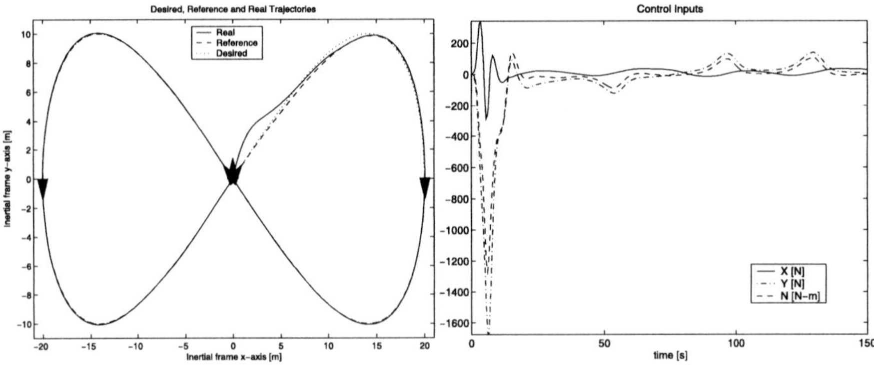

Fig. 2. Octomorphic trajectory and corresponding control command, $\tau(t)$.

5.2 AUV

The control algorithm will now be applied to the most commonly used propulsion system available on AUVs as well as surface vessels: a thruster, used for propulsion, and a rudder for steering, or, equivalently, a vectored thruster ([14, 15]). The propulsion system considered provides two independent control commands, while the vehicle has three degrees of freedom. The vehicle will thus have fewer independent actuators than degrees of freedom, making it *underactuated*. The corresponding dynamic model is characterized by unstable zero dynamics, as mentioned in [7, 14, 15], and the system is said to be *nonminimum phase*. Furthermore, $m = 2$ and the available control command $\tau \in \mathbb{R}^2$ appearing in (9) is of the form

$$\tau \triangleq \begin{bmatrix} \tau_1 \ \tau_2 \end{bmatrix}^{\mathrm{T}}, \tag{65}$$

where τ_1 and τ_2 correspond to the surge and the sway force, respectively.

Tracking Errors

The error in position in the EFF and BFF are respectively defined by

$$e_{\mathrm{p}}(x_{1\mathrm{r}}, \eta_{\mathrm{s}}) \triangleq \eta_{\mathrm{s}} - x_{1\mathrm{r}}, \tag{66}$$

$$\tilde{e}(x_{1\mathrm{r}}, \eta) \triangleq J_{\mathrm{s}}^{-1}(\psi) e_{\mathrm{p}}(x_{1\mathrm{r}}, \eta_{\mathrm{s}}) = \begin{bmatrix} \tilde{e}_1(x_{1\mathrm{r}}, \eta_{\mathrm{s}}) \ \tilde{e}_2(x_{1\mathrm{r}}, \eta_{\mathrm{s}}) \end{bmatrix}^{\mathrm{T}}, \tag{67}$$

where $e_{\mathrm{p}}(x_{1\mathrm{r}}, \eta_{\mathrm{s}})$ is the error in the EFF, $\tilde{e}(x_{1\mathrm{r}}, \eta)$ is this same error in position, but projected in the BFF, $\eta_{\mathrm{s}} \triangleq \begin{bmatrix} x_{\mathrm{N}} \ y_{\mathrm{E}} \end{bmatrix}^{\mathrm{T}}$ is the actual position of the vehicle, $x_{1\mathrm{r}} \in \mathbb{R}^2$ is the position of the reference system that the vehicle is tracking, and

$$J_{\mathrm{s}}(\psi) \triangleq \begin{bmatrix} \cos\psi & -\sin\psi \\ \sin\psi & \cos\psi \end{bmatrix}. \tag{68}$$

Using (7) and (67), we obtain the following first and second time derivatives for $\tilde{e}(t)$:

$$\dot{\tilde{e}}(x, x_r) = \nu_{\mathrm{s}} - J_{\mathrm{s}}^{-1}(\psi)\dot{x}_{1\mathrm{r}} + rS\tilde{e}, \tag{69}$$

$$\ddot{\tilde{e}}(x, x_r, \tau, \hat{r}) = \dot{\nu}_{\mathrm{s}}(x, \tau) - J_{\mathrm{s}}^{-1}(\psi)\left(\dot{x}_{2\mathrm{r}} + rS\dot{\eta}_{\mathrm{r}}\right) + \dot{r}S\tilde{e}(x_{1\mathrm{r}}, \eta) + rS\dot{\tilde{e}}(x, x_r), \tag{70}$$

where $\nu_{\mathrm{s}} \triangleq \begin{bmatrix} u \ v \end{bmatrix}^{\mathrm{T}}$, and S is the following skew-symmetric matrix,

$$S \triangleq \begin{bmatrix} 0 & 1 \\ -1 & 0 \end{bmatrix}. \tag{71}$$

Note that the control action $\tau(t)$ appears explicitly in the expression of $\ddot{\tilde{e}}(t)$ (70) through $\dot{\nu}_{\mathrm{s}}(t)$. The distance between the vehicle and its desired position is defined as

$$e_{\mathrm{d}}(x_{1\mathrm{r}}, \eta_{\mathrm{s}}) \triangleq \|\tilde{e}(x_{1\mathrm{r}}, \eta_{\mathrm{s}})\| = \|e_{\mathrm{p}}(x_{1\mathrm{r}}, \eta_{\mathrm{s}})\|. \tag{72}$$

The time derivative of such a distance is given by

$$\dot{e}_{\mathrm{d}}(x_{\mathrm{r}}, x) = \frac{1}{e_{\mathrm{d}}(x_{1\mathrm{r}}, \eta_{\mathrm{s}})} \tilde{e}^{\mathrm{T}}(x_{1\mathrm{r}}, \eta_{\mathrm{s}}) \dot{\tilde{e}}(x_{1\mathrm{r}}, \eta_{\mathrm{s}}). \tag{73}$$

Next, we define the error $\beta(x_{1\mathrm{r}}, \eta_{\mathrm{s}}) \in (-\pi, \pi]$, which is the angle between the longitudinal axis of the vehicle and the direction of the desired position [1]. It can be computed as follows:

$$\beta(x_{1\mathrm{r}}, \eta_{\mathrm{s}}) \triangleq \tan^{-1}\left(\frac{\tilde{e}_2(x_{1\mathrm{r}}, \eta_{\mathrm{s}})}{\tilde{e}_1(x_{1\mathrm{r}}, \eta_{\mathrm{s}})}\right), \tag{74}$$

and its time derivative is given by

$$\dot{\beta}(x_{\mathrm{r}}, x) = \frac{1}{e_{\mathrm{d}}^2(x_{1\mathrm{r}}, \eta_{\mathrm{s}})} \tilde{e}^{\mathrm{T}}(x_{1\mathrm{r}}, \eta_{\mathrm{s}}) S \dot{\tilde{e}}(x_{1\mathrm{r}}, \eta_{\mathrm{s}}). \tag{75}$$

We will consider the tracking error $e_1(x_{1\mathrm{r}}, \eta_{\mathrm{s}}) = \left[e_{\mathrm{d}}(x_{1\mathrm{r}}, \eta_{\mathrm{s}})\ \beta(x_{1\mathrm{r}}, \eta_{\mathrm{s}}) \right]^{\mathrm{T}}$. As $\beta(x_{1\mathrm{r}}, \eta_{\mathrm{s}})$ converges to zero, the vehicle will orient itself toward its desired position.

We can note that $\beta(x_{1\mathrm{r}}, \eta_{\mathrm{s}})$ is not defined for $e_{\mathrm{d}}(x_{1\mathrm{r}}, \eta_{\mathrm{s}}) = 0$, which implies that we can not guarantee that $e_1(x_{1\mathrm{r}}, \eta_{\mathrm{s}}) = 0$ if and only if $\eta_{\mathrm{s}} = x_{1\mathrm{r}}$. We are, however, still able to apply Theorem 2, but we will prove ultimate boundedness of the error dynamics rather than asymptotic stability. In particular, keeping in mind that perfect tracking is not achievable in the case of a nonminimum phase system [16], we will control the vehicle in such a fashion that $e_1(x_{1\mathrm{r}}, \eta_{\mathrm{s}})$ converges toward a compact set.

Control Command

The time derivative of $e_1(x_{1\mathrm{r}}, \eta_{\mathrm{s}})$ is given by (22), with

$$Q_1(x_{1\mathrm{r}}, \eta_{\mathrm{s}}) = \begin{bmatrix} 1 & 0 \\ 0 & \frac{1}{e_{\mathrm{d}}} \end{bmatrix} J_s^{-1}(\beta),$$

$$q_2(x, x_{\mathrm{r}}, \chi) = \nu_{\mathrm{s}} - J_s^{-1}(\psi)\dot{x}_{1\mathrm{r}} + \chi, \tag{76}$$

where χ is a known exogenous signal whose time derivative is given by

$$T\dot{\chi}(x_{1\mathrm{r}}, \eta, \chi) = rS\tilde{e}(x_{1\mathrm{r}}, \eta_{\mathrm{s}}) - \chi, \tag{77}$$

where T is a diagonal matrix with positive constant elements. We note that $\det[Q_1(x_{1\mathrm{r}}, \eta_{\mathrm{s}})] \neq 0$. We now consider the following Lyapunov function candidate $V_s(e_1)$:

$$V_s(e_1) = e_{\mathrm{d}} \sin^2\left(\frac{\beta}{2}\right) + \frac{1}{2}e_{\mathrm{d}}^2. \tag{78}$$

Note that

$$\frac{dV_s(e_1)}{de_1} = \left[\sin^2\left(\tfrac{\beta}{2}\right) + e_{\mathrm{d}} \quad \tfrac{1}{2}e_{\mathrm{d}}\sin(\beta)\right], \tag{79}$$

and that $\frac{dV_s(e_1)}{de_1} = 0$ if and only if $e_1 = 0$.

The velocity error given by (28) is of the form

$$e_2(x, x_{\mathrm{r}}, \chi) = \nu_{\mathrm{s}} - J_{\mathrm{s}}^{-1}(\psi)\dot{x}_{1\mathrm{r}} + \chi + e_{\mathrm{d}}G_1 J_s(\beta)\begin{bmatrix}\sin^2\left(\tfrac{\beta}{2}\right) + e_{\mathrm{d}} - a \\ \tfrac{1}{2}e_{\mathrm{d}}\sin(\beta)\end{bmatrix}, \tag{80}$$

where the dependencies of β and e_{d} from x_{r} and x were omitted, $a > 0$ is an arbitrary constant that will measure the maximum allowable tracking error. As mentioned in [16], perfect tracking for a nonminimum phase system is not achievable. Accordingly, the control design objective for such a system should not be perfect tracking, but bounded-error tracking. Therefore, the expression of e_2 given by (28) is altered by introducing the $-a$ term. This introduces a position error corresponding to the distance between the vehicle and its desired position, which avoids unstable vehicle behavior.

We now consider the new Lyapunov function candidate

$$V^*(e_1, e_2) = V_s(e_1) + \frac{1}{2}e_2^{\mathrm{T}}G_2 e_2. \tag{81}$$

Using the control command (23) with

$$\Theta^* = (B\Lambda_1)^{-1}\Lambda_2, \tag{82}$$

$$w(x, x_{\mathrm{r}}, \chi, \hat{r}) = \frac{\partial q_2(x, x_{\mathrm{r}}, \chi)}{\partial \eta} J(\eta)\nu - J_{\mathrm{s}}^{-1}(\psi)\dot{x}_{\mathrm{r}} + T^{-1}\left(rS\tilde{e}(x_{1\mathrm{r}}, \eta_{\mathrm{s}}) - \chi\right)$$

$$-\dot{q}_{2\mathrm{des}}(x, x_{\mathrm{r}}, \chi) + e_2(x, x_{\mathrm{r}}, \chi) + G_2^{-1}J_s(\beta)\begin{bmatrix}\sin^2\left(\tfrac{\beta}{2}\right) + e_{\mathrm{d}} \\ \tfrac{1}{2}e_{\mathrm{d}}\sin(\beta)\end{bmatrix}, \tag{83}$$

$$\delta^* = (B\Lambda_1)^{-1}f(x), \tag{84}$$

where

$$\Lambda_1 \triangleq \begin{bmatrix}1 & 0 & 0 \\ 0 & 1 & 0\end{bmatrix}, \qquad \Lambda_2 \triangleq \begin{bmatrix}1 & 0 \\ 0 & 1 \\ 0 & 0\end{bmatrix}, \tag{85}$$

we obtain the following time derivative for the above Lyapunov function candidate:

$$\dot{V}^*(e_1, e_2) = -e_2^{\mathrm{T}}G_2 e_2 - e_{\mathrm{d}}\left[\sin^2\left(\tfrac{\beta}{2}\right) + e_{\mathrm{d}} \tfrac{1}{2}\sin(\beta)\right]G_1\begin{bmatrix}\sin^2\left(\tfrac{\beta}{2}\right) + e_{\mathrm{d}} - a \\ \tfrac{1}{2}\sin(\beta)\end{bmatrix}. \tag{86}$$

Recognizing that $V^*(e_1, e_2) > 0$ and $\dot{V}^*(e_1(t), e_2(t)) < 0$, $t \geq 0$, provided that $e_{\mathrm{d}} \geqslant a$, it follows from Theorem 1 that the error dynamics are ultimately bounded.

Introducing the parameter estimates, the control command is modified according to (37) in Theorem 3,

$$\tau(t) = -\Lambda_1 \left[\Theta(t)w(x(t), x_{\mathrm{r}}(t), \hat{r}(t)) + W(t)\sigma(x(t)) \right], \qquad t \geq 0.$$

The parameter estimates (38) and (39) become

$$\dot{\Theta} = \Lambda_2 G_2^{\mathrm{T}} e_2(x, x_{\mathrm{r}}, \chi)w^{\mathrm{T}}(x, x_{\mathrm{r}}, \chi, \hat{r})\Gamma_1 - \sigma_1\Theta, \tag{88}$$

$$\dot{W} = \Lambda_2 G_2^{\mathrm{T}} e_2(x, x_{\mathrm{r}}, \chi)\sigma^{\mathrm{T}}(x)\Gamma_2 - \sigma_2 W. \tag{89}$$

It follows that the dynamics of the error and parameter estimates are ultimately bounded, in accordance with Theorem 3.

Circular Trajectory Simulation Results

The first maneuver we will attempt is a counterclockwise circle of radius 10m at a velocity of 1m/s, with the following initial conditions:

$$\eta(0) = \begin{bmatrix} 0 & 0 & 0 \end{bmatrix}^{\mathrm{T}}, \qquad \nu(0) = \begin{bmatrix} 0 & 0 & 0 \end{bmatrix}^{\mathrm{T}}. \tag{90}$$

The reference model initial conditions are

$$x_{\mathrm{r1}}(0) = \begin{bmatrix} 0.4 & 0 \end{bmatrix}^{\mathrm{T}}, \qquad x_{\mathrm{r2}}(0) = \begin{bmatrix} 0 & 0 \end{bmatrix}^{\mathrm{T}}. \tag{91}$$

The natural frequency and damping matrices of the reference system are set at $0.2I_2$ and $0.9I_2$, respectively. The initial conditions chosen for the Θ and W estimates are

$$\Theta(0) = \begin{bmatrix} 15 & 0 \\ 0 & 50 \end{bmatrix}, \qquad W(0) = \begin{bmatrix} -5 & 0 & -5 & 0 & 40 & 0 \\ 0 & 10 & 0 & 10 & 0 & 10 \end{bmatrix}. \tag{92}$$

Furthermore, $\Gamma_1 = I_2$, $\Gamma_2 = I_6$, $a = 0.4$, and $G_1 = G_2 = I_2$. The dynamic model of the vehicle corresponds to the Silent Quick Unmanned Intelligent Diver [14]. The values for M, $C(\nu)$, and $D(\nu)$ are given in [14]. Finally, the initial position of the desired trajectory is

$$\eta_{\mathrm{ds}}(0) = \begin{bmatrix} 7.0711 & 7.0711 \end{bmatrix}^{\mathrm{T}}.$$

As shown in Figure 3, the tracking performances are excellent.

Octomorphic Trajectory Simulation Results

For our second maneuver, we consider an octomorphic trajectory. The initial conditions are the same as for the previous example, except for the desired trajectory

$$\eta_{\mathrm{d}}(0) = \begin{bmatrix} 0 & 0 \end{bmatrix}^{\mathrm{T}}.$$

The result of this simulation are displayed in Figure 4. The tracking performances are very good.

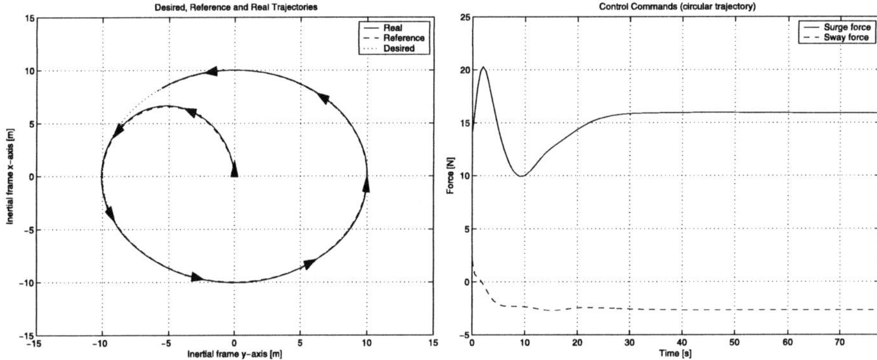

Fig. 3. Circular trajectory and corresponding control command, $\tau(t)$.

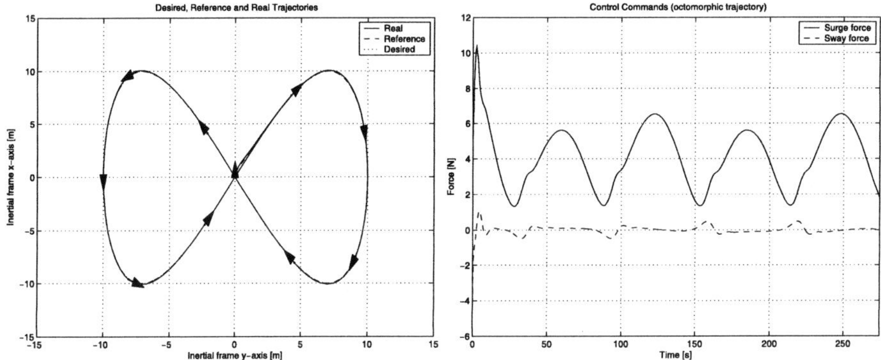

Fig. 4. Octomorphic trajectory and corresponding control command, $\tau(t)$.

6 Conclusion

A NN-MRAC algorithm was developed that uses Lyapunov stability theory to guarantee an ultimately bounded tracking error. The single layer neural network combines with the parameter update laws to eliminate the need to know any of the system dynamics, including its structure. This adds portability to the controller, which was demonstrated by its implementation on two different marine vehicles. Numerical simulations performed for an ASV showed excellent tracking performance despite a strong dependence on the reference system dynamics during the transient region. Equivalent results were seen for the AUV, where the complexity was heightened by the nonminimum phase properties of the system.

References

1. Aguiar A, Pascoal A (2001) Regulation of a nonholonomic autonomous underwater vehicle with parametric modeling uncertainty using Lyapunov functions. 4178–4183. In: Proc. 40th IEEE Conf. Dec. Contr., Orlando, FL
2. Do K, Jiang Z.-P, Pan J (2002) Universal controllers for stabilization and tracking of underactuated ships. Sys. Contr. Lett., 47:299–317
3. Do K, Jiang Z.-P, Pan J (2003) Underactuated ship global tracking without measurement of velocities. In: Proc. 2003 IEEE Am. Contr. Conf., Denver, CO
4. Encarnacão PMM (2002) Nonlinear path following control systems for ocean vehicles. PhD Thesis, Instituto Superior Technico, Lisbon, Portugal
5. Fang Y, Zergeroglu E, de Queiroz M, Dawson D (2004) Global output feedback control of dynamically positioned surface vessels: an adaptive control approach. Mechatronics, 14(4):341–356
6. Fantoni I, Lozano R, Mazenc F, Pettersen K (2000) Stabilization of a nonlinear underactuated hovercraft. International Journal of Robust and Nonlinear Control, 10(8):645–654
7. Fossen T, Godhavn J.-M, Berge S, Lindegaard K.-P (1998) Nonlinear control of underactuated ships with forward speed compensation. 121–126., In: Proc. IFAC NOLCOS98, Enschede, Netherlands
8. Fossen TL (1999) Guidance and control of ocean vehicles. John Wiley & Sons Ltd., Chichester, England
9. Godhavn J.-M (1996) Nonlinear control of underactuated surface vessels, 975–980. In: Proc. 35th IEEE Conf. Dec. Contr. Kobe, Japan
10. Hayakawa T, Haddad WM, Hovakimyan N, Chellaboina V (June 2003) Neural network adaptive control for nonlinear nonnegative dynamical systems, 561–566. In: Proc. Amer. Contr. Conf., Denver, CO
11. Krstić M, Kanellakopoullos I, Kokotović P (1995) Nonlinear and adaptive control design. John Wiley and Sons, New York, NY
12. LaSalle J, Lefschets S (1961) Stability by Liapunov's direct method with applications. Academic Press, Reading, MA
13. Lewis FL, Yesildirek A, Liu K (1999) Neural network control of robot manipulators and nonlinear systems. CRC Press, London, UK
14. Morel Y (2002) Design of an adaptive nonlinear controller for an autonomous underwater vehicle equipped with a vectored thruster. Master's thesis, Florida Atlantic University, Department of Ocean Engineering
15. Morel Y, Leonessa A (2003) Adaptive nonlinear tracking control of an underactuated nonminimum phase model of a marine vehicle using ultimate boundedness. In: Proc. 42th IEEE Conf. Dec. Contr., Maui, HI
16. Slotine J, Li W (1991) Applied nonlinear control. Prentice Hall, Englewood Cliffs, NJ
17. Toussaint G, Basar T, Bullo F (2000a) H^∞-optimal tracking control techniques for nonlinear underactuated systems, 2078–2083. In: Proc. 39th IEEE Conf. Dec. Contr., Sydney, Australia
18. Toussaint G, Basar T, Bullo F (2000b) Tracking for nonlinear underactuated surface vessels with generalized forces, 355–360. In: Proc. of the IEEE International Conference on Control Applications, Anchorage, AK
19. VanZwieten TS (2003) Dynamic simulation and control of an autonomous surface vehicle. Master's thesis, Florida Atlantic University, Department of Ocean Engineering

Part V

Manufacturing Systems

Projection and Aggregation in Maxplus Algebra

Guy Cohen,[1] Stéphane Gaubert,[2] and Jean-Pierre Quadrat[2]

[1]ENPC, 6-8, avenue Blaise Pascal Cité Descartes, Champs-sur-Marne 77455
Marne-La-Vallée Cedex 2, France
`Guy.Cohen@enpc.fr`
[2]INRIA-Rocquencourt, B.P. 105, 78153 Le Chesnay Cedex, France
`{Stephane.Gaubert,Jean-Pierre.Quadrat}@inria.fr`

Summary. In maxplus algebra, linear projectors on an image of a morphism B parallel to the kernel of another morphism C can be built under transversality conditions of the two morphisms. The existence of a transverse to an image or a kernel of a morphism is obtained under some regularity conditions. We show that those regularity and transversality conditions can be expressed linearly as soon as the space to which $\mathrm{Im}(B)$ and $\mathrm{Ker}(C)$ belong is free and its order dual is free. The algebraic structure $\overline{\mathbb{R}}_{\max}^n$ has these two properties. Projectors are constructed following a previous work. Application to aggregation of linear dynamical systems is discussed.

1 Introduction

The work of Kokotović, Delebecque, and Quadrat [9] on the aggregation of Markov chains, which we have tried to extend to the maxplus algebra context, has motivated our research on the construction of maxplus linear projectors [5, 6]. This construction presents some difficulties in this new algebraic context. In this paper, we survey some facts given in these previous papers and clarify some links with module theory.

The practical motivation to study projectors in the context of maxplus algebra is the curse of dimensionality in dynamic programming, which is the main restriction to the application of this technique. Indeed, the dynamic programming equation can be written linearly in maxplus algebra. It is therefore tempting to try to adapt the standard linear techniques (for example, [9]) of linear system aggregation. This is possible and has been done in [15]. But some difficulties appear during this adaptation.

The maxplus analogs of linear space are an idempotent semimodule. As for modules, not all idempotent semimodules have a basis; it is possible that there does not exist a set of generators such that any vector of the semimodule has a unique set of coordinates. When this is the case, we say that the semimodule is free. Moreover, like in module theory, any surjective linear operator does not admit a linear inverse. When this is the case, the range space is called projective. Dually, when any injective linear application admits a linear inverse, we say that the domain is injective. To be able to build the analog of a linear projector on an image parallel to a kernel, we need a transversality condition, which is the existence and uniqueness of the intersection of the kernel and the image (more precisely, the kernel introduced in the following defines a fibration and transversality means that each kernel fiber always intersects the image in a unique point). The transversality condition can be checked by means of linear algebra as soon as the ambient semimodule has some properties of projectiveness and injectiveness.

In a previous work [6], we have obtained necessary and sufficient conditions for the existence of a linear projector, but the links with module theory were not given. In this chapter, we extend this result using assumptions of freeness on the ambient space and its order dual. This dual space is no longer defined as the set of continuous linear forms, but is rather derived from the definition of a scalar product using residuation. In so doing, the bidual space is equal to the initial space. Apart from this improvement of the assumptions, the construction of the linear projector is the same and is recalled here.

There exist few works on these questions. We can find some results on semimodules in [10, 16]. The application of linear projection to aggregation and coherency has been studied in [15]. We recall here some of these results, in particular, the notion of lumpability of dynamic programming equations.

2 Ordered Algebraic Structures

2.1 Structure Definitions

A *semiring* is a set $\mathcal{D}$ equipped with two operations $\oplus$ and $\otimes$ such that $(\mathcal{D}, \oplus)$ is a commutative monoid whose zero is denoted by ϵ, $(\mathcal{D}, \otimes)$ is a monoid whose unit is denoted by e, $\otimes$ is distributive with respect to $\oplus$, the zero is absorbing, that is, $\epsilon \otimes a = a \otimes \epsilon = \epsilon$. A *dioid* is an idempotent semiring: $a \oplus a = a$. A *semifield* is a semiring in which nonzero elements have an inverse.

Example 1. The maxplus semifield $\mathbb{R}_{\max}$ is the set $\mathbb{R} \cup \{-\infty\}$ equipped with $\oplus = \max$ and $\otimes = +$. It is an idempotent semifield.

A *semimodule* is the analog of a module but defined on a semiring instead of a ring.

Example 2. $\mathbb{R}_{\max}^n$ is an idempotent semimodule.

Example 3. The set of $n \times n$ matrices with entries in $\mathbb{R}_{\max}$ is a dioid for the operations:

$$(A \otimes B)_{ik} = \max_j \{A_{ij} + B_{jk}\}, \quad (A \oplus B)_{ij} = \max\{A_{ij}, B_{ij}\}.$$

In a dioid, the operation $\oplus$ induces an order relation: $a \geqslant b \iff a = a \oplus b$, $\forall a, b \in \mathcal{D}$. Then $a \vee b = a \oplus b$. We say that a dioid $\mathcal{D}$ is *complete* if any arbitrary subset $B \subset \mathcal{D}$ has a supremum and if the product distributes with respect to suprema. When the supremum of a set B belongs to B, we denote it by $\top B$.

Example 4. $\mathbb{R}_{\max}$ is not complete but $\overline{\mathbb{R}}_{\max} = \mathbb{R}_{\max} \cup \{+\infty\}$ is complete.

A complete dioid is a complete lattice. Indeed, we can define the infimum of a set $A \subset \mathcal{D}$ by $\top \{d \in \mathcal{D} \mid d \leqslant a, \ \forall a \in A\}$. When the infimum of A belongs to A, we denote it by $\perp A$.

2.2 Residuated Mappings

If $\mathcal{D}$ and $\mathcal{C}$ are complete lattices, we say that a map $f : \mathcal{D} \to \mathcal{C}$ is *isotone* if $a \geqslant b$ implies $f(a) \geqslant f(b)$. We say that f is *lower semicontinuous (lsc)* if f commutes with arbitrary suprema. We say that f is *residuated* if the maximal element $f^\sharp(y) \triangleq \top \{x \in \mathcal{D} : f(x) \leqslant y\}$ exists for all $y \in \mathcal{C}$. The function $f^\sharp$ is called the *residual* of f. We say that a map $g : \mathcal{C} \mapsto \mathcal{D}$ is *dually residuated* if $g^\flat(x) \triangleq \perp \{y \in \mathcal{C} : g(y) \geqslant x\}$ exists for all $x \in \mathcal{C}$.

It is shown in [1] that f is residuated iff f is lsc with $f(\epsilon) = \epsilon$. Moreover, f is residuated iff f is isotone and there exists an isotone map $g : \mathcal{C} \to \mathcal{D}$ such that $f \circ g \leqslant I_\mathcal{C}$ and $g \circ f \geqslant I_\mathcal{D}$.

The residual has the following properties: $f \circ f^\sharp \circ f = f$; $f^\sharp \circ f \circ f^\sharp = f^\sharp$; f is injective iff $f^\sharp \circ f = I_\mathcal{D}$ iff $f^\sharp$ is surjective; f is surjective iff $f \circ f^\sharp = I_\mathcal{C}$ iff $f^\sharp$ is injective; $(h \circ f)^\sharp = f^\sharp \circ h^\sharp$; $f \leqslant g$ iff $g^\sharp \leqslant f^\sharp$; $(f \oplus g)^\sharp = f^\sharp \wedge g^\sharp$; $\left(f^\sharp \wedge g^\sharp\right)^\flat = f \oplus g$.

Example 5. The function $f : x \in \overline{\mathbb{R}}_{\max}^2 \mapsto x_1 \wedge x_2 \in \overline{\mathbb{R}}_{\max}$ is isotone but not residuated. The function $f : x \in \overline{\mathbb{R}}_{\max}^2 \mapsto x_1^2 \oplus x_2^2 \triangleq \max(2x_1, 2x_2) \in \overline{\mathbb{R}}_{\max}$ is isotone, additive but nonlinear and residuated.

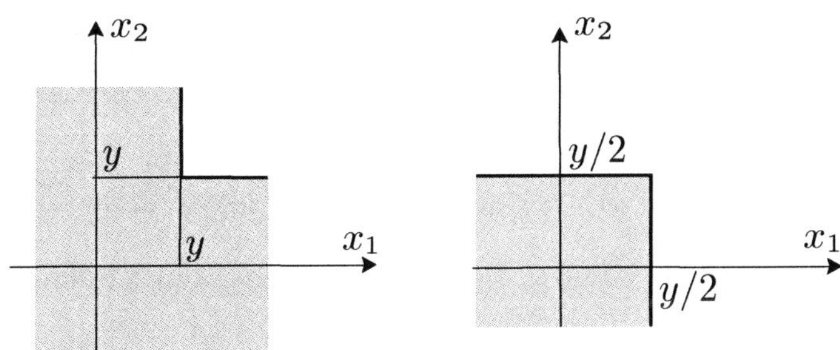

Fig. 1. $x_1 \wedge x_2 \leqslant y$ and $x_1^2 \oplus x_2^2 \leqslant y$.

Example 6. If A is a $m \times n$ matrix with entries in $\overline{\mathbb{R}}_{\max}$, the morphism $x \in \overline{\mathbb{R}}_{\max}^m \mapsto Ax \in \overline{\mathbb{R}}_{\max}^n$ is residuated: the maximal element of the set $\{x \in \overline{\mathbb{R}}_{\max}^n \mid Ax \leqslant b\}$ exists, it is denoted $A \backslash b$ with $(A \backslash b)_j = \bigwedge_i A_{ij} \backslash b_i$. Indeed,

$$Ax \leqslant b \Leftrightarrow \bigoplus_j A_{ij} x_j \leqslant b_i, \ \forall j \Leftrightarrow A_{ij} x_j \leqslant b_i, \ \forall i, j$$

$$\Leftrightarrow x_j \leqslant A_{ij} \backslash b_i, \ \forall i, j \Leftrightarrow x_j \leqslant \bigwedge_i A_{ij} \backslash b_i, \ \forall j \ .$$

Thus $A \backslash b = (-A^t) \odot b$ where $\odot$ denotes the minplus matrix product. In the same way, we can compute $A \backslash B$, B/A, $C \backslash A/B$ for matrices with compatible dimensions.

For example, $A \backslash B$ is the largest matrix such that $AX \leqslant B$ and, if there exists X such that $AX = B$, we have $A(A \backslash B) = B$. Moreover $B \mapsto A \backslash B$ is an $\wedge$ morphism.

3 Idempotent Semimodules

In this section we extend the notion of projective module and injective module to projective and injective idempotent semimodule.

3.1 Projective Idempotent Semimodules

Definition 1. *A complete idempotent semimodule $\mathcal{P}$ is projective if for all complete idempotent semimodules $\mathcal{U}$ and $\mathcal{X}$ and for all morphisms $A : \mathcal{P} \to \mathcal{X}$ and $B : \mathcal{U} \to \mathcal{X}$ such that* $\mathrm{Im}(A) \subset \mathrm{Im}(B)$ *there exists a morphism $R : \mathcal{P} \to \mathcal{U}$ such that $A = BR$.*

This is, of course, the analog of the classical notion in module theory. The following proposition shows that free complete idempotent semimodules are projective and the factor R can be computed by residuation.

Proposition 1. *Given the free idempotent complete semimodule $\mathcal{P}$, the two idempotent complete semimodules $\mathcal{U}$, $\mathcal{X}$, the morphism $B : \mathcal{U} \to \mathcal{X}$, and the morphism $A : \mathcal{P} \to \mathcal{X}$ such that* $\mathrm{Im}(A) \subset \mathrm{Im}(B)$, *we have*

$$A = B(B \backslash A) \ . \tag{1}$$

The existence of a morphism R such that $A = BR$ follows from the classical argument [1, Th. 8.6] The maximal R is equal to $(B \backslash A)$. In the case of matrices, $(B \backslash A)$ has been computed as in Example 6.

Example 7. $\mathcal{X} = \mathrm{Im}(B)$ subsemimodule of $\overline{\mathbb{R}}_{\max}^2$ with $B = \begin{bmatrix} 1 & e \\ e & 1 \end{bmatrix}$ is not free.

The vector $\begin{bmatrix} 1 \\ e \end{bmatrix}$ can also be written $\begin{bmatrix} 1 \\ e \end{bmatrix} \oplus (-1) \begin{bmatrix} e \\ 1 \end{bmatrix}$ and therefore does not have a unique set of coordinates on the generating family $\left\{ \begin{bmatrix} 1 \\ e \end{bmatrix}, \begin{bmatrix} e \\ 1 \end{bmatrix} \right\}$. However, $\mathrm{Im}(B)$ is projective because B has a generalized inverse B^g, meaning that $BB^gB = B$. We can take: $B^g = \begin{bmatrix} -1 & -2 \\ -2 & -1 \end{bmatrix}$.

3.2 Injective Idempotent Semimodules

To discuss injective idempotent semimodules, it is necessary to extend the notion of morphism kernel. Indeed, since in a semiring the general linear equation $Ax = A'x$ cannot be reduced to $A''x = \epsilon$, the standard kernel definition is not very useful. Instead, we introduce the fibration of the domain by the injectivity classes of a morphism.

Definition 2. *Given two semimodules $\mathcal{X}$ and $\mathcal{Y}$ and a morphism $C : \mathcal{X} \to \mathcal{Y}$, we define the equivalence relation $\mathrm{Ker}(C)$ on $\mathcal{X}$ by $x \sim x'$ modulo $\mathrm{Ker}(C)$ if $Cx = Cx'$.*

We say that $\mathrm{Ker}(C) \subset \mathrm{Ker}(A)$ when the fibration induced by $\mathrm{Ker}(C)$ is a subfibration of that of $\mathrm{Ker}(A)$, that is, when $Cx = Cx'$ implies $Ax = Ax'$. With this kernel definition, we can extend the notion of injective modules to injective semimodules.

Definition 3. *A complete idempotent semimodule $\mathcal{I}$ is* injective *if for all complete idempotent semimodules $\mathcal{Y}$ and $\mathcal{X}$ and for all morphisms $A : \mathcal{X} \to \mathcal{I}$ and $C : \mathcal{X} \to \mathcal{Y}$ satisfying $\mathrm{Ker}(C) \subset \mathrm{Ker}(A)$ there exists a morphism $L : \mathcal{Y} \to \mathcal{I}$ such that $A = LC$.*

To understand the duality between projective and injective idempotent semimodules, it is useful to introduce a duality on idempotent semimodules based on residuation.

If $\mathcal{X}$ is a complete idempotent semimodule on the idempotent semiring $\mathcal{D}$, we call *(order) dual* of $\mathcal{X}$, the semimodule on the semiring $\mathcal{D}$, denoted $\mathcal{X}^\sharp$, with underlying set $\mathcal{X}$, addition $(x \in \mathcal{X}, x' \in \mathcal{X}) \mapsto (x \wedge x') \in \mathcal{X}$ (where $\wedge$ is relative to the natural order on $\mathcal{X}$ induced by $\oplus$) and action $(\lambda \in \mathcal{D}, x \in \mathcal{X}) \mapsto x/\lambda \in \mathcal{X}$. The semimodule property of $\mathcal{X}^\sharp$ comes easily from the residuation properties.

Proposition 2 ([7] Prop. 4). *For all complete idempotent semimodules $\mathcal{X}$, $(\mathcal{X}^\sharp)^\sharp = \mathcal{X}$.*

Associated with a semimodule morphism $C : \mathcal{X} \to \mathcal{Y}$, the residuation operation defines the semimodule morphism $C^\sharp : \mathcal{Y}^\sharp \to \mathcal{X}^\sharp$ with which we can characterize the kernel inclusions.

Proposition 3. *Given two idempotent complete semimodules $\mathcal{X}$, $\mathcal{Y}$, two morphisms A and C from $\mathcal{X}$ to $\mathcal{Y}$, we have*

$$\mathrm{Ker}(C) \subset \mathrm{Ker}(A) \Leftrightarrow \mathrm{Im}(A^\sharp) \subset \mathrm{Im}(C^\sharp) .$$

Proof. If $\mathrm{Im}(A^\sharp) \subset \mathrm{Im}(C^\sharp)$, any point in $\mathrm{Im}(A^\sharp)$ does not move when projected on $\mathrm{Im}(C^\sharp)$, which translates into the equality $C^\sharp C A^\sharp = A^\sharp$. By pre- and postcomposition with A of both sides of this equality, we get that

$$AC^\sharp C A^\sharp A = AA^\sharp A = A .$$

We have $AC^\sharp C A^\sharp A \geqslant AC^\sharp C \geqslant A$ because both $A^\sharp A$ and $C^\sharp C$ are greater than the identity. Finally, equality holds throughout and we have proved that $A = AC^\sharp C$, from which it is clear that $\mathrm{Ker}(C) \subset \mathrm{Ker}(A)$.

Conversely, it is easily checked that $C^\sharp C x$ is equivalent to x modulo $\mathrm{Ker}(C)$. Moving in the same class modulo $\mathrm{Ker}(C)$ implies moving in the same class modulo $\mathrm{Ker}(A)$ from the assumption. Therefore $A = AC^\sharp C$. Then, pre- and postcomposition with $A^\sharp$ shows that

$$A^\sharp A C^\sharp C A^\sharp = A^\sharp A A^\sharp = A^\sharp$$

and $A^\sharp A C^\sharp C A^\sharp \geqslant C^\sharp C A^\sharp \geqslant A^\sharp$, which shows that equality holds throughout. Therefore $A^\sharp = C^\sharp C A^\sharp$, which shows that $\mathrm{Im}(A^\sharp) \subset \mathrm{Im}(C^\sharp)$.

Theorem 1. *An idempotent complete semimodule is injective if and only if its (order) dual is projective.*

Proof. If $\mathcal{I}^\sharp$ is projective and the morphisms $A : \mathcal{X} \to \mathcal{I}$ and $C : \mathcal{X} \to \mathcal{Y}$ satisfy $\mathrm{Ker}(C) \subset \mathrm{Ker}(A)$, then $\mathrm{Im}(A^\sharp) \subset \mathrm{Im}(C^\sharp)$ and because $\mathcal{I}^\sharp$ is projective there exists a morphism $L^\sharp : \mathcal{I}^\sharp \to \mathcal{X}^\sharp$ such that $A^\sharp = C^\sharp L^\sharp$, which implies $A = LC$ thanks to the residuation properties.

Conversely, if $\mathcal{I}$ is injective, consider the morphisms $A^\sharp : \mathcal{I}^\sharp \to \mathcal{X}^\sharp$ and $C^\sharp : \mathcal{X}^\sharp \to \mathcal{Y}^\sharp$ such that $\mathrm{Im}(A^\sharp) \subset \mathrm{Im}(C^\sharp)$, which implies $\mathrm{Ker}(C) \subset \mathrm{Ker}(A)$, which implies $A = LC$ since $\mathcal{I}$ is injective and therefore $A^\sharp = C^\sharp L^\sharp$, which shows that $\mathcal{I}^\sharp$ is projective.

Proposition 4. *Given a complete idempotent semimodule $\mathcal{I}$ with a free dual, two complete idempotent semimodules $\mathcal{Y}$, $\mathcal{X}$, and two morphisms $A : \mathcal{X} \to \mathcal{I}$, $C : \mathcal{X} \to \mathcal{Y}$ with $\mathrm{Ker}(C) \subset \mathrm{Ker}(A)$ we have*

$$A = (A/C)C . \tag{2}$$

It is useful to remark that the dual of a free semimodule is not always free.

Example 8. $\overline{\mathbb{R}}_{\max}^n$ is free and its dual $\overline{\mathbb{R}}_{\min}^n$ is free and therefore these two semimodules are projective and injective.

Example 9. $\overline{\mathbb{R}}_{\max}^+ = (\mathbb{R}^+ \cup \{-\infty, +\infty\}, \max, +)$ is a complete idempotent semimodule on $\overline{\mathbb{R}}_{\max}^+$ which is free, its basis is $\{0\}$, but its dual is not free.

4 Projectors

In the following, the sets $\mathcal{U}$, $\mathcal{X}$, and $\mathcal{Y}$ will be complete idempotent semimodules. Moreover, we will suppose that $\mathcal{X}$ and its dual are free so that we can test the image or kernel inclusions by (1) or (2). Based on those assumptions, following [6], we build linear projectors on subsemimodules of $\mathcal{X}$ that can be described as images of regular morphisms.

We say that B is *regular* if there exists a generalized inverse B^g satisfying $B = BB^gB$. Then there exists a largest generalized inverse equal to $B\backslash B/B$. Therefore we have

$$B = B(B\backslash B/B)B \tag{3}$$

if and only if B is regular. Indeed, the existence of B^g implies the existence of X such that $B = BXB$ and therefore the largest X such that $B \geqslant BXB$ (which is equal to $B\backslash B/B$) satisfies the equality.

Moreover, with every generalized inverse, we can associate a generalized reflexive inverse. Indeed, $B^r \triangleq BB^gB$ is an inverse that satisfies

$$B = BB^rB, \quad B^r = B^rBB^r,$$

which are the relations that define the reflexive inverses.

4.1 Projector on Im(B)

Proposition 5. *There exists a linear projector Q on* $\mathrm{Im}(B)$ *iff B is regular.*

By linear projector we mean a projector that is a morphism of complete idempotent semimodules.

Proof. If Q is a linear projector on $\mathrm{Im}(B)$, we have $QB = B$. Since B and Q have the same image, by (1) we have $Q = B(B\backslash Q)$, which implies $B = B(B\backslash Q)B$. Therefore we can take $B^g = B\backslash Q$ as generalized inverse. Thus B is regular.

Conversely if B^g is a generalized inverse of B, then $Q = BB^g$ is a linear projector on $\mathrm{Im}(B)$.

To prepare the transition with the next section, it is useful to give other forms to the projector Q.

Proposition 6. *If B is regular and B^r is a generalized reflexive inverse of B, we have*

$$Q = BB^r = (B/(B^rB))B^r = B((B^rB)\backslash B^r) \,.$$

Proof. From $B = BB^rB$, we deduce from the residuation properties that $B = (B/(B^rB))B^rB$. Similarly from $B^r = B^rBB^r$, we deduce $B^r = B^rB((B^rB)\backslash B^r)$. Let P defined by $P \triangleq (B/(B^rB))B^r$ we have $P = (B/(B^rB))B^rB((B^rB)\backslash B^r) = B((B^rB)\backslash B^r)$.

Moreover $PB = B$, $B^r P = B^r$, and $P^2 = (B/(B^r B))B^r B((B^r B)\backslash B^r) = (B/(B^r B))B^r = P$, therefore P is a projector on $\text{Im}(B)$ and since $B^r P = B^r$ the projection is parallel to $\text{Ker}(B^r)$.

Finally, $Q = BB^r = (B/(B^r B))B^r BB^r B((B^r B)\backslash B^r) = (B/(B^r B))B^r = P$.

4.2 Projector on Im(B) parallel to Ker(C)

To define a projector on $\text{Im}(B)$ parallel to $\text{Ker}(C)$, we need a *transversality* condition, that is, each equivalent class of $\text{Ker}(C)$ intersects $\text{Im}(B)$ in exactly one point (for all x there exists a unique x' such that $Cx = Cx'$ with $x' : x' = Bu$). The following theorem gives a test for transversality.

Theorem 2 ([6] Th. 8 and 9). *The three following assertions are equivalent:*

1. $\text{Ker}(CB) = \text{Ker}(B)$ *and* $\text{Im}(CB) = \text{Im}(C)$, *that is,*

$$B = (B/(CB))CB, \quad C = CB((CB)\backslash C) .$$

2. There exists a linear projector P on $\text{Im}(B)$ *parallel to* $\text{Ker}(C)$:

$$PB = B, \ CP = C, \ P^2 = P = (B/(CB))C = B((CB)\backslash C) .$$

3. $\text{Im}(B)$ *is transverse to* $\text{Ker}(C)$.

Let us only show that the first proposition implies the third one.

The existence of the intersection follows from $C = CB((CB)\backslash C)$; indeed, $Cx = CB((CB)\backslash C)x$, therefore $y = B((CB)\backslash C)x$ belongs to $\text{Im}(B)$ and is in the same $\ker(C)$-class as x.

The uniqueness of the intersection follows from $B = (B/(CB))CB$; indeed, $CBu = CBu'$ implies $(B/(CB))CBu = (B/(CB))CBu' = Bu = Bu'$.

Example 10. With $\mathcal{U} = \mathcal{X} = \mathcal{Y} = \mathbb{R}^2_{\max}$ and

$$B = C = \begin{bmatrix} e & -1 \\ -1 & e \end{bmatrix} ,$$

we have $B = B/B$ and it follows that $P = B$. More generally, for any E, as soon as $B = E/E$ we have $B = B/B = B^2 = B\backslash B = B\backslash B/B = B^r = P$.

Example 11. For $\mathcal{X} = \mathbb{R}^2_{\max}$, $\mathcal{U} = \mathcal{Y} = \mathbb{R}_{\max}$ and B and C, given by

$$B = \begin{bmatrix} e \\ e \end{bmatrix} , \quad C = \begin{bmatrix} e & -1 \end{bmatrix} ;$$

the projector is given in Figure 3.

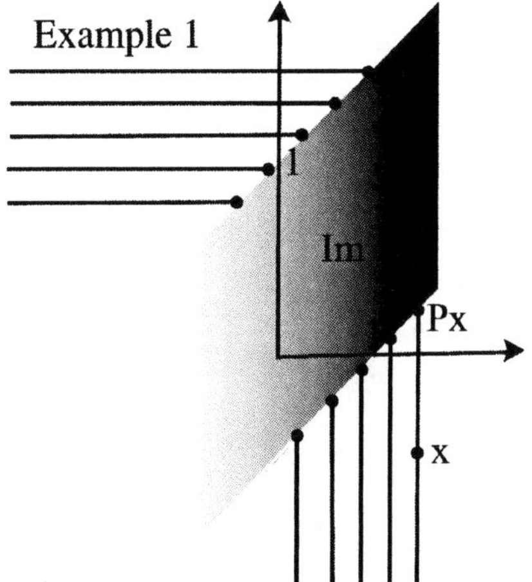

Fig. 2. Projection on $\mathrm{Im}(B)$ parallel to $\mathrm{Ker}(B)$.

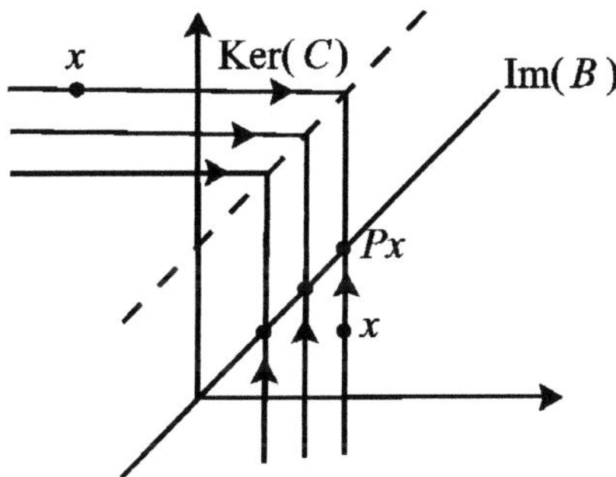

Fig. 3. Projection on $\mathrm{Im}(B)$ parallel to $\mathrm{Ker}(C)$.

Example 12. For the following B,

$$B = \begin{bmatrix} a & e & e \\ e & a & e \\ e & e & a \end{bmatrix},$$

we have $B\backslash B$

$$B\backslash B = \begin{bmatrix} e & -|a| & -|a| \\ -|a| & e & -|a| \\ -|a| & -|a| & e \end{bmatrix}.$$

With some calculation we can show that

$$B^{\sharp}B \begin{bmatrix} x \\ y \\ z \end{bmatrix} = (B\backslash B) \begin{bmatrix} x \\ y \\ z \end{bmatrix} \oplus \chi \begin{bmatrix} y \wedge z \\ x \wedge z \\ x \wedge y \end{bmatrix},$$

where $\chi = e$ if $a < e$ and $\chi = \epsilon$ otherwise.

Then we see that $B^\sharp B = (B \backslash B)$ when $a \geqslant e$. We can verify that the case $a \geqslant e$ is precisely the case where the matrices B are regular.

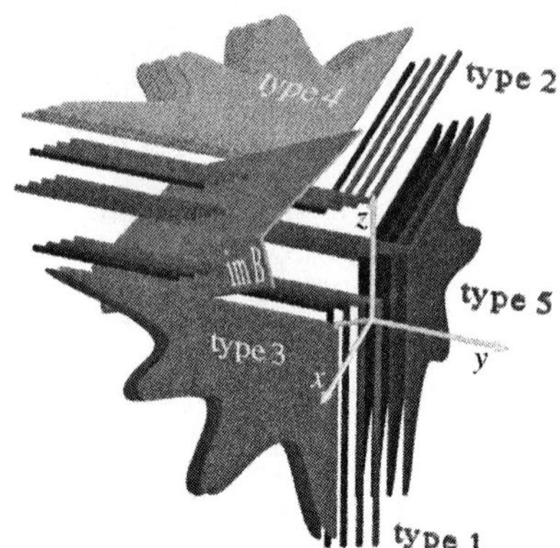

Fig. 4. Projection on $\mathrm{Im}(B)$ parallel to $\mathrm{Ker}(C)$ in dimension 3.

Example 13. The projector shown in Figure 4 is defined with the following B and C:

$$B = \begin{bmatrix} 0 & 1 \\ 0.5 & 0 \\ 2 & 1 \end{bmatrix}, \quad C = \begin{bmatrix} 0 & 0 & 0 \\ 2 & 1 & 0 \end{bmatrix}.$$

We see on this example that the fiber shapes can be different. In this example, we have five kinds of fiber having three different shapes.

5 Aggregation and Coherency

Thanks to the projectors defined in the previous section, the results about aggregation and coherency given in [9] can be extended to the case of maxplus algebra. This means that the aggregation tools used in the theory of linear systems can be applied to the aggregation of dynamic programming problems or Hamilton-Jacobi equations. In this section, we recall some results given in [15].

Given $\mathcal{X}$, a complete free idempotent semimodule with free dual, we consider the endomorphism $A : \mathcal{X} \to \mathcal{X}$ and the dynamic system $X_{n+1} = AX_n$. We say that A is *aggregable* by $C : \mathcal{X} \to \mathcal{Y}$ (regular morphism on the idempotent complete semimodule $\mathcal{Y}$) if there exists A_C such that $CA = A_C C$. Then $Y_n \triangleq CX_n$ satisfies the aggregate dynamics $Y_{n+1} = A_C Y_n$.

Proposition 7. *If C is regular, A is aggregable by C iff there exists B such that the projector on $\mathrm{Im}(B)$ parallel to $\mathrm{Ker}(C)$ satisfies $PA = PAP$.*

Proof. Since C is regular, there exists B and P with $P = B\left((CB)\backslash C\right) = (B/(CB))\,C$.

- Sufficiency $(PA = PAP \Rightarrow CA = A_C C)$:

$$CA = CPA = CPAP = CAP = CA\,(B/\,(CB))\,C = [CA\,(B/\,(CB))]\,C\,,$$

  and we have $A_C = CA(B/(CB))$.
- Necessity $(PA = PAP \Leftarrow CA = A_C C)$:

$$PA = (B/\,(CB))\,CA = (B/\,(CB))\,A_C C = (B/\,(CB))\,A_C CP$$
$$= (B/\,(CB))\,CAP = PAP\,.$$

In a similar way, we say that B, assumed regular, is *coherent* with A if there exists A_B such that $AB = BA_B$. In this case, if $X_0 = BU_0$, then $X_n = BU_n$ is defined by the aggregate dynamics $U_{n+1} = A_B U_n$.

Proposition 8. *If B is regular, B is coherent with A iff there exists C such that the P projector on $\mathrm{Im}(B)$ parallel to $\mathrm{Ker}(C)$ satisfies $AP = PAP$.*

To show an analogy with the lumpability of Markov chains [12], we can specialize the previous results to the case when C is defined as the characteristic function of a partition.

Let us suppose that $\mathcal{X} = \overline{\mathbb{R}}_{\min}^n$. Consider a partition $\mathcal{U} = \{J_1, \ldots, J_p\}$ of $F = \{1, \cdots, n\}$ and its characteristic matrix:

$$U_{iJ} = \begin{cases} e & \text{if } i \in J, \\ \epsilon & \text{if } i \notin J, \end{cases} \quad \forall i \in F,\ \forall J \in \mathcal{U}\,.$$

If $w \in \overline{\mathbb{R}}_{\min}^n$ is a cost (which is analogous to a probability, that is, if w satisfies $gw = e$ with g a row vector with all entries equal to e), the conditional cost with respect to $\mathcal{U}$ is defined by

$$\left(w^{\mathcal{U}}\right)_{iJ} = \frac{w_j}{\bigoplus_{j \in J} w_j}, \quad \forall j, J.$$

Clearly we have

$$w^{\mathcal{U}} = WUS^{-1}\,, \quad \text{with } S \triangleq U^t WU,\ W = \mathrm{diag}(w)\,.$$

If A is the transition cost of a Bellman chain (that is, $gA = g$), we say that A is $\mathcal{U}$-*lumpable* if A is aggregable with $C = U^t$.

If A admits a unique invariant cost $Aw = w$, and if A is $\mathcal{U}$-lumpable, we can take $P = BC$ with $B = w^{\mathcal{U}}$ for projection on $\mathrm{Im}(B)$ parallel to $\mathrm{Ker}(C)$ since $CB = U^t WUS^{-1} = SS^{-1} = I$ (the identity matrix I is the matrix with a diagonal of e and ϵ elsewhere).

In this case, looking at the meaning of $CA = A_C C$, we have the following.

Proposition 9 ([15] Th. 20). *A is lumpable iff:*

$$\bigoplus_{k \in K} a_{kj} = \bar{a}_{KJ}, \quad \forall j \in J,\ \forall J, K \in \mathcal{U}\,.$$

References

1. Blyth TS (1990) Module theory. Oxford Science Publications, Oxford
2. Blyth TS, Janowitz MJ (1972) Residuation theory. Pergamon Press, Oxford
3. Cao ZQ, Kim KH, Roush FW (1984) Incline Algebra and Applications, Hellis Horwood, New York
4. Baccelli F, Cohen G, Olsder GJ, Quadrat JP (1992) Synchronization and linearity, Wiley, New York, and `http://www-rocq.inria.fr/metalau/cohen/SED/book-online.html`
5. Cohen G, Gaubert S, Quadrat JP (1996) Kernels, images and projections in dioids. Proceedings of WODES'96, Edinburgh, and `http://www-rocq.inria.fr/metalau/quadrat/kernel.pdf`
6. Cohen G, Gaubert S, Quadrat JP (1997) Linear projectors in the max-plus algebra, 5th IEEE-Mediterranean Conf. Paphos, Cyprus, and `http://www-rocq.inria.fr/metalau/quadrat/projector.pdf`
7. Cohen G, Gaubert S, Quadrat JP (2004) Duality and separation theorems in idempotent semimodules, Linear Algebra and its Applications 379:395–422 and arXiv:math.FA/0212294
8. Cunninghame-Green RA (1979) Minimax Algebras. Lecture Notes in Economics and Math. Systems, Springer-Verlag, New York
9. Delebecque F, Kokotović P, Quadrat JP (1984) Aggregation and coherency in networks and Markov chains. Int. Jour. of Control 40(5):939–952, and `http://www-rocq.inria.fr/metalau/quadrat/Aggregation.pdf`
10. Golan JS (1992) The theory of semirings with applications in mathematics and theretical computer science 54, Longman Sci & Tech., Harlow, UK
11. Kim KH (1982) Boolean matrix theory and applications, Macel Dekker, New York
12. Kemeny JG, Snell JL, Knap AW (1976) Finite Markov chains, Springer-Verlag, New York
13. Kelly FP (1976). Reversibility and stochastic networks. Wiley, New York
14. Maslov V, Samborskii S (1992) editors, Idempotent analysis Vol. 13 of Adv. in Sov. Math. American Mathematical Society, Providence, RI
15. Quadrat JP (1997) Min-plus linearity and statistical mechanics. Markov Processes and Related Fields, N.3, `http://www-rocq.inria.fr/metalau/quadrat/MecaStat.pdf`
16. Wagneur E (1991) Moduloids and pseudomodules, dimension theory, Discrete Math. 98:57–73

Other related papers and information about maxplus algebra are available at: `http://maxplus.org`.

A Switched System Model for the Optimal Control of Two Symmetric Competing Queues with Finite Capacity

Mauro Boccadoro and Paolo Valigi

Dipartimento di Ingegneria Elettronica e dell'Informazione
Università di Perugia, Perugia, Italia
{boccadoro,valigi}@diei.unipg.it

Summary. An optimal scheduling problem for a two-part type symmetric manufacturing system subject to nonnegligible setup times and characterized by finite buffer capacity is addressed. The solution method relies on i) restricting the possible control policies to those that respect some general necessary conditions of optimality, ii) exploiting such properties to introduce a two-dimensional sampled version of the original system, and iii) mapping such a planar model onto an equivalent scalar one. This modeling approach makes it possible to derive the analytical solution, which exhibits a simple feedback structure.

1 Introduction

This work is motivated by dynamic scheduling problems for a class of manufacturing systems, sometimes referred to as switched server queuing networks [1]–[4]. Such systems are usually modeled accounting for nonnegligible setup times to switch among different production modes, as well as finite buffers capacity (see, e.g., [5],[6]).

Here we consider a fluid approximation model for a two-part type, make-to-order system. The steady state optimal solution for such a system is well known and is given by the limit cycle closest to the origin [7]. To give a complete solution to the problem, the transient behavior has to also be taken into account, providing an optimal path to approach such a limit cycle. This issue has been addressed in [8], where the optimal solution is derived by means of a dynamic programming algorithm. Considering infinite capacity systems, in [7] heuristic rules are proposed to derive the optimal switching curves; [9] introduces a suitable cost index that ecompasses the optimization of the transient behavior; in [10], the problem of reaching the optimal limit cycle in finite time is studied. Optimal control problems of manufacturing systems have been successfully addressed by making use of hybrid system modeling (e.g., [1], [11], [12]). In this chapter, following the same approach, the optimal solution to the problem addressed is derived in two stages: the first to find the optimal continuous input and the second to find the optimal switching time sequence. Similar decomposition techniques have been used in [15], for finite switching sequence, and in [12] for a slightly different class of systems.

Here the timing optimization problem (second stage) is addressed by introducing a sampled discrete event model for the dynamics of the system (originally proposed in [13]). This makes it possible to analytically characterize the optimal feedback control policy for the transient optimization problem addressed. Loose similarities to the sampling approach developed here can be found in [14], which deals with the optimization of trajectories converging to an equilibrium point of some stable subsystem. In our case, the trajectory of the system must be steered to a steady state limit cycle, so that the system presents an infinite number of switchings, characterized by a predefined sequence.

The chapter is organized as follows. Section 2 presents the problem formulation. In Section 3 we introduce the modeling approach. Section 4 provides the optimal solution and an illustrative example.

2 Problem formulation

Consider a make-to-order manufacturing system producing two-part types with nonnegligible setup times and having finite demand buffer capacities (see Figure 1). According to Gershwin's hierarchical structure [16], the machine can be considered completely reliable at the level of setup scheduling, and a fluid approximation model is adopted.

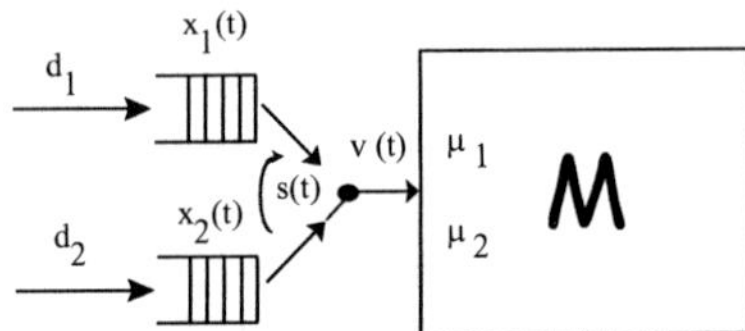

Fig. 1. The manufacturing system.

For part type i, $i = 1, 2$, let x_i be the content of buffer i, with capacity L_i; d_i the demand rate, assumed constant; $v(t)$ the instantaneous production rate, $0 \leq v \leq \mu_i$. The dynamics of the system are given by

$$\dot{x}_i = d_i 1(x_i \neq L) - s_i(t)v(t), \quad i = 1, 2, \tag{1}$$

where $1(\cdot)$ denotes the indicator function, and $s \in \{[0,0]', [1,0]', [0,1]'\}$ machine setup: $s = [1,0]'$ for production of part type 1, $s = [0,1]'$ for production of part type 2, and $s = [0,0]'$ during a setup change. Eq. (1) describes a switched system with *continuous state* given by $\mathbf{x} := (x_1, x_2) \in \mathcal{X} := [0, L_1] \times [0, L_2]$, and *discrete states* s. The system switches from any of the production modes to the setup change mode when an exogenous control signal is received, whereas the system remains in the state $s = [0, 0]'$ for a time interval τ_{ij}, the setup time from part type i to part type $j \neq i$. A setup change event is always feasible in the production modes and not feasible in the setup change mode. Notice that the presence of the indicator function in (1) yields additional dynamic behaviors of the system.

The control inputs to the system are the continuous variable $v(\cdot)$ and the sequence of setup-change instants $\{q_1, q_2, \ldots\}$. From the assumption of nonnegligible setup times, and due to the structure of the event-driven dynamics, it follows that Zenoness (i.e., infinitely many switchings in finite time) is avoided.

For this class of systems, considering infinite buffers and a linear cost index

$$J_s = \lim_{T \to \infty} \frac{1}{T} \int_0^T (c_1 x_1 + c_2 x_2) dt, \tag{2}$$

the optimal steady state solution is given by the limit cycle closest to the origin [7].

Here we consider the problem of finding the optimal transient control policy that brings the system to such a steady state. This kind of situation arises, for example, when the system restarts operating normally after recovering from a failure or performing maintenance. In conformity with Gershwin's hierarchical structure we make the following assumption.

Assumption 1. *Times between failures/maintenance are large enough that the transient evolution of the system converges to the steady state solution.*

The optimal performance of the system are defined in terms of a cost index, which depends on the transient behavior of the system as in [9], where infinite buffers are considered, and [8]:

$$J_t = \lim_{T \to \infty} \inf \int_0^T \left(g[\mathbf{x}(t)] - \bar{J} \right) dt, \tag{3}$$

where $g(\mathbf{x})$ is a positive instantaneous cost function associated with buffer content $\mathbf{x} := (x_1, x_2)$, and $\bar{J}$ is the average cost of the optimum limit cycle. Minimization of J_t yields the optimal transient trajectory reaching the limit cycle. Considering finite buffer capacities we associate a cost to rejected demands of the two-part types. We assume the instantaneous cost $g(\mathbf{x})$ to be the sum of such a rejection cost plus a linear backlog cost, i.e.,

$$g(\mathbf{x}) = \sum_{i=1}^{2} c_i x_i + R_i \mathbf{1}(x_i = L) d_i, \tag{4}$$

where c_i, R_i denote the holding cost and the cost per unit of lost demand, respectively, for part type i.

In the following we study a symmetric system.

Assumption 2. $L_1 = L_2 = L$, $\mu_1 = \mu_2 = \mu$, $d_1 = d_2 = d$, $c_1 = c_2 = c$, $R_1 = R_2 = R$, $\tau_{12} = \tau_{21} = \tau$.

To make the problem nontrivial we also make the following assumptions.

Assumption 3. *The system is stable, that is $\mu - 2d > 0$.*

Assumption 4. *The limit cycle is contained in the state space $\mathcal{X}_L$, that is,*
$$L \geq L_{min} := \frac{2(\mu - d)}{\mu - 2d} d\tau.$$

The problem studied in this chapter can be formulated as follows.

Problem 1. For the manufacturing system described by the dynamics (1), under Assumptions 1–4, find the initial setup choice and the control policy minimizing J_t (3,4), for any initial condition $\mathbf{x}_0 \in \mathcal{X}_L$.

In the following we present a solution approach to this problem based on the introduction of a sampled version of the system dynamics (1) and of the transitory cost index (3).

3 A sampled model

We begin this section by presenting necessary optimality conditions for the transient policies.

Lemma 1. *The system always works at full capacity, i.e., $v(t) = \mu$, when the system is in any of the discrete states $s = [1, 0]'$, $[0, 1]'$.*

Proof. We first consider that the production rate can never be smaller than the (constant) demand rate d: such a choice would obviously result in a suboptimal solution. Let (t_a, t_b) be a time interval during which the system is in one of the production modes, and denote $\tilde{v}(t) \in [d, \mu]$, $t \in (t_a, t_b)$ the production rate exploited for such an interval, and $x_i(t_a) = a$, $x_i(t_b) = b$. Clearly $b \leq a$. It is easy to see that the production rate function $v^*(t) = \mu$, $t \in (t_a, t_a + \frac{b-a}{\mu})$ and $v^*(t) = d$, $t \in (t_a + \frac{b-a}{\mu}, t_b)$ achieves a smaller value for J_t. Hence, optimal production rate functions are always characterized by a first part of full production capacity followed by a second part of "just-in-time" production. Now consider a generic transient trajectory and denote by q_i, $i = 1, 2, \ldots$ the time instants at which a setup change is performed. Assume that a control policy $\tilde{\pi}$ yields a production rate $\tilde{v}$, which is not full capacity. By the considerations made above, there exists an index k such that $\tilde{v}(t) = \mu$ for (q_{k-1}, t_k) and $\tilde{v}(t) = d$ for (t_k, q_k). Consider a control policy π^* that performs setup at time t_k, produces at full rate in the interval $(q_k + \tau d, t_{k+1})$ and at rate $v^*(t) = d$ in the interval (t_{k+1}, q_{k+1}), where t_{k+1} is such that the two production rate functions $\tilde{v}(.)$ and $v^*(.)$ in the interval $(q_k + \tau d, q_{k+1})$ yield the same value for $\mathbf{x}(q_{k+1})$ (see Figure 2). Also in this case it is easy to verify that π^* achieves a smaller value for J_t than $\tilde{\pi}$.[1] Let $\Delta t_i = q_i - t_i$, $i = k, k+1$; it results that $\Delta t_{k+1} = \frac{d}{\mu-d} \Delta t_k$. By the stability Assumption 3, $\Delta t_{k+1} < \Delta t_k$.

Iterating this argument we can construct a full capacity control policy (since $\lim_{i\to\infty} \Delta t_i = 0$) characterized by the same sequence of setup change instants, achieving a lower transient cost (3) than $\tilde{\pi}$.

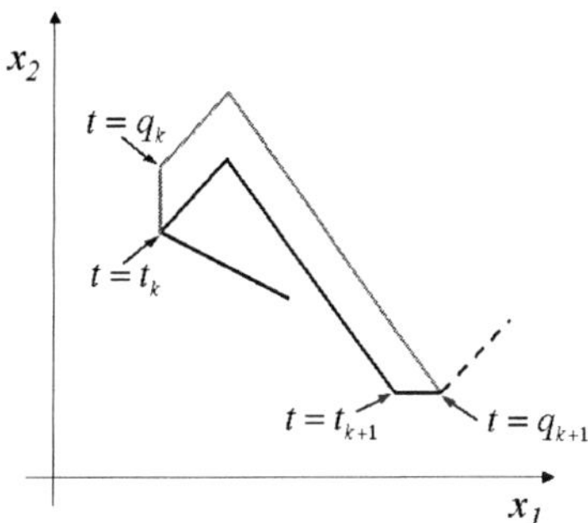

Fig. 2. Trajectories resulting from the policies $\tilde{\pi}$ and π^* described in Theorem 1.

Define $\mathcal{X}_l$ as the open subset of $\mathcal{X}$ given by $\mathcal{X}_l := (0, l) \times (0, l)$, $l := L - \tau d$, τd being the increase in the buffer content of both part types during a setup.

Lemma 2. *A setup change performed at a point* $\mathbf{x} \in \mathcal{X}_l$ *is not optimal.*

Proof. It is a simple extension of Theorem 2 in [8].

The system controlled by the policies satisfying the necessary conditions of optimality stated in Lemmas 1 and 2 will be referred to as the *feedback control system* (FCS). The use of such control policies yields modifications in the finite state machine model of the system. Namely, the discrete dynamics of the "plant" are characterized by an exogenous event, the setup change, feasible in the whole state space, whereas for the FCS such an event becomes feasible as the trajectory in the continuous state space $\mathcal{X}$ hits the boundaries of $\mathcal{X}_l$. This amounts to an endogenous event, occurring at time instants and in correspondence of points in the state space, which we call *decision times*, and *decision points* (d.p.), respectively (see Figure 3). D.p.s belong either to the set $\partial \mathcal{X}_l := \{\mathbf{x} \in \mathcal{X} \mid x_1 = l\} \cup \{\mathbf{x} \in \mathcal{X} \mid x_2 = l\}$, or to the set $\partial \mathcal{X}_0 := \{\mathbf{x} \in \mathcal{X} \mid x_1 = 0\} \cup \{\mathbf{x} \in \mathcal{X} \mid x_2 = 0\}$. In view of Lemma 1, the only true decision variables for the FCS are the setup change time instants q_i, which can be expressed in terms of a delay $u_i \geq 0$ with respect to decision times t_i, i.e., $q_i = t_i + u_i$. A strictly positive value for u_i corresponds to rejecting du_i demands, yielding a faster approach to the limit cycle, while incurring in a rejection cost Rdu_i. The sequence of nonnegative values $\{u_1, u_2, \ldots\}$ represent the sequence of *control actions* of the FCS.

---

[1] A visual inspection could help the intuition of this fact.

Remark 1. Notice that when a trajectory hits $\partial \mathcal{X}_0$, by Lemmas 1 and 2 setup change has to be suddenly performed, so that the corresponding optimal control action is given by $u = 0$.

3.1 Planar sampled model

The dynamic behavior of the FCS can be completely characterized by the sequence of d.p.s and corresponding control actions. Thus we introduce a discrete map $\mathbf{T} : (\partial \mathcal{X}_0 \cup \partial \mathcal{X}_l) \times \mathcal{U} \mapsto \partial \mathcal{X}_0 \cup \partial \mathcal{X}_l$, where $\mathcal{U}$ is the set of feasible controls, describing a one-step transition from the ith d.p. $\mathbf{x}_i$ and the corresponding control action u_i, to the next d.p. $\mathbf{x}_{i+1}$, i.e.,

$$\mathbf{x}_{i+1} = \mathbf{T}(\mathbf{x}_i, u_i). \tag{5}$$

The whole trajectory in the continuous state space $\mathcal{X}$ from any d.p. to the next shall be referred to as *(transient) semicycle*. We define *sampled system* the system governed by the dynamics (5) and *sampled trajectory* a sequence of d.p.s $\{\mathbf{x}_1, \mathbf{x}_2, \ldots\}$.

To derive a sampled cost function for this model, first consider a sampled trajectory starting from a d.p.

The integral in the cost index (3) exhibits a periodic steady state behavior, and its minima are achieved in correspondence of decision instants $t_i, i = 1, 2, \ldots$ Hence, expressing the cost index as

$$J_t = \sum_{i=0}^{\infty} \left(\int_{t_i}^{t_{i+1}} \left(g[\mathbf{x}(s)] - \bar{J} \right) ds \right), \tag{6}$$

the *lim inf* operation can be neglected. Taking into account the stationarity of g, it is easy to see that any integral in the right-hand side of (6) is merely a function of $\mathbf{x}_i$ and $\mathbf{x}_{i+1}$:

$$\int_{t_i}^{t_{i+1}} \left(g(\mathbf{x}(s)) - \bar{J} \right) ds =: \kappa(\mathbf{x}_i, \mathbf{x}_{i+1}). \tag{7}$$

For the same reasons, the function κ can be extended to any pair of points that belong to the same semicycle. Define the *sampled cost function* $\mathbf{C} : (\partial \mathcal{X}_0 \cup \partial \mathcal{X}_l) \times U \mapsto \Re$ as

$$\mathbf{C}(\mathbf{x}, u) := \kappa \left(\mathbf{x}, \mathbf{T}(\mathbf{x}, u) \right), \tag{8}$$

giving the transient cost of the generic semicycle in terms of its initial d.p. $\mathbf{x}$ and of the performed control action u. Notice that $\mathbf{C}(\mathbf{x}, u)$ expresses the difference between a) the overall costs incurred in the transient semicycle $\mathbf{x} \to \mathbf{T}(\mathbf{x}, u)$, whose duration is $t_s = t_{i+1} - t_i$, and b) the quantity $\frac{t_s}{t_{CL}} J_{CL}$, where J_{CL} and t_{CL} are the cost and duration of the limit cycle ($\bar{J} = J_{CL}/t_{CL}$). The quantity $\frac{t_s}{t_{CL}} J_{CL}$ can be seen as a measure of the progress toward the limit cycle, hence, the cost function $\mathbf{C}$ implicitly accounts for the trade-off between such a progress and the cost incurred in any semicycle. In [7] the setup curves for an infinite capacity system have been derived, making use of an heuristic objective function that expresses in a different way such a trade-off.

The cost index (3), for a generic initial state $\mathbf{x}_0 \in \mathcal{X}$ can now be expressed in a sampled form as

$$J_t = \kappa(\mathbf{x}_0, \mathbf{x}_1) + \sum_{i=1}^{\infty} \mathbf{C}(\mathbf{x}_i, u_i), \tag{9}$$

where $\mathbf{x}_1$ denotes the first d.p. reached from the initial condition.

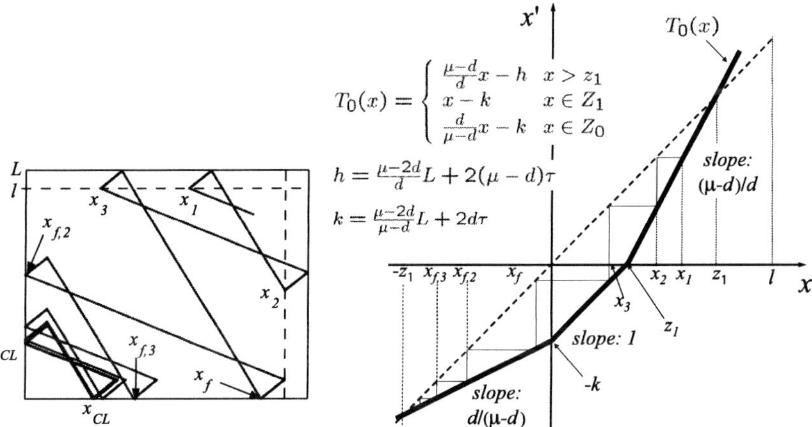

Fig. 3. A transient trajectory converging to limit cycle: planar representation (left figure) and correspondent scalar sampled dynamics (right figure).

3.2 Scalar sampled model

The sampled system dynamics can be expressed in a more compact form taking into account that d.p.s lie on the one-dimensional set $\partial\mathcal{X}_l \cup \partial\mathcal{X}_0$. As a consequence, scalar versions of map $\mathbf{T}$ and cost function $\mathbf{C}$ can be obtained.

Formally, this can be achieved by introducing a relation $\mathcal{R}$ between d.p.s: $\mathbf{a} \, \mathcal{R} \, \mathbf{b} \Leftrightarrow \mathbf{C}(\mathbf{a}, u) = \mathbf{C}(\mathbf{b}, u)$, which is an equivalence relation. The symmetry Assumption 2 implies that given two d.p.s $\mathbf{a}$ and $\mathbf{b}$, $\mathbf{a} \neq \mathbf{b}$, $\mathbf{a} \, \mathcal{R} \, \mathbf{b}$, if and only if $\mathbf{a}$ and $\mathbf{b}$ are symmetric with respect to the line $x_1 = x_2$. Therefore, the quotient set is given by the union of only two of the four segments constituting $\partial\mathcal{X}_0 \cup \partial\mathcal{X}_l$, for example, $\mathcal{X}^+ \cup \mathcal{X}^-$, where $\mathcal{X}^+ := \{(x, l) : x \in [0, l]\}$ and $\mathcal{X}^- := \{(x, 0) : x \in [0, l]\}$.[2] Let $\mathcal{X}_R := [-l, l]$ and define $M : (\mathcal{X}^+ \cup \mathcal{X}^-) \to \mathcal{X}_R$

---

[2] By means of the canonical projection the state space is folded along the line $x_1 = x_2$ so that $x_1 = 0$ and $x_1 = l$ collapse into $x_2 = 0$ and $x_2 = l$, respectively.

as follows:

$$M((x,l)) = x \qquad (x,l) \in \mathcal{X}^+$$
$$M((x,0)) = x - l \ (x,0) \in \mathcal{X}^- . \tag{10}$$

The composition of mapping M with the canonical projection $P : \partial\mathcal{X}_0 \cup \partial\mathcal{X}_l \mapsto \mathcal{X}^+ \cup \mathcal{X}^-$ allows us to represent d.p.s by the scalar values $x = M(P(\mathbf{x}))$ (see Figure 3) and makes $\mathcal{X}_R$ the scalar state space.

Accordingly, the transient evolution of the system will be described by "scalar" discrete map $T : \mathcal{X}_R \times \mathcal{U} \mapsto \mathcal{X}_R$ and transient cost function $C : \mathcal{X}_R \times \mathcal{U} \mapsto \Re^+$, defined as follows:

$$T := M \circ P \circ \mathbf{T} \circ M^{-1} \tag{11}$$

$$C := \mathbf{C} \circ M^{-1} . \tag{12}$$

We extend the discrete map T from domain $\mathcal{X}_R \times \mathcal{U}$ to domain $\mathcal{X}_R \times \mathcal{U}^*$ in the following recursive manner:

$$T(x, U_0) := x; \quad T(x, U_{i+1}) := T(T(x, U_i), u_{i+1}), \tag{13}$$

where $\mathcal{U}^*$ is the set of finite sequences of control actions $U_i := \{u_1, \ldots, u_i\}$, including the empty sequence $U_0 := \emptyset$. Analogously, define $C(x, U_k) := \sum_{i=1}^{k} C(T(x, U_{i-1}), u_i)$, representing the overall cost incurred by the sampled trajectory from x to $T(x, U_k)$. Moreover, let $T_0(x) := T(x, 0)$, $T_0^i(x) := T_0(T_0^{i-1}(x))$ and $C_0(x) := C(x, 0)$, describing the evolution and cost function of the autonomous FCS. Finally, observe that the set of feasible control actions at a d.p. x is state dependent and given by $[0, x/(\mu - d)]$.

The scalar model makes possible an alternative formulation of the cost incurred in a semicycle cost (8), namely, as given by the sum of two contributions, one due to the control variable, and the other representing the cost of an autonomous evolution starting from a properly shifted initial d.p.

Assume a semicycle starts at d.p. $\mathbf{x}$ and, under a control action u, terminates at $\mathbf{x}'$ passing through point $\mathbf{z}$, in correspondence of which setup change ends (see Figure 4). Let $\mathbf{y}$ be the d.p. from which an autonomous semicycle leads to $\mathbf{x}'$ and let x, y be the scalars for $\mathbf{x}$ and $\mathbf{y}$. Split the costs incurred in the semicycles from $\mathbf{x}$ to $\mathbf{x}'$ as $C(x, u) = \kappa(\mathbf{x}, \mathbf{z}) + \kappa(\mathbf{z}, \mathbf{x}')$ and from $\mathbf{y}$ to $\mathbf{x}'$ as $C(y, 0) = \kappa(\mathbf{y}, \mathbf{z}) + \kappa(\mathbf{z}, \mathbf{x}')$. Clearly $C(x, u) = \kappa(\mathbf{x}, \mathbf{z}) - \kappa(\mathbf{y}, \mathbf{z}) + C(y, 0)$. Defining $S(x, y) := \kappa(\mathbf{x}, \mathbf{z}) - \kappa(\mathbf{y}, \mathbf{z})$ we have

$$C(x, u) = S(x, y) + C_0(y) . \tag{14}$$

This leads us to consider things under the following point of view: we can assume that a control action has the "virtual" effect of making the system start an autonomous evolution from a d.p. $y = x - (\mu - d)u$, thus incurring the cost $C_0(y)$ instead of $C(x, u)$ but "paying" for this the price $S(x, y)$.

It results that

$$S(x, y) := \frac{\frac{R}{c}d + L - L_{min}}{v_1}(x - y) + \frac{1}{2v_1}(x^2 - y^2). \tag{15}$$

We remark the following property of function S:

$$S(x', x''') = S(x', x'') + S(x'', x''') \quad \text{for any} \quad x' \geq x'' \geq x''', \tag{16}$$

which can be readily verified from (15).

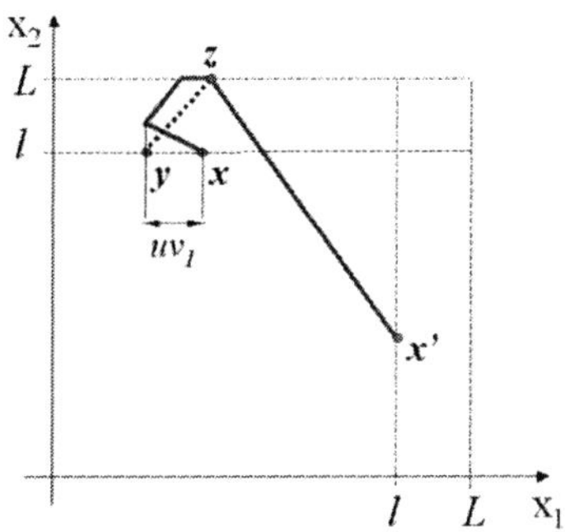

Fig. 4. A control action u performed at d.p. $\mathbf{x}$, brings eventually the semicycle at the same point that is reached by an autonomous semicycle that starts from a d.p. $y = x - v_1 u$.

3.3 Reformulation of the optimal control problem

We now introduce a partition of the state space of the scalar sampled system. Let $z_i \in \mathcal{X}_R$ be the point from which the autonomous system reaches the state $z_0 := 0$ in i steps, i.e., such that $T_0^i(z_i) = 0$, and let $z_\infty := \lim_{i \to \infty} z_i$. It turns out that the fixed points of map T are $-z_\infty$ and z_∞, representing, in the proposed scalar formulation, the d.p.s of the stable and of an unstable limit cycle, indeed, $\forall x \in [-l, z_\infty)$ $\lim_{i \to \infty} T_0^i(x) = -z_\infty$ (see Figure 3).

Define the regions $Z_i := (z_{i-1}, z_i]$, $i = 1, 2, \ldots$; $Z_0 := [-l, z_0]$, and $Z_\infty := [z_\infty, l]$. Observe that for any initial condition in Z_i the corresponding autonomous evolution of the FCS reaches Z_0 exactly in i steps. The scalar state space $\mathcal{X}_R$ can be partitioned as follows: $\mathcal{X}_R = \bigcup_{i \in \mathbb{N}} Z_i \cup Z_\infty$.

According to Remark 1, once a scalar sampled trajectory reaches the region Z_0, say at x_f, the subsequent evolution of the FCS is optimal and completely determined; indeed, 1) all subsequent d.p.s are known and given by $T_0(x_f)$, $T_0^2(x_f)$, etc., and 2) the cost incurred from x_f until convergence to the limit cycle is given by

$$C_\infty(x_f) := \sum_{i=0}^{\infty} C_0(T_0^i(x_f)). \tag{17}$$

Hence a point $x_f \in Z_0$ can be interpreted as a final state with cost $C_\infty(x_f)$, since the optimal cost-to-go is known and independent of the control variable

u. This implies that a feedback solution to the optimal control problem has to be sought only for the d.p.s of the set $\mathcal{X}_R \setminus Z_0$, with $x_f \in Z_0$ being a free terminal state.

To give the expression of the scalar discrete map T, we also introduce a partition of the set $\mathcal{X}_R \times \mathcal{U}$. Starting from a d.p. x, if a control action u is performed such that $x - u(\mu - d) \in Z_1$, then $T(x, u) \in Z_0$, whereas, if $x - u(\mu - d) > z_1$, $T(x, u) > z_0$. Defining $\Gamma_0 := (x, u) : T(x, u) \in Z_0$ and $\Gamma_1 := (x, u) : T(x, u) > z_0$, we have

$$T(x, u) = \begin{cases} a_2 x + g_2 u + b_2, & (x, u) \in \Gamma_1 \\ a_1 x + g_1 u + b_1, & (x, u) \in \Gamma_0 \\ a_0 x + b_0, & x \in Z_0, \end{cases} \tag{18}$$

where $a_2 = \frac{v_1}{d}$, $g_2 = -\frac{v_1^2}{d}$, $b_2 = -\frac{v_2}{d} L + 2v_1 \tau$; $a_1 = 1$, $g_1 = -v_1$, $b_1 = -\frac{v_2}{v_1} L + 2d\tau$; $a_0 = \frac{d}{v_1}$, $b_0 = -\frac{v_2}{v_1} L + 2d\tau$ and $v_1 = \mu - d$, $v_2 = \mu - 2d$.

Remark 2. Assume that the values of two d.p.s x and y are given: the problem of finding a control action u such that $T(x, u) = y$, if it exists, is equivalent to finding the inverse of map T with respect to u. By the form (18) of T, the expression of such an inverse depends on the value of the terminal d.p. y: if $y > z_0 = 0$, the pair $(x, u) \in \Gamma_1$, while if $y \leq z_0$, the pair $(x, u) \in \Gamma_0$, so that either the first or the second of (18) have to be inverted, accordingly.

Note also that due to the stability Assumption 3, v_1 and v_2 are positive quantities, $a_2 > 1$, $a_0 < 1$, and

$$\begin{aligned} T(x, u) < x \ \ \forall u \in [0, x/(\mu - d)] \ \ & \text{if } x < z_\infty \\ T_0(x) > x \ \ & \text{if } x > z_\infty . \end{aligned} \tag{19}$$

We are now able to reformulate Problem 1, *restricted to d.p.s as initial states*, in the following way.

Problem 2. Given $x_0 \in \mathcal{X}_R$, find a control sequence U such that $T(x, U) \in Z_0$, minimizing

$$J_U(x_0) = C(x_0, U) + C_\infty(T(x_0, U)) . \tag{20}$$

4 Solution method

Denote a scalar sampled trajectory that brings the system from a d.p. x to a d.p. y in k steps employing a control sequence U_k, by $x \xrightarrow{U_k} y$. Define the class of first step rejection (FSR) policies, those yielding control sequences having all null components but the first one, i.e., sequences of the form $\{u, 0, \ldots, 0\}, u \geq 0$, and denote by $\mathcal{U}_\mathcal{F}$ the set of FSR control sequences. In a similar fashion, last step rejection (LSR) policies are characterized by control sequences of the form $\{0, \ldots, 0, u\}$. The acronyms FSR and LSR will be also used to indicate trajectories and control sequences featuring FSR and LSR policies.

It turns out that FSR policies are optimal for all trajectories joining two d.p.s, and we present a sketch of the rationale followed in the derivation of this result, which is presented in detail in the following.

First, two steps trajectories are studied. The total cost of a two-step trajectory, for fixed initial and final d.p.s, can be expressed as a continuous concave function of one (control) variable, hence having its minima at the boundary of its domain. This implies that the optimal policy is either FSR or a policy rejecting demand only on the last step (LSR policy).

Next, comparing the cost of FSR and LSR two-step trajectories, it is possible to derive an indicator function giving the best decision in terms of the system parameters. The study of such a function shows that it is not possible to have optimality of LSR policies for any consistent values of the system parameters.

From this result for two-step trajectories, an induction argument yields the complete result.

For the sake of simplicity, in the following we will assume c to have a unitary value so that the rejection cost R implicitly denotes the rate between rejection and backlog costs R/c.

4.1 Two-step trajectories

First we show that two-step minimum cost trajectories are either FSR or LSR.

Lemma 3. *Consider a two-step sampled trajectory from a d.p. x to a d.p. y.[3] Then, depending on the sign of a function $\chi(R, x, y)$, linear in its arguments, the two-step optimal control policy is either FSR or LSR.*

Proof. We search for

$$\min_{U_2} C(x, U_2) \tag{21}$$

$$\text{subject to:} \quad T(x, U_2) = y. \tag{22}$$

Denote $U_2 = \{u_1, u_2\}$; taking into account that constraint (22) is linear in u_1, u_2, and solving for u_2, we derive a function $\varphi(x, y, u_1)$ such that $C(x, U_2) = \varphi(x, y, u_1)$, with admissible values of u_1 belonging to the interval $[0, u_{1max}(x, y)]$, where $u_{1max}(x, y)$ denotes the maximum admissible value of u_1 for the pair of d.p.s x, y, i.e., $T_0(T(x, u_{1max})) = y$. Similarly, solving (22) for u_1 yields $C(x, U_2) = \varphi'(x, y, u_2)$ with $u_2 \in [0, u_{2max}(x, y)]$. By inspection of (15), it is easy to see that φ and φ' are linear in R. Also, it turns out that both φ and φ' are concave quadratic functions of u_1 and u_2, respectively,

---

[3]When speaking of initial and terminal d.p.s of a two-step trajectory x and y, we will always assume they are feasible for two-step trajectories (i.e., $x, y : y \leq T_0^2(x)$).

so that the optimal two-step control sequence is either $U_F = \{u_{1max}, 0\}$ or $U_L = \{0, u_{2max}\}$, and, in any case, $T(x, U_F) = T(x, U_L) = y$. Denote by $u_1^*(x, y)$, which is a linear function of x and y, the value at which $\varphi(x, y, u_1)$ attains its maximum. Then $u_{1max}(x, y)$ minimizes $\varphi(x, y, u_1)$ if and only if $2u_1^*(x, y) \leq u_{1max}(x, y)$, a condition that can be expressed as

$$\chi(R, x, y) \geq 0, \tag{23}$$

for a suitable linear function χ.

Note that all the quantities and functions appearing in the preceding lemma depend on the region in which lies the final d.p. y of the two-step trajectory $x \to y$. This is a consequence of Remark 2 and also of the fact that different expressions of the autonomous cost $C_0(x)$ given in (39) of the Appendix have to be used in computing (21,22). In particular, we will denote χ^+ and χ^- such two different expressions of function χ, for $y > 0$, and $y \leq 0$, respectively.

Remark 3. Optimality is only relevant for those two-step trajectories $x \to y$ that feature a positive intermediate d.p. $x' = T(x, u_1)$: indeed, if $x' \leq 0$ (that is, x' lies in the "terminal" region of $\mathcal{X}_R$), then the second step of such a trajectory should be autonomous (see Section 3.3), and the optimal control sequence would result in, trivially, an FSR one.

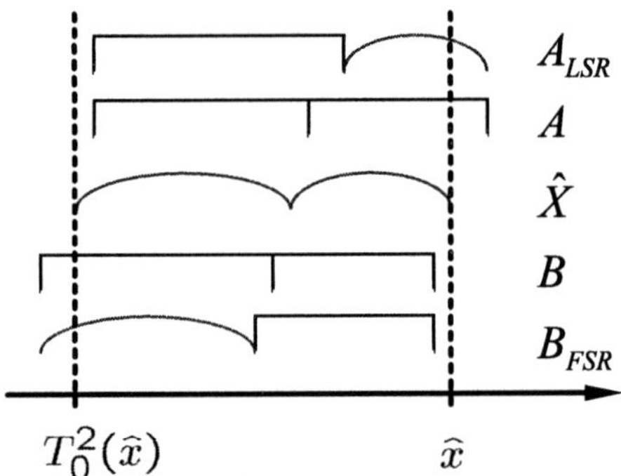

Fig. 5. Properties of optimal two-step nonterminal trajectories close to point $\hat{x}$. Autonomous steps are curved, nonautonomous steps right-angled. $\hat{X}$ is the trajectory: $\hat{x} \to T_0^2(\hat{x})$; A (B) generic two-step trajectory terminating at a d.p. greater than $T_0^2(\hat{x})$ (starting from a d.p. smaller than $\hat{x}$), A_{LSR} (B_{FSR}) optimal trajectory sharing the same extremal d.p.s of A (B).

The sign of function χ in (23) determines whether an optimal control sequence for a two-step trajectory $x \to y$ is FSR or LSR (depending on the sign of the final d.p. y either χ^+ or $\chi-$ should be checked). It is useful to derive a threshold value for d.p.s characterizing the boundary between optimal FSR and LSR policies.

Remark 4. Functions χ^+ and χ^- decrease with the coordinates of the starting and terminal point x and y of a two-step trajectory. This implies, for the case of function χ^+, that if the optimal two-step trajectory for the pair $\bar{x}, \bar{y}$, $\bar{y} > 0$ is FSR (i.e., $\chi^+(\bar{x}, \bar{y}) \geq 0$), then $\forall x, y : x \leq \bar{x}, 0 < y \leq \bar{y}$, the optimal trajectory from x to y is also FSR. Similarly, if for some $\bar{x} \to \bar{y}$, $\bar{y} > 0$, function χ^+ indicates that the optimal two-step trajectory is LSR (i.e., $\chi^+(\bar{x}, \bar{y}) < 0$), then $\forall x, y : x > \bar{x}, y > \bar{y}$ the optimal two-step trajectory is LSR. Analogous considerations can be made about two-step trajectories with $y \leq 0$, characterized by function χ^-.

Now consider the solution $\hat{x}^+$ of $\chi^+(R, \hat{x}^+, T_0^2(\hat{x}^+)) = 0$, which is unique, being $T_0^2(x)$ affine in x, and assume that $z_2 < \hat{x}^+ < z_\infty$. In this case, by (19) $T_0^2(\hat{x}^+) < \hat{x}^+$, and the segment $[T_0^2(\hat{x}^+), \hat{x}^+]$ is the shortest one among all the segments generated by two-step trajectories starting in $\hat{x}^+$, being defined on the basis of a two-step autonomous evolution. It follows that two-step trajectories $x \to y : y > 0$ starting from d.p.s $x : x \leq \hat{x}^+ < z_\infty$, yield, by Remark 4, $\chi^+(x, y) \geq 0$, which implies that the optimal trajectory is FSR. Hence, an LSR policy is candidate optimal if its initial d.p. lies above $\hat{x}^+$; indeed, an LSR policy is the optimal one if in addition to $x > \hat{x}^+$ it is also true that $y > T_0^2(\hat{x}^+)$. Again, a similar argument leads us to identify in the value $\hat{x}^-$ (the solution of $\chi^-(\hat{x}^-, T_0^2(\hat{x}^-)) = 0$) the smallest possible starting d.p. of a two-step trajectory, arriving to some d.p. $y \leq 0$, for which an LSR policy is candidate optimal.

4.2 Optimality of FSR trajectories

The aim of this section is to prove the optimality of FSR trajectories for any number of steps and of generic initial and final d.p.s. To this end, we compare two-step LSR trajectories, which are candidate optimal, with single-step trajectories having the same extremal d.p.s, and prove that such single-step trajectories are always of lower cost than their LSR correspondent. By the reformulation of the cost function (8) that we provided in Subsection 3.3, a single-step trajectory $x \to y$ incurs a lower cost than a two-step LSR $x \to y$,[4] when

$$S(x, z) + C_0(z) < C_0(x) + S(x', z) + C_0(z), \tag{24}$$

where $x' = T_0(x)$, $z : T_0(z) = y$, $z < x'$. By Remark 3, $x > z_1$, and $z > z_0$.

If $x > z_\infty$, the existence of a single-step trajectory $x \to y$ and (19) imply $y \leq x' = T_0(x)$; hence $z \leq x < x'$, which implies $S(x', z) = S(x', x) + S(x, z)$, so that condition (24) is equivalent to

$$C_0(x) + S(x', x) > 0, \tag{25}$$

which, evidently, is always true.

---

[4]In this paragraph x and y will always indicate the initial and final d.p. of any trajectory considered.

If the initial d.p. lies in the region $(z_1, z_\infty]$, then $x' \leq x$, which yields $S(x, z) = S(x, x') + S(x', z)$. Defining $\phi(x) := S(x, T_0(x)) - C_0(x)$, condition (24) is equivalent to

$$\phi(x) < 0. \tag{26}$$

Function ϕ is a continuous quadratic function of x, with domain $(z_1, z_\infty]$: the lower bound z_1 of such a domain is due to the considerations in Remark 3. It results that ϕ is convex and yields a negative value at the right extremum of its domain, i.e., $\phi(z_\infty) < 0$, and therefore, if $\phi(x^*) < 0$, for some x^*, a single-step trajectory is better than its two-step LSR counterpart for any $x \geq x^*$. Based on these considerations we state the following.

Lemma 4. *There exist two values $L_\emptyset^+$ and $L_\emptyset^-$ such that two-step LSR trajectories are suboptimal under the conditions:*

$$\begin{aligned}
L \leq L_\emptyset^+, \ \ y > 0 \\
L \leq L_\emptyset^-, \ \ y \leq 0.
\end{aligned} \tag{27}$$

Proof. By the considerations made above about the key points $\hat{x}$, LSR policies are suboptimal in region (z_1, z_∞) under the following conditions:

$$\begin{aligned}
\hat{x}^+ > z_\infty \\
\hat{x}^- > z_\infty.
\end{aligned} \tag{28}$$

Each condition[5] in (28) yields an upper bound for the values of parameter R for which LSR trajectories $x \to y$ are candidate optimal, i.e., there exist two values R_U^+ and R_U^- such that

$$\begin{aligned}
R \geq R_U^+, \ \ y > 0, \\
R \geq R_U^-, \ \ y \leq 0,
\end{aligned} \tag{29}$$

imply suboptimality of LSR two-step trajectories.

Note also that a candidate optimal two-step LSR is worse than its single-step counterpart under the following conditions:

$$\begin{aligned}
\phi(\hat{x}^+) < 0 \\
\phi(\hat{x}^-) < 0.
\end{aligned} \tag{30}$$

The values $\hat{x}^+ \ \hat{x}^-$ are affine functions of the parameter R so that the expressions $\phi(\hat{x}^+)$ and $\phi(\hat{x}^-)$ above are quadratic functions of R. By substituting in each of (30) the minimum possible value for $\hat{x}^+$ and $\hat{x}^-$, i.e., z_2, and z_1, respectively, we obtain lower bound values for R, i.e., there exist two values R_L^+ and R_L^- such that

---

[5]Conditions (28) are equivalent to $\min_{\xi, \eta} \chi^+(R, \xi, \eta) > 0$, $\min_{\xi, \eta} \chi^-(R, \xi, \eta) > 0$, which are also easily recognized as sufficient conditions for the optimality of FSR policies in the region (z_1, z_∞).

$$R \leq R_L^+, \ y > 0,$$
$$R \leq R_L^-, \ y \leq 0, \tag{31}$$

imply suboptimality of LSR two-step trajectories.

Therefore, two-step LSR trajectories can never be optimal if $R_L^- \leq R_U^-$ (for $y \leq 0$) and $R_L^+ \leq R_U^+$ ($y > 0$). In fact, these conditions are verified under (27).

Lemma 5. *There exist two values L_G^+ and L_G^- such that two-step LSR trajectories are suboptimal under the conditions*

$$L > L_G^+, \ y > 0$$
$$L > L_G^-, \ y \leq 0. \tag{32}$$

Proof. Sufficient conditions for the suboptimality of two-step LSR trajectories are given by $\max_R \phi(\hat{x}^+(R)) < 0$ and $\max_R \phi(\hat{x}^-(R)) < 0$, which yield (32).

The two lemmas above lead to the following lemma.

Lemma 6. *Optimal two-step trajectories are FSR.*

Proof. Under the stability Assumption 3 it results that $L_G^+ < L_\emptyset^+$ and $L_G^- < L_\emptyset^-$, hence LSR trajectories are always suboptimal for any possible "size" of the state space. $\square$

We are now able to state that optimal trajectories are FSR.

Theorem 1. $\forall x, y \in \mathcal{X}_R$ *such that* $\exists \, U : T(x, U) = y$, *let*

$$U^* := \arg \min_{U:T(x,U)=y} C(x, U),$$

then

$$U^* \in \mathcal{U}_F.$$

Proof. If a k-step ($k > 2$) optimal trajectory were not FSR, then it would contain some nonautonomous step, say the jth, which is not the first (hence $j > 1$). In such a case we could construct a better trajectory substituting the two-step subtrajectory $x_{j-1} \to x_{j+1}$ with its FSR correspondent, according to Lemma 6. If the trajectory resulting from this substitution process were not FSR yet, we could iterate the same argument, finally establishing a contradiction. $\square$

Remark 5. Consider a transient trajectory starting in the region Z_∞ and converging to the stable limit cycle. Observe that, as a consequence of the above Theorem 1, the first control action has to bring such a trajectory inside the "stable" region $[-l, z_\infty)$, otherwise the trajectory could not be both FSR and convergent to $-z_\infty$.

4.3 Derivation of the optimal policy

Consider an initial d.p. $x \in Z_k$, so that the autonomous sampled trajectory from x to the first point in Z_0 is $\{x, T_0(x), \ldots, T_0^k(x)\}$. The overall cost of the autonomous sampled trajectory starting from x and converging to the limit cycle is given by the quadratic cost function (see the Appendix)

$$\sum_{i=1}^{k} C_0(T_0^{i-1}(x)) + C_\infty(T_0^k(x)) = B_{2,k}x^2 + B_{1,k}x + B_{0,k} . \tag{33}$$

The above extends the function C_∞, defined in (17), to the domain $[-l, z_\infty)$, i.e., $C_\infty : [-l, z_\infty) \mapsto \Re$.

We are now able to find the solution of Problem 2, since by Theorem 1 it is sufficient to perform a single control action. Its optimal value can be deduced by minimizing the sum of the cost of the first step and of the overall cost of the autonomous trajectory from the second step onward, which is given by the function $H_x : (0, \min\{x, z_\infty\}) \mapsto \Re$, defined as $H_x(y) = S(x, y) + C_\infty(y)$, in terms of the variable $y = x - (\mu - d)u$, representing the level of the buffer currently being served, reached from level x by performing a control action u. Hence, from an initial d.p. x, the optimal control action is characterized in terms of the buffer level $\tilde{y}$ satisfying

$$\tilde{y} = \arg\min_{y} H_x(y) \tag{34}$$

$$\text{subject to} : \ y \in [0, x],$$

$$y < z_\infty ,$$

where the last constraint is due to Remark 5. It turns out that function H_x is continuous but not differentiable at the "key" points z_i. Therefore, the minimization problem has to be referred to the theory of subdifferentials [17]; under this framework, the condition for which $\tilde{y}$ is a minimum of $H_x(y)$ is stated as $0 \in \partial H_x(\tilde{y})$, $\partial H_x(\tilde{y})$ being the subdifferential of H_x at $\tilde{y}$. Since H_x is a scalar function it results that

$$\partial H_x(\bar{y}) = \left[\lim_{y \uparrow \bar{y}} \frac{\partial H_x(y)}{\partial y} , \ \lim_{y \downarrow \bar{y}} \frac{\partial H_x(y)}{\partial y} \right] . \tag{35}$$

For those values such that H_x is differentiable, the above subdifferential is the singleton $\frac{\partial H_x}{\partial y}(\bar{y})$. The extrema of the interval in (35) only depend on the system parameters and can be easily computed in closed form, so that the computation of the optimal control action is straightforward. This amounts to solving Problem 2.

The optimal solution enjoys an interesting threshold feedback structure. Consider an initial d.p. x, and let

$$\arg\min S(x, y) + C_\infty(y) = \tilde{y}.$$

Starting from $x' > x$ instead, by property (16) of function S, it results that

$$\arg\min S(x',y) + C_\infty(y) = S(x',x) + S(x,y) + C_\infty(y) = S(x,y) + C_\infty(y) = \tilde{y},$$

since $S(x',x)$ is a constant independent of y. Hence, for any $x' > x > y$

$$\arg\min_y H_{x'}(y) = \arg\min_y H_x(y) . \tag{36}$$

The solution of Problem 2 is easily characterized, in light of (36), by computing $y^* = \arg\min H_{z_\infty}(y)$, since, by Remark 5, demand rejection must be performed at least until the system reaches z_∞, so that $y^* < z_\infty$ (indeed, $\lim_{y \to z_\infty} H_{z_\infty}(y) \to \infty$). By the considerations made above

$$\arg\min_y H_x(y) = \min\{x, y^*\} . \tag{37}$$

The solution of Problem 1 can be easily derived from the above solution for Problem 2. Let $\mathbf{x}_1$ ($\mathbf{x}_2$) be the first d.p., mapped into x_1 (x_2), reached from $\mathbf{x}_0$ by choosing to begin production with setup 1 (2), respectively. Then the optimal initial setup is given by

$$m^* = \arg\min_{m=1,2}\{\kappa(\mathbf{x}_0, \mathbf{x}_m) + H^*_{x_m}\}, \tag{38}$$

where $H^*_x = \min_y H_x(y)$. Hence, the solution to the transient optimal control problem addressed in this chapter is given by the optimal initial setup m^* given by (38), and, according to the results given in Section 3, and (37), by the following (feedback) threshold policy.

Policy 1. *At any d.p. above y^*, delay a setup change until the buffer currently being served reaches y^*; at any d.p. below y^*, perform setup immediately.*

We remark that any system, as defined by its parameter values, features a typical "structural" value y^*, which completely characterizes its optimal transient policies.

4.4 Example

As an example, consider the system characterized by the following parameter values: $L = 50$, $\mu = 3$, $d = 1$, $\tau = 4.5$ (so that $l = 45.5$ and $z_\infty = 32$), and $R/c = 200$. Let $\mathbf{x}_0 = (45.5, 38)$. The first d.p.s reached from such an initial value are $\mathbf{x}_1 = (30.5, 45.5)$ and $\mathbf{x}_2 = (45.5, 38)$ ($\mathbf{x}_0$ is itself a d.p.). By applying (38) for the scalar values $x_1 = 30.5$ and $x_2 = 38$ mapping $\mathbf{x}_1$ and $\mathbf{x}_2$, we get $m^* = 1$; also, $y^* = \arg\min_y H_{32}(y) = 16$: the optimal policy is to start producing part type 1 and to perform a control action rejecting demand until the first buffer reaches the level $y^* = 16$. Thereafter, the system follows an autonomous evolution, i.e., performs setup changes in correspondence of d.p.s.

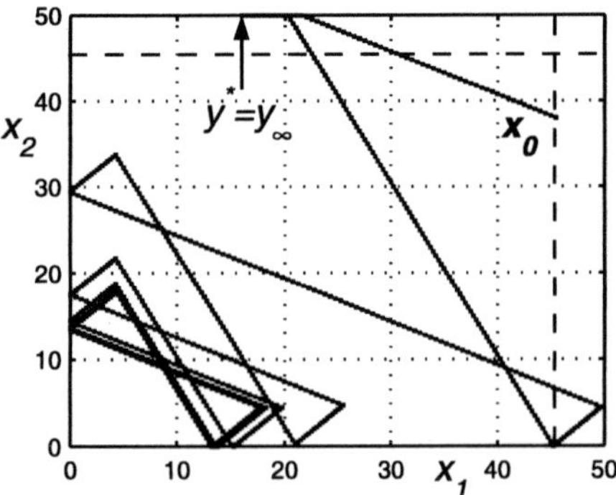

Fig. 6. The optimal transient trajectory for the example system. $R/c = 200$.

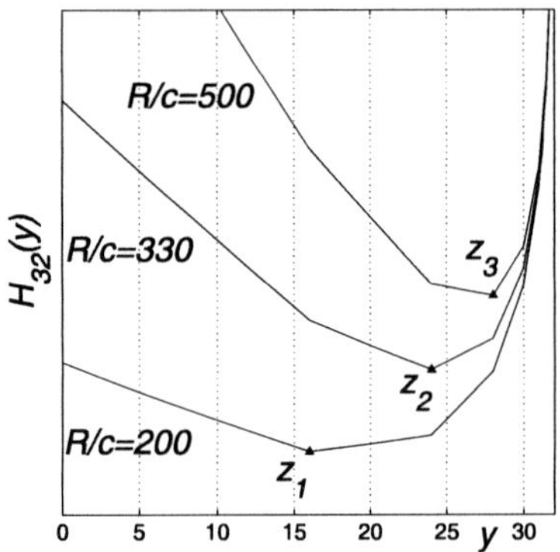

Fig. 7. Function H_{32} for the example system.

If parameter R/c is increased to 330 and 500, the optimal initial setup is still $s = [1, 0]'$, but y^* increases to 24 and 28, respectively. In all these cases, such minima are achieved for $y = z_1$, z_2, z_3, corresponding to the corners of function $H_{32}(y)$ (see Figure 7). The optimal trajectory for the case $R/c = 200$ is illustrated in Figure 6.

A Appendix

A.1 Autonomous cost function

The autonomous cost function C_0 is quadratic function of x, with different expressions for different subsets of the domain. In particular, in Z_1, C_0 is affine in x:

$$C_0(x) = \begin{cases} \alpha_2 x^2 + \beta_2 x + \gamma_2, & x \in (z_1, l - \tau d] \\ \beta_1 x + \gamma_1, & x \in Z_1 \\ \alpha_0 x^2 + \beta_0 x + \gamma_0, & x \in Z_0. \end{cases} \tag{39}$$

A.2 Derivation of $C_\infty(x)$, $x \in Z_0$

$$C_\infty(x) = \sum_{n=0}^{\infty} \alpha_0 \big(T_0^n(x)\big)^2 + \beta_0 T_0^n(x) + \gamma_0 =$$

$$\sum_{n=0}^{\infty} \alpha_0 a_0^{2n} x^2 + \left(2\alpha_0 b_0 \frac{1-a_0^n}{1-a_0} + \beta_0\right) a_0^n x + \alpha_0 b_0^2 \left(\frac{1-a_0^n}{1-a_0}\right)^2 + \beta_0 b_0 \frac{1-a_0^n}{1-a_0} + \gamma_0$$

$$= A_2 x^2 + A_1 x + A_0 \,, \tag{40}$$

where we used the fact that for a k-step autonomous trajectory for which it always applies one of the three of (18), we have

$$T_0^n(x) = a^n x + \frac{1-a^n}{1-a} b \,, \tag{41}$$

with a and b the generic coefficients a_i, b_i, for some $i \in \{0,1,2\}$.

A.3 Derivation of $C_\infty(x)$, $x \in (0, z_\infty]$

We focus on the derivation of the sole $B_{2,k}$ and $B_{1,k}$ of (33), since those are the only relevant coefficients for the determination of the minimum. Assume $\sum_{i=1}^{0} f(.) = 0$ (this convention permits us to generalize the derivation presented for the case $x \in Z_1$, i.e., $k = 1$):

$$\sum_{i=1}^{k} C_0(T_0^{i-1}(x)) + C_\infty(T_0^k(x)) = \sum_{i=1}^{k-1} C_0(T_0^{i-1}(x)) + C_0(T_0^{k-1}(x)) + C_\infty(T_0^k(x)).$$

Let $\Delta_k = \frac{1-a_2^k}{1-a_2} b_2$.

$$\sum_{i=1}^{k-1} C_0(T_0^{i-1}(x)) = \sum_{i=1}^{k-1} \alpha_2 a_2^{2(i-1)} x^2 + (2\alpha_2 \Delta_{i-1} + \beta_2) a_2^{i-1} x + \beta_2 \Delta_{i-1} + \gamma_2$$

$$C_0(T_0^{k-1}(x)) = \beta_1(a_2^{k-1} x + \Delta_{k-1}) + \gamma_1$$

$$C_\infty(T_0^k(x)) = A_2 \left(a_1(a_2^{k-1} x + \Delta_{k-1}) + b_1\right)^2 + A_1 \left(a_1(a_2^{k-1} x + \Delta_{k-1}) + b_1\right) + A_0$$

$$= A_2(a_1 a_2^{k-1})^2 x^2 + a_1 a_2^{k-1}\left(2A_2(a_1 \Delta_{k-1} + b_1) + A_1\right) x + (\text{const.}).$$

Therefore,

$$B_{2,k} = \sum_{i=1}^{k-1} \alpha_2 a_2^{2(i-1)} + A_2(a_1 a_2^{k-1})^2$$

$$B_{1,k} = \sum_{i=1}^{k-1} (2\alpha_2 \Delta_{i-1} + \beta_2) a_2^{k-1} + a_2^{2(k-1)} \beta_1 + a_1 a_2^{k-1}\left(2A_2(a_1 \Delta_{k-1} + b_1) + A_1\right).$$

References

1. Savkin AV (1998) Regularizability of complex switched server queueing networks modelled as hybrid dynamical systems. Systems and Control Letters 35(5):291–299
2. Perkins J, Humes, CJ, Kumar P (1994) Distributed scheduling of flexible manufacturing systems: stability and performance. IEEE Transactions on Robotics and Automation 10(2):133–141
3. Khmelnitsky E, Caramanis M (1998) One-machine n-part-type optimal setup scheduling: analytical characterization of switching surfaces. IEEE Trans. on Automatic Control 43(11):1584–1588
4. Kim E, Oyen MV (2000) Finite-capacity multi-class production scheduling with setup times. IIE Transactions 32:807–818
5. Martinelli F, Valigi P (2002) Impact of finite buffers on the optimal scheduling of a single-machine two-part-type manufacturing system. IEEE Trans. on Automatic Control 47(10):1705–1710
6. Suk JB, Cassandras CG (1991) Optimal scheduling of two competing queues with blocking. IEEE Trans. on Automatic Control 36:1086–1091
7. Connolly S, Dallery Y, Gershwin SB (1992) A real-time policy for performing setup changes in a manufacturing system. 764–770. In: Proc. of the 31st Conf. on Decision and Control (CDC '92), Tucson, AZ
8. Del Gaudio M, Martinelli F, Valigi P (2001) A scheduling problem for two competing queues with finite capacity and nonnegligible setup times. 2355–2360. In: 40th IEEE Conf. on Decision and Control (CDC '01), Orlando, FL
9. Hu J, Caramanis M (1992) Near optimal set-up scheduling for flexible manufacturing systems. 192–201. In: Proceedings of the Third International Conference on Computer Integrated Manufacturing, Troy, NY
10. Bai SX, Elhafsi M (1994) Optimal feedback control of a manufacturing system with setup changes. 191–196. In: Proceedings of the 4th International Conference on Computer Integrated Manufacturing and Automation Technology
11. Cassandras C, Pepyne D, Wardi Y (2001) Optimal control of a class of hybrid systems. IEEE Trans. on Automatic Control 46(3):398–415
12. Gokbayrak K, Cassandras C (2000) A hierarchical decomposition method for optimal control of hybrid systems. 1816–1821. In: 39th IEEE Conf. on Decision and Control (CDC 2000), Sydney, Australia
13. Boccadoro M, Valigi P (2003) A modelling approach for the dynamic scheduling problem of manufacturing systems with nonnegligible setup times and finite buffers. In: 42nd IEEE Conf. on Decision and Control (CDC '03), Maui, HI
14. Giua A, Seatzu C, Van Der Mee C (2001) Optimal control of switched autonomous linear systems. 2472–2477. In: 40th IEEE Conf. on Decision and Control (CDC '01), Orlando, FL
15. Xu X, Antsaklis P (2000) A dynamic programming approach for optimal control of switched systems. 1822–1827. In: 39th IEEE Conf. on Decision and Control (CDC 2000), Sydney, Australia
16. Gershwin SB (1989) Hierarchical flow control: a framework for scheduling and planning discrete events in manufacturing systems. Proc. of IEEE, Special Issue on Dynamics of Discrete Event Systems 77(1):195–209
17. Clarke F, Ledyaev Y, Stern R, Wolenski P (1998) Nonsmooth analysis and control theory. Springer-Verlag, New York

Cooperative Inventory Control

Dario Bauso,[1] Raffaele Pesenti,[1] and Laura Giarré[2]

[1]Dipartimento di Ingegneria Informatica, Università di Palermo, Viale delle Scienze, Palermo, Italia
`dariobauso@libero.it,pesenti@unipa.it`
[2]DIAS, Dipartimento di Ingegneria dell'Automazione, Università di Palermo, Viale delle Scienze, Palermo, Italia
`giarre@unipa.it`

Summary. In multi-retailer inventory control the possibility of sharing setup costs motivates communication and coordination among the retailers. We solve the problem of finding suboptimal distributed reordering policies that minimize setup, ordering, storage, and shortage costs incurred by the retailers over a finite horizon. Neuro-dynamic programming (NDP) reduces the computational complexity of the solution algorithm from exponential to polynomial on the number of retailers.

1 Introduction

In a competitive environment, decision makers may find it convenient to coordinate their strategies to share costs for using resources, services, or facilities. For this reason, there is a vast literature on game theory devoted to *mechanism design* [24], i.e., the definition of game rules or incentive schemes that induce self-interested players to coordinate their strategies so that they converge to Pareto optimal equilibria [2].

The main contribution of this chapter is the design of a *consensus protocol* (see, e.g., [23]) where each player exchanges a limited amount of information with a subset of other players. We cast this protocol within the *minimal information paradigm* [13] to reduce each player's data exposure to the competitors.

Our results apply to repeated noncooperative games in which, at each stage, the payoff of the players is a monotonic function of the strategies of the others. Possible examples are the so-called *externality games* [15] and *cost-sharing games* [29]. Some previous results provided in [4] are a starting point for the protocol design.

We consider a multi-retailer inventory application. The players, namely different competing retailers, aim to coordinate joint orders to share fixed transportation costs. The reader is referred to [27] for a general introduction to multi-retailer inventory problems with particular emphasis on coordination among noncooperative retailers. Recent more specific examples are [5, 8, 9, 14, 30]. The role of information is discussed, e.g., in [11, 12]. The modeling of inventory problems as noncooperative games is discussed in [1, 17, 19, 28].

Each day a stochastic demand materializes at each node. Unfulfilled demand is backlogged. Retailers observe their own inventory level, communicate, and make decisions whether to reorder from warehouse to fulfill the expected demand. Ordered quantities plus inventory at hand may not exceed storage capacity at each store. Reordering occurs by means of a single track which serves all the retailers. Setup costs are shared among all retailers who reorder, also called *active retailers*. This motivates a certain coordination of reordering policies. The system under concern is depicted in Figure 1.

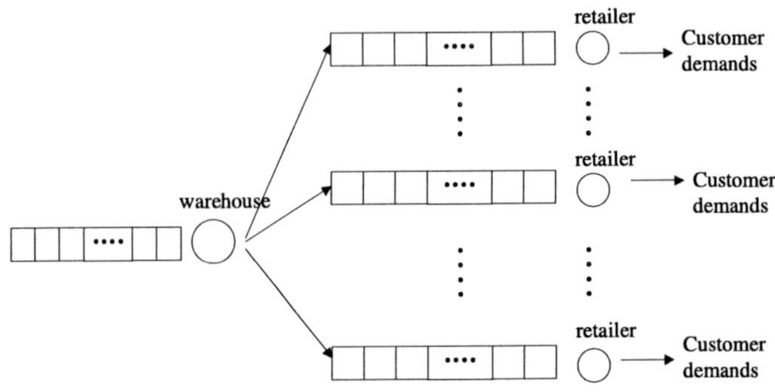

Fig. 1. One-warehouse multi-retailer inventory system.

Alternative ways to achieve coordination among retailers are to i) centralize control at the supplier through vendor-managed inventory [10, 18], or ii) allow side payments and formulate the multi-retailer inventory problems as cooperative games with transferable utility [16, 20, 21]. However, in a competitive environment both central coordination and side payments are difficult to implement. For this reason, differently from [20, 21], we also exclude the presence of any intermediary to whom the retailers may disclose some private information. Decentralization of policies under partial information is the main focus in [14]. In [30] the authors analyze the benefits of information sharing on the performance of the entire chain. In [1] issues are discussed, regarding the use of different kinds of penalties, transfer prices, and cost-sharing schemes to improve the coordination of policies optimized on a local basis.

In a static context, i.e., for fixed day and fixed inventory levels, we introduced in [4] a distributed consensus protocol [22] for estimating the number of active retailers and coordinating the reordering policies. Each retailer is assumed to choose a fixed *threshold policy*, with threshold l_i on the number of active retailers. In other words one retailer defines its intention to reorder only if at least $l_i - 1$ other retailers are willing to do the same. We proved that consensus on the number of active retailers is asymptotically globally reached and coordination is the same to that obtained if the decision making process is centralized, namely, any retailer has access to the thresholds of all other retailers and chooses whether to reorder. The proposed distributed protocol has the advantage that the retailers do not communicate their threshold policy to reach consensus on the number of active retailers.

This chapter extends the aforementioned results to a dynamic inventory control context, i.e., where inventory levels change each day. We show that the threshold policies assumed in [4] are strictly connected to the well-known (s, S) incentive [25, 26]. In some cases, we prove that a optimal policy, for each ith retailer, is to order only in conjunction with at least other $l_i - 1$ retailers. We prove also that the threshold l_i can be computed locally by the ith retailer depending on the current inventory level and expected demand. This is possible by implementing a distributed neuro-dynamic programming (NDP) algorithm polynomial on the number of retailers, which avoids the curse of dimensionality and reduces errors due to model uncertainties.

This chapter is organized as follows. In Section 2 we develop a hybrid model for the cooperative inventory control problem. In Section 3 we prove that the cost function is K-convex and hence can be efficiently computed in a reduced number of points. We show also that threshold policies on the number of active retailers are optimal. In Section 4 we present the NDP algorithm and in Section 5 we provide conclusions.

2 Hybrid Model

In this section we present a novel hybrid model for the multi-retailer inventory system (see, e.g., Figure 2). In particular, in Subsection 2.1, we model the n decoupled inventory subsystems. In Subsection 2.2, we model the information flow among the subsystems. In Subsection 2.3, we introduce the structure of the local controllers and formally state the problem.

2.1 System Dynamics

Consider a network $\mathcal{G} = (\mathcal{V}, \mathcal{E})$; each retailer is a node $v_i \in \mathcal{V}$, where $i \in \Gamma := \{1, 2, ..., n\}$, and each communication link is an edge $e = (v_i, v_j) \in \mathcal{E}$; $i, j = 1, 2, ..., n$. Let $n = |\mathcal{V}|$, where $|S|$ indicates the cardinality of the set S. The model input u_i^k is the quantity of inventory ordered by the ith retailer at each stage $k = 0, 1, ..., N - 1$. We model with ω_i^k the stochastic demand faced by the ith retailer.

The ith inventory subsystem is a finite-state discrete-time model, that for all $i \in \Gamma$ takes on the form

$$x_i^{k+1} = x_i^k + u_i^k - \omega_i^k.$$

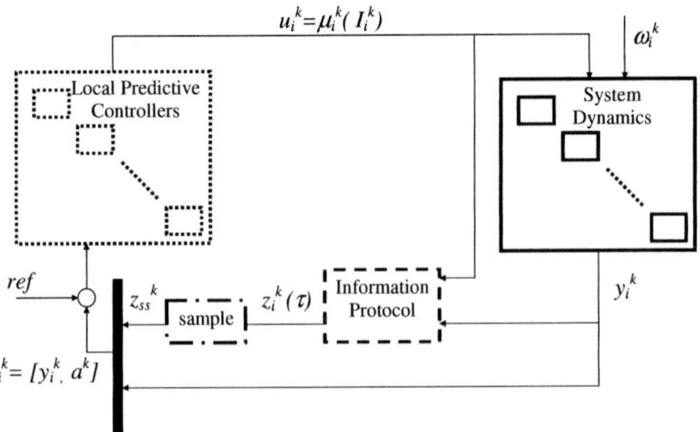

Fig. 2. Block diagram of the closed-loop inventory system.

The inventory at hand plus inventory ordered may not exceed storage capacity as is expressed in the following equation:

$$x_i^k + u_i^k \leq C_{store}.$$

The ith output y_i^k, referred to as *sensed information*, is

$$y_i^k = x_i^k,$$

i.e., each retailer observes only his inventory level.

2.2 Consensus Protocols

The information flow is managed through a *distributed* protocol $\Pi = \{(f_i, h_i, \phi_i) : \text{for all } i \in \mathcal{V}\}$

$$\dot{z}_i^k(\tau) = f_i(z_j^k(\tau), \text{ for all } j \in N_i), \, 0 \leq \tau \leq T, \tag{1}$$
$$z_i^k(0) = h_i(y_i^k), \tag{2}$$
$$a^k = \phi_i(z_{ss}^k), \tag{3}$$

where

- $f_i : \Re^n \to \Re$ describes the dynamics of the transmitted information of the ith node as a function of the information both available at the node itself and transmitted by the other nodes, as expressed in (1);
- $h_i : \mathcal{Z} \to \Re$ generates a new transmitted information vector given his output at the stage k, as described in (2);
- $\phi_i : \Re \to \mathcal{Z}$ estimates, based on current information, the aggregate info (3).

Here N_i is the neighborhood of the ith retailer, $N_i = \{j \in \Gamma : (v_i, v_j) \in \mathcal{E}\} \cup \{i\}$, i.e., the set of all the retailers j that are connected to i and i itself and

$$z_{ss}^k = \lim_{\tau \to T^-} z_i(kT + \tau), \quad \text{for all } i \in \Gamma, \tag{4}$$

represents the steady-state value assumed by $z_i^k(\tau)$ within the interval $[kT, (k+1)T]$.

We refer the reader to [22] for studies on the convergence of consensus protocols. For given scenario, defined by the full state vector, $x^k = \{x_i^k, \text{ for all } i \in \Gamma\}$, the converging value of the transmitted information a_i^k, plus the sensed information y_i^k, constitute the *partial information* vector $I_i^k = [y_i^k, a_i^k]$ available to the ith retailer.

2.3 Local Predictive Controllers

The local controllers compute the following cost over a finite horizon

$$J_i(\hat{I}_i^k, u_i^k) = \mathrm{E}\left\{g_i(\hat{I}_i^N) + \sum_{\hat{k}=k}^{N-1}(\alpha^{\hat{k}}g_i(\hat{I}_i^{\hat{k}}, u_i^{\hat{k}}))\right\}, \tag{5}$$

where $\hat{I}_i^k$ is the predicted information and α^k is the discount factor at stage k. The stage cost $g_i(\hat{I}_i^k, u_i^k, k)$ is defined as

$$\begin{aligned} g_i(\hat{I}_i^k, u_i^k, k) = \tfrac{K}{a^k}\delta(u_i^k) + cu_i^k + p\mathrm{E}\{\max(0, -\hat{y}_i^{k+1})\} \\ + h\mathrm{E}\{\max(0, \hat{y}_i^{k+1})\}, \end{aligned} \tag{6}$$

where K represents the setup cost, c is the purchase cost per unit stock, p is the penalty on storage, h the penalty on shortage, and $\delta(u_i(k))$ is zero if the ith retailer does not reorder, and one if he reorders. In (6) we assume that the setup cost is equally shared among the active retailers.

As will be clear later on, the idea of the solution algorithm is to use a simulation-based tunable predictor of the form

$$\hat{I}_i^{k+1} = \begin{bmatrix} \hat{y}_i^{k+1} \\ \hat{a}_i^{k+1} \end{bmatrix} = \begin{bmatrix} x_i^k + u_i^k - \hat{\omega}_i^k \\ \psi_i(a_i^k, u_i^k) \end{bmatrix}. \tag{7}$$

We report hereafter the formalization of the problem under consideration: given a set of retailers reviewed as dynamic agents of a network with topology $\mathcal{G} = (\mathcal{V}, \mathcal{E})$.

Problem (Local Controller Synthesis) *For each ith retailer, determine the reordering policy $u_i^k = \mu(I_i^k)$ that minimizes the N-stage individual payoff defined in (5).*

Subproblem (Protocol Design) *Determine a distributed protocol Π that maximizes the set of active retailers A_Π.*

3 Dynamic Programming Approach

In this section, we prove that the inventory must be ordered in quantity to fulfill exactly the expected demand for the upcoming days, as summarized in Theorem 1. We provide an intuitive explanation of such a result.

Let $K^k = \frac{K}{a^k}$ be the setup cost charged to each retailer that reorders at stage k, and $d_i^k = x_i^k + u_i^k$ the instantaneous inventory position, i.e., the inventory level just after the order has been issued. Then we claim as follows.

- If the setup cost K^k decreases with time (in the future more retailers are interested in reordering) retailers place short-term orders. Optimal policies are multi-period policies (s^k, S^k), with a unique lower and upper threshold (see, e.g., Figure 3).
- On the contrary, if the setup cost K^k increases with time (in the future fewer retailers are interested in reordering), retailers place long-term orders. Optimal policies are multi-period policies (s^k, S^k) with multiple thresholds at different inventory levels (see, e.g., Figure 4).

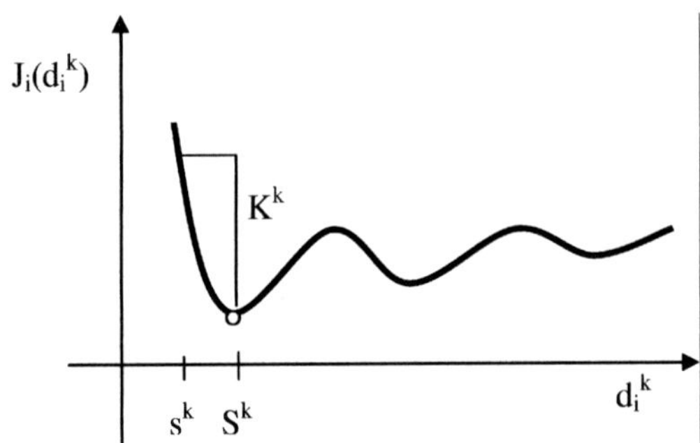

Fig. 3. Intuitive plot of the cost when the setup cost decreases with time: single thresholds (s^k, S^k).

3.1 Searching for Structure: K-Convex Analysis

To show that the individual objective functions J_i; $i \in \Gamma$ have at most N local minima, we first apply the dynamic programming (DP) algorithm (8)–(9) to minimize the cost (5). The Bellman's equation is then rearranged by defining a new function $H(\cdot)$ that verifies the $\overline{K}_i$-convexity property, where $\overline{K}_i$ is the maximum setup cost incurred by the ith retailer over the horizon. Exploiting

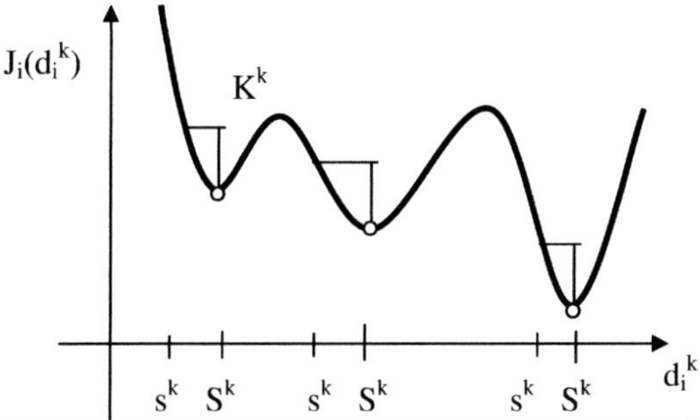

Fig. 4. Intuitive plot of the cost when the setup cost increases with time: multiple thresholds (s^k, S^k).

the definition of the inventory position d_i and the setup cost K^k, we rewrite the stage cost (6) as

$$g_i(d_i^k, a^k) = K^k \delta(u_i^k) + cu_i^k + pE\{\max(0, -(d_i^k - \omega_i^k))\}$$
$$+ hE\{\max(0, d_i^k - \omega_i^k))\}.$$

By applying the dynamic programming algorithm, we have

$$J_i^N(I_i^N) = 0, \tag{8}$$

$$J_i^k(I_i^k) = \min_{u_i^k \in U} [g_i(d_i^k, a^k) + \alpha^{k+1} E\{J_i^{k+1}(I_i^{k+1})\}]. \tag{9}$$

Let us define the new function

$$G_i^k(d_i^k, a^{k+1}) = cd_i^k + E\{p\max(0, -(d_i^k - \omega_i^k))$$
$$+ h\max(0, d_i^k - \omega_i^k) + J_i^{k+1}(I_i^{k+1})\},$$

and rewrite the Bellman's equation (9) as follows:

$$J_i^k(I_i^k) = -c_i x_i^k + \min_{d_i^k \geq x_i^k} [K^k + G_i^k(d_i^k, a^{k+1}), G_i^k(x_i^k, a^{k+1})]. \tag{10}$$

Note that if we can show that J_i^{k+1} is K^k-convex, then G_i^k is also K^k-convex and the Bellman's equation (10) has a unique minimizer.

Indeed, it has been proven in [6] that K^k-convexity of $G_i^k(d_i, a^{k+1})$ implies K^k-convexity of $J_i^k(I_i^k)$.

This represents a sufficient condition that guarantees optimality of multi-period (s_i^k, S_i^k) order-up-to policies.

We recall that s_i^k represents the minimum threshold on inventory level below which retailers reorder to restore level S_i^k.

Let us remember that S_i^k minimizes $G_i^k(\cdot, a^{k+1})$ and threshold s_i^k verifies

$$G_i^k(s_i^k, a^{k+1}) = G_i^k(S_i^k, a^{k+1}) + K^k.$$

Now let us call $\underline{s}_i^k$ the threshold that corresponds to the assumption that the ith retailer is charged the whole setup cost; namely we have $K_i^k = K$; $i \in \Gamma$. In the same way, let us define as $\overline{s}_i$ the threshold computed as if all retailers would share equally the setup costs; thus, each retailer is charged a setup cost $K_i^k = \frac{K}{n}$, namely one nth of the entire cost K. We now render explicit the dependence of threshold s_i^k on setup cost K_i^k by defining the function $s_i^k(\frac{K}{a^k})$ for which it holds that $\underline{s}_i^k \leq s_i^k(\cdot) \leq \overline{s}_i^k$.

In the following, we consider a^k a parameter and show that the individual objective function $J_i^k(x_i)$; $i \in \Gamma$, which is generically nonconvex, has all local minima coincide with the demand summed over one or more days.

Theorem 1. *Solutions of the Bellman's equation (9) are at most $N - k$ different multi-period policies (s_i^k, S_i^k), where $S_i^k \in \{\sum_{j=k}^{M} \omega_i^j; M = k, k+1, ..., N\}$ and threshold s_i^k verifies $G_i^k(s_i^k, a^{k+1}) = G_i^k(S_i^k, a^{k+1}) + K^k$. Policy is associated to different intervals of inventory levels.*

Proof. The essential idea is that the cost is piecewise linear. This is evident in the Bellman's equation where the cost J_i^k is the summation of a piecewise linear stage cost g_i^k (with unique global minimum at ω_i^k) and a piecewise linear future cost (with potential local minima at $\omega_i^k + S_i^{k+1}$).

An immediate consequence of the above theorem is that the set of feasible decisions is finite and each element represents the exact ordered quantity to fulfill the expected demand for the upcoming 1, 2, ..., N days.

3.2 Threshold Reordering Policies

The aim now is to show that Nash equilibrium reordering policies have a threshold structure on the number of retailers interested in reordering. To see this, we first introduce a preliminary lemma on single-stage inventory control and reinterpret the concept of threshold (s, S) in a way more suitable for a multi-retailer scenario. In particular we change a threshold on inventory level s into a threshold l on "how many retailers are interested in reordering."

Lemma 1. (Single-Stage Optimization) *For each inventory level x_i there exists a threshold $l_i \in \{1, 2, ..., n\}$, such that the reordering policy*

$$\mu_i(I_i) = \begin{cases} S_i - x_i & \text{if } a \geq l_i \\ 0 & \text{if } a < l_i \end{cases} \tag{11}$$

is a Nash equilibrium for the single-stage formulation of the multi-retailer inventory control problem.

Proof. From Theorem 1, if $N = 1$, we have a unique multi-period policy (s_i, S_i). This means that retailers make decisions according to

$$\mu_i(I_i) = \begin{cases} S_i - x_i & \text{if } x_i < s_i \\ 0 & \text{if } x_i \geq s_i. \end{cases} \qquad (12)$$

For given x_i, the idea is to find the minimum value of l_i that verifies the condition $x_i < s_i$. This is straightforward for the two limit cases of "low" and "high" inventory level, namely $x_i < \underline{s}_i$, and $x_i \geq \overline{s}_i$, respectively. It is left to prove (12) for the intermediate case $\underline{s}_i \leq x_i \leq \overline{s}_i$ (see proof in [3]).

As evident from (11) the single-stage formulation of the multi-retailer inventory control problem leads to reordering policies with a threshold structure. Results from Lemma 1 can be extended to the multi-stage formulation.

Theorem 2. (Multi-Stage Optimization) *For each inventory level x_i^k there exists a threshold $l_i^k \in \{1, 2, ..., n\}$, such that the reordering policy*

$$\mu(I_i^k) = \begin{cases} S_i^k - x_i^k & \text{if } a^k \geq l_i^k \\ 0 & \text{if } a^k < l_i^k \end{cases} \qquad (13)$$

is a Nash equilibrium for the multi-stage formulation of the multi-retailer inventory control problem.

Proof. The structure of the proof is the same as for the single-stage inventory problem in Lemma 1. Note, however, that from Theorem 1 we now have at most $N - k$ different multi-period policies (s_i^k, S_i^k), each one associated to a different interval of inventory levels. The trick of the proof is to repeat the argument above for each interval.

We then conclude that optimizing the multi-retailer inventory control problem over a multi-stage horizon leads to Nash equilibrium reordering policies with threshold structure on the number of active retailers.

3.3 Local Estimation via Consensus Protocols

In this subsection, we discuss the solution of the subproblem on protocol design. The focus is on consensus protocols to estimate the number of active retailers a^k. Indeed, given the vector $l = \{l_i\}$, collecting the optimal thresholds, each retailer makes the decision "do not reorder" if its local estimation is lower than its threshold, as expressed in (13). We assume that the transmitted information is the current estimate of the percentage of retailers who are interested in reordering. The current estimate $z_i(\cdot)$ is reinitialized to $\{0, 1\}$ at the beginning of each time interval $[kT, kT+1]$ based on the current inventory level x_i^k. In particular, if the ith inventory level is "low," i.e., the corresponding threshold l_i does not exceed the network size n, then the retailer is willing to reorder. He has no information yet except its observed inventory level; thus, he assumes that all other retailers are in the same circumstances (spatially invariant assumption) and set $z_i^k = 1$, indicating that everyone is interested in reordering. On the contrary, if the inventory level is "high" (l_i exceeds n), he is not willing to join the group to order and set $z_i^k = 0$, indicating that no one needs to reorder. Thus we can write

$$z_i^k = \begin{cases} 0 \ l_i(x_i^k) > n \\ 1 \ \text{otherwise.} \end{cases} \tag{14}$$

Then each retailer updates the estimate on-line on the basis of new estimate data received from neighbors. At any time t_i whenever the number of retailers interested in reordering a^k goes below his threshold l_i, the ith retailer communicates his decision to "give up" to reorder by activating an exogenous impulse signal $\delta_i(t - t_i)$. This exogenous impulse can be activated only one time (once you exit the group you are no longer allowed to rejoin it) and only when the all local estimates have reached consensus on a final value. This occurs every t_f, where t_f is an estimate of the worst case possible settling time of the protocol dynamics.

Given (14), an average-consensus protocol leads all local estimates to converge to the max a^k (see [4]). The continuous-time average-consensus protocol takes on the form

$$\begin{cases} h_i(x_i^k) &= l_i(x_i^k) \leq n \\ f_i(z^k(\tau)) &= -L_{i\bullet} z^k(\tau) + \delta_i(t - t_i) \cdot u_i^k \\ \phi(z_i^k(\tau)) &= n(\lim_{t \to T^-} z_i^k(\tau)), \end{cases}$$

where L is the Laplacian matrix of the communication network topology and t_i is in turn the time instant where the current estimate converges to a value below the threshold. It can be defined by the following logic conditions:

$$t_i : s.t. [l_i(x_i^k) > n] \text{ OR } [(l_i(x_i^k) \leq n)\text{AND}(nz_i(t_i) < l_i) \\ \text{AND } (t_i = kt_f, \ k \in \mathcal{N})].$$

We refer the reader to [4] for details on the optimality of the protocol above.

4 NDP Solution Algorithm

In this section, we cast the hybrid model within the framework of NDP.

4.1 Consensus on Features a_i^k

To review the features as a compact description of the behavior of the other retailers, we consider i) the NDP architecture based on feature extraction displayed in Figure 5 (see, e.g., [7]) and ii) the block diagram of the hybrid model displayed in Figure 6.

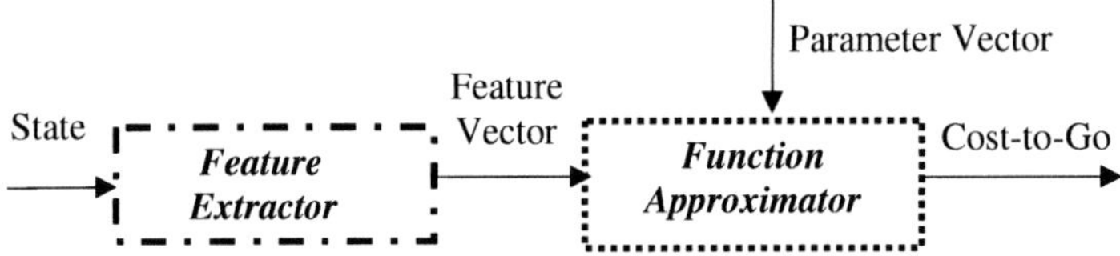

Fig. 5. The information flow management uses consensus protocols to extract the features.

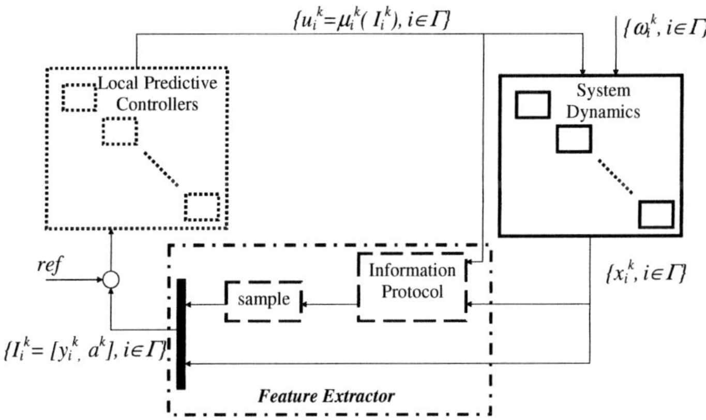

Fig. 6. Block diagram of the closed-loop system.

The full-state vector of the hybrid model x^k becomes, in the approximation architecture, the input to the feature extractor. The information flow management block can be reviewed as the feature extractor. The full-state vector reduces to the partial information vector $I_i^k = [y_i^k, a_i^k]$ available to the ith retailer. Each local controller implements a function approximator, which receives the partial information vector and returns the individual cost-to-go $\widetilde{J}_i^k(I_i^k, r)$ over the horizon.

4.2 Linear Architecture

We assume that the probability distribution over all potential values assumed by a^k propagates according to the linear dynamics $a^{k+1} = a^k \Psi^k$ where $\Psi^k = \{\psi_{ij}^k, i, j \in \Gamma\}$. In this case we have i) a matrix of weights r that coincides with the transition probability matrix of the predictor, namely, $r = \Psi = \{\Psi^k, k = 1, 2, ..., N\}$, and ii) *basis functions* $\widetilde{J}_i^{k+1}(I_i^{k+1}, a^{k+1})$ representing different future costs associated to different a^{k+1}.

The approximation architecture linearly parameterizes the future costs associated to all possible behaviors of the other retailers over the horizon. This can be described as

$$\sum_{a^{k+1}=1}^{|\mathcal{Z}|} \Psi_{a^k,\,a^{k+1}}^{k}\, \widetilde{J}_i^{k+1}(I_i^{k+1},\, a^{k+1}) = \psi_{a^k\bullet}^{k}\, \widehat{J}_i^{\,k+1}(I_i^{k+1},\, \bullet)^{T},$$

where $\psi_{a^k\bullet}^{k}$ is the row of the transition probabilities from a^k to all possible a^{k+1}, and $\widehat{J}_i^{\,k+1}(I_i^{k+1},\, \bullet)^{T}$ is the transposed row of the associated future costs.

4.3 The NDP Algorithm

This algorithm is organized in two parts. In the first part the retailers compute the set of admissible decisions U_i^k and reachable states R_i^k over the horizon. The second part presents three steps.

1. *Policy improvement.* For given prediction Ψ, we improve the policy via the stochastic Bellman's equation backward in time:

$$\mu_i^k(I_i^k) = \mathrm{argmin}_{u_i^k \in U_i^k(x_i^k)} \left[g_i(I_i^k,\, u_i^k,\, k) + \alpha^{k+1}\psi_{a^k\bullet}\, \widehat{J}_i^{\,k+1}(I_i^{k+1},\, \bullet) \right].$$

2. *Value iteration.* The improved policy is valued through repeated quasi-Monte Carlo simulations. Active exploration guarantees that initial states are sufficiently spread over the local minima. During the value iteration we compute and store the number of times a transition Ψ_{ij} occurs during the repeated finite length simulations. At the end of each simulation, the protocol runs over the horizon and returns the training set for the next step.

3. *Temporal difference.* We use the training set to update the transition probabilities of the predictor.

The tree steps are iteratively repeated until convergence of policies.

Lemma 2. *Each iteration of the NDP algorithm, for given initial state x^0, has computational complexity polynomial on the number of retailers, i.e., $O(n^2 N R^2)$.*

Proof. The proof starts by considering that the complexity of the algorithm depends essentially on the complexity of the second part. Here we write the Bellman's equation considering the set of feasible decisions U_i^k, for each retailer $i \in \Gamma$, for each stage $k = 1, 2, ..., N$ and for each decomposed state $I_i^k \in (R_i^k \times \Gamma)$. Thus, complexity is $O(n^2 N R^2)$.

Assuming that convergence is achieved in a finite number of iterations, the temporal difference algorithm returns stochastic Nash equilibrium policies, paths, and costs-to-go. Further efforts are still to be made to investigate the convergence conditions of this algorithm.

Table 1. Expected demand for the upcoming ten days.

ω_1	4	8	6	5	7	8	4	5	6	8
ω_2	0	0	1	7	8	0	6	2	1	4
ω_3	0	3	2	0	3	1	1	3	3	0

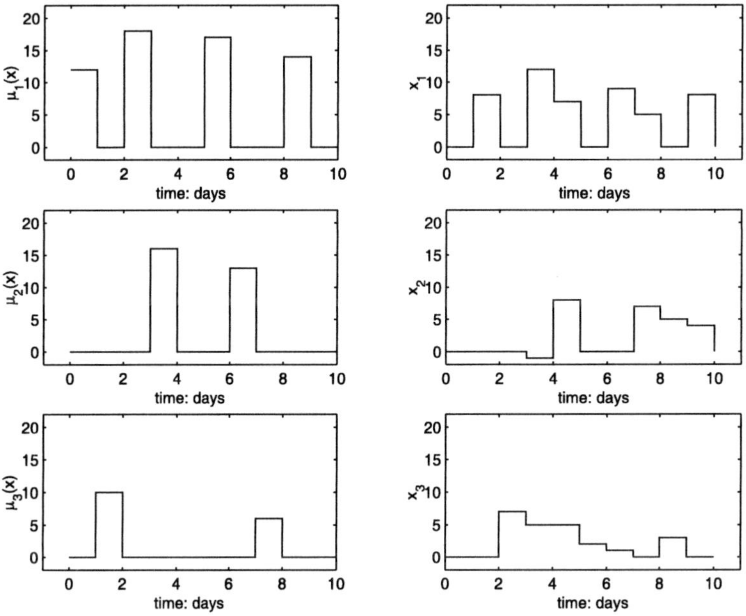

Fig. 7. Uncoordinated reordering policies.

Example 1. Let us consider a group of three retailers and parameters $K = 24$, $p = 8$, $h = 1$, and $c = 2$. Retailers face a stochastic Poissonian demand with expected values over the horizon of ten days as in Table 1.

At the first iteration, no communication has occurred among the retailers and the "policy improvement" returns the uncoordinated reordering policies displayed in Figure 7.

The "value iteration" consists of 12 simulations of the inventory system under the improved reordering policies. The set of initial states is a stochastic sequence extracted from a Poissonian distribution with mean value, respectively, equal to 25, 10, and 6 for the first, second, and third retailer. Indeed, we know from deterministic simulation results that J_1 has potential local minima at 18, 23, 30, J_2 at 1, 8, 16, and J_3 at 8, 10 as displayed in Figure 9 (solid and dotted lines). Here the costs associated to the first, second, third and fourth policy improvements when demand is deterministic are represented by four lines of different type. At the end of each simulation the retailers run a consensus protocol returning a^k over the horizon. Based on this new aggregate information, during the "temporal difference" the retailers update the transition probabilities of the predictor and a new iteration starts. In this example, the algorithm eventually converges to a Nash equilibrium in six iterations returning a coordinated distribution of reorders over the horizon as shown in Figure 8. We see from Figure 9 that the costs-to-go at the fourth and fifth iteration draw much near to the cost-to-go of the deterministic problem. We may conclude that the NDP algorithm possesses satisfying learning capabilities.

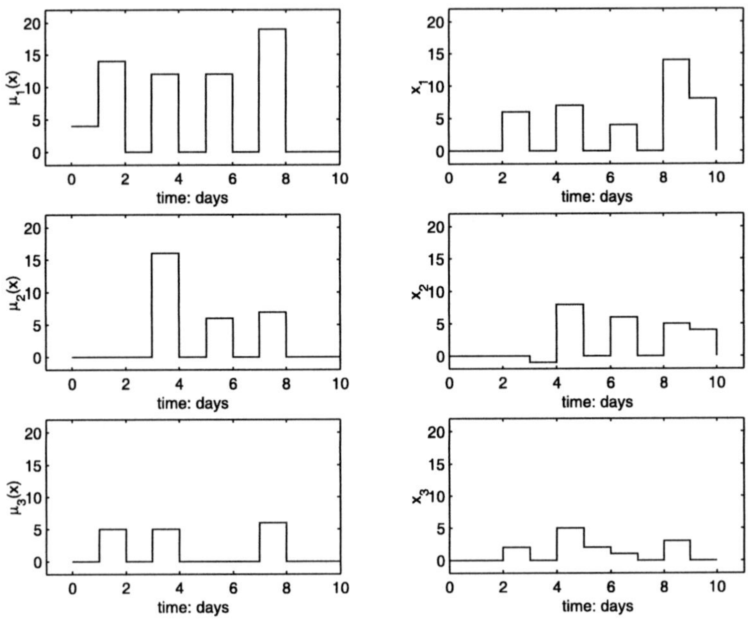

Fig. 8. Coordinated reordering policies.

5 Conclusion and Future Research

In this chapter we propose an NDP approach to coordinate the reordering policies of a group of retailers. Coordination is motivated by the possibility of sharing setup cost when orders are placed in conjunction. We develop a hybrid model to describe the inventory subsystems and the information flow. We designed consensus protocols for the information flow. Finally we presented a scalable and suboptimal NDP algorithm. The main question is how consensus protocol can achieve distributed convergence to Pareto optimal Nash equilibria for a class of repeated noncooperative games under incomplete information. A possible solution is given by considering games with monotonic payoffs

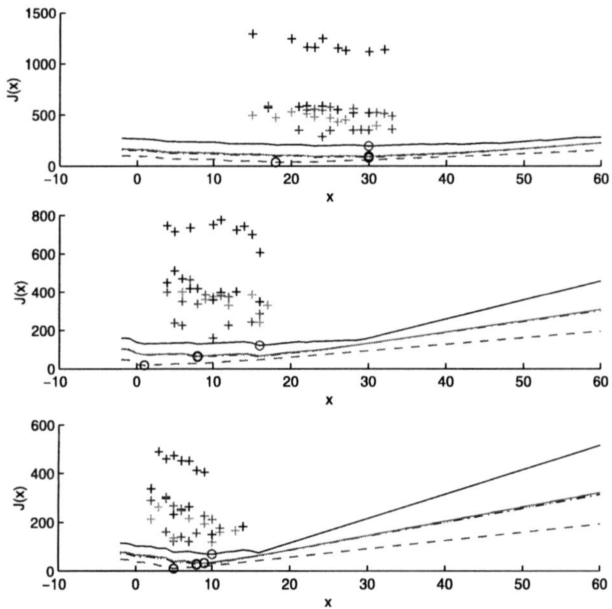

Fig. 9. Costs vs inventory: deterministic (lines) and stochastic demand (crosses).

and specializing them to multi-retailer inventory problems, where transportation or setup costs are shared among all retailers reordering from a common warehouse. In future works, the authors will prove the existence and the stability of Nash equilibria and their convergence properties. Further work in this direction would involve the study of information protocols and decision mechanisms in presence of stochastic processes.

References

1. Axsäter S (2001) A framework for decentralized multi-echelon inventory control. IIE Transactions 33(1):91–97
2. Basar T, Olsder G (1995) Dynamic noncooperative game theory, 2nd ed., Academic Press, London
3. Bauso D (2003) Cooperative control and optimization: a neuro-dynamic programming approach. PhD thesis, Università di Palermo, Dipartimento di Ingegneria dell'Automazione e dei Sistemi
4. Bauso D, Giarré L, Pesenti R (2003) Distributed consensus protocols for coordinated buyers, 588–592. In: Proc. of the 42nd Conference on Decision and Control, Maui, HI

5. Berovica DP, Vinter RB (2004) The application of dynamic programming to optimal inventory control. IEEE Trans. on Automatic Control 49(5):676–685

6. Bertsekas D (1995) Dynamic programming and optimal control, 2nd ed., Athena, Belmont, MA

7. Bertsekas D, Tsitsiklis JN (1996) Neuro-dynamic programming. Athena, Belmont, MA

8. Blanchini F, Miani S, Ukovich W (2000) Control of production-distribution systems with unknown inputs and systems failures. IEEE Trans. on Automatic Control 45(6):1072–1081

9. Cachon G (2001) Stock wars: inventory competition in a two-echelon supply chain with multiple retailers. Operations Research 49(5):658–674

10. Cetinkaya S, Lee CY (2000) Stock replenishment and shipment scheduling for vendor-managed inventory systems. Management Science 46(2):217–232

11. Chen F, Drezner Z, Ryan JK, Simchi-Levi D (2000) Quantifying the bullwhip effect in a simple supply chain: the impact of forecasting, lead times, and information. Management Science 46(3):436–443

12. Corbett CJ (2001) Stochastic inventory systems in a supply chain with asymmetric information: cycle stocks, safety stocks, and consignment stock. Operations Research 49:487–500

13. Fax A, Murray RM (2002) Information flow and cooperative control of vehicle formations In: The 15th IFAC World Congress, Barcelona, Spain

14. Fransoo JC, Wouters MJF, de Kok TG (2001) Multi-echelon multi-company inventory planning with limited information exchange. Journal of the Operational Research Society 52(7):830–838

15. Friedman EJ (1996) Dynamics and rationality in ordered externality games. Games and Economic Behavior 16:65–76

16. Hartman B, Dror M, Shaked M (2000) Cores of inventory centralization games. Games and Economic Behavior 31(1):26–49

17. Jorgensen S, Kort PM (2002) Optimal pricing and inventory policies: centralized and dencentralized decision making. European Journal of Operational Research 138:578–600

18. Kleywegt A, Nori V, Savelsbergh M (2002) The stochastic inventory routing problem with direct deliveries. Transportation Science 36(1):94–118

19. Lee H, Whang S (1999) Decentralized multi-echelon supply chains: incentives and information. Management Science 45(5):633–640

20. Meca A, Garcia-Jurado I, Borm P (2003) Cooperation and competition in inventory games. Mathematical Methods of Operations Research 57(3):481–493

21. Meca A, Timmer J, Garcia-Jurado I, Borm P (2004) Inventory games. European Journal of Operational Research 156:127–139

22. Olfati-Saber R, Murray R (2003) Consensus protocols for networks of dynamic agents, 951–956. In: American Control Conference, Denver, CO

23. Olfati-Saber R, Murray R (2004) Consensus problems in networks of agents with switching topology and time-delays. IEEE Trans. on Automatic Control 49(9):1520–1533

24. Osborne M, Rubinstein A (1994) A course in game theory. MIT Press, Cambridge, MA

25. Scarf HE (1995) The optimality of (s, S) policies in the dynamic inventory problem. Stanford University Press, Stanford, CA

26. Scarf HE (2002) Inventory theory. Operations Research 50(1):189–191
27. Silver E, Pyke D, Peterson R (1998) Inventory management and production planning and scheduling. John Wiley & Sons, New York
28. Wang Q (2001) Coordinating independent buyers in a distribution system to increase a vendor's profits. Manufacturing and Service Operations Management 3(4):337–348
29. Watts A (2002) Uniqueness of equilibrium in cost sharing games. Journal of Mathematical Economics 37:47–70
30. Yu Z, Yan H, Cheng T (2002) Modelling the benefits of information sharing-based partnerships in a two-level supply chain. Journal of the Operational Research Society 53(4):436–446

Part VI

Networked Control Systems

Communication Logic Design and Analysis for Networked Control Systems*

Yonggang Xu and João P. Hespanha

Center for Control Engineering and Computation
Dept. of Electrical and Computer Engineering
University of California
Santa Barbara, CA 93106, USA
{yonggang,hespanha}@ece.ucsb.edu

Summary. This chapter addresses the control of spatially distributed processes via communication networks with a fixed delay. A distributed architecture is utilized in which multiple local controllers coordinate their efforts through a data network that allows information exchange. We focus our work on linear time-invariant processes disturbed by Gaussian white noise and propose several logics to determine when the local controllers should communicate. Necessary conditions are given under which these logics guarantee boundedness and the trade-off is investigated between the amount of information exchanged and the performance achieved. The theoretical results are validated through Monte Carlo simulations. The resulting closed loop systems evolve according to stochastic differential equations with resets triggered by stochastic counters. This type of stochastic hybrid system is interesting on its own.

1 Introduction

The architectures for feedback control of spatially distributed processes generally fall in one of the three classes, *centralized, decentralized,* or *distributed.* Centralized architectures yield the best performance because they pose the least constraints on the structure of the controller, whereas decentralized architectures are the simplest to implement. We pursue here distributed architectures, as they provide a range of compromise solutions between the two extremes. The communication among local controllers is supported by a data network that allows information between local controllers to be exchanged at discrete time instants.

Our objective is to understand the trade-off between the amount of information exchanged and the performance achieved. Several results can be

---

*This research was supported by the National Science Foundation under the grant numbers CCR-0311084 and ECS-0242798.

found in the literature on how to reduce communication in networked control systems (NCS). The problem of stabilization with finite communication bandwidth is introduced by [21, 22] and further pursued by [14, 18, 11, 6, 12]. An estimation problem is investigated in [21] under the constraint that observations must be coded digitally and transmitted over a channel with finite capacity. The corresponding stabilization problem under similar limitations is addressed in [22]. [18, 14] and [6] determine the minimum bandwidth (measured in discrete symbols per second) needed to stabilize a linear process. In all these references a digital communication channel is assumed so that any information transmitted has to be quantized. [3] focuses on the quantization aspect and shows that the memoryless quantization scheme that minimizes the product of quantization density times sampling rate follows a logarithmic rule.

We depart from the work summarized above in that we only penalize the number of times that information is exchanged. This is motivated by the fact that in most widely used communication protocols there is a fixed overhead incurred by sending a packet over the network, which is not reduced by decreasing the number of data-bits. For example, a fixed-size ATM (asynchronous transfer mode) cell consists of a 5-byte header and a 48-byte information field, whereas an Ethernet frame has a 14-byte or 22-byte header and a data field that must be at least 46 bytes long. In either case, one "pays" the same price for sending a single bit or 48/46 bytes of data.

Several practical issues motivate us to reduce packet rate in NCS. Higher data traffic may induce longer communication delay and more data dropouts, which are undesirable in real-time systems [10]. In sensor network applications, an important criterion in assessing communication protocols is energy efficiency, and the primary source of energy consumption in the nonmobile wireless settings is the radio [1]. A smart communication scheduling method can extend the battery lifetime and therefore reduce the sensor network deployment costs.

The systems of interest are spatially distributed processes whose dynamics are decoupled but for which the control objectives are not, e.g., the control of a group of autonomous aircraft flying in a geometric formation far enough from each other so that their dynamics are decoupled. However, many of these ideas could be extended to coupled processes.

Each process with an associated local controller is viewed as a *node*. The overall control system consists of certain number of nodes connected via a communication network. Figure 1 depicts the internal structure of the ith node. It consists of a *local process*, a *local controller*, a *bank of estimators* and a *communication logic*. The *synchronized estimators* are used by the local controller to replace the state of remote processes not available locally. They are simply computational models of the remote processes and the reason to call them "synchronized" will become clear shortly. These estimators run open

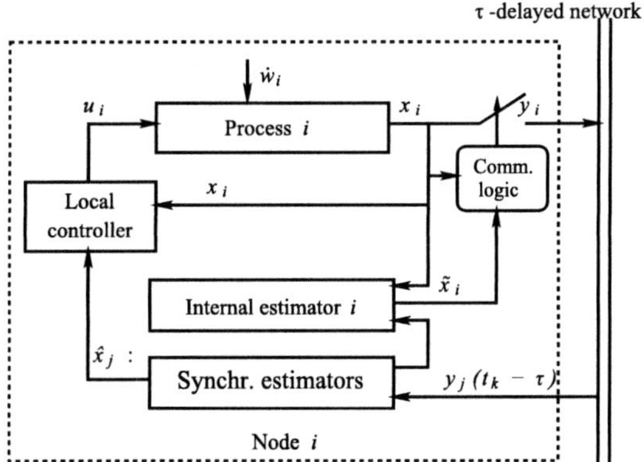

Fig. 1. One of the nodes in a networked control system.

loop most of the time but are sometimes reset to "correct" values when state measurements are received through the network. These resets do not necessarily occur periodically and it is the responsibility of each node to decide when to broadcast to the network the state of its local process. The communication logic makes use of an *internal estimator* to determine how well other nodes can "predict" the state of its local process and decides when to broadcast it. In general, the communication network introduces delay and therefore these data only become available to the other nodes some time later. This type of architecture is proposed by Yook, Tilbury and Soparkar [24] for the control of discrete-time distributed systems over delay-free networks.

Several algorithms can be used by the communication logic to determine *when* the state of the local process should be broadcast. The quality of a communication logic should be judged in terms of the performance it can achieve for a given message broadcast rate. We measure performance in terms of the statistical moments of the estimation errors associated with the synchronized estimators, which provide a measure of the penalty introduced by the fact that the state measurements of the remote processes are not available locally.

One simple algorithm that can be used by the communication logic consists of broadcasting messages *periodically*. However, as we shall see, this is not optimal because data may be transmitted with little new information. In [24], it is proposed that a node should broadcast the true value of the state of its local process when it differs from the estimate known to the remaining nodes by more than a given threshold. For the linear discrete-time case, they showed that this scheme results in a system that is BIBO stable. The relation between the threshold level and the message exchange rate is investigated through simulation in the context of examples.

We proposed new communication logics that can be analyzed to determine stability as well as the trade-off between communication (in terms of average message exchange rates) and performance. We start by considering *stochastic communication logics* for which the probability of a node broadcasting a message is a function of the current estimation error. *Deterministic communication logics* similar to the ones proposed in [24] are also considered. We will see that the latter can be viewed as limiting cases of the former. We also simulate different communication logics, including periodic, stochastic and deterministic, and compare their performances.

The stochastic communication logics are based on doubly stochastic Poisson processes (DSPPs) [2]. In essence, the state of the local process is broadcast according to a Poisson process whose rate depends on the estimation error. This type of stochastic hybrid system seems to be interesting on its own. For stochastic communication logics, our stability analysis uses tools from jump diffusion processes. Deterministic logics are analyzed in the context of first exit time problems.

In Section 2, the control-communication architecture is formally described for the case of two linear time-invariant processes. Stochastic communication logics are analyzed in Section 3, first for delay-free networks and later for networks that introduce a delay of τ time units. Deterministic logics are addressed in Section 4 for delay-free systems. Simulation results are presented in Section 5 for a second-order leader-follower problem. We also provide trade-off curves showing the average communication rate versus the variance of the estimation error for an unstable process. Section 6 contains conclusions and directions for future work.

2 Networked control system model

In this section we propose an estimator-based architecture for distributed control. For simplicity of presentation we consider only two nodes like the ones in Figure 1.

2.1 Estimator-based control architecture

The processes are assumed to be linear time-invariant with an exogenous disturbance input,

$$\dot{\mathbf{x}}_i = A_i \mathbf{x}_i + B_i \mathbf{u}_i + \sigma_i \dot{\mathbf{w}}_i$$
$$\mathbf{y}_i = \mathbf{x}_i + \boldsymbol{\zeta}_i \qquad\qquad \forall i \in \{1,2\},$$

where $\mathbf{x}_i \in \mathbb{R}^{n_i}$ denotes the state, $\mathbf{u}_i \in \mathbb{R}^{m_i}$ the control input, $\mathbf{y}_i \in \mathbb{R}^{n_i}$ the state measurement, $\dot{\mathbf{w}}_i$ the ℓ_i-dimensional standard Gaussian white noise, and $\boldsymbol{\zeta}_i$ the zero-mean measurement noise and/or quantization errors. The two noise processes are assumed independent and all matrices are real and of appropriate dimensions. The process $\boldsymbol{\zeta}_i$ is also assumed stationary with known probability distribution $\mu(\cdot)$.

We assume given state-feedback control laws

$$\mathbf{u}_i = K_{i1}\mathbf{x}_1 + K_{i2}\mathbf{x}_2, \qquad \forall i \in \{1,2\}, \tag{1}$$

which would provide adequate performance in a centralized configuration, i.e., if the states of both processes were available to both local controllers. In a centralized configuration, the closed loop system would be

$$\begin{aligned}
\dot{\mathbf{x}}_1 &= (A_1 + B_1 K_{11})\mathbf{x}_1 + B_1 K_{12}\mathbf{x}_2 + \sigma_1 \dot{\mathbf{w}}_1 \\
\dot{\mathbf{x}}_2 &= (A_2 + B_2 K_{22})\mathbf{x}_2 + B_2 K_{21}\mathbf{x}_1 + \sigma_2 \dot{\mathbf{w}}_2.
\end{aligned} \tag{2}$$

Since the state of the ith process is not directly available at the jth node ($j \neq i$, $i, j \in \{1,2\}$), we build at the node j an estimate $\hat{\mathbf{x}}_i$ of the state $\mathbf{x}_i$. In the distributed architecture, the centralized laws (1) are replaced by

$$\begin{aligned}
\mathbf{u}_1 &= K_{11}\mathbf{x}_1 + K_{12}\hat{\mathbf{x}}_2 \\
\mathbf{u}_2 &= K_{21}\hat{\mathbf{x}}_1 + K_{22}\mathbf{x}_2.
\end{aligned} \tag{3}$$

The distributed control laws (3) result in closed-loop dynamics given by

$$\begin{aligned}
\dot{\mathbf{x}}_1 &= (A_1 + B_1 K_{11})\mathbf{x}_1 + B_1 K_{12}\hat{\mathbf{x}}_2 + \sigma_1 \dot{\mathbf{w}}_1 \\
\dot{\mathbf{x}}_2 &= (A_2 + B_2 K_{22})\mathbf{x}_2 + B_2 K_{21}\hat{\mathbf{x}}_1 + \sigma_2 \dot{\mathbf{w}}_2,
\end{aligned} \tag{4}$$

to be contrasted with (2). To understand the effect of the distributed architecture on the performance of the closed loop system, we write the closed loop dynamics (4) in terms of the estimation errors:

$$\begin{aligned}
\dot{\mathbf{x}}_1 &= (A_1 + B_1 K_{11})\mathbf{x}_1 + B_1 K_{12}\mathbf{x}_2 + \sigma_1 \dot{\mathbf{w}}_1 + B_1 K_{12}\mathbf{e}_2, \\
\dot{\mathbf{x}}_2 &= (A_2 + B_2 K_{22})\mathbf{x}_2 + B_2 K_{21}\mathbf{x}_1 + \sigma_2 \dot{\mathbf{w}}_2 + B_2 K_{21}\mathbf{e}_1,
\end{aligned}$$

where the estimation error is defined as $\mathbf{e}_i := \hat{\mathbf{x}}_i - \mathbf{x}_i$. Comparing these equations with (2), we observe that the penalty paid for a distributed architecture is expressed by the additive "disturbance" terms $B_i K_{ij}\mathbf{e}_j$. The estimation error term $\mathbf{e}_i$ is the focus of our investigation.

2.2 Estimators

Since remote state information is not directly available, each node needs to construct *synchronized state estimates* to be used in (3), based on the data received via the network, which is assumed to introduce a delay of τ time units. Moreover, each node also needs to send its own state to the remote nodes to allow them to construct their synchronized estimates. Each node's *communication logic* is responsible for determining when data transmission should take place and makes this decision based on an *internal estimate* of its own state. The difference between this internal state estimate and the measured state provides a criterion to judge the quality of the synchronized estimates currently being used at the remote nodes.

To simplify the presentation, we only write the equations for the synchronized state estimators inside node 2 and the internal estimator used by the communication logic inside node 1. These estimators are relevant to investigate the impact of the rate at which measurements are sent from node 1 to node 2. The flow of data in the opposite direction is determined by completely symmetric structures.

Node 2's synchronized state estimates at some time t are based on all information received from node 1 up to time t:

$$\left\{ \mathbf{y}_1(\mathbf{t}_j) : \mathbf{t}_j \leq t - \tau \right\} \tag{5}$$

where $0 =: \mathbf{t}_0 < \mathbf{t}_1 < \mathbf{t}_2 < \cdots$ are the times at which node 1 sends its state measurement $\mathbf{y}_1(t_k)$ to node 2. The corresponding minimum-variance estimate is given by a Kalman filter for (4) with discrete measurements [13]. For simplicity we shall assume that the measurement noise $\boldsymbol{\zeta}_i(t)$ is negligible, in which case the filter takes a particularly simple form because one does not need to propagate the covariance matrix and the optimal estimator is given by the following open loop "computational model":

$$\dot{\hat{\mathbf{x}}}_1 = (A_1 + B_1 K_{11})\hat{\mathbf{x}}_1 + B_1 K_{12}\hat{\mathbf{x}}_2, \tag{6}$$

which, upon receiving $\mathbf{y}_1(\mathbf{t}_k)$ at time $\mathbf{t}_k + \tau$, is updated according to

$$\hat{\mathbf{x}}_1(\mathbf{t}_k + \tau) = \mathbf{z}_1(\mathbf{t}_k + \tau) := \exp\{(A_1 + B_1 K_{11})\tau\}\mathbf{y}_1(\mathbf{t}_k) +$$
$$+ \int_{\mathbf{t}_k}^{\mathbf{t}_k + \tau} \exp\{(A_1 + B_1 K_{11})(\mathbf{t}_k + \tau - r)\}B_1 K_{12}\hat{\mathbf{x}}_2(r)dr. \tag{7}$$

To implement (6), node 2 also needs to compute $\hat{\mathbf{x}}_2$ based on the information that it has been sending to node 1. This is done using equations completely symmetric to (6)–(7):

$$\dot{\hat{\mathbf{x}}}_2 = (A_2 + B_2 K_{22})\hat{\mathbf{x}}_2 + B_2 K_{21}\hat{\mathbf{x}}_1 \tag{8}$$

$$\hat{\mathbf{x}}_2(\bar{\mathbf{t}}_k + \tau) = \mathbf{z}_2(\bar{\mathbf{t}}_k + \tau) := \exp\{(A_2 + B_2 K_{22})\tau\}\mathbf{y}_2(\bar{\mathbf{t}}_k) +$$
$$+ \int_{\bar{\mathbf{t}}_k}^{\bar{\mathbf{t}}_k + \tau} \exp\{(A_2 + B_2 K_{22})(\bar{\mathbf{t}}_k + \tau - r)\}B_2 K_{21}\hat{\mathbf{x}}_1(r)dr, \tag{9}$$

where each $\bar{\mathbf{t}}_k$ denotes a time at which node 2 sends the measurement $\mathbf{y}_2(\bar{\mathbf{t}}_k)$ to node 1, and $\bar{\mathbf{t}}_k + \tau$ the time at which this measurement is expected to arrive at its destination. Although node 2 has $\mathbf{x}_2$ always available, in building this estimate, it only does the discrete updates of $\hat{\mathbf{x}}_2$ τ time units after it sends the state measurement $\mathbf{y}_2$ to node 1, because only at that time will node 1 be able to update its estimate. By construction, the estimators inside node 2 defined by (6)–(9) will always remain equal to the corresponding estimators inside node 1. For this reason, we call them *synchronized state estimators*.

2.3 Estimation error processes

The estimator equations (6)–(7) can be formally written as the following jump diffusion process:

$$d\hat{\mathbf{x}}_1(t) = A\hat{\mathbf{x}}_1(t)dt + B_1 K_{12}\hat{\mathbf{x}}_2(t)dt + \big(\mathbf{z}_1(t) - \hat{\mathbf{x}}_1(t)\big)d\mathbf{N}_1(t - \tau), \qquad (10)$$

where $A := A_1 + B_1 K_{11}$ and $\mathbf{N}_1(t)$ is an integer counting process that is constant almost everywhere except at the times t_k, $k \geq 0$ when it is increased by 1. Moreover, at any time t when the measurement $\mathbf{y}_1(t - \tau) = \mathbf{x}_1(t - \tau) + \boldsymbol{\zeta}_1(t - \tau)$ is received, we have that

$$\mathbf{z}_1(t) - \hat{\mathbf{x}}_1(t) = \exp\{A\tau\}\boldsymbol{\zeta}_1(t - \tau) + \exp\{A\tau\}\mathbf{x}_1(t - \tau)$$

$$+ \int_{t-\tau}^{t} \exp\{A(t - r)\}B_1 K_{12}\hat{\mathbf{x}}_2(r)dr - \mathbf{x}_1(t) - \mathbf{e}_1(t)$$

$$= \boldsymbol{\eta}_1(t) - \mathbf{e}_1(t), \qquad (11)$$

where

$$\boldsymbol{\eta}_1(t) = \exp\{A\tau\}\boldsymbol{\zeta}_1(t - \tau) - \int_{t-\tau}^{t} \exp\{A(t - r)\}\sigma_1 d\mathbf{w}_1(r), \qquad (12)$$

with the integral defined in the Itô sense [16, 9]. It is straightforward to show that the stochastic moments of $\boldsymbol{\eta}_1(t)$ are finite for any delay of τ.

From (4), (10), and (11) we conclude that the estimation error $\mathbf{e}_1$ satisfies

$$d\mathbf{e}_1(t) = A\mathbf{e}_1(t)dt - \sigma_1 d\mathbf{w}_1(t) + \big(\boldsymbol{\eta}_1(t) - \mathbf{e}_1(t)\big)d\mathbf{N}_1(t - \tau). \qquad (13)$$

Periodic communication is not optimal to reduce network utilization because a node does not need to send its measured state to the network if the other nodes have a good estimate of it. An optimal communication logic problem is solved in [23] for discrete-time systems, in which it is shown that the optimal communication decision for node 1 is a function of the estimation error associated with an additional estimate of its local state $\mathbf{x}_1$ that should be updated in a delay-free fashion right after data is transmitted [i.e., without waiting τ time units in the discrete update (7)], even though the network may exhibit significant delay. This *internal estimate* $\tilde{\mathbf{x}}_1$ is constructed inside node 1 as follows:

$$d\tilde{\mathbf{x}}_1(t) = A\tilde{\mathbf{x}}_1(t)dt + B_1 K_{12}\hat{\mathbf{x}}_2(t)dt + \big(\mathbf{y}_1(t) - \tilde{\mathbf{x}}_1(t)\big)d\mathbf{N}_1(t), \qquad (14)$$

in which $\mathbf{N}_1(t)$ is determined by node 1's communication logic. Inspired by the results in [23], we consider communication logics that base their decision on the internal estimation error $\tilde{\mathbf{e}}_i := \tilde{\mathbf{x}}_i - \mathbf{x}_i$, whose dynamics are given by

$$d\tilde{\mathbf{e}}_1(t) = A\tilde{\mathbf{e}}_1(t)dt - \sigma_1 d\mathbf{w}_1(t) + \big(\boldsymbol{\zeta}_1 - \tilde{\mathbf{e}}_1(t)\big)d\mathbf{N}_1(t). \qquad (15)$$

For simplicity of notation, in the following we drop the subscript $_1$ in all the signals and rewrite (15) and (13) as follows:

$$d\tilde{\mathbf{e}}(t) = A\tilde{\mathbf{e}}(t)dt - \sigma d\mathbf{w}(t) + \big(\boldsymbol{\zeta} - \tilde{\mathbf{e}}(t)\big)d\mathbf{N}(t) \tag{16}$$

$$d\mathbf{e}(t) = A\mathbf{e}(t)dt - \sigma d\mathbf{w}(t) + \big(\boldsymbol{\eta}(t) - \mathbf{e}(t)\big)d\mathbf{N}(t - \tau), \tag{17}$$

in which the integer process $\mathbf{N}(t)$ is determined by the communication logic based on $\tilde{\mathbf{e}}(t)$. For networks with negligible delay, (16) and (17) are identical and we can simply write

$$d\mathbf{e}(t) = A\mathbf{e}(t)dt - \sigma d\mathbf{w}(t) + (\boldsymbol{\zeta} - \mathbf{e}(t))d\mathbf{N}(t). \tag{18}$$

The reader is reminded that the equations above have analogous counterparts for all nodes in the network.

2.4 Communication measure

The "communication cost" of a particular communication logic is measured in terms of the *communication rate*, defined to be the asymptotic rate that messages are sent, i.e.,

$$R := \lim_{k \to \infty} \mathbf{E}\left[\frac{k}{\mathbf{t}_k}\right].$$

Define $\mathbf{T}_k := \mathbf{t}_k - \mathbf{t}_{k-1}$ to be the *intercommunication time* between the $(k-1)$th and the kth messages. If all the $\mathbf{T}_k$ are *i.i.d.*, it is straightforward to show that

$$R = \lim_{k \to \infty} \mathbf{E}\left[\frac{k}{\sum_{i=1}^{k} \mathbf{T}_k}\right] = \frac{1}{\mathbf{E}[\mathbf{T}_k]}. \tag{19}$$

For several communication logics we proceed to investigate the relation between performance, measured in terms of the statistical moments of the estimation error $\mathbf{e}$, and communication cost, measured in terms of the communication rate R.

3 Stochastic communication logics

The idea behind stochastic communication logics is for each node to broadcast at an average rate that depends on the current value of the internal estimation error $\tilde{\mathbf{e}}$, as in (16). To this effect, we define $\mathbf{N}(t)$ to be a DSPP whose increments are associated with message exchanges. The instantaneous rate at which increments occur is a function of the estimation error $\tilde{\mathbf{e}}(t)$. In particular we take $\mathbf{N}(t)$ to be a DSPP with *intensity* $\lambda\big(\tilde{\mathbf{e}}(t)\big)$, which has the property that

$$\mathbf{E}\big[\mathbf{N}(t) - \mathbf{N}(s)\big] = \mathbf{E}\left[\int_s^t \lambda(\tilde{\mathbf{e}}(r))dr\right], \qquad \forall t > s \geq 0,$$

where $\lambda : \mathbb{R}^n \to [0, \infty)$ is called the *intensity function*. For this type of communication logic, the communication rate R is given by

$$R = \lim_{t \to \infty} \frac{\mathbf{E}[\mathbf{N}(t) - \mathbf{N}(0)]}{t} = \lim_{t \to \infty} \frac{\int_0^t \mathbf{E}\big[\lambda(\tilde{\mathbf{e}}(r))\big]dr}{t}, \tag{20}$$

which shows that when $\mathbf{E}[\lambda(\tilde{\mathbf{e}}(t))]$ converges as $t \to \infty$, its limit is precisely the communication rate R.

We start by considering the delay-free case (18) and provide sufficient conditions for stochastic stability for both constant and state-dependent intensity functions. These results are later generalized for the case of a τ time units delay network expressed by (16)–(17).

3.1 Infinitesimal generators

For the stability analysis of (18), it is convenient to consider its *infinitesimal generator*. Given a twice continuously differentiable function $V : \mathbb{R}^n \to \mathbb{R}$, the generator $\mathcal{L}$ of a jump diffusion process $\mathbf{e}$ is defined by

$$(\mathcal{L}V)(e) := \lim_{t \to s} \frac{\mathbf{E}\big[V(\mathbf{e}(t))|\mathbf{e}(s) = e\big] - V(e)}{t - s}, \qquad \forall e \in \mathbb{R}^n, \ t > s \geq 0, \tag{21}$$

where $\mathbf{E}\big[V(\mathbf{e}(t)|\mathbf{e}(s) = e\big]$ denotes the expectation of $V(\mathbf{e}(t))$ given $\mathbf{e}(s) = e$ [15, 5]. The generator for the jump diffusion process described by (18) is given by

$$\mathcal{L}V(e) = \frac{\partial V(e)}{\partial e} \cdot Ae + \frac{1}{2}\mathrm{tr}\left[\sigma' \frac{\partial^2 V(e)}{\partial e^2}\sigma\right]$$
$$+ \lambda(e)\left(\int V(\zeta)\, d\mu(\zeta) - V(e)\right), \tag{22}$$

where $\frac{\partial V(e)}{\partial e}$ and $\frac{\partial^2 V(e)}{\partial e^2}$ denote the gradient vector and Hessian matrix of V, respectively [9]. Setting $e = \mathbf{e}(t)$ in (21) and taking expectation, one obtains

$$\frac{\mathrm{d}}{\mathrm{d}t}\mathbf{E}[V(\mathbf{e}(t))] = \mathbf{E}[(\mathcal{L}V)(\mathbf{e}(t))], \tag{23}$$

from which the stochastic stability properties of the process $\mathbf{e}(t)$ can be deduced by appropriate choices of V.

3.2 Constant intensity

Consider a constant intensity function $\lambda(e) = \gamma$ for the DSPP. From (20), the corresponding communication rate is $R = \gamma$. The following statements hold.

Theorem 1. *Let* **e** *be the jump diffusion process defined by* (18) *with* $\lambda(e) = \gamma$, $\forall e$.

1. *If* $\gamma > \Re\{\lambda_i(A)\}$, *for every eigenvalue* $\lambda_i(A)$ *of* A, *then* $\mathbf{E}[\mathbf{e}(t)]$ *converges to zero exponentially fast.*
2. *If* $\gamma > 2\,m\,\Re\{\lambda_i(A)\}$, *for every eigenvalue* $\lambda_i(A)$ *of* A *and some* $m \geq 1$, *then* $\mathbf{E}[(\mathbf{e}(t) \cdot \mathbf{e}(t))^m]$ *is bounded.*
3. *If* $\gamma > 2\,\Re\{\lambda_i(A)\}$, *for every eigenvalue* $\lambda_i(A)$ *of* A, *and* P, Q *are* $n \times n$ *positive definite matrices and* c *a positive constant such that*

$$P\left(A - \frac{\gamma}{2}I\right) + \left(A - \frac{\gamma}{2}I\right)' P \leq -Q, \qquad Q \geq cP,$$

then $\mathbf{E}[\mathbf{e}(t) \cdot \mathbf{e}(t)]$ *is uniformly bounded and*

$$\lim_{t \to \infty} \mathbf{E}[\mathbf{e}(t) \cdot P\mathbf{e}(t)] \leq \frac{\gamma \rho^2 + \theta}{c}, \tag{24}$$

where $\rho^2 := \int \zeta \cdot P\zeta \, d\mu(\zeta)$, *and* $\theta := \mathrm{tr}(\sigma' P\sigma)$.

To prove this theorem, we need the following lemma, which relates the expectations of different moments of a positive random variable.

Lemma 1. *Given a scalar random variable* $\mathbf{x}$ *that is nonnegative with probability one, a positive constant* δ, *and positive integers* $k > \ell > 0$, $\mathbf{E}[\mathbf{x}^k] \geq \delta^\ell \mathbf{E}[\mathbf{x}^{k-\ell}] - \delta^k$.

Proof (Lemma 1). Suppose $\mathbf{x}$ has distribution $\mu(\mathbf{x})$. For every $\delta > 0$, the following inequalities hold:

$$\mathbf{E}[\mathbf{x}^k] \geq \int_{x \geq \delta} x^k \, d\mu(x) \geq \delta^\ell \int_{x \geq \delta} x^{k-\ell} \, d\mu(x)$$

$$= \delta^\ell \left(\int_{x \geq 0} x^{k-\ell} \, d\mu(x) - \int_{x < \delta} x^{k-\ell} \, d\mu(x) \right) \geq \delta^\ell (\mathbf{E}[\mathbf{x}^{k-\ell}] - \delta^{k-\ell}). \qquad \square$$

Proof (Theorem 1). To prove 1, put $V(e) = e$. From (22),

$$\mathcal{L}V(e) = Ae - \gamma e$$

because $\int V(\zeta) d\mu(\zeta) = 0$. Therefore (23) takes the form

$$\frac{\mathrm{d}}{\mathrm{d}t} \mathbf{E}[\mathbf{e}(t)] = (A - \gamma I)\mathbf{E}[\mathbf{e}(t)],$$

and statement 1 follows since $A - \gamma I$ is stable. To prove 2, let P and Q be $n \times n$ positive-definite matrices and c a positive constant such that

$$P\left(A - \frac{\gamma}{2m}I\right) + \left(A - \frac{\gamma}{2m}I\right)' P \leq -Q, \qquad Q \geq cP.$$

Such matrices exist as long as $\gamma > 2\,m\,\max\{\Re[eig(A)]\}$. For $m \geq 1$, define

$$V(e) := (e \cdot Pe)^m. \tag{25}$$

We conclude from (22) that

$$\begin{aligned}
\mathcal{L}V(e) &= m(e \cdot Pe)^{m-1} e \cdot (PA + A'P)e + \lambda(e)\rho^{2m} - \lambda(e)V(e) \\
&\quad + 2m(m-1)(e \cdot Pe)^{m-2} e \cdot P\sigma\sigma'Pe + m(e \cdot Pe)^{m-1}\theta \\
&= m(e \cdot Pe)^{m-1} e \cdot \left[P\left(A - \frac{\gamma}{2m}I\right) + \left(A - \frac{\gamma}{2m}I\right)' P \right] e + \gamma\,\rho^{2m} \\
&\quad + 2m(m-1)(e \cdot Pe)^{m-2} e \cdot P\sigma\sigma'Pe + m(e \cdot Pe)^{m-1}\theta \\
&\leq -m(e \cdot Pe)^{m-1} e \cdot Qe + \gamma\,\rho^{2m} \\
&\quad + 2m(m-1)(e \cdot Pe)^{m-2} e \cdot P\sigma\sigma'Pe + m(e \cdot Pe)^{m-1}\theta \\
&\leq -cmV(e) + \gamma\,\rho^{2m} + m\big(2c_2(m-1) + \theta\big)(e \cdot Pe)^{m-1},
\end{aligned}$$

where $\rho^{2m} := \int (\zeta \cdot P\zeta)^m d\mu(\zeta)$ and $c_2 > 0$ is such that $P\sigma\sigma'P \leq c_2 P$. From this and (23), we conclude that

$$\frac{d}{dt}\mathbf{E}[V(e)] \leq -cm\mathbf{E}[V(e)] + \gamma\,\rho^{2m} + m\big(2c_2(m-1) + \theta\big)\mathbf{E}[(e \cdot Pe)^{m-1}].$$

Given some $\delta > 0$, from Lemma 1,

$$\mathbf{E}[(e \cdot Pe)^{m-1}] \leq \frac{1}{\delta}\mathbf{E}[V(e)] + \delta^{m-1},$$

and therefore

$$\frac{d}{dt}\mathbf{E}[V(e)] \tag{26}$$

$$\begin{aligned}
&\leq -cm\mathbf{E}[V(e)] + \gamma\rho^{2m} + m\big(2c_2(m-1) + \theta\big)\left(\frac{1}{\delta}\mathbf{E}[V(e)] + \delta^{m-1}\right) \\
&= -m\left(c - \frac{2c_2(m-1) + \theta}{\delta}\right)\mathbf{E}[V(e)] + \gamma\rho^{2m} + m\delta^{m-1}\big(2c_2(m-1) + \theta\big).
\end{aligned}$$

For sufficiently large δ, $c - \frac{2c_2(m-1)+\theta}{\delta} > 0$ and the boundedness of $\mathbf{E}[V(e)]$ and consequently that of $\mathbf{E}[(e \cdot e)^m]$ follows.

To prove 3, we rewrite (26) for $m = 1$ and obtain

$$\frac{d}{dt}\mathbf{E}[V(e)] \leq -\left(c - \frac{\theta}{\delta}\right)\mathbf{E}[V(e)] + \gamma\rho^2 + \theta.$$

Applying the Comparison Lemma [7], we conclude that

$$\lim_{t \to \infty} \mathbf{E}[V(e)] \leq \frac{\gamma\rho^2 + \theta}{c - \frac{\theta}{\delta}},$$

from which (24) follows as we make $\delta \to \infty$. $\qquad\square$

3.3 Error-dependent intensity

We now consider an intensity for the DSPP that depends on the current estimation error. The rationale is that a larger estimation error should more rapidly lead to a message exchange. We consider intensities of the form

$$\lambda(e) = (e \cdot Pe)^k, \qquad \forall e \in \mathbb{R}^n, \tag{27}$$

where P is some positive-definite matrix and k a positive integer.

Theorem 2. *Let* $\mathbf{e}$ *be the jump diffusion process defined by* (18) *with intensity* (27). *For every* $k > 0$, *the communication rate and all finite moments of* $\mathbf{e}(t)$ *are bounded.*

Proof (Theorem 2). Choose c_1 sufficiently large so that $A - \frac{c_1}{2}I$ is asymptotically stable. Then there exists a matrix $P > 0$ such that

$$P\left(A - \frac{c_1}{2}I\right) + \left(A - \frac{c_1}{2}I\right)P < 0,$$

i.e., $PA + A'P < c_1 P$. Moreover $P\sigma\sigma'P \le c_2 P$ for sufficiently large $c_2 > 0$.

We start by proving that the mth moment of $\mathbf{e}(t)$ is bounded for $m > k$. Let V be as in (25) and $\rho^{2m} := \int (\zeta \cdot P\zeta)^m d\mu(\zeta)$. From (22) we obtain

$$
\begin{aligned}
\mathcal{L}V(e) &= m(e \cdot Pe)^{m-1} e \cdot (PA + A'P)e + \lambda(e)\rho^{2m} - \lambda(e)V(e) \\
&\quad + 2m(m-1)(e \cdot Pe)^{m-2} e \cdot P\sigma\sigma'Pe + m(e \cdot Pe)^{m-1}\theta \\
&= m(e \cdot Pe)^{m-1} e \cdot (PA + A'P)e + \rho^{2m}(e \cdot Pe)^k - (e \cdot Pe)^{m+k} \\
&\quad + 2m(m-1)(e \cdot Pe)^{m-2} e \cdot P\sigma\sigma'Pe + m(e \cdot Pe)^{m-1}\theta \\
&\le c_1 m(e \cdot Pe)^m + \rho^{2m}(e \cdot Pe)^k - (e \cdot Pe)^{m+k} \\
&\quad + m\big(2c_2(m-1) + \theta\big)(e \cdot Pe)^{m-1}.
\end{aligned}
$$

From this and (23), we conclude that

$$
\begin{aligned}
\frac{\mathrm{d}}{\mathrm{d}t}\mathbf{E}[V(\mathbf{e})] &\le c_1 m\mathbf{E}[V(\mathbf{e})] + \rho^{2m}\mathbf{E}[(\mathbf{e} \cdot P\mathbf{e})^k] - \mathbf{E}[(\mathbf{e} \cdot P\mathbf{e})^{m+k}] \\
&\quad + m\big(2c_2(m-1) + \theta\big)\mathbf{E}[(\mathbf{e} \cdot P\mathbf{e})^{m-1}].
\end{aligned}
$$

Given some $\delta_1, \delta_2, \delta_3 > 0$, we conclude from Lemma 1 that

$$\mathbf{E}[(\mathbf{e} \cdot P\mathbf{e})^k] \le \frac{\mathbf{E}[V(\mathbf{e})]}{\delta_1^{m-k}} + \delta_1^k,$$

$$\mathbf{E}[(\mathbf{e} \cdot P\mathbf{e})^{m+k}] \ge \delta_2^k \mathbf{E}[V(\mathbf{e})] - \delta_2^{m+k},$$

$$\mathbf{E}[(\mathbf{e} \cdot P\mathbf{e})^{m-1}] \le \frac{1}{\delta_3}\mathbf{E}[V(\mathbf{e})] + \delta_3^{m-1},$$

therefore

$$\frac{\mathrm{d}}{\mathrm{d}t}\mathbf{E}[V(\mathbf{e})] \leq c_1 m \mathbf{E}[V(\mathbf{e})] + \rho^{2m}\left(\frac{\mathbf{E}[V(\mathbf{e})]}{\delta_1^{m-k}} + \delta_1^k\right) - \left(\delta_2^k \mathbf{E}[V(\mathbf{e})] - \delta_2^{m+k}\right)$$

$$+ m\left(2c_2(m-1) + \theta\right)\left(\frac{\mathbf{E}[V(\mathbf{e})]}{\delta_3} + \delta_3^{m-1}\right)$$

$$\leq \left(c_1 m + \frac{\rho^{2m}}{\delta_1^{m-k}} - \delta_2^k + m\frac{2c_2(m-1) + \theta}{\delta_3}\right)\mathbf{E}[V(\mathbf{e})]$$

$$+ \rho^{2m}\delta_1^k + \delta_2^{m+k} + m\left(2c_2(m-1) + \theta\right)\delta_3^{m-1}.$$

For sufficiently large δ_2,

$$c_1 m + \frac{\rho^{2m}}{\delta_1^{m-k}} - \delta_2^k + \frac{m\left(2c_2(m-1) + \theta\right)}{\delta_3} < 0,$$

and the boundedness of $\mathbf{E}[V(\mathbf{e})]$ and consequently of $\mathbf{E}[(\mathbf{e}\cdot\mathbf{e})^m]$ follows.

To prove the boundedness of the mth moment of $\mathbf{e}(t)$ for $m \leq k$, we use Lemma 1 to bound

$$\mathbf{E}[(\mathbf{e}\cdot\mathbf{e})^m] \leq \frac{\mathbf{E}[(\mathbf{e}\cdot\mathbf{e})^{k+1}]}{\delta_4^{k+1-m}} + \delta_4^m, \tag{28}$$

where $\delta_4 > 0$. Since the boundedness of the $(k+1)$th moment has already been established, we conclude that the mth moment is also bounded for $m \leq k$.

$\square$

3.4 τ-delayed network

We now extend the stochastic stability results to networks with τ time units delay. For constant intensity functions, the estimation error process in (17) is driven by a constant intensity Poisson process $\mathbf{N}(t - \tau)$ and Theorem 1 still holds. It turns out that, for intensity functions like (27), a result analogous to Theorem 2 but for delayed networks can also be proved.

Theorem 3. *Let $\tilde{\mathbf{e}}$ and $\mathbf{e}$ be the jump diffusion processes defined by (16) and (17) with the intensity of the DSPP $\mathbf{N}(t)$ given by $\lambda\big(\tilde{\mathbf{e}}(t)\big)$, with*

$$\lambda(\tilde{e}) = (\tilde{e}\cdot P\tilde{e})^k, \qquad \forall\tilde{e} \in \mathbb{R}^n,$$

for some $k > 0$. The communication rate and all the finite moments of both $\tilde{\mathbf{e}}(t)$ and $\mathbf{e}(t)$ are bounded.

For a given time $t \geq 0$, let $\mathbf{s}(t)$ be the time at which the communication logic sends the last measurement before or at t, i.e.,

$$\mathbf{s}(t) := \max\left\{r \leq t : d\mathbf{N}(r) > 0\right\}.$$

The random variable $\mathbf{s}(t)$ is a *stopping time* [15], which is independent of any event after time t. Since no data are sent during the time interval $(\mathbf{s}(t), t]$, the processes $\mathbf{N}$ and $\tilde{\mathbf{e}}$ in (16) have no jumps on this interval. Consequently, because of the network delay of τ, the remote node does not receive any data on $(\mathbf{s}(t) + \tau, t + \tau]$ and during this interval the process $\mathbf{e}$ in (17) has no jumps.

Proof (Theorem 3). Since the process $\tilde{\mathbf{e}}$ is not affected by the delay, the boundedness of all moments of this process as well as the communication rate follows directly from the proof of Theorem 2. It remains to prove boundedness of all moments of the estimation error $\mathbf{e}$ associated with the synchronized state estimators. To this effect, we consider the function $V : \mathbb{R}^n \to \mathbb{R}$ defined in (25) and show that for an arbitrary time t, $\mathbf{E}\big[V(\mathbf{e}(t + \tau))\big]$ is bounded. Note that once we prove boundedness of $\mathbf{E}[V(\mathbf{e})]$, the boundedness of $\mathbf{E}[(\mathbf{e} \cdot \mathbf{e})^m]$ follows. At time $\mathbf{s}(t) + \tau$, $\mathbf{e}$ is reset to $\boldsymbol{\eta}(\mathbf{s}(t) + \tau)$ defined by (12), and therefore

$$
\mathbf{e}(\mathbf{s}(t) + \tau) = \exp\{A\tau\}\boldsymbol{\zeta} - \int_{\mathbf{s}(t)}^{\mathbf{s}(t)+\tau} \exp\{A(\mathbf{s}(t) + \tau - r)\}\sigma\, d\mathbf{w}(r). \tag{29}
$$

On the other hand, since the process $\tilde{\mathbf{e}}$ is reset to $\boldsymbol{\zeta}$ at time $\mathbf{s}(t)$ and this process has no jumps on $(\mathbf{s}(t), t]$, we conclude from (16) that

$$
\tilde{\mathbf{e}}(t) = \exp\{A(t - \mathbf{s}(t))\}\boldsymbol{\zeta} - \int_{\mathbf{s}(t)}^{t} \exp\{A(t - r)\}\sigma\, d\mathbf{w}(r),
$$

which is equivalent to

$$
\boldsymbol{\zeta} = \exp\{A(-t + \mathbf{s}(t))\}\tilde{\mathbf{e}}(t) + \int_{\mathbf{s}(t)}^{t} \exp\{A(\mathbf{s}(t) - r)\}\sigma\, d\mathbf{w}(r). \tag{30}
$$

Using (30) to eliminate $\boldsymbol{\zeta}$ in (29), we conclude that

$$
\mathbf{e}(\mathbf{s}(t) + \tau) = \exp\{A(\tau + \mathbf{s}(t) - t)\}\tilde{\mathbf{e}}(t)
$$
$$
- \int_{t}^{\mathbf{s}(t)+\tau} \exp\{A(\mathbf{s}(t) + \tau - r)\}\sigma\, d\mathbf{w}(r). \tag{31}
$$

Moreover, since the process $\mathbf{e}$ has no jumps on $(\mathbf{s}(t) + \tau, t + \tau]$, we conclude from (17) that

$$
\mathbf{e}(t + \tau) = \exp\{A(t - \mathbf{s}(t))\}\mathbf{e}(\mathbf{s}(t) + \tau) - \int_{\mathbf{s}(t)+\tau}^{t+\tau} \exp\{A(t + \tau - r)\}\sigma\, d\mathbf{w}(r).
$$

From this and (31), we obtain

$$
\mathbf{e}(t + \tau) = \exp\{A\tau\}\tilde{\mathbf{e}}(t) - \int_{t}^{t+\tau} \exp\{A(t + \tau - r)\}\sigma\, d\mathbf{w}(r).
$$

Since $V(a + b) \leq 2^{2m}V(a) + 2^{2m}V(b)$, $\forall a, b \in \mathbb{R}^n$, we conclude that

$$V(\mathbf{e}(t+\tau)) \leq 2^{2m}V\big(\exp\{A\tau\}\tilde{\mathbf{e}}(t)\)$$

$$+ 2^{2m}V\left(\int_t^{t+\tau} \exp\{A(t+\tau-r)\}\sigma\,d\mathbf{w}(r)\right).$$

Since the process that appears in the integral is a Gaussian white noise, independent of $\mathbf{s}(t)$, we conclude that there exist finite constants c_5, c_6 such that

$$\mathbf{E}\big[V(\mathbf{e}(t+\tau))\big] \leq c_5\mathbf{E}\big[V(\tilde{\mathbf{e}}(t))\big] + c_6.$$

The boundedness of $\mathbf{E}\big[V(\mathbf{e}(t+\tau))\big]$ then follows from the already established boundedness of $\mathbf{E}[V(\tilde{\mathbf{e}}(t))]$. $\qquad\square$

4 Deterministic communication logics

We now consider communication logics that utilize deterministic rules but restrict our attention to delay-free networks, whose estimation error satisfies (18). The communication logic monitors a continuous positive and radially unbounded *communication index* $S : \mathbb{R}^n \to \mathbb{R}^+$ and forces a node to broadcast its state when $S(\mathbf{e}) \geq 1$. In particular, a message broadcast occurs at time $\mathbf{t}_k$ when $\lim_{t\uparrow t_k} S(\mathbf{e}(t)) \geq 1$. To avoid chattering, the post-reset value $\boldsymbol{\zeta}(t_k)$ should satisfy $S\big(\boldsymbol{\zeta}(t_k)\big) < 1$ with probability one. This type of resetting guarantees that $\mathbf{e}(t)$ is bounded, since

$$\mathbf{e}(t) \in \mathcal{D} := \{e \in \mathbb{R}^n | S(e) \leq 1\}, \qquad \forall t \geq 0, \tag{32}$$

with probability one.

To determine the communication rate, suppose that a message exchange occurred at time $\mathbf{t}_{k-1}$ and $\mathbf{e}(\mathbf{t}_{k-1})$ was reset to $\boldsymbol{\zeta}(t_{k-1})$. From $\mathbf{t}_{k-1}$ to the next reset time $\mathbf{t}_k$, $\mathbf{e}(t)$ is a pure diffusion process

$$\dot{\mathbf{e}} = A\mathbf{e} - \sigma\dot{\mathbf{w}}. \tag{33}$$

Given $\boldsymbol{\zeta}(t_{k-1})$, define $\mathbf{T}_k(\boldsymbol{\zeta})$ to be the intercommunication time, i.e.,

$$\mathbf{T}_k(\boldsymbol{\zeta}) = \inf\{t - \mathbf{t}_{k-1} \geq 0 : \mathbf{e}(t) \in \partial\mathcal{D}, \mathbf{e}(\mathbf{t}_{k-1}) = \boldsymbol{\zeta}\},$$

where $\mathbf{e}(t)$ is governed by (33) for $t \geq \mathbf{t}_{k-1}$ and $\partial\mathcal{D}$ denotes the boundary of $\mathcal{D}$. The random variable $\mathbf{T}_k(\boldsymbol{\zeta})$ is called the *first exit time* of $\mathbf{e}(t)$ from $\mathcal{D}$. It is, in general, not easy to obtain the distribution of $\mathbf{T}_k(\boldsymbol{\zeta})$ in closed form, but its expected value can be obtained from Dynkin's equation. In particular, defining $g(\boldsymbol{\zeta}) := \mathbf{E}[\mathbf{T}_k(\boldsymbol{\zeta})]$, it is known that $g(\boldsymbol{\zeta})$ is a solution to the following boundary value problem:

$$\frac{\partial g(\zeta)}{\partial \zeta} \cdot A\zeta + \frac{1}{2}\mathrm{tr}\left[\sigma'\frac{\partial^2 g(\zeta)}{\partial \zeta^2}\sigma\right] = -1, \tag{34}$$

$$\forall \zeta \in \mathcal{D}, \quad g(\zeta) = 0, \quad \forall \zeta \in \partial\mathcal{D},$$

where $\frac{\partial g(\zeta)}{\partial \zeta}$ and $\frac{\partial^2 g(\zeta)}{\partial \zeta^2}$ denote the gradient vector and Hessian matrix of g, respectively [17]. Once $g(\zeta)$ is known, the expected intercommunication time $\mathbf{T}_k$ can be obtained from

$$\mathbf{E}[\mathbf{T}_k] = \mathbf{E}[g(\zeta_{k-1})] = \int g(\zeta)d\mu(\zeta),$$

and the communication rate follows from (19):

$$R = \frac{1}{\int g(\zeta)d\mu(\zeta)}.$$

In practice, (34) needs to be solved numerically. Since $\mathcal{D}$ is compact, (32) provides an upper bound on $\mathbf{e}(t)$ and consequently on its statistical moments. To obtain tighter bounds one can use Kolmogorov's forward equation with appropriate boundary conditions to compute the probability density function of the error $\mathbf{e}(t)$. However, this method is computationally intensive for higher-order systems.

5 Simulation results

In this section we validate the theoretical results through Monte Carlo simulations. All the simulations are done in Matlab/Simulink. The DSPP $\mathbf{N}(t)$ is realized by a single binomial test. Specifically, for a fixed time step h, a message exchange is triggered at time $t := k\,h$, $k \in \mathbb{N}$, if a binomial test characterized by a probability of success $p = 1 - e^{-h\lambda(\mathbf{e}(t))}$ succeeds. Convergence results for similar procedures can be found in [4] and references therein.

5.1 Leader-follower

A leader-follower problem is used to illustrate the distributed control architecture with different communication logics. The two processes have identical dynamics and are disturbed by uncorrelated white Gaussian noise processes. The dynamics of the leading and following vehicles are given by

$$\text{leader:} \qquad \dot{\mathbf{x}}_1(t) = A\mathbf{x}_1(t) + Br(t) + \sigma\dot{\mathbf{w}}_1(t) \tag{35}$$

$$\text{follower:} \qquad \dot{\mathbf{x}}_2(t) = A\mathbf{x}_2(t) + Bu_2(t) + \sigma\dot{\mathbf{w}}_2(t), \tag{36}$$

where each state $\mathbf{x}_i$ contains the position and velocity of one of the vehicles, u_i are the controls, r is an external reference, each $\dot{\mathbf{w}}_i(t)$ is standard Gaussian white noise, and $A = \begin{bmatrix} 0 & 1 \\ 0 & -0.5 \end{bmatrix}$, $\sigma = \begin{bmatrix} 1 & 0 \\ 0 & 1 \end{bmatrix}$, $B = \begin{bmatrix} 0 & 1 \end{bmatrix}'$. The follower's control objective is to follow the leader's position. The reference r is also known by the follower. The open loop state estimator for the leader's state is given by

$$\dot{\hat{\mathbf{x}}}_1 = A\hat{\mathbf{x}}_1 + Br, \qquad\qquad \hat{\mathbf{x}}_1(\mathbf{t}_k) = \mathbf{y}_1(\mathbf{t}_k) := \mathbf{x}_1(\mathbf{t}_k) + \boldsymbol{\zeta}_k,$$

where the $\mathbf{t}_k$ denote the times at which the leader broadcasts its state $\mathbf{x}(\mathbf{t}_k)$ to the follower, and the $\boldsymbol{\zeta}_k$ are zero-mean uniformly distributed random vectors over an interval of about 5% of the maximum estimation error. The follower uses the following controller:

$$u_2 = -K(\mathbf{x}_2 - \hat{\mathbf{x}}_1),$$

where $K = \begin{bmatrix} 32.6 & 8.07 \end{bmatrix}$ is obtained from a LQR design.

Table 1 summarizes the communication rates and the variances of both the estimation and the tracking errors for three communication logics: periodic, DSPP with quadratic intensity $\lambda(e)$, and deterministic with a quadratic communication index $S(e) := e \cdot Pe$, where $P = \begin{bmatrix} 1 & 0 \\ 0 & 0.1 \end{bmatrix}$. For fair comparison, the parameters are selected to achieve communication rates approximately equal to 0.2 for all logics. We see that the deterministic logic outperforms both the DSPP logic and the periodic communication.

Table 1. Communication rate versus variance of the estimation and tracking errors.

Logics	Parameters	Comm. rate	Est. err. var.	Trck. err. var.
Determ.	$S(e) \leq 0.070$	0.19	0.011	0.017
DSPP	$\lambda(e) := 0.5\frac{e \cdot Pe}{0.070}$	0.22	0.029	0.037
Period.	$period = 5$	0.20	0.037	0.042

Figure 2 shows sample trajectories of the position tracking and estimation errors for a 20-second period, in which $r(t)$ is a sinusoid. The communication instants are indicated by markings in the horizontal lines at the bottom of the left figure. Under similar communication rates, both the deterministic and the DSPP logics show advantage over that of periodic communication, as they both exhibit lower error variances. Aperiodic transmission in the stochastic updating rules requires data to be time-stamped.

5.2 Rate-variance curves

To study the trade-off between communication rate and estimation error variance, we consider the remote state estimator of a first-order unstable process $d\mathbf{x} = \mathbf{x}dt + d\mathbf{w}$. This corresponds to a jump diffusion process defined by (18) with $A = 1$, $\sigma = 1$. The results presented refer to a simulation time of 1000

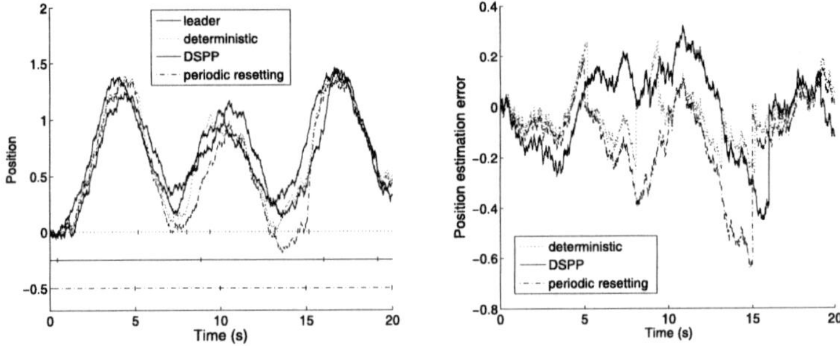

Fig. 2. Leader and follower positions (left) and leader estimation error (right) obtained with the different logics: deterministic, DSPP, and periodic. The message exchange time instants are indicated with bars in the left plot.

seconds. The system's instability presents an added challenge to a distributed architecture.

Figure 3 (left) depicts the trade-off between the communication rate and the variance of the estimation error for four different communication logics: periodic, DSPP with constant intensity, DSPP with quadratic intensity, and deterministic with quadratic communication index. The curves are obtained by varying the parameters that define these logics. For a given communication rate, the DSPP logic with constant intensity results in the largest error, whereas the deterministic logic results in the smallest. The communication rate obtained with the DSPP logic for the quadratic $\lambda(e)$ is significantly smaller than the upper bound provided by (28), which for this example numerically equals 1.

Figure 3 (right) provides a comparison between deterministic and DSPP logics. The deterministic logics have a communication index of form $S(e) := \frac{e^2}{\Delta} \leq 1$, and the different points on the curve are generated by changing Δ. The DSPP logics have intensities of the form $\lambda(e) = (\frac{e^2}{\Delta})^k$, where Δ is a positive parameter and $k \in \{1, 2, 3, 4, 5\}$. For large k, $\lambda(e)$ essentially provides a barrier at $e^2 = \Delta$, which acts as the bound in the deterministic logics. It is therefore not surprising to see that as k increases, the DSPP logics converge to the deterministic logics. As proved for discrete systems, the deterministic curve provides optimal trade-off between communication cost and control performance [23].

6 Conclusion and future work

Deterministic and stochastic communication logics are proposed to determine when local controllers should communicate in a distributed control architec-

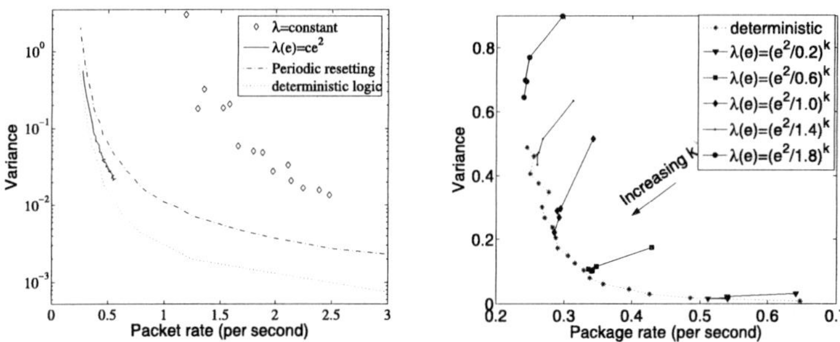

Fig. 3. Communication rate versus variance of the estimation error: for different communication logics (left) and for deterministic and polynomial-intensity DSPP logics (right).

ture. Using tools from jump diffusion processes and the Dynkin's equation, we investigated conditions under which these logics guarantee boundedness as well as the trade-off between the amount of information exchanged and the performance achieved. Monte Carlo simulations confirm that these communication logics can save communication resources over periodic schemes.

In this work, a linear certainty equivalence controller structure (3) is assumed, which may not achieve optimal control performance in the distributed settings. In fact, the counterexample in [19] shows that optimal LQG controllers for distributed linear processes are in general nonlinear. Our problem falls under the class of delayed sharing information patterns for which a general separation theorem for controller and estimator design does not appear to exist [8, 20]. We are currently investigating if some form of separation can hold for this specific problem under consideration.

Other future work includes studying the impact of modeling errors on the system's performance as well as the impact of a nonideal networks that drop packets.

References

1. Coleri S, Puri A, Varaiya P (2003) Power efficient system for sensor networks. 837–842. In: Eighth IEEE International Symposium on Computers and Communication, Kemer–Antalya, Turkey
2. Cox DR (1955) Some statistical methods connected with series of events. Journal of the Royal Statistical Society 17(2):129–164
3. Elia N, Mitter SK (2001) Stabilization of linear systems with limited information. IEEE Trans. on Automat. Contr. 46(9):1384–1400
4. Glasserman P, Merener N (2003) Numerical solution of jump-diffusion LIBOR market models. Finance and Stochastics 7:1–27
5. Hespanha JP (2004) Stochastic hybrid systems: applications to communication networks. In: Alur R, Pappas GJ (eds), Hybrid systems: computation and

control, number 2993 in Lect. Notes in Comput. Science, 387–401. Springer-Verlag, Berlin

6. Hespanha JP, Ortega A, Vasudevan L (2002) Towards the control of linear systems with minimum bit-rate. In: Proc. of the Int. Symp. on the Math. Theory of Networks and Systems, University of Notre Dame, France

7. Khalil HK (1996) Nonlinear systems. Prentice-Hall, Upper Saddle River, NJ

8. Kurtaran B (1979) Corrections and extensions to "decentralized stochastic control with delayed sharing information pattern." IEEE Trans. on Automat. Contr. AC-24(4):656–657

9. Kushner H (2001) Heavy traffic analysis of controlled queueing and communication networks, vol. 47 of Applications of Mathematics. Springer-Verlag, Berlin

10. Lian FL (2001) Analysis, design, modeling, and control of networked control systems. PhD thesis. University of Michigan, Ann Arbor, MI

11. Liberzon D (2002) A note on stabilization of linear systems using coding and limited communication. 836–841. In: Proc. of the 41st Conf. on Decision and Contr., Las Vegas, NV

12. Liberzon D, Hespanha JP (2005) Stabilization of nonlinear systems with limited information feedback IEEE Trans. on Automatic Control 50(6):910–915

13. Matveev AS, Savkin AV (2003) The problem of state estimation via asynchronous communication channels with irregular transmission times. IEEE Trans. on Automat. Contr. 48(4):670–676

14. Nair GN, Evans RJ (2000) Communication-limited stabilization of linear systems. 1005–1010. In: Proc. of the 39th Conf. on Decision and Contr.

15. Oksendal B (2000) Stochastic differential equations: an introduction with applications. Springer-Verlag, Berlin

16. Revuz D, Yor M (1999) Continuous martingales and brownian motion. Springer-Verlag, Berlin

17. Schuss Z (1980) Theory and applications of stochastic differential equations. Wiley Series in Probability and mathematical statistics, John Wiley and Sons, New York

18. Tatikonda S (2000) Control under communication constrains. PhD thesis, MIT, Cambridge, MA

19. Witsenhausen HS (1968) A counterexample in stochastic optimum control. SIAM J. Contr. 6(1):131–147

20. Witsenhausen HS (1971) Separation of estimation and control for discrete time systems. Proceedings of the IEEE 59(11):1557–1566

21. Wong WS, Brockett RW (1997) Systems with finite communication bandwidth constraints–part I: state estimation problems. IEEE Trans. on Automat. Contr. 42(9):294-299

22. Wong WS, Brockett RW (1999) Systems with finite communication bandwidth constraints–II: stabilization with limited information feedback. IEEE Trans. on Automat. Contr. 44(5):1049–1053

23. Xu Y, Hespanha JP (2004) Optimal communication logics for networked control systems. In: Proc. of the 43rd Conf. on Decision and Contr., Atlantis, Bahamas

24. Yook JK, Tilbury DM, , Soparkar NR (2002) Trading computation for bandwidth: reducing communication in distributed control systems using state estimators. IEEE Trans. Contr. Syst. Technol. 10(4):503–518

Networked Decentralized Control of Multirate Sampled-Data Systems

Roberto Ciferri, Gianluca Ippoliti, and Sauro Longhi

Dipartimento di Ingegneria Informatica, Gestionale e dell'Automazione
Università Politecnica delle Marche
Via Brecce Bianche, 60131 Ancona, Italia
`{r.ciferri,g.ippoliti}@diiga.univpm.it,sauro.longhi@univpm.it`

Summary. The vast progress in network technology over the past decade has certainly influenced the area of control systems. Nowadays, it is becoming more common to use networks in systems, especially in those that are large scale and physically distributed or that require extensive cabling. A network-based decentralized control of multirate sampled-data systems approach is introduced and analyzed as an effectiveness solution to the control problem of large-scale continuous-time plant. The main idea is that of information exchange among the various input-output channels of a complex system throughout a local area network. The chapter analyzes the stabilization problem for decentralized control of a large-scale continuous-time plant with different sampling rates in the input-output plant channels. Existence conditions and synthesis procedures are introduced and discussed. The results on the analysis and control of periodic discrete-time systems are used for finding such solutions. The chapter ends with some remarks on the possible extensions to the robust stabilization for the decentralized control of multirate sampled-data systems.

1 Introduction

The problem of stabilizing a linear time-invariant multivariable large-scale system can be solved by using a conventional approach based on a classical centralized controller, or by an innovative and more effective methodology based on the use of several local *decentralized controllers*. In many contexts, it is possible to find useful applications of decentralized control. Interesting applications of these innovative solutions can be found in the field of team decision and adjustment processes, in the field of decentralization in economic systems [9, 23], or in the field of decentralized control to frequency and voltage regulation problems associated with power sharing in multiarea electrical power pools [12].

During the last few decades, interest in the study of decentralized control systems has increased and the stabilization problem of a large-scale plant with independent decentralized controllers has been investigated. In [33], the problem of stabilizing a linear time-invariant multivariable system by using several local feedback control laws is considered. The existence condition of a solution is stated in terms of the notion of *fixed modes* of a decentralized control system. A generalization of this problem has been analyzed in [10] with direct implications to the decentralized robust servomechanism problem. In [8], the effects of decentralized feedback on the closed loop properties of jointly controllable, jointly observable k-channel linear systems are analyzed. Channel interactions within such systems are described by means of suitably defined direct graphs and the concept of a *complete system* is introduced. In [1], algebraic characterizations for the existence of fixed modes of a linear closed loop system with decentralized feedback control are presented.

Digital solutions to the decentralized control problem have also been studied. In [20], the decentralized stabilization problem for linear, discrete-time, periodically time-varying plants using periodic controllers is analyzed. The main tool used is the technique of lifting a periodic system to a time-invariant one via extensions of the input and the output spaces. Moreover, multirate decentralized controllers have been proposed in [30, 28, 19], also with asynchronous sampling mechanism [18]. In a multirate control scheme the plant channels are sampled with different rates and the corresponding controllers are time-invariant digital systems working with different sampling rates.

In a conventional interpretation of decentralized control, each one of the several control stations that characterize the decentralized scheme observes only local measured outputs and controls only local control inputs. However, all the controllers are involved in controlling the same large system. In a more enlarged vision, it is possible to imagine a decentralized scheme in which the measured outputs can be available for all the local controllers, making up for lack of structural properties and performing a more accurate control. The measured output can be transmitted to the all local controllers by means of a local network.

The vast progress in network technology over the past decades has greatly influenced the area of control systems, especially those that are large scale and physically distributed or that need extensive cabling. In such systems, networks assure efficiency of communication and flexibility of the systems. At lowest level of hierarchy of networks is the control network, the so-called *fieldbus* [11], which connects sensors and actuators to control devices with a single network cable, whereas usually point-to-point links have been used, performing a great cost reduction of interfaces and a growth in the communication rate, so that the use of control networks has become a more common solution in industry [27]. The main drawback of these solutions is given by the *limitation on bandwidth*. However, data sent over control networks differ from those encountered in networks for data communications purposes [27]. In data networks, large sets of data messages are transmitted occasionally at high data rates for short intervals of time. Data in control networks are continuously sent out at relatively constant data rates and there are crucial real-time requirements to achieve certain control performance. Hence, specific communication methods or protocols are introduced for achieving these aspects [27]. A consequence of limited bandwidth is that control functions often need to be distributed over the systems. Control of distributed systems is examined in the area of decentralized control [11, 32] and the idea of information exchange among local controllers has been also considered (see [24] and its references). However, there are only a few works considering realistic, practical models of networks.

An interesting idea of control with network communications is analyzed by Ishii and Francis [17]. Their work deals with the time-sequencing aspect of bandwidth, by proposing a design problem of dynamic local controllers that are connected to each other by a network.

The problem of limited bandwidth in data control transmission for a large-scale continuous-time plant with networked decentralized multirate sampled-data control system is analyzed in this chapter. In particular, the proposed solution to the *networked decentralized multirate control problem* is based on the idea of *information exchange* among the several channels of the system by using a local area network [3, 4]. A generalization of solutions developed in [30], for a single-rate sampling in each input-output plant channel, and in [28], for discrete-time plants, has been investigated by examining the case of a continuous-time plant, with each input and each measured output of a plant channel updated with different time intervals and sampled with different sampling rate, respectively [3]. A continuous-time plant with such a multirate sampling mechanism can be efficiently modeled by a periodic discrete-time system (see, e.g., [7, 22]). Therefore, the results on the analysis and control of periodic discrete-time systems can be used for finding solutions to different multirate control problems. In particular, concepts and tools typical of the algebraic approach, developed for the class of *periodic systems* in [14, 15], are here specified and adapted to the class of multirate sampled-data systems, in order to identify conditions for the existence of a solution to the networked decentralized multirate control problem, in terms of the original continuous-time plant and to develop appropriate design procedures. Sufficient conditions for the existence of the networked decentralized multirate control problem are given in terms of the continuous-time plant. The multirate control mechanism does not compromise the stabilization of a large-scale continuous-time plant with decentralized independent controllers; the multirate sampling mechanism and the use of a local network enlarge the class of plants to be stabilized with decentralized controllers [3, 4].

The decentralized multirate control problem and some review of decentralized control system properties and concepts are introduced in Section 2. The networked decentralized multirate control system and the relative stabilization problem are presented in Section 3. In Section 4 the time-invariant representation of the extended multirate sampled-data system is recalled and the existence conditions of the introduced problem are analyzed. Some possible solutions to the networked decentralized multirate control problem are investigated in Section 5. Concluding remarks and direction for future research activities end the chapter.

2 Preliminaries and Notation

In this section, the *decentralized multirate control problem* is introduced in order to approach the more complex problem defined for control networked systems, which will be presented in the following sections. Some decentralized control system properties and concepts are reviewed for the development of some conditions for the solution of the above-mentioned problem [3]. Consider a *linear continuous time-invariant plant* Σ^c, characterized by σ input-output channels and described by

$$\dot{x}^c(t) = A^c\, x^c(t) + \sum_{i=1}^{\sigma} B_i^c\, u_i^c(t) \tag{1}$$

$$y_j^c(t) = C_j^c\, x^c(t), \qquad j = 1, \ldots, \sigma, \tag{2}$$

where $x^c(t) \in \mathbb{R}^{n^c}$ is the state, $u_i^c(t) \in \mathbb{R}^{p_i}, i = 1, \ldots, \sigma$, are the control inputs, $y_j^c(t) \in \mathbb{R}^{q_j}, j = 1, \ldots, \sigma$, are the measured outputs. The *stabilization problem* of Σ^c with a *decentralized continuous-time control system*, constituted by σ independent controllers with input $y_i^c(\cdot)$ and output $u_i^c(\cdot), i = 1, \ldots, \sigma$, has a solution if and only if Σ^c is stabilizable and detectable and all the $2^\sigma - 2$ complementary subsystems are *weakly complete*, i.e., Σ^c has no unstable *decentralized fixed modes* (DFM) (see, e.g., [33, 1]).

In the study of decentralized control problems for linear time-invariant systems, the notion of DFM plays an important role. The DFMs of a system are a subset (possibly empty) of the system's open loop eigenvalues that have the property that they remain fixed, independent of any linear time-invariant controller (subject to the given decentralized control information constraint), which may be applied to the system. The weakly completeness of complementary subsystems can be verified making use of the notation introduced in [28]. Denoting by $\mathcal{I} := \{i_1, \ldots, i_\mu\}$ and $\mathcal{J} := \{j_1, \ldots, j_\nu\}$ two arbitrary nonempty subsets of $\{1, \ldots, \sigma\}$ such that $\mathcal{I} \cap \mathcal{J} = \emptyset$ and $\mathcal{I} \cup \mathcal{J} = \{1, \ldots, \sigma\}$ and defining

$$B_{\mathcal{I}}^c := \begin{pmatrix} B_{i_1}^c & B_{i_2}^c & \cdots & B_{i_\mu}^c \end{pmatrix}, \qquad C_{\mathcal{J}}^c := \begin{pmatrix} C_{j_1}^c & C_{j_2}^c & \cdots & C_{j_\nu}^c \end{pmatrix}^T, \tag{3}$$

the complementary subsystem $\Sigma_{\mathcal{I}\mathcal{J}}^c$ has the form

$$\dot{x}^c(t) = A^c\, x^c(t) + B_{\mathcal{I}}^c\, u_{\mathcal{I}}^c(t), \tag{4}$$

$$y_{\mathcal{J}}^c(t) = C_{\mathcal{J}}^c\, x^c(t). \tag{5}$$

System Σ^c has no unstable decentralized fixed modes if and only if for all $\mathcal{I}$ and $\mathcal{J}$ the following condition is verified:

$$rank \begin{pmatrix} A^c - \lambda I & B^c_{\mathcal{I}} \\ C^c_{\mathcal{J}} & 0 \end{pmatrix} \geq n^c \tag{6}$$

for each unstable eigenvalue λ of A^c, i.e., all the $2^\sigma - 2$ complementary subsystems are *weakly complete*.

To solve the stabilization problem by a decentralized control system, which will be dealt with in the following section, consider a multirate control scheme of system Σ^c, where each component of output $y^c_j(t)$ of channel j is sampled with a period $Z^j_h T_s$, $h = 1, \ldots, q_j$, $j = 1, \ldots, \sigma$, and each component of control input $u^c_i(t)$ of channel i is connected with a zero-order hold circuit, whose hold interval is $N^i_h T_s$, $h = 1, \ldots, p_i$, $i = 1, \ldots, \sigma$. Denote with ω the least common multiple of the integers N^i_h, $h = 1, \ldots, p_i$, $i = 1, \ldots, \sigma$, and Z^j_h, $h = 1, \ldots, q_j$, $j = 1, \ldots, \sigma$. Without loss of generality, it is assumed that the greatest common divisor of the integers N^i_h, $h = 1, \ldots, p_i$, $i = 1, \ldots, \sigma$, and Z^j_h, $h = 1, \ldots, q_j$, $j = 1, \ldots, \sigma$, is equal to 1 and all the samplers and hold circuits are synchronized at time $t = 0$ [3].

The corresponding discrete-time state space model Σ of the multirate sampled-data system is characterized by σ input-output channels and is given by the series connection of discrete-time systems $\widetilde{\Sigma}_i$, $i = 1, \ldots, \sigma$, with Σ^d, which are periodic with period ω (or, briefly, ω-periodic).

The ω-periodic system $\widetilde{\Sigma}_i$, $i = 1, \ldots, \sigma$, describes the mechanism of the *zero-order hold circuits* of the channel i and has the following form:

$$\widetilde{x}_i((k+1)T_s) = \overline{S}_i(k)\, \widetilde{x}_i(kT_s) + S_i(k)\, u_i(kT_s) \tag{7}$$

$$u^c_i(kT_s) = \overline{S}_i(k)\, \widetilde{x}_i(kT_s) + S_i(k)\, u_i(kT_s), \tag{8}$$

where $k \in \mathbb{Z}$, $\widetilde{x}_i(kT_s) \in \mathbb{R}^{p_i}$ is the state, $u_i(kT_s) \in \mathbb{R}^{p_i}$ is the input of channel i of Σ, $\overline{S}_i(k) := (I_{p_i} - S_i(k))$, I_{p_i} denotes the identity matrix of dimension p_i, and $S_i(\cdot)$ is an ω-periodic matrix given by

$$S_i(k) := diag\{\sigma^i_m(k),\ m = 1, \ldots, p_i\}, \tag{9}$$

$$\sigma^i_m(k) := \begin{cases} 1, & k = jN^i_m, \\ 0, & k \neq jN^i_m, \end{cases} \qquad j \in \mathbb{Z},\ m = 1, \ldots, p_i. \tag{10}$$

The ω-periodic system Σ^d with σ input-output channels, which represents the *sampled-data system* associated with Σ^c, has the following form:

$$x^c((k+1)T_s) = e^{A^c T_s} x^c(kT_s) + \sum_{i=1}^{\sigma} B^d_i\, u^c_i(kT_s) \tag{11}$$

$$y_j(kT_s) = T_j(k)\, C^c_j\, x^c(kT_s), \qquad j = 1, \ldots, \sigma, \tag{12}$$

where $k \in \mathbb{Z}$, $B^d_i := \int_0^{T_s} e^{A^c(T_s - \theta)} B^c_i d\theta$ and $T_j(\cdot)$ is an ω-periodic matrix given by

$$T_j(k) := diag\{\tau^j_m(k),\ m = 1, \ldots, q_j\}, \tag{13}$$

$$\tau_m(k) := \begin{cases} 1, & k = iZ^j_m, \\ 0, & k \neq iZ^j_m, \end{cases} \qquad i \in \mathbb{Z},\ m = 1, \ldots, q_j. \tag{14}$$

Then, the *ω-periodic discrete-time model Σ of the multirate sampled-data system* is given by

$$x((k+1)T_s) = A(k)\,x(kT_s) + \sum_{i=1}^{\sigma} B_i(k)\,u_i(kT_s) \tag{15}$$

$$y_j(kT_s) = C_j(k)\,x(kT_s), \qquad j = 1,\ldots,\sigma, \tag{16}$$

where $k \in \mathbb{Z}^+$,

$$x(kT_s) := \left(\widetilde{x}_1(kT_s)\,\widetilde{x}_2(kT_s) \cdots \widetilde{x}_\sigma(kT_s)\,x^c(kT_s) \right)^T \in \mathbb{R}^n \tag{17}$$

with $n := n^c + \sum_{i=1}^{\sigma} p_i$, is the state, and the ω-periodic matrices $A(\cdot)$, $B(\cdot)$, and $C(\cdot)$ have the following form:

$$A(k) = \begin{pmatrix} \overline{S}_1(k) & 0 & \cdots & 0 & 0 \\ 0 & \overline{S}_2(k) & \cdots & 0 & 0 \\ \vdots & \vdots & \ddots & \vdots & \vdots \\ 0 & 0 & \cdots & \overline{S}_\sigma(k) & 0 \\ B_1^d\overline{S}_1(k) & B_2^d\overline{S}_2(k) & \cdots & B_\sigma^d\overline{S}_\sigma(k) & e^{A^c T_s} \end{pmatrix} \tag{18}$$

$$B_i(k) = \left(0 \cdots 0\,S_i(k)^T\,0 \cdots 0\,(B_i^d S_i(k))^T \right)^T \tag{19}$$

$$C_j(k) = \left(0\,0 \cdots 0\,T_j(k)C_j^c \right). \tag{20}$$

3 Networked Decentralized Control Problem

Now given the plant Σ^c and the set of sampling and hold circuits corresponding to the σ channels, consider a control network scheme characterized by linear discrete-time local controllers $\mathcal{C}_i$, for $i = 1,\ldots,\sigma$, making use of not only measured output $y_i(\cdot)$ of local channel i, but also of the other outputs connected by the local network, as shown in Figure 1. The idea is that each controller takes information from all the σ channels, in order to avoid the possible lack of the structural properties required for the decentralized solution to the stabilization problem of Σ [3].

The different sampling periods of the σ channels and the time delay of the data transmission over the network make impossible, for each controller $\mathcal{C}_i$, with $i \in \{1,\ldots,\sigma\}$, the real-time acquisition of measured outputs $y_j(\cdot)$, for $j = 1,\ldots,\sigma$ and $j \neq i$. Controller $\mathcal{C}_i$, at sampling time t_i, acquires the last sample $y_j(t_j)$ of output channel j, with $t_j < t_i$, $j = 1,\ldots,\sigma$ and $j \neq i$.

Example 1. Given a 2-channel system whose sampling and holding intervals are $T_1 = 2T_s$ and $T_2 = 3T_s$ (Figure 2). The local controller $\mathcal{C}_1$ acquires as its own inputs the following sequences of measured outputs:

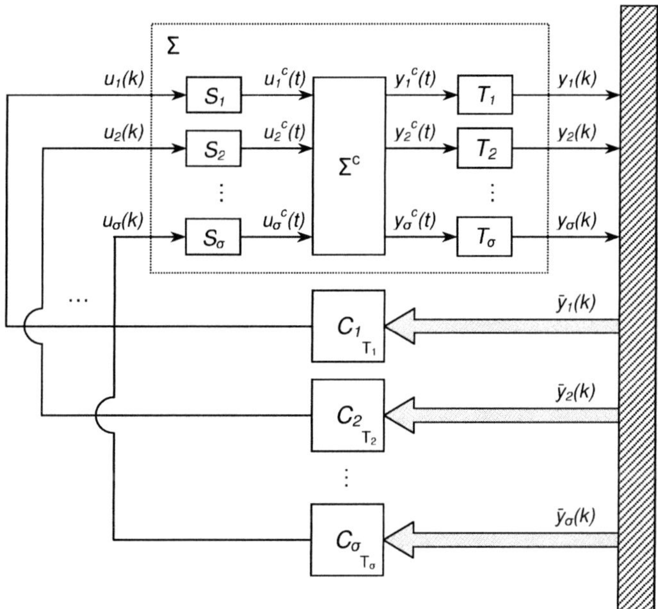

Fig. 1. Networked decentralized multirate control architecture.

$$y_1(0),\ y_1(2),\ y_1(4),\ y_1(6),\ y_1(8),\ y_1(10),\ \ldots$$
$$*,\quad y_2(0),\ y_2(3),\ y_2(3),\ y_2(6),\ y_2(9),\ \ldots$$

coming from both the channels and according to the sampling and holding interval T_1; while the local controller $\mathcal{C}_2$, operating on channel 2, acquires as its own inputs the following sequences of measured outputs:

$$y_2(0),\ y_2(3),\ y_2(6),\ y_2(9),\ y_2(12),\ y_2(15),\ \ldots$$
$$*,\quad y_1(2),\ y_1(4),\ y_1(8),\ y_1(10),\ y_1(14),\ \ldots$$

coming from both the channels with sampling and holding interval T_2.

As shown in this example, some samples of the sequence belonging to the other channels than the considered one are lost. However, more complete information is guaranteed to be available to each channel, as compared to the typical lack of exchange information of a control system which does not make use of a local network.

Also in the case of a switch box solution [17], the number of samples that integrates the information for each channel is less than the networked scheme proposed here.

The delay on the acquisition is modeled keeping memory of the samples of each channel for a time delay $\delta := max_{i,j=1,\ldots,\sigma,\,k=1,\ldots,\omega-1}\big(\delta_{ij}(k)\big)$, where $\delta_{ij}(k)$ is the ω-periodic time-shift, at time k, between the sampling time at

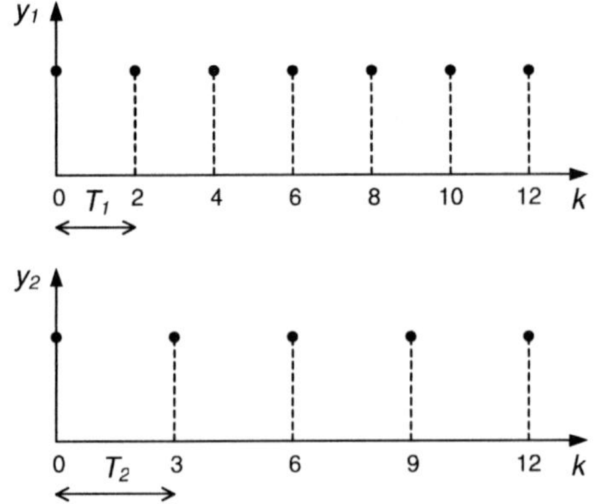

Fig. 2. Sample sequences of a two-channel system.

channel i and the sampling time at channel j, for $i, j = 1, \ldots, \sigma$ [3]. The *networked extended ω-periodic system* $\overline{\Sigma}$, whose representation allows the modeling of this mechanism, is given by

$$\bar{x}((k+1)T_s) = \bar{A}(k)\,\bar{x}(kT_s) + \sum_{i=1}^{\sigma} \bar{B}_i(k)\,u_i(kT_s) \tag{21}$$

$$\bar{y}_j(kT_s) = \bar{C}_j(k)\,\bar{x}(kT_s), \quad j = 1, \ldots, \sigma, \tag{22}$$

where the extended state and outputs are described by

$$\bar{x}(kT_s) := \begin{pmatrix} x(kT_s) \\ x^c((k-1)T_s) \\ \vdots \\ x^c((k-\delta)T_s) \end{pmatrix}, \qquad \bar{y}_j(kT_s) := \begin{pmatrix} y_j(kT_s) \\ y_1(l_1^k T_s) \\ \vdots \\ y_{j-1}(l_{j-1}^k T_s) \\ y_{j+1}(l_{j+1}^k T_s) \\ \vdots \\ y_\sigma(l_\sigma^k T_s) \end{pmatrix}, \tag{23}$$

with $k = hN_j$, for $h \in \mathbb{Z}^+$, $i, j = 1, \ldots, \sigma$ and $i \neq j$, and

$$l_i^k := \begin{cases} k - N_i, & k = hN_i, \\ N_i[k/N_i], & k \neq hN_i, \end{cases} \tag{24}$$

where $[\cdot]$ is the integer part function; besides, $\bar{x}(kT_s) \in \mathbb{R}^{\bar{n}}$, with $\bar{n} := (\delta + 1)n^c + \sum_{i=1}^{\sigma} p_i$, $\bar{y}_j(kT_s) \in \mathbb{R}^{\bar{q}}$, with $\bar{q} := \sum_{j=1}^{\sigma} q_j$, and

$$\bar{A}(k) := \begin{pmatrix} \overline{S}_1(k) & 0 & \cdots & 0 & 0 & 0 & \cdots & 0 & 0 \\ 0 & \overline{S}_2(k) & \cdots & 0 & 0 & 0 & \cdots & 0 & 0 \\ \vdots & \vdots & \ddots & \vdots & \vdots & \vdots & \ddots & \vdots & \vdots \\ 0 & 0 & \cdots & \overline{S}_\sigma(k) & 0 & 0 & \cdots & 0 & 0 \\ B_1^d\overline{S}_1(k) & B_2^d\overline{S}_2(k) & \cdots & B_\sigma^d\overline{S}_\sigma(k) & e^{A^c T_c} & 0 & \cdots & 0 & 0 \\ 0 & 0 & \cdots & 0 & I_{n^c} & 0 & \cdots & 0 & 0 \\ 0 & 0 & \cdots & 0 & 0 & I_{n^c} & \cdots & 0 & 0 \\ \vdots & \vdots & \ddots & \vdots & \vdots & \vdots & \ddots & \vdots & \vdots \\ 0 & 0 & \cdots & 0 & 0 & 0 & \cdots & I_{n^c} & 0 \end{pmatrix} \tag{25}$$

$$\bar{B}_i(k) := \begin{pmatrix} 0 & \cdots & 0 & S_j(k)^T & 0 & \cdots & 0 & (B_i^d S_i(k))^T & 0 & \cdots & 0 \end{pmatrix}^T, \tag{26}$$

$$\bar{C}_j(k) := \begin{pmatrix} 0 \cdots 0 & T_i(k)C_j^c & 0 \\ 0 \cdots 0 & 0 & \widetilde{C}_1(k) \\ \vdots \ddots \vdots & \vdots & \vdots \\ 0 \cdots 0 & 0 & \widetilde{C}_{j-1}(k) \\ 0 \cdots 0 & 0 & \widetilde{C}_{j+1}(k) \\ \vdots \ddots \vdots & \vdots & \vdots \\ 0 \cdots 0 & 0 & \widetilde{C}_\sigma(k) \end{pmatrix} \tag{27}$$

with

$$\widetilde{C}_i(k) := \begin{pmatrix} \widetilde{T}_1(l_i^k)C_i^c & \widetilde{T}_2(l_i^k)C_i^c & \cdots & \widetilde{T}_\delta(l_i^k)C_i^c \end{pmatrix} \tag{28}$$

$$\widetilde{T}_m(l_i^k) := \begin{cases} 1, & l_i^k = k - m, \\ 0, & l_i^k \neq k - m, \end{cases} \qquad i = 1,\ldots,\sigma, \quad m = 1,\ldots,\delta. \tag{29}$$

Remark 1. Note that at time $k \neq hN_j$, with $h \in \mathbb{Z}$ and $j \in \{1,\ldots,\sigma\}$, output $\bar{y}_j(kT_s)$ is not considered, according to the assumption that the time-base of channel j is given by $T_j = N_j T_s$.

The stabilization problem of a networked decentralized multirate control system is here defined and existence conditions are introduced in the next section in terms of rank test of the system matrix of the original linear continuous time-invariant plant. Making use of the introduced notation the considered control problem is stated here.

Problem 1. Networked Decentralized Multirate Control Problem (NDMCP) The stabilization problem of Σ^c by a networked decentralized multirate control system consists of finding for each input-output channel a linear discrete-time periodic local controller $\mathcal{C}_i$ with period ω and of the form

$$\widehat{x}_i((h+1)N_iT_s) = F^i(h)\widehat{x}_i(hN_iT_s) + G^i(h)\bar{y}_i(hN_iT_s) \tag{30}$$

$$u_i(hN_iT_s) = P^i(h)\widehat{x}_i(hN_iT_s) + Q^i(h)\bar{y}_i(hN_iT_s), \tag{31}$$

such that the ω-periodic closed loop system given by $\bar{\Sigma}$ and independent controllers $\mathcal{C}_i$, for $i = 1, \ldots, \sigma$, is asymptotically stable.

Remark 2. Observe that, for the implementation of the digital periodic controller $\mathcal{C}_i$, it is possible to consider the time-base of the corresponding channel i as equal to $T_i = N_i T_s$. In this sense, $\mathcal{C}_i$ operates only at instants multiple of T_i, i.e., the measured output $\bar{y}_i(h N_i T_s) = \bar{y}_i(h T_i)$, with $h \in \mathbb{Z}$, is collected only for $k = h T_i$. This means that the period of $\mathcal{C}_i$ may be more easily implemented as equal to w/N_i, with the consequence of a less complex structure.

4 Main Results

To analyze the existence conditions for a solution to the introduced problem, the time-invariant representation of the extended multirate sampled-data system $\bar{\Sigma}$ is here recalled. The state transition matrix of $\bar{\Sigma}$ is expressed by

$$\bar{\Phi}(k, k_0) := \bar{A}(k-1)\,\bar{A}(k-2)\,\ldots\,\bar{A}(k_0), \tag{32}$$

with $k > k_0$, $k, k_0 \in \mathbb{Z}^+$, and $\bar{\Phi}(k, k) := I_{\bar{n}}$ for all $k \in \mathbb{Z}^+$. For any initial time $k_0 \in \mathbb{Z}^+$, the output response of the ω-periodic system $\bar{\Sigma}$, for $k \geq k_0$, to a given initial state $\bar{x}(k_0)$ and control functions $u_i(\cdot)$, $i = 1, \ldots, \sigma$, can be expressed throughout the time-invariant associated system of $\bar{\Sigma}$ at time k_0, denoted by $\bar{\Sigma}^{k_0}$ [14]. For an arbitrary time k, the *time-invariant state space representation* of system $\bar{\Sigma}^k$ is

$$\bar{x}^k(h+1) = \bar{E}^k\,\bar{x}^k(h) + \sum_{i=1}^{\sigma} \bar{J}_i^k\,u_i^k(h) \tag{33}$$

$$\bar{y}_j^k(h) = \bar{L}_j^k\,\bar{x}^k(h) + \sum_{i=1}^{\sigma} \bar{M}_{ji}^k u_i^k(h), \qquad j = 1, \ldots, \sigma, \tag{34}$$

where

$$\bar{E}^k := \bar{\Phi}(k+w, k), \quad \bar{J}_i^k := \left(\bar{\Delta}_i^k(0)\,\bar{\Delta}_i^k(1) \cdots \bar{\Delta}_i^k(\omega-1) \right) \tag{35}$$

$$\bar{L}_j^k := \begin{pmatrix} \bar{\Gamma}_j^k(0) \\ \bar{\Gamma}_j^k(1) \\ \vdots \\ \bar{\Gamma}_j^k(\omega-1) \end{pmatrix}, \quad \bar{M}_{ji}^k := \begin{pmatrix} \bar{\Theta}_{ji}^k(0,0) & \cdots & \bar{\Theta}_{ji}^k(0, \omega-1) \\ \vdots & \ddots & \vdots \\ \bar{\Theta}_{ji}^k(\omega-1, 0) & \cdots & \bar{\Theta}_{ji}^k(\omega-1, \omega-1) \end{pmatrix}, \tag{36}$$

with

$$\bar{\Delta}_i^k(\ell) := \bar{\Phi}(k+w, k+\ell+1)\bar{B}_i(k+\ell) \tag{37}$$

$$\bar{\Gamma}_j^k(\ell) := \bar{C}_j(k+\ell)\bar{\Phi}(k+\ell, k) \tag{38}$$

$$\bar{\Theta}_{ji}^k(\ell, r) := \begin{cases} 0, & \ell \leq r, \\ \bar{C}_j(k+\ell)\bar{\Phi}(k+\ell, k+r+1)\bar{B}_i(k+r), & \ell > r, \end{cases} \tag{39}$$

for $\ell, r = 0, 1, \ldots, \omega - 1, i, j = 1, \ldots, \sigma$.

It is easy to see that if

$$u_i^k(h) = \begin{pmatrix} u_i(k + h\omega) \\ u_i(k + 1 + h\omega) \\ \vdots \\ u_i(k + \omega - 1 + h\omega) \end{pmatrix} \quad and \quad \bar{y}_j^k(h) = \begin{pmatrix} \bar{y}_j(k + h\omega) \\ \bar{y}_j(k + 1 + h\omega) \\ \vdots \\ \bar{y}_j(k + \omega - 1 + h\omega) \end{pmatrix} \tag{40}$$

$i, j = 1, \ldots, \sigma$, then for $\bar{x}^k(0) = \bar{x}(k)$, it results that $\bar{x}^k(h) = \bar{x}(k + h\omega)$ for all $h \in \mathbb{Z}^+$. Thus, $\bar{y}_j^k(\cdot), j = 1, \ldots, \sigma$, gives $\bar{y}_j(\cdot)$ in a lifted form over each period, provided that $u_i^k(\cdot), i = 1, \ldots, \sigma$, coincides with the lifted form of $u_i(\cdot)$ over each period. Moreover, the characteristic polynomial of $\bar{E}^k$ (the monodromy matrix) is independent of k, and, by the periodicity of $\overline{\Sigma}$, it characterizes the stability of $\overline{\Sigma}$ [14]. For this reason, the eigenvalues of $\bar{E}^k$ are called the eigenvalues of $\overline{\Sigma}$.

The existence conditions of a solution to the decentralized problem of periodic systems have been stated in [20]. These results can be used for introducing the existence conditions of NDMCP in terms of periodic representation of the multirate sampled-data system $\overline{\Sigma}$. In order to introduce such conditions, the following notation are needed. A complementary subsystem $\overline{\Sigma}_{\mathcal{I}\mathcal{J}}$ of $\overline{\Sigma}$, associated to the sets $\mathcal{I} := \{i_1, \ldots, i_\mu\}$ and $\mathcal{J} := \{j_1, \ldots, j_\nu\}$, with $\mathcal{I} \cap \mathcal{J} = \emptyset$ and $\mathcal{I} \cup \mathcal{J} = \{1, \ldots, \sigma\}$, has the form

$$\bar{x}((k+1)T_s) = \bar{A}(k)\bar{x}(kT_s) + \bar{B}_{\mathcal{I}}(k)u_{\mathcal{I}}(kT_s) \tag{41}$$

$$\bar{y}_{\mathcal{J}}(kT_s) = \bar{C}_{\mathcal{J}}(k)\bar{x}(kT_s), \tag{42}$$

where

$$\bar{B}_{\mathcal{I}}(k) := \begin{pmatrix} \bar{B}_{i_1}(k) \ \bar{B}_{i_2}(k) \ \cdots \ \bar{B}_{i_\mu}(k) \end{pmatrix} \quad and \quad \bar{C}_{\mathcal{J}}(k) := \begin{pmatrix} \bar{C}_{j_1}(k) \\ \bar{C}_{j_2}(k) \\ \vdots \\ \bar{C}_{j_\nu}(k) \end{pmatrix}, \tag{43}$$

with $\bar{B}_i(k)$ and $\bar{C}_j(k)$ stated by (27). The time-invariant representation $\overline{\Sigma}_{\mathcal{I}\mathcal{J}}^k$ of $\overline{\Sigma}_{\mathcal{I}\mathcal{J}}$ has the same structure of $\overline{\Sigma}^k$ with matrices $\bar{J}_{\mathcal{I}}^k, \bar{L}_{\mathcal{J}}^k$ and $\bar{M}_{\mathcal{J}\mathcal{I}}^k$ defined as $\bar{J}^k, \bar{L}^k$ and $\bar{M}^k$, with $\bar{B}(k)$ and $\bar{C}(k)$ substituted by $\bar{B}_{\mathcal{I}}(k)$ and $\bar{C}_{\mathcal{J}}(k)$ respectively.

Lemma 1. *[20] The NDMCP admits a solution if and only if*

(i) system $\overline{\Sigma}$ is stabilizable and detectable, i.e., for an arbitrary $k \in \mathbb{Z}^+$ and for all z outside the open unitary disk,

$$rank \begin{pmatrix} \bar{E}^k - zI_{\bar{n}} \ \bar{J}^k \end{pmatrix} = \bar{n}, \qquad rank \begin{pmatrix} \bar{E}^k - zI_{\bar{n}} \\ \bar{L}^k \end{pmatrix} = \bar{n}; \tag{44}$$

(ii) the $2^\sigma - 2$ complementary subsystems $\bar{\Sigma}_{\mathcal{I}\mathcal{J}}$ are weakly complete, i.e., for all $\mathcal{I}$ and $\mathcal{J}$, for an arbitrary $k \in \mathbb{Z}^+$ and for all z outside the open unitary disk,

$$rank \begin{pmatrix} \bar{E}^k - zI_{\bar{n}} & \bar{J}^k_{\mathcal{I}} \\ \bar{L}^k_{\mathcal{J}} & \bar{M}^k_{\mathcal{J}\mathcal{I}} \end{pmatrix} \geq \bar{n}. \tag{45}$$

Denoting

$$B^c_{\mathcal{I}} := \begin{pmatrix} B^c_{i_1} & B^c_{i_2} & \cdots & B^c_{i_\mu} \end{pmatrix} \tag{46}$$

the existence conditions of NDMCP can be stated in terms of the given continuous time-invariant system Σ^c if the sampling rates are chosen appropriate to system Σ^c.

Theorem 1. *Given a continuous-time plant Σ^c that is stabilizable and detectable, the NDMCP has a solution if*

(i) every pair $(\lambda^c_a, \lambda^c_b)$ of distinct eigenvalues of A^c, with $\mathrm{Re}\,[\lambda^c_a] = \mathrm{Re}[\lambda^c_b] \geq 0$, has

$$\mathrm{Im}[\lambda^c_a - \lambda^c_b] \neq \pm \frac{2h\pi}{\omega T_s} \tag{47}$$

for all $h \in \mathbb{Z}^+$;
(ii) the $2^\sigma - 2$ conditions

$$rank \begin{pmatrix} A^c - \lambda I & B^c_{\mathcal{I}} \\ C^c & 0 \end{pmatrix} \geq n^c \tag{48}$$

are verified for all $\mathcal{I}$ and for each unstable eigenvalue λ of A^c.

For the sake of brevity the proof of this theorem is here omitted. This proof has been performed making use of Lemma 1, of results on the analysis of linear periodic discrete-time systems [15], of the Jordan form of matrix A^c, and of elementary operations on the matrices of conditions (44) and (45).

The conditions of Theorem 1 are not related to the multirate mechanism, but only to the least common multiple of the sampling and holding intervals. The condition (i) of Theorem 1 preserves the stabilizability and determinability of system $\bar{\Sigma}$ and the fulfillment of condition (ii) of Lemma 1 if condition (ii) of Theorem 1 is verified.

Example 2. Consider a linear continuous time-invariant plant Σ^c, characterized by $\sigma = 2$ input-output channels where

$$A^c = \begin{pmatrix} 2 & 0 & 1 \\ 0 & 0 & -1 \\ 1 & -2 & 0 \end{pmatrix}, \qquad B^c_1 = \begin{pmatrix} -2 \\ 0 \\ 0 \end{pmatrix}, \qquad B^c_2 = \begin{pmatrix} 0 \\ 1 \\ -1 \end{pmatrix}$$

$$C^c_1 = \begin{pmatrix} 1 & -1 & 0 \end{pmatrix}, \qquad C^c_2 = \begin{pmatrix} 0 & 1 & 1 \end{pmatrix}$$

and characterized by different sampling and updating periods, $T_1 = 2T_s$ and $T_2 = 3T_s$. It is possible to show that the conditions for the existence of a simple decentralized multirate control system designed without information exchange between the two output channels are not verified, missing a solution for the decentralized control problem [3]. On the contrary, the conditions of Theorem 1 for the existence of a solution to the NDMCP are verified and a decentralized control multirate system can be implemented, by appropriately choosing time-base T_s.

For the plant Σ^c considered in the Example 2, making use of the synthesis procedure introduced in the following section, it is possible to design two independent periodic controllers, $\mathcal{C}_1$ and $\mathcal{C}_2$, which guarantee the asymptotic stability of the closed loop system.

5 Networked Decentralized Controller Design Techniques

Solutions to the NDMCP are investigated in this section. A classical synthesis technique is proposed and discussed. Other approaches for robust solutions to the considered problem are discussed in the last section.

Networked decentralized control is an appealing concept, because it offers an essential reduction in communication requirements without significant, if any, loss of performance. Each subsystem is controlled on the basis of all the available information, coming from all the channels belonging to the entire system, with states of one subsystem being seldom used in control of another subsystem. This attractive property disappears when states of the subsystems are accessible at the local level. In building observers or estimators to reconstruct the states, it is necessary to force the exchange of states among the subsystems and this violates the information structure constraints of decentralized control laws. A logical way out is to consider static or dynamic output-feedback satisfying simultaneously the global (decentralized) and a local (output) information constraints.

In this section, a classical synthesis technique, making use of pole-placement and minimal periodic realization techniques, is deeply analyzed. The design of the local controllers $\mathcal{C}_i$, for $i = 1, \ldots, \sigma$, can be performed throughout a three-step technique that initially involves the time-invariant representation $\overline{\Sigma}^k$, for an arbitrary time k, associated to the multirate sampled-data extended system $\overline{\Sigma}$. The idea is that if a set of local controllers can be designed for the time-invariant case, it is possible to find a correspondent realization in the multirate discrete-time case.

Algorithm 1. *The synthesis procedure is now specified in three steps.*

(STEP 1) Making use of classical algorithms, compute time-invariant decentralized controllers $\overline{\mathcal{C}}_i^0$ for the stabilization of the time-invariant system $\overline{\Sigma}^0$.

(STEP 2) Making use of the algorithm proposed in [6], compute an ω-periodic realization $\overline{\mathcal{C}}_i$ associated to $\overline{\mathcal{C}}_i^{\,0}$.

(STEP 3) Compute the discrete-time system $\mathcal{C}_i$ with sampling rate $N_i T_c$ corresponding the ω-periodic system $\overline{\mathcal{C}}_i$.

Remark 3. Note that STEP 3 of this algorithm can be applied only in some cases in which a minimal single-rate realization $\mathcal{C}_i$ can be associated with the ω-periodic one $\overline{\mathcal{C}}_i$, computed in the previous step. The analysis of conditions that let achieve this minimal single-rate realization have to be investigated in future research.

The first step of the design procedure above can make use of results on pole assignment for the decentralized control for linear time-invariant systems that have been developed by several authors (see, e.g., [33, 8, 25, 10]). The time-invariant representation $\overline{\Sigma}^k$ of a multirate sampled-data extended system $\overline{\Sigma}$, obtained from the original continuous time-invariant Σ^c, is considered. This lifted system $\overline{\Sigma}^k$ can be analyzed and manipulated independently from the correspondent original sampled-data one. This means that $\overline{\Sigma}^k$ may be considered, for an arbitrary fixed $k \in \mathbb{Z}$, as any one linear time-invariant system and the synthesis results on this class of systems can be applied making use of techniques such as that presented by Corfmat and Morse [8] or by Davison and Chang [10]. One possible approach to the spectrum assignment problem is to apply to $\overline{\Sigma}^k$ local nondynamic controllers of the form

$$u_i^k(h) = F_i^k \bar{y}_i^k(h), \qquad i \in \{1, \ldots, \sigma\}, \tag{49}$$

the objective being to select F_i^k such that the resulting closed loop system $\overline{\Sigma}_{cl_1}^k$ is both controllable and observable through a single channel. If this can be accomplished, then standard techniques may be used to construct a second local dynamic control system

$$\bar{\eta}_j^k(h+1) = \bar{R}_j^k \bar{\eta}_j^k(h) + \bar{U}_j^k \bar{y}_j^k(h) \tag{50}$$

$$u_j^k(h) = \bar{V}_j^k(h)\bar{\eta}_j^k(h) + \bar{W}_j^k \bar{y}_j^k(h), \tag{51}$$

which, when applied to a channel $j \in \{1, \ldots, \sigma\}$ of $\overline{\Sigma}_{cl_1}^k$, results in a closed loop system $\overline{\Sigma}_{cl_2}^k$ with prescribed spectrum Λ.

Example 3. Consider a linear continuous time-invariant plant Σ^c, characterized by $\sigma = 2$ input-output channels where

$$A^c = \begin{pmatrix} 1 & 0 & 1 \\ 0 & 0 & 1 \\ 0 & -1 & 0 \end{pmatrix}, \qquad B_1^c = \begin{pmatrix} -2 \\ 0 \\ 1 \end{pmatrix}, \qquad B_2^c = \begin{pmatrix} 0 \\ 1 \\ 1 \end{pmatrix}$$

$$C_1^c = \begin{pmatrix} 1 & -1 & 0 \end{pmatrix}, \qquad C_2^c = \begin{pmatrix} 0 & 1 & 1 \end{pmatrix},$$

and with different sampling and updating periods, $T_1 = 2T_s$ and $T_2 = 3T_s$. The period ω is 6 and the maximum relative delay between samples of output channel 1 and samples of output channel 2 is $T_\delta = 3T_s$ (i.e., $\delta = 3$). According to (47), the time-base T_s has no constraint on its own value and $T_s = 0.06\,\mathrm{sec}$ is chosen. The conditions of Theorem 1 for the existence of a solution to the NDMCP are verified and a decentralized control multirate system can be implemented. The monodromy matrix $\bar{E}^0$ of the corresponding time-invariant representation in networked configuration at time $k = 0$, is given by

$$
\bar{E}^0 =
\left(
\begin{array}{c|ccc|c}
0_{2\times2} & \multicolumn{3}{c|}{0_{2\times3}} & 0_{2\times9} \\
\hline
& 1.4333 & -0.0741 & 0.4419 & \\
0_{3\times2} & 0 & 1.0655 & -0.3678 & 0_{3\times9} \\
& 0 & -0.3678 & 1.0655 & \\
\hline
& 1.3499 & -0.0502 & 0.3547 & \\
& 0 & 1.0453 & -0.3045 & \\
& 0 & -0.3045 & 1.0453 & \\
& 1.2712 & -0.0314 & 0.2737 & \\
0_{9\times2} & 0 & 1.0289 & -0.2423 & 0_{9\times9} \\
& 0 & -0.2423 & 1.0289 & \\
& 1.1972 & -0.0173 & 0.1982 & \\
& 0 & 1.0162 & -0.1810 & \\
& 0 & -0.1810 & 1.0162 & \\
\end{array}
\right).
$$

The non-null eigenvalues of this monodromy matrix are those of the block associated to rows $3, 4, 5$ and columns $3, 4, 5$, therefore the pole-placement technique can be applied only to the part of the entire system that has this block as dynamic matrix, in order to shift the unstable modes. A nondynamic output feedback $\overline{C}_1^0$, with transfer function $\bar{F}_1$, for channel 1, is designed:

$$
u_1^0(h) = \bar{F}_1^0 \, \bar{y}_1^0(h), \quad \bar{F}_1^0 = diag\{(43\,0), (43\,0), (1\,1), (1\,1), (1\,1), (1\,1)\},
$$

which is able to guarantee that the resulting closed loop system $\overline{\Sigma}_{cl_1}^0$ is both controllable and observable through the single channel 2. Standard techniques of pole-placement are used to design the second local controller $\overline{C}_2^0$ for channel 2, of the form (50) and (51) with

$$
\bar{R}_2^0 = 10^3 \cdot
\begin{pmatrix}
-0.5126 & 0.8421 & -0.7178 \\
-0.6247 & 1.0232 & -0.8682 \\
0.5232 & -0.8651 & 0.74363
\end{pmatrix},
$$

$$
\bar{U}_2^0 =
\begin{pmatrix}
-21.9417 & 0 & 0 & 0 & 0 & 21.9294 & 8.8524 & 0 & 0 & 0 & 0 \\
-26.8490 & 0 & 0 & 0 & 0 & 26.8342 & 11.6405 & 0 & 0 & 0 & 0 \\
21.8269 & 0 & 0 & 0 & 0 & -22.8982 & -9.6316 & 0 & 0 & 0 & 0
\end{pmatrix}
$$

$$\bar{V}_2^0 = \begin{pmatrix} -83.1187 & 136.7669 & -116.7358 \\ 0 & 0 & 0 \\ 0 & 0 & 0 \\ 32.9629 & -80.8966 & 101.7515 \\ 0 & 0 & 0 \\ 0 & 0 & 0 \end{pmatrix}, \quad \bar{W}_2^0 = 0.$$

The closed loop system $\overline{\Sigma}_{cl_2}^0$, obtained by applying the two computed controllers $\bar{C}_1^0$ and $\bar{C}_2^0$ to the whole system $\overline{\Sigma}^0$, has the following eigenvalues $p_i = -0.3481$ for $i \in \{1,2,3,4\}$, $p_j = 0$ for $j \in \{5,6,\cdots,17\}$, which are all inside the open unitary disk $\mathbb{D}$.

The aim of the second step of the synthesis algorithm is to define a periodic minimal realization for the decentralized control system designed in the previous step with a pole-placement technique.

This problem has been considered with particular interest in recent years, for the increasing attention devoted to analysis and control of discrete-time linear periodic systems and for reasons related to the widespread use of digital schemes for control purposes, which have led research on this topic to explore the possibility of enhancing the properties of linear time-invariant plants with the use of periodic controllers [21]. As just recalled, the lifting isomorphism between periodic systems and time-invariant ones can exploit the theory of time-invariant systems for the control of periodic ones, provided that results achieved can be easily reinterpreted in a periodic framework. More precisely, the design procedure of a periodic controller, carried out in the context of time-invariant systems, has to incorporate the constraints on periodic realizability of the achieved time-invariant controller. Therefore, a main role in the design of periodic controller is played by the issue of determining a periodic state space model from input-output maps. This represents the *periodic realization problem* that the second step of the above-described synthesis algorithm tries to solve.

It is well known that in the discrete-time case the reachable and/or observable subspaces may have time-varying dimensions. Therefore it is natural, in order to consistently solve the problem of periodic minimal realization, to allow for state space descriptions having time-varying dimensions. A minimal realization with time-invariant dimension is called *minimal uniform realization* [6]. Necessary and sufficient conditions for the solvability of the periodic realization problem are given by Colaneri and Longhi [6] in terms of the existence of a minimal (reachable and observable at any time instant) periodic realization, which is generally described by periodic difference equations whose matrices have time-varying dimensions. Such an algorithm is based on the structural properties of systems with time-invariant dimensions [15] and with time-varying dimensions [13]. The algorithm for computing a periodic minimal realization of $\bar{C}_i$ associated to $\bar{C}_i^0$ is stated in [6].

Example 4. Consider the time-invariant solutions of Example 3. For time-invariant controller $\overline{\mathcal{C}}_1^0$, with nondynamic output feedback $\bar{F}_1^0$, a periodic realization $\overline{\mathcal{C}}_1$ is provided by a 3-periodic matrix $\bar{F}_1(h)$, with respect to channel 1 time-base $T_1 = 2T_s$ [6-periodic matrix, with respect to global time-base T_s], of the form

$$\bar{F}_1(kT_s) = (43\ 0),\ k = 0, 1, \quad \bar{F}_1(kT_s) = (1\ 1),\ k = 2, 3, 4, 5.$$

For time-invariant controller $\overline{\mathcal{C}}_2^0$, a periodic realization $\overline{\mathcal{C}}_2$ is provided by a 2-periodic matrix, with respect to channel 2 time-base $T_2 = 3T_s$ [6-periodic matrix, with respect to global time-base T_s] of the form

$$\eta((k+1)T_s) = \bar{R}_2(kT_s)\,\eta(kT_s) - \bar{U}_2(kT_s)\,\bar{y}_2(kT_s)$$
$$u_2(kT_s) = \bar{R}_2(kT_s)\,\eta(kT_s),$$

with $\xi(\cdot) \in \mathbb{R}^3$, $u_2(\cdot) \in \mathbb{R}$, $\bar{y}_2(\cdot) \in \mathbb{R}^2$ and the 2-periodic [6-periodic] matrices

$$\bar{R}_2(kT_s) = 10^3 \cdot \begin{pmatrix} -0.5126 & 0.8421 & -0.7178 \\ -0.6247 & 1.0232 & -0.8682 \\ 0.5232 & -0.8651 & 0.7436 \end{pmatrix}, \quad k = 0, 1, 2,$$

$$\bar{R}_2(kT_s) = I_3, \quad k = 3, 4, 5,$$

$$\bar{U}_2(kT_s) = \begin{pmatrix} 21.9417\ 0\ 0\ 0\ 0\ 0 \\ 26.8490\ 0\ 0\ 0\ 0\ 0 \\ -21.8269\ 0\ 0\ 0\ 0\ 0 \end{pmatrix}, \quad k = 0, 1, 2,$$

$$\bar{U}_2(kT_s) = \begin{pmatrix} -21.9294 & -8.8524\ 0\ 0\ 0\ 0 \\ -26.8342 & -11.6405\ 0\ 0\ 0\ 0 \\ 22.8982 & 9.6316\ \ 0\ 0\ 0\ 0 \end{pmatrix}, \quad k = 3, 4, 5,$$

$$\bar{V}_2(kT_s) = \begin{pmatrix} -83.1187 & 136.7669 & -116.7358 \\ 0 & 0 & 0 \\ 0 & 0 & 0 \end{pmatrix}, \quad k = 0, 1, 2,$$

$$\bar{V}_2(kT_s) = \begin{pmatrix} -6.6245 & 6.4280 & 1.2476 \\ 0 & 0 & 0 \\ 0 & 0 & 0 \end{pmatrix}, \quad k = 3, 4, 5.$$

The periodical realization of controller $\overline{\mathcal{C}}_2$ is a minimal uniform realization.

As emphasized in Remark 3, STEP 3 of Algorithm 1 has to be deeply explored to get digital time-invariant controllers for the sampled-data plant Σ.

6 Concluding Remarks

The problem of stabilizing a large-scale continuous-time plant characterized by different sampling and updating intervals for each input-output channel of the plant has been analyzed.

A preliminary result for the development of a networked decentralized digital control system is introduced. Existence conditions of a solution to the NDMCP have been achieved and formulated in terms of rank conditions on the system matrix describing the continuous-time plant and of choice of sampling frequencies of the sampling mechanism. The NDMCP admits a solution if the sampling frequency is not pathological (relative to the dynamic matrix A^c), i.e., the sampling frequencies preserve the stabilizability and detectability of the continuous-time plant Σ^c. Besides, the existence conditions are not related to multirate mechanism, but only to the least common multiple of the sampling and hold intervals. The multirate sampling mechanism and the use of a local network enlarge the class of plants to be stabilized with decentralized control. In fact, the possibilities for stabilizing a large-scale continuous plant by a decentralized digital control system are improved throughout the output data exchange by using local networks. Each local controller of the digital control scheme can make use of the local sampled-data output measures and of someone or of all output measures of the other channels to avoid the lack of structural properties.

A classical technique for the design of the local independent controllers that constitute the decentralized scheme is analyzed. To the best of our knowledge, no general stabilization techniques exist for decentralized multirate systems. The decentralized stabilizing controller can be designed by applying the pole-placement procedure in [8] to a lifted system. In so doing, a time-invariant controller is obtained, which can be given an ω-periodic reformulation suitable for implementation.

The application of classical techniques, like the pole-placement for the design of output-feedback dynamic independent controllers used here, does not manage to solve the great problems of robustness and reliability that arise in networked systems. Recently, robust solutions have been developed and specific solutions are stated in [29, 31, 16].

The necessity of exploring the chance to design a networked decentralized multirate control system capable of assuring robustness and reliability to the controlled plant will be investigated in future works by analyzing some interesting possible approaches for solving robust problems:

- a stable proper fractional one, for the parametrization of all solutions of NDMCP, applied to the time-invariant representation associated to the multirate system [25];
- a periodic polynomial one, for the design of local periodic controllers [2, 5], capable of assuring a real-time behaviour to the entire decentralized control systems and, in so doing, to follow possible changes of the parameter of the system;
- a linear matrix inequalities one, for periodic discrete-time periodic systems [26], capable of synthesizing local output-feedback controllers with periodic dynamics.

All these techniques could represent a possible direction for future research activities and, clearly, need a more extensive investigation, for the analysis of the performances of a networked decentralized control in terms of robustness and/or parameter uncertainties.

Moreover, another interesting open issue concerning control problem for large-scale multirate sampled-data systems making use of networks is the necessity of managing communication on the network and manipulating control data in a correct manner. It is necessary to plan a methodology that lets us put in a precise correspondence the measures drawing from output channels with data produced by the local controllers and addressed to control inputs so that the control action could be applied to the controlled process in the right instant. The ideal case is that of a real-time control, but it struggles with the several causes producing different delays on networks. This problem becomes more difficult to solve in the case of data exchange among the output channels, as proposed here, because the multirate sampling and holding introduce a further degree of complexity in the problem. In this sense, the study of the probable asynchronism of data sampling and data updating among the various channels of a digital networked decentralized control scheme, due to physical and/or technological constraints, represents a further fundamental direction of investigation.

References

1. Anderson B, Clements D (1981) Algebraic characterization of fixed modes in decentralized control. Automatica 17:703–712
2. Ciferri R, Colaneri P, Longhi S (2001) CAD tools for control design in linear periodic discrete-time systems subject to input constraints. 193–198. In: Proc. of the IFAC Workshop on Periodic Control Systems, Cernobbio, Italy
3. Ciferri R, Longhi S (2002) Decentralized control of multirate sampled-data systems In: Proc. of the 41th IEEE Conference on Decision and Control, Orlando, FL
4. Ciferri R, Longhi S (2003) Decentralized control networks of multirate sampled-data systems In: Proc. of the 11th Mediterranean Conference on Control and Automation, Rhodes, Greece
5. Colaneri P, Kučera V, Longhi S (2003) Polynomial approach to the control of SISO periodic systems subject to input constraint. Automatica 39:1417–1424
6. Colaneri P, Longhi S (1995) The realization problem for linear periodic systems. Automatica 31:775–779
7. Colaneri P, Scattolini R, Shiavoni N (1990) Stabilization of multirate sampled-data linear systems. Automatica 26:377–380
8. Corfmat J, Morse A (1976) Decentralized control of linear multivariable systems. Automatica 12:479–495
9. Dantzig GB, Wolfe P (1960) Decomposition principle of linear programs. Operation Res. 8(1):101–111
10. Davison E, Chang TN (1990) Decentralized stabilization and pole assignment for general proper systems. IEEE Trans. Automat. Contr. 35:652–664

11. Decotignie JD, Pleinevaux P (1993) A survey on industrial communication networks. Annals of Telecommunications 48:435–448
12. Elgerd O (1971) Electric energy systems theory: an introduction. McGraw-Hill, New York
13. Gohberg I, Kaashoek MA, Lerer L (1992) Minimality and realization of discrete time-varying systems. Operator Theory: Advanced Applications 56:261–296
14. Grasselli O, Longhi S (1988) Zeros and poles of linear periodic multivariable discrete-time systems. Circuit System Signal Processing 7:361–380
15. Grasselli O, Longhi S (1991) The finite zero structure of linear periodic discrete-time systems. International Journal of Systems Science 22:1785–1806
16. Ikeda GZM, Fujisaki Y (2001) Decentralized $\mathcal{H}_\infty$ controller design: a matrix inequality approach using a homotopy method. Automatica 37:565–572
17. Ishii H, Francis B (2002) Stabilization with control networks. Automatica 38:1745–1751
18. Ito H (1997) Worst-case performance and stability of multirate sampled-data systems with nonsynchronous decentralized controllers. 778–783. In: Proc. of the American Control Conference, Albuquerque, NM
19. Jian X, Lei T (2000) Pole assignment of multirate digital control systems with decentralized structure In: Proc. of the International Workshop on Autonomous Decentralized Systems, Chengdu, China
20. Khargonek P, Özgüler A (1994) Decentralized control and periodic feedback. IEEE Trans. Automat. Contr. 36:877–882
21. Khargonekar PP, Poolla K, Tannenbaum A (1985) Robust control of linear time-invariant plants using periodic compensation. IEEE Trans. Automat. Contr. 30:1088–1096
22. Longhi S (1994) Structural properties of multirate sampled-data systems. IEEE Trans. Automat. Contr. 39:692–696
23. Marshak J, Radner R (1971) The economic theory of teams. Yale University Press, New Haven, CT
24. Ooi J, Verbout S, Ludwig J, Wornell G (1997) A separation theorem for periodic sharing information patterns in decentralized control. IEEE Trans. Automat. Contr. 42:1546–1550
25. Özgüler A (1990) Decentralized control: a stable proper fractional approach. IEEE Trans. Automat. Contr. 35:1109–1117
26. Park P, Ko J (2002) A dynamic output feedback controller for discrete time linear periodic systems: LMI approach. 193–198. In: Proc. of the 41st IEEE Conference on Decision and Control, Orlando, FL
27. Raji R (1994) Smart networks for control. IEEE Spectrum 31(6):49–55
28. Scattolini R, Schiavoni N (1996) Decentralized control of multirate systems subject to exogenous signals. IEEE Trans. Automat. Contr. 41:1540–1544
29. Seo J, Jo CH, Lee SH (1999) Decentralized $\mathcal{H}_\infty$-controller design. Automatica 35:865–876
30. Sezer M, Siljak D (1990) Decentralized multirate control. IEEE Trans. Automat. Contr. 35:60–65
31. Shen TLJ, Chiu MS (2001) Independent design of robust partially decentralized controllers. Journal of Process Control 11:419–428
32. Siljak D (1991) Decentralized control of complex systems. Academic Press, Boston, MA
33. Wang S, Davison E (1973) On the stabilization of decentralized control systems. IEEE Trans. Automat. Contr. AC-18:473–478

Finite-Time Stability for Nonlinear Networked Control Systems

Silvia Mastellone, Peter Dorato, and Chaouki T. Abdallah

Department of Electrical and Computer Engineering
University of New Mexico
Albuquerque, NM 87131, USA
{silvia,peter,chaouki}@ece.unm.edu

Summary. Finite-time stability of nonlinear networked control systems is studied in this chapter. Focusing on packet dropping, a deterministic model for networked control systems is realized by incorporating the network dynamics. This links the fields of control of networks and networked control systems.

1 Introduction

In several recent works, the problem of networked control systems (NCS) has been posed and partially investigated, see [1]–[23]. This problem deals with controlling a system remotely via a communication network (as represented in Figure 1); as such, instantaneous and perfect signals between controller and plant are not achievable. This casts classical control problems into a setting that provides control solutions to remotely located systems such as assembling space structures, exploring hazardous environment, and executing tele-surgery.

Within this new setting we are able to overcome the necessity of collocated control and processes, thus overcoming many of the spatial restrictions. NCSs, however, do not exist without new challenging sets of problems. In fact, the networks introduce delays of time-varying and possibly random nature, packet losses that degrade the performance of the system and possibly destabilize it, and limited bandwidth that compromises our otherwise achievable control objective. Most of classical control theory is based on the assumption that the controller, system, and sensors are collocated so the aforementioned problems were not apparent. A challenging aspect of the networked setting is that we need to compensate for the effects of the network to retain stability and performance of the system under study. Many models have been proposed to study the effects of the network, and in this chapter we aim to provide a novel model that links the effects of the network to the traditional control design.

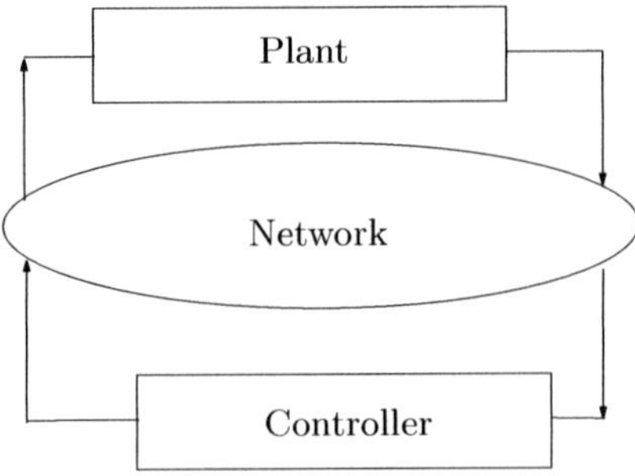

Fig. 1. NCS.

Another novel aspect of this chapter is that, unlike the current trends that study the Lyapunov stability of networked systems, we use the concept of finite-time stability where specific bounds are desired on the performance of the system and the study is restricted to a finite interval of time. This issue appears in several problems where we are interested in the system's behavior only over a specific, finite-time interval.

The remainder of this chapter is organized as follows. Section 2 states the general problem and in particular describes the model used to control a nonlinear plant, assuming a model of the original plant available on the controller's side of the network. The state of the plant is sent through the network and is therefore subject to packet dropping. On the other side of the network, when a state is received it is used to update the model and the controller, or else the state provided by the model is used to update the controller. In both cases the controller is attempting to stabilize the closed loop plant. In Section 3 we provide a model description of the networks used. For such models we describe how packets are dropped and thus complete our model of the networked control system.

In Section 4 the closed loop system resulting from the NCS and the network is considered and finite-time stability is investigated for this case. An example is shown in Section 5.

2 Problem Formulation

In [16] a model for the networked control of linear time-invariant systems was proposed. The network is modeled as a sampler placed between the plant and sensors, on one side, and the controller, on the other side of the network. Utilizing an approximate model of the process at the controller's side, the controller may be able to maintain stability while receiving only periodic updates of the actual state of the plant. Whenever a new update is received, the model is initialized with the new information. This idea was utilized in [8], where the system evolved in discrete time, and state updates were either received or dropped at each sampling time due to the effects of the network. The characterization of such a dropout is achieved through the use of a Markov chain that takes on values of 0 or 1 depending on whether a sample was lost or received, respectively. Recently in [17], the model for a continuous-time plant and a network modeled with a fixed-rate sampler was extended to bounded yet random sample times driven by a Markov chain.

Our objective in this section is to propose a similar framework to the discrete-time result of [8], by extending the problem to a nonlinear setting, i.e., our plant and model used for state estimation are both nonlinear. After formulating the control problem in the nonlinear discrete-time setting with generic packet dropout, we then describe a particular class of NCS and describe its properties.

We consider the nonlinear discrete-time plant described by the following:

$$x_{k+1} = f(x_k) + g(x_k)u_k \,, \tag{1}$$

where $x_k \in \mathbb{R}^n$, and $f, g : \mathbb{R}^n \to \mathbb{R}^n$ are two sufficiently smooth vector functions, and $u_k \in \mathbb{R}$ is a scalar input.

As depicted in Figure 2, a discrete-time model-based NCS [16] contains a plant and a model with the network residing between the sensors of the plant and the model and actuators.

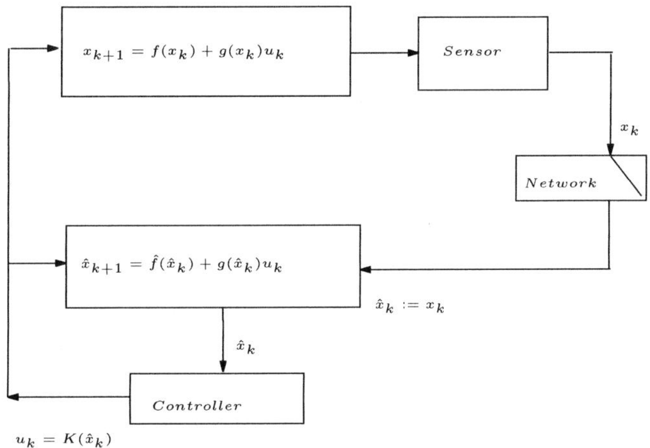

Fig. 2. Model-based NCS.

The network is modeled as a two-value variable sequence θ_k (assumed for now to be generic), where a measurement is dropped if $\theta_k = 0$, and a measurement is received when $\theta_k = 1$. Due to our inability to receive an update of the plant's state at each discrete instant of time, we use an inexact plant model on the controller side to provide us with a state estimate when packets are dropped. Such a model is given by

$$\hat{x}_{k+1} = \hat{f}(\hat{x}_k) + \hat{g}(\hat{x}_k)u_k\,, \tag{2}$$

in which $\hat{x}_k \in \mathbb{R}^n$, and $\hat{f}, \hat{g}$ are two smooth vector functions that map $\mathbb{R}^n$ into $\mathbb{R}^n$.

To carry out the analysis, we define the estimation error as $e_k = x_k - \hat{x}_k$ and augment the state vector x_k with e_k so that the closed loop state vector is given by $z_k = \left(x_k^T;\, e_k^T\right)^T$, $z_k \in \mathbb{R}^{2n}$. The closed loop system evolves according to

$$z_{k+1} = \begin{pmatrix} f(x_k) \\ (f(x_k) - \hat{f}(x_k)) + (1 - \theta_k)(\hat{f}(x_k) - \hat{f}(\hat{x}_k)) \end{pmatrix}$$
$$+ \begin{pmatrix} g(x_k)u_k \\ (g(x_k) - \hat{g}(x_k))K(\hat{x}_k) + (1 - \theta_k)(\hat{g}(x_k) - \hat{g}(\hat{x}_k))u_k \end{pmatrix}. \tag{3}$$

In the above model, $\theta_k \in \{0, 1\}$ is a receiving sequence (or equivalently, $\varphi_k = (1 - \theta_k)$ is a dropping sequence) that indicates the reception ($\theta_k = 1$) or the loss ($\theta_k{=}0$) of the packet containing the state measurement x_k. We assume that at each step time k a state is sent across the network in one packet. If a packet is received, it is used as an initial condition for the next time step in the model, otherwise the previous state of the model is used. Note that $u_k = K(\hat{x})$ is a scalar state-feedback input. Note how the stability of the plant depends on the rate of packets lost, the accuracy of the model, and the initial conditions for the model and the plant. We then classify the NCS errors as follows:

1. Model structure errors

$$e_{f1}(x_k) = f(x_k) - \hat{f}(x_k), \qquad e_{g1}(x_k) = g(x_k) - \hat{g}(x_k). \tag{4}$$

   These are the errors between the plant and the model evaluated at the plant's state and are therefore dependent on the system's structure.

2. State-dependent errors

$$e_{f2}(x_k, \hat{x}_k) = \hat{f}(x_k) - \hat{f}(\hat{x}_k), \qquad e_{g2}(x_k, \hat{x}_k) = \hat{g}(x_k) - \hat{g}(\hat{x}_k). \tag{5}$$

   These represent the errors between the model evaluated at the plant's state and at its own state, i.e., the error introduced by the difference in the states.

3. Structure and state-dependent errors

$$e_{f3}(x_k, \hat{x}_k) = f(x_k) - \hat{f}(\hat{x}_k), \qquad e_{g3}(x_k, \hat{x}_k) = g(x_k) - \hat{g}(\hat{x}_k), \tag{6}$$

   which include both model structure and state-dependent errors.

With the new notation, the system (3) becomes

$$z_{k+1} = \begin{pmatrix} f(x_k) + g(x_k)u_k \\ e_{f1}(x_k) + e_{g1}(x_k)u_k + (1 - \theta_k)(e_{f2}(x_k, \hat{x}_k) + e_{g2}(x_k, \hat{x}_k)u_k) \end{pmatrix}.$$

Based on the value of θ_k we have two possible situations.

1. For $\theta_k = 1$, the closed-loop system becomes

$$z_{k+1} = \begin{pmatrix} f(x_k) + g(x_k)u_k \\ e_{f1}(x_k) + e_{g1}(x_k)u_k \end{pmatrix}. \tag{7}$$

2. For $\theta_k = 0$, we have

$$z_{k+1} = \begin{pmatrix} f(x_k) + g(x_k)u_k \\ e_{f3}(x_k, \hat{x}_k) + e_{g3}(x_k, \hat{x}_k)u_k \end{pmatrix}. \tag{8}$$

For the remainder of this work we use the following compact form to represent the closed loop system and to highlight the fact that θ_k represents packet dropouts:

$$z_{k+1} = H_1(z_k) + H_2(z_k)(1 - \theta_k), \ k \geq 0, \tag{9}$$

with

$$H_1(z_k) = F_1(z_k) + G_1(z_k)u_k, \qquad H_2(z_k) = F_2(z_k) + G_2(z_k)u_k, \tag{10}$$

$$F_1(z_k) = \begin{pmatrix} f(x_k) \\ e_{f1}(x_k) \end{pmatrix}, \qquad F_2(z_k) = \begin{pmatrix} 0 \\ e_{f2}(x_k, \hat{x}_k) \end{pmatrix}, \tag{11}$$

$$G_1(z_k) = \begin{pmatrix} g(x_k) \\ e_{g1}(x_k) \end{pmatrix}, \qquad G_2(z_k) = \begin{pmatrix} 0 \\ e_{g2}(x_k, \hat{x}_k) \end{pmatrix}, \tag{12}$$

in which $H_i, F_i, G_i \in \mathbb{R}^{2n}$, $i = 1, 2$, are vector functions that map $\mathbb{R}^{2n}$ into $\mathbb{R}^{2n}$.

Moreover, we assume that the control law $u_k = K(\hat{x}_k)$ stabilizes, in some sense, the plant in the case of full-state availability.

In the following we denote a model-based NCS as MB-NCS.

2.1 Bounded NCS

Next we define a particular class of NCS for which we characterize the accuracy of the model in representing the plant's dynamics and describe how the model discrepancy affects the NCS structure.

Definition 1. *Class $C_{B\text{-}NCS}$* **NCS**
An MB-NCS of the form (9) belongs to a class $C_{B\text{-}NCS}$ with the bounds $(B_f, B_g, B_{efi}, B_{egi}; B_{h_i})$, $i = 1, 2$ if for all $k = 0, \ldots, N$ and for all $x_k \in S$,

where S is a given subset of $\mathbb{R}^n$, the system structure and error norms are bounded as follows:

$$\|f(x_k)\| \le B_f, \qquad\qquad \|g(x_k)u(\hat{x}_k)\| \le B_g(\hat{x}_k) \qquad (13)$$
$$\|e_{f1}(x_k)\| \le B_{ef1}, \qquad\qquad \|e_{f2}(x_k,\hat{x}_k)\| \le B_{ef2}(\hat{x}_k)$$
$$\|e_{g1}(x_k)u(\hat{x}_k)\| \le B_{eg1}(\hat{x}_k), \qquad \|e_{g2}(x_k,\hat{x}_k)u(\hat{x}_k)\| \le B_{eg2}(\hat{x}_k),$$

where B_f, B_{ef1} are constant bounds and $B_g(\hat{x}_k)$, $B_{ef2}(\hat{x}_k)$, $B_{eg1}(\hat{x}_k)$, $B_{eg2}(\hat{x}_k)$ are bounds that depend on the model state. Such NCS are called bounded model-based NCS (B-MB-NCS).

The above definition describes the class of NCS for which it is possible to define bounds on the plant and the NCS errors and where such bounds depend only on the model's state.

Next we state a lemma that describes properties of class $C_{B\text{-}NCS}$. In particular the lemma describes how bounds on the norm of the B-MB-NCS errors imply bounds on the weighted norm of the NCS dynamics, i.e., on $\|z_k\|_M = z_k^T M(k) z_k$.

Lemma 1. *Consider the NCS (9) and $M(k) > 0$, $(2n \times 2n)$ time-varying real-valued matrix,*

$$M(k) = \left[\begin{array}{c|c} m_1(k) & m_2(k) \\ \hline m_3(k) & m_4(k) \end{array}\right], \quad m_i(k) \in \mathbb{R}^{n\times n}, \quad m_2(k)^T = m_3(k). \qquad (14)$$

Also assume the system belongs to class $C_{B\text{-}NCS}$. Then the following bounds hold on the norm of the NCS dynamics weighted by $M(k)$ for $i,j = \{1,2\}$, $j \ne i$, $k = 0,\ldots,N$ and for all $x_k \in S$, where $S \subset \mathbb{R}^n$,

$$H_i^T M(k+1) H_j \le B_{H_{i,j}}(\hat{x}_k), \qquad H_i^T M(k+1) H_i \le B_{H_i}(\hat{x}_k), \qquad (15)$$

and where the bounds on the vector functions are related to the bounds on the errors as follows:

$$\begin{aligned}
B_{H_1}(\hat{x}_k) &= (B_f + B_g(\hat{x}_k))\lambda_{max}(m_1(k+1)) + (B_{ef1} + B_{eg1}(\hat{x}_k)) \quad (16)\\
&\quad (\|m_3(k+1)\| + \|m_2(k+1)\|)(B_f + B_g(\hat{x}_k)) + \\
&\quad (B_{ef1} + B_{eg1}(\hat{x}_k))\lambda_{max}(m_4(k+1)) \\
B_{H_{1,2}}(\hat{x}_k) &= (B_{ef2}^T(\hat{x}_k) + B_{eg2}^T(\hat{x}_k))\lambda_{max}(m_4(k+1)) \qquad\qquad\qquad (17)\\
B_{H_2}(\hat{x}_k) &= (B_{ef1} + B_{eg1}(\hat{x}_k))(\|m_4(k+1)\|)(B_{ef2} + B_{eg2}(\hat{x}_k)) \quad (18)\\
&\quad (B_f + B_g(\hat{x}_k))(\|m_4(k+1)\|)(B_{ef2} + B_{eg2}(\hat{x}_k)).
\end{aligned}$$

The proof can be found in [14], Theorem 2.1.

The above lemma states that if in an NCS the norms of the plant and of the NCS errors are bounded by constants, or by a model's state bound in the finite interval of time $[0, N]$ (see (13)), then there exists a bound on the weighted norm of the NCS dynamics in the interval of time $[0, N]$, and moreover this bound depends on the errors bounds. Assuming the NCS is such that the above bounds on the errors hold, then it is possible to bound the weighted norm defined by the matrix M of the B-MB-NCS dynamics. In particular, bounds defined on the vector function H_1 do not depend on packet dropping, whereas those on H_2 do, see (15).

Lemma 2. *Consider the NCS (9) and $M(k) > 0$ matrix, and denote $\| x \|_M = x^T M x$, then for all $x_k \in S \subset \mathbb{R}^n, \forall k = 0, \ldots, N$*

$$||z_k||_M \geq \lambda_{min}\{M\}B_z(\hat{x}). \tag{19}$$

Also assume the system belongs to class $C_{B\text{-}NCS}$ then for all $x_k \in S \subset \mathbb{R}^n$ and Euclidean norm $||.||$

$$||x_k|| \geq B_f + B_g(\hat{x}_k) = B_x(\hat{x}) \tag{20}$$

$$||e_k|| \geq B_{ef1} + B_{eg1}(\hat{x}_k) + B_{ef2}(\hat{x}_k) + B_{eg2}(\hat{x}_k) = B_e(\hat{x}) \tag{21}$$

$$||z_k|| \geq B_x(\hat{x}) + B_e(\hat{x}) = B_z(\hat{x}). \tag{22}$$

Proof. For the first part of the lemma observe that

$$||z_k||_M \geq \lambda_{min}\{M\}||z_k|| = \lambda_{min}\{M\}(||x_k|| + ||e_k||) \geq$$
$$\lambda_{min}\{M\}(B_x(\hat{x}) + B_e(\hat{x})) = \lambda_{min}\{M\}B_z(\hat{x}). \tag{23}$$

The second part trivially follows from the system definition.

3 Network Control and Models for Packet Dropout

Communication networks and their complex dynamics have been studied by several researchers, see, for example, [20, 25, 19]. Due to the growth in the size and complexity of the Internet, and with the advent of industrial networks, an understanding of the organization and efficiency of communication networks has become necessary.

As communication between two systems takes place across a network, several problems arise including delays and loss of information due to limited bandwidth and congestion. Considering the bandwidth as a fixed resource, to avoid the loss of information and delays, an efficient use of such resources is required. Congestion control represents an important aspect of the problem. As an example, in [20] the network is modeled as a dynamical system and the congestion control problem is reformulated as an optimization problem, which is the characterization of the equilibrium conditions from the point of view of fairness, efficiency in resource usage. Here we define and model a simplified network, and with it we model the dropping sequence for NCS defined previously, as a deterministic event, driven by the network dynamics. To explore the causes of packet dropping we start by defining the network setting in which the packet drops take place and exploring some of their properties. Among several models and descriptions of communication networks provided in the literature, we choose to redefine the network in a simpler framework.

Definition 2. *Communication Network*

A communication network is a couple (L, S), where L is a set of n_L links, and S is a set of n_S nodes that can potentially perform as sources or sinks of traffic.

Each link is a transmission medium whose capacity, also referred to as bandwidth or data rate, is measured in packets per second, where a packet is the information carrier. Each source so_i has an associated rate $r_i(k)$ that is a function of time, and denotes the number of packets per second sent to the sink si_i.

In Figure 3 a communication network is depicted: in particular the network may be bidirectional or undirected, and the nodes can be either sources or sinks, i.e., they can send or receive packets.

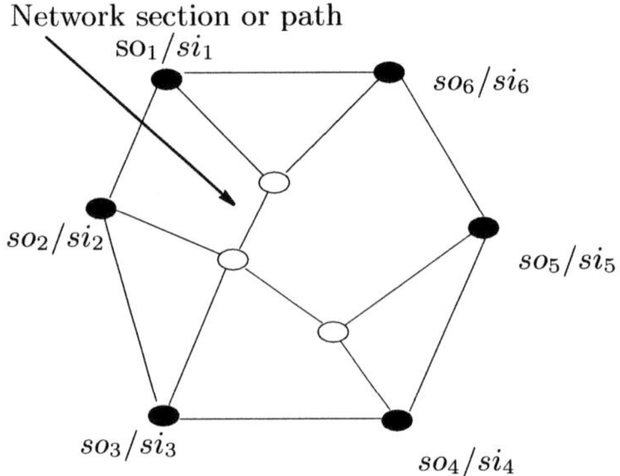

Fig. 3. Undirected or bidirectional network; nodes are sources and sinks.

Definition 3. *Network Section*

Consider a network (L, S), a couple (L_s, S_s), in which $L_s \subset L$, $S_s \subset S$ is called a network section of the network (L, S). Moreover, L_s is called a path and is a set of n_l links, and S_s is a set of n_s nodes that access the path.

Consider a network (L, S) composed of n_L links l_j, $j = 1, \ldots, n_L$ in which n_S sources access sending information to n_L sinks. Each source so_i, $i = 1, \ldots, n_s$ sends information to the sink si_i encoded in packets through the network with time rate $r_i(k)$, $k = 0, 1, 2, \ldots,$. Also each link l_i has an associated fixed bandwidth capacity C_i and at each time k we have a corresponding leftover capacity $c_i(k)$, $0 \le c_i(k) \le C_i$ that represents the amount of packets per second it can support. Let each link l_i be used by n_s sources, each sending at a rate $r_j(k)$ and therefore the global rate at the ith link is $G_i(k) = \sum_j^{n_s} r_j(k)$. We are interested in the section network (L_{si}, S_{si}) that is being used by the system, where L_{si} is the path, or set of $n_{l,i}$ links, associated with each source-sink (so_i, si_i). We define the following possible state for the network.

Definition 4. *Congested Link*

A link l_i is congested at time k if the amount of packets sent through it exceed its leftover capacity, i.e., $G_i(k) > c_i(k)$ for $k : 1, \ldots, N$.

Definition 5. *Congested Network Section*

A network section (L_s, S_s) is congested at time k if at least one link is congested.

If a link is congested then it starts dropping packets. After a certain period of time, the congestion disappears as a consequence of the sources reducing their rate so that $c_i(k) \geq G_i(k)$. A sink does not receive packets as a consequence of congestion in one of the links in the path associated with it.

3.1 Deterministic Model for Packet Dropout

With the described framework, we are now ready to provide a deterministic model for the packet dropout depending on the network dynamics. In particular we are interested in the network section that includes the path that a packet is going to follow. This path is composed of a number of n_l links, and with each link is associated an actual traffic, depending on the number and rate of sources that are accessing the path, and on the link physical capacity.

We want to study how the loss of packets affects the stability of the overall system by including the network dynamics in the model. In particular, this will allow us to explicitly relate the stability of the system to the capacity of the links involved in the path used by the system, and to the rate of the sources that are accessing such a path. This relation gives us the possibility of eventually designing for the stability of the system by controlling the rate of the sources accessing the path.

Let (L, S) be a network in which each source s_i has an associated rate $r_i(k)$ that is a function of time at which it sends packets through a set $L_i \subset L$ of links. So through every link l_j a total rate that is the sum of all the rates of n_s sources is given by

$$R_j(k) = \sum_{i=1}^{n_s} r_i(k). \tag{24}$$

Moreover, each link will have a capacity function proportional to the total rate that will indicate the level of occupation of the link

$$G_j(k) = K_l R_j(k), \quad j = 1, \ldots, n_l. \tag{25}$$

As stated previously, a link has a limiting capacity beyond which it will drop packets. In particular there is a critical level of leftover capacity $c_i(k)$ above which the link will accommodate packets, and below which it will start dropping them. The packet drop will be modeled by the binary value variable θ_k, as discussed earlier. With the described setting we have

$$\theta_k = \prod_{j=1}^{n_l} \left[\frac{sign(c_i - G_j(k)) + 1}{2} \right], \tag{26}$$

where the function $sign : \mathbb{R} \to \{-1, 1\}$ is defined as follows:

$$sign(a) = \begin{cases} 1 & a \geq 0 \\ -1 & a < 0. \end{cases} \tag{27}$$

With the provided framework we are now able to study the stability of the following dynamical nonlinear time-varying system:

$$z_{k+1} = H_1(z_k) + H_2(z_k)\varphi_k \tag{28}$$
$$= (H_1(z_k) + H_2(z_k)) \left[1 - \prod_{j=1}^{n_l} \left[\frac{sign(c_i(k) - K \sum_{j=1}^{n_s} r_i(k)) + 1}{2} \right] \right],$$

where $G_j(k)$ is given by (25), and where $r_i(k)$ are the known sequence of rates for sources accessing the path.

This model of NCS is a discrete-time, time-varying dynamical system that incorporates the system state z_k, and the network dynamics $c_i(k), r_j(k)$. The network is an integral part of the overall system, therefore achieving our goal.

4 Nonlinear Model-Based NCS: Deterministic Finite-Time Stability

As discussed previously, several studies have been conducted in modeling and controlling NCS and, on the other hand, in control of network dynamics. However, to the best of our knowledge, the network model itself has not yet been directly incorporated into the NCS model, but only through the effects that arise as a result of the network's conditions.

In this section, we consider packet dropping as a deterministic event and make use of the possibility of knowing when a packet drop may occur. To accomplish this, the network dynamics must be incorporated in the control loop as was done in Section 3, providing a more accurate model of packet dropping. The model asserts that the dropping of packets containing sensor measurements is due to network congestion.

As mentioned earlier, congestion is caused by sources that transmit at an aggregate rate exceeding the channel capacity. It may therefore be possible to control packet dropping by appropriately managing the sources' rates in specific paths. Since the stability of NCS depends on the controller performance, model accuracy, and the amount of packets dropped, we can design for FTS, not only through direct state feedback, but also by implementing a network controller to reduce packet dropping by controlling path capacities and source rates. Therefore, the model studied in this chapter is extended to include network dynamics, such as link capacities and source rates.

4.1 Extended Finite-Time Stability

We introduce next a novel concept not discussed in earlier works. But first we recall some known framework. We focus on discrete-time dynamical systems described by

$$x_{k+1} = f(x_k),\ x \in \mathbb{R}^n,\ x(0) = x_0, \tag{29}$$

where $x \in \mathbb{R}^n$ is the system state, and $f : \mathbb{R}^n \to \mathbb{R}^n$ is a sufficiently smooth vector function. We are interested in studying the state trajectory of the system in a finite time interval; in other words we want to guarantee that specific bounds on the state are maintained in this finite-time interval.

Definition 6. *Finite-Time Stability*

The system (29) is finite-time stable (FTS) with respect to $(\alpha, \beta, N, \|.\|)$ with $\alpha < \beta$ if every trajectory x_k starting in $\|x_0\| \leq \alpha$ satisfies the bound $\| x_k \| < \beta$ for all $k = 1, \ldots, N$.

Some results on FTS for nonlinear discrete-time systems may be found in [15]. However, in this chapter we will focus on an extended concept of FTS. Consider the case in which the state norm may exceed the bound β, but only for a finite number of consecutive steps, after which it needs to contract again below the bound β. The rationale for this is to consider for the deterministic case an equivalent concept to a stochastic one, where the possibility of exceeding the bound for some time is allowed. The proposed extension fits many real situations such as the example of driving a car in a tunnel, where we do not want to hit the tunnel walls, but in case the car is robust enough, we may hit the walls for short periods of time. We formalize such a concept with the following definition.

Definition 7. *Extended FTS*

The nonlinear discrete-time system (29) is extended FTS (EFTS) with respect to $(\alpha, \beta; N, N_o)$, if one of the following holds:

1. for some $k \in [0, N]$ either the state norm never exceeds the bound β, which is

$$\{\|x_k\| < \beta : k \in [0, N]|\ \|x_0\| \leq \alpha\}, \tag{30}$$

or, if there exists a time j at which the state norm exceeds β, then in at most N_o steps the state norm will converge back behind β, i.e.,

2. *if*

$$\{\exists j \in [0, N] : ||x_j|| > \beta, \Rightarrow \min_{j+1 \leq i \leq j+N_o+1} ||x_i|| \leq \beta\}, \ N_o < N, \quad (31)$$

where N_o is the number of consecutive steps the system state is allowed to exceed the finite-time bound.

Definition 8. *Attracted System*

A discrete-time system of the form (29) is an attracted system with respect to $(\alpha_1, \beta, \alpha_2, N, N_r)$, $\alpha_1 \leq \beta \leq \alpha_2$ if it is FTS with respect to (α_1, β, N) and contracting with respect to (α_2, β, N_r), i.e.,

$$||x_0|| \leq \alpha_1 \Rightarrow ||x_k|| \leq \beta, \ k = [0, N]$$
$$\alpha_2 \geq ||x_0|| \geq \beta \Rightarrow ||x_k|| \leq \beta, \ k = [N_r, N].$$

Theorem 1. *Consider a system (29) and assume it is attracted with respect to $(\alpha_1, \beta, \alpha_2, N, N_r)$ where the region $[-\beta, \beta]$ is a global region of attraction for the state. Also assume N_r is the number of steps needed for the state to contract into the ball of radius β from a distance α_2. Then the system is EFTS with respect to $(\alpha_1, \beta, N, N_r + 1)$.*

Proof. In the case of $||x_0|| \leq \alpha_1$ the assumption that the system is contractive implies that $||x_k|| \leq \beta, \ k = [0, N]$ from which FTS follows and therefore EFTS.

In the case of $\alpha_2 \geq ||x_0|| \geq \beta$ we have $||x_k|| \leq \beta, \ k = [N_r, N]$, which implies $||x_{N_r}|| \leq \beta$ and therefore $min_{0 \leq j \leq N_r+1}||x_j|| \leq \beta$, which means it is EFTS with respect to $(\alpha_1, \beta, N, N_r + 1)$.

Let us now go back to the NCS setting. We consider the deterministic MB-NCS described in (9). Recall the dropping sequence $\varphi_k = (1 - \theta_k) \in \{0, 1\}$ defined as follows: assume the state of the system is sent across a network path T of n_l links, each of which has a maximum allowed packet rate (or capacity $c_i(k)$), and a global capacity $G_i(k)$ determined by the source rates at time k, then the dropping sequence is given by $\varphi_k = 1 - \prod_{j=1}^{n_l} \frac{sign(c_j(k) - G_j(k)) + 1}{2}$.

The NCS described in (28) is a deterministic system, and we are interested in investigating its stability over a finite time in the event of packet dropping. A deterministic definition of EFTS is given next, which also allows bounds to be exceeded, but over limited intervals.

Definition 9. *EFTS-NCS*

The NCS (28) is EFTS with respect to $(\alpha_x, \beta_x; \alpha_z, \beta_z; N, N_o)$, if the following conditions hold:

1. *the system is FTS with respect to (α_x, β_x, N), if no packet dropping occurs, i.e., $\varphi_k = 0$*

$$\{z_k^T z_k < \beta_z : k \in [0, N] | z_0^T z_0 \leq \alpha_z\}; \quad (32)$$

2. for $\varphi_k = 1$, and some $k \in [0, N]$ either

$$\{z_k^T z_k < \beta_z : k \in [0, N] | z_0^T z_0 \leq \alpha_z\} \tag{33}$$

or if

$$\left\{ \exists j \in [0, N] : z_j^T z_j > \beta_z, \Rightarrow \min_{j+1 \leq i \leq j+N_o+1} x_i^T x_i \leq \beta_x \right\}, \ N_o < N, \tag{34}$$

where N_o is the number of consecutive steps the system state is allowed to exceed the finite-time bound due to packet dropping.

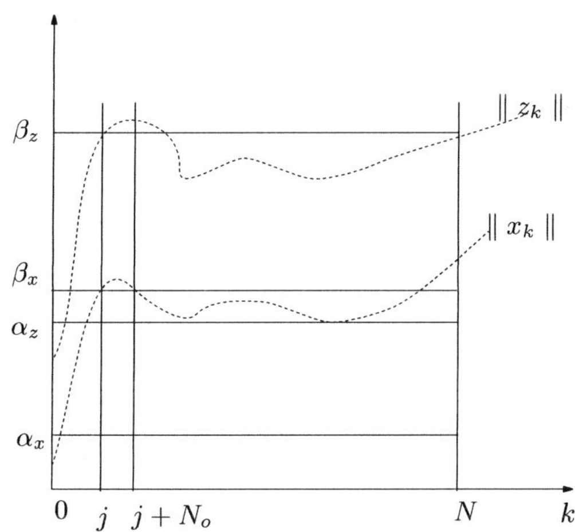

Fig. 4. Extended definition of FTS for NCS: the global state norm $\|z_k\|$ may exceed the bound β_z and $\|x_k\|$ the bound β_x as long as the plant state norm $\|x_k\|$ contracts in a finite number of steps N_o, back to β_x.

In particular, FTS for NCS is redefined so that if packet dropping occurs, the system state may exceed the bound β_x for a fixed finite number of consecutive steps N_o. Note that the above definition requires the knowledge of future states to ensure FTS at each step. We will also redefine quadratic EFTS in case it is desired to bound a given quadratic function of the state.

Definition 10. *Quadratically EFTS-NCS*
The NCS (28) is quadratically EFTS with respect to $(\gamma_x, \gamma_{x0}; \gamma_z, \gamma_{z0}; N, N_o, M)$, if for the choice of quadratic Lyapunov functions $V_z(z_k, k) = z_k^T M(k) z_k$, $V_x(x_k, k) = x_k^T m_1(k) x_k$ and $V_{\hat{x}}(\hat{x}_k, k) = \hat{x}_k^T m_4(k) \hat{x}_k$, in which $M(k) = M^T(k)$ is a $2n \times 2n$ time-varying matrix, with $m_1(k) > 0$, $m_4(k) > 0$, we have

1. for $\varphi_k = 0$

$$\{V_z(z_k, k) < \gamma_z : k \in [0, N] | V_z(z_0, 0) \leq \gamma_{z0}\}; \qquad (35)$$

2. for $\varphi_k = 1$ either

$$\{V_z(z_k, k) < \gamma_z : k \in [0, N] | V_z(z_0, 0) \leq \gamma_{z0}\}; \qquad (36)$$

or

$$\left\{ \exists j \in [0, N] : V_z(z_j, j) > \gamma_z, \Rightarrow \min_{j+1 \leq i \leq j+N_o+1} V_x(x_i, i) \leq \gamma_x \right\}.$$

Theorem 2. *Every NCS that is quadratically EFTS with respect to the parameters $(\gamma_x, \gamma_{x0}; \gamma_z, \gamma_{z0}; N, N_o, M)$ is also EFTS with respect to $(\alpha_x, \beta_x; \alpha_z, \beta_z; N, N_o)$. Also $\gamma_x = \delta_1 \beta_x$, $\gamma_{x0} = \delta_2 \alpha_x$, $\gamma_z = \delta_1 \beta_z$, $\gamma_{z0} = \delta_2 \alpha_z$.*

Proof. The proof easily follows by considering the fact that $\delta_1 \|z_k\|^2 \leq V_z(z_k, k) \leq \delta_2 \|z_k\|^2$, $\delta_1(k) = \lambda_{min}\{M(k)\}$, $\delta_2(k) = \lambda_{max}\{M(k)\}$ are the minimum and maximum eigenvalues of $M(k)$, respectively.

4.2 EFTS Analysis

In this section, we consider sufficient conditions that will guarantee EFTS for the NCS. In the new setting, if the NCS state exceeds the bound specified at time j, then, in order to predict the future values of the state, it is required to have an estimate of the plant state for the successive $N_o + 1$ steps. This is presented in the following theorem by using the model to predict future states.

The sets of bounded states are denoted as $S_{\gamma_z} = \{z_k : V_z(z_k, k) \leq \gamma_z\}$, $S_{\gamma_x} = \{x_k : V_x(x_k, k) \leq \gamma_x\}$, $S_{\gamma_{\hat{x}}} = \{\hat{x}_k : V_{\hat{x}}(\hat{x}_k, k) \leq \gamma_{\hat{x}}\}$.

Theorem 3. *Consider the class $C_{B\text{-}NCS}$ NCS (28), and the state prediction using the model*

$$\hat{x}_{k+(j+1)} = \hat{f}(\hat{x}_{k+j}) + \hat{g}(\hat{x}_{k+j}) u_{k+j}, \quad k+1 \leq j \leq k+1+N_o, \qquad (37)$$

and assume for all $x_k \in S_{\gamma_x}$ and $\forall k = 1, \ldots, N$ that

$$\Delta V_z \leq \Delta V_{B_z} = B_{H_2}(\hat{x}_k)\varphi_k^2 + 2(B_{H_{1,2}}(\hat{x}_k))\varphi_k + B_{H_1}(\hat{x}_k) - \hat{x}_k^T M(k)\hat{x}_k$$

$$\Delta V_x \leq \Delta V_{B_x} = B_{H_1}(\hat{x}_k) - \lambda_{min}\{M\} B_z(\hat{x}), \qquad (38)$$

then if for $V_z(0) \leq \gamma_z 0$, $V_{\hat{x}}(0) \leq \gamma_{\hat{x}0}$, $\exists \rho_i > 0$, $\rho_i' > 0$ such that

$$[\rho_k V_z(z_k, k) - \Delta V_{B_z}(z_k, k))] \geq 0 \qquad (39)$$

$$\frac{\gamma_z}{\gamma_{z0}} \geq \sup_{0 \leq k \leq N} \prod_{j=0}^{k-1} (1 + \rho_j)$$

$$[\rho_k V_z(z_k, k) - \Delta V_{B_z}(z_k, k))] \leq 0 \tag{40}$$

$$\min_{k+1 \leq i \leq k+N_o+1} [\rho_i' V_x(\hat{x}_i, i) - \Delta V_{B_x}(\hat{x}_i, i)] \tag{41}$$

$$\frac{\beta_{\hat{x}}}{\alpha_{\hat{x}}} \geq \sup_{0 \leq k \leq N} \prod_{j=0}^{k-1} (1 + \rho_j') \tag{42}$$

and finally

$$B_e(\hat{x}_k) + \beta_{\hat{x}} \leq \beta_x, \quad \forall \hat{x}_k \in S_{\gamma_{\hat{x}}}, \tag{43}$$

then the NCS (28) is EFTS with respect to $(\alpha_x, \beta_x; \alpha_z, \beta_z; N, N_o)$

Proof. From condition (39) we have that if

$$[\rho_k V_z(z_k, k) - \Delta V_{B_z}(z_k, k)] \geq 0, \tag{44}$$

then using the fact that the NCS belongs to class C_{B-NCS} together with the result in [14], Theorem 4.1, we can show the EFTS for the NCS. Let us study the case in which $[\rho_k V_z(z_k, k) - \Delta V_{B_z}(z_k, k)] \leq 0$, then inequality (41) reduces to

$$\min_{j+1 \leq i \leq j+N_o+1} [\rho_k V_x(\hat{x}_{i)}, i) - \Delta V_{B_x}(\hat{x}_i, i))] \geq 0,$$

from which it follows that there exists a $j + 1 \leq i \leq N_o + 1$ for which

$$[\rho_i V(\hat{x}_i, i) - \Delta V(\hat{x}_i, i)] \geq 0, \tag{45}$$

which, when combined with condition (42) and theorem 4.1 in [14], implies FTS for the model state $\hat{x}$ with respect to $(\alpha_{\hat{x}}, \beta_{\hat{x}}, 1)$. Also consider the following:

$$||x_k|| = ||x_k - \hat{x}_k + \hat{x}_k|| \leq ||e_k|| + ||\hat{x}_k|| \leq B_e(x_k) + ||x_k||, \tag{46}$$

then considering the condition (43), and the FTS of $\hat{x}_k$, from which it follows $||x_k|| \leq \beta_x$ for at least one $k \in [j + 1 \leq i \leq N_o + 1]$ and moreover EFTS for the NCS.

5 Example

Recalling from [4] the Brockett integrator and considering the discrete-time version of it, we investigate in a deterministic setting how packets losses affect the closed loop EFTS of the system. Consider the discrete version of the Brockett integrator:

$$x_1(k+1) = x_1(k) + u_1(k) \tag{47}$$
$$x_2(k+1) = x_2(k) + u_2(k) \tag{48}$$
$$x_3(k+1) = x_3(k) + (x_1(k)u_2(k) + x_2(k)u_1(k)) \tag{49}$$

and the following model that approximates the integrator

$$\hat{x}_1(k+1) = -23\hat{x}_1(k) - 17u_1(k) \tag{50}$$
$$\hat{x}_2(k+1) = -19\hat{x}_2(k) + 3.33u_2(k) \tag{51}$$
$$\hat{x}_3(k+1) = -5\hat{x}_3(k) - 8(\hat{x}_1(k)u_2(k) - 7\hat{x}_2(k)u_1(k)). \tag{52}$$

We study EFTS with respect to $(\alpha_z = 1, \beta_z = 3, \alpha_x = 0.6, \beta_x = 1.5, N = 10, N_o = 2)$. Let us use the controller

$$u(k) = - \begin{bmatrix} e^{-ak} & 0 \\ 0 & e^{-bk} \end{bmatrix} y(k), \tag{53}$$

with parameters $a = 1.3$, $b = 0.7$. Then the conditions of theorem 3 are satisfied if full information is available, i.e., $\varphi_k = 1$, $\forall k = 0, \ldots 10$.

In order to simulate the system, we consider the path used to the NCS composed of three links l_1, l_2, l_3, each with limit capacity $c_i(k)$. The links are used by five sources $s_1, \ldots, s_5$ so that we get $(l1; s1, s4)$, $(l_2; s_1, s_3)$, $(l_3; s_2, s_3, s_5)$. Meanwhile the sources send at the following respective rates:

$$r_1(k) = 1(sin(k) + 1), \qquad r_2(k) = 3(cos(k) + 1),$$
$$r_3(k) = 1.7exp^{-k}, \qquad r_4(k) = 8(cos(k) + 1), \qquad r_5(k) = 9exp^{-k},$$

from which we can calculate the global rates at each link as $G_i(k) = \sum_i r_i(k)$.

We study the closed loop behavior of the NCS as the limit rate of the link, and therefore the amount of packets dropped vary. Starting from initial conditions $x_i(0) = \hat{x}_i(0) = 0.3$, $i = 1, 2, 3$, we first consider a fixed limit capacity $c_i = c = 17$ *packets/second* that will lead to a dropping sequence $\{\varphi_k\}$ of all zeros, that is, all the packets are received (and therefore a receiving sequence $\{\theta_k\}$ of all ones). Figure 5 shows the evolution of the system state over time.

If we lower the limit capacity to $c = 13$ packets/second, the receiving sequence becomes $\theta = [0\,1\,1\,1\,1\,0\,0\,1\,1\,1\,1\,0\,0\,1\,1]$, for which EFTS conditions are still satisfied, as shown in Figure 6. For $c = 1$ we obtain a dropping sequence of all ones and the state dynamics are depicted in Figure 7. Finally in Figure 8, we show the norms for the three values of capacities for $x_i(0) = \hat{x}_i(0) = 0.3$.

References

1. Azimi-Sadjadi B (2003) Stability of networked control systems in the presence of packet losses. In: IEEE Conference on Decision and Control, Maui, HI

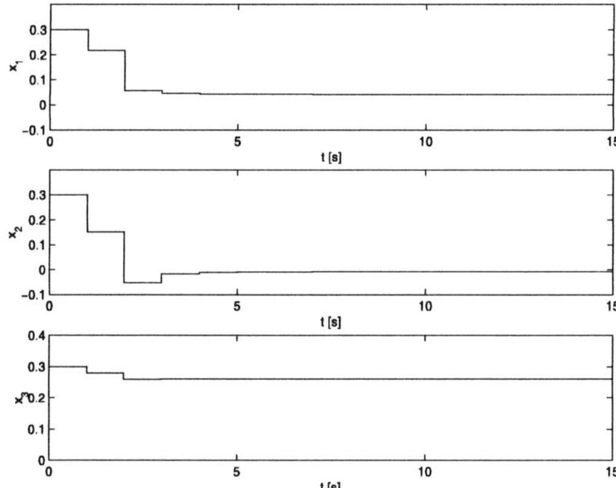

Fig. 5. Brockett integrator controlled through the network with exponential class (*b*) controller with $a_1 = 1.3$, $a_2 = 0.7$, capacity $c = 17$.

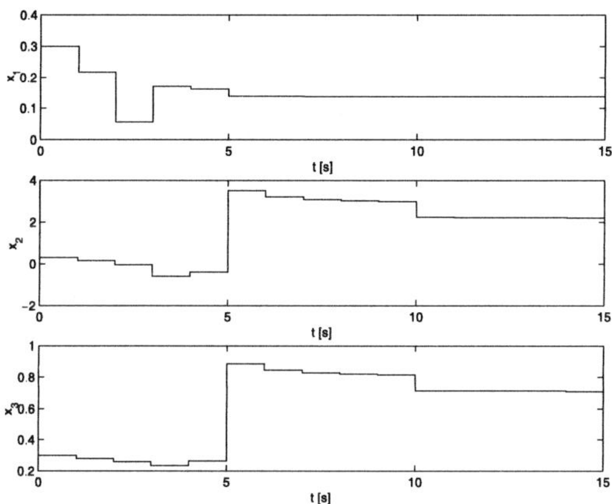

Fig. 6. Brockett integrator controlled through the network with exponential class (*a*) controller with $a_1 = 1.3$, $a_2 = 0.7$, static capacity $c = 13$.

2. Brandt J, Hein W (2001) Polymer materials in joint surgery. In: Grellmann W, Seidler S (eds) Deformation and fracture behavior of polymers. Engineering materials. Springer-Verlag, Berlin
3. Brockett RW (1997) Minimum attention control. 2628–2632. In: Proceedings of the IEEE Conference on Decision and Control, San Diego, CA
4. Brockett RW (1983) Asymptotic stability and feedback stabilization. 181–191. In Millman RS, Sussmann HJ (eds) Differential Geometry Control Theory, Birkhäuser, Boston, MA
5. Che M, Grellmann W, Seidler S (1997) Crack resistance behavior of polyvinylchloride. J. of Appl Polym Sci 64(6):1079–1090

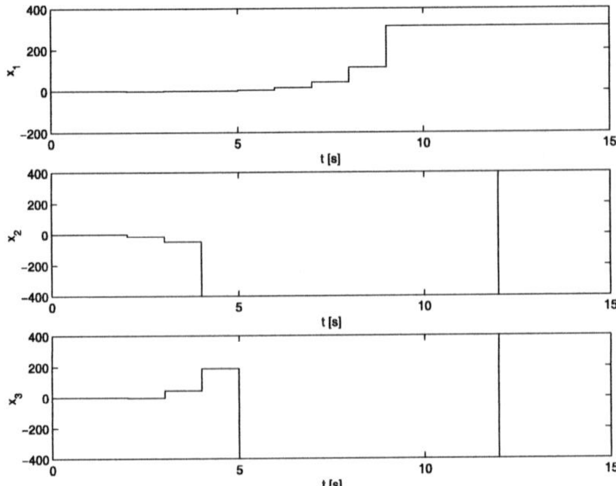

Fig. 7. Brockett integrator controlled through the network with exponential class (a) controller with $a_1 = 1.3$, $a_2 = 0.7$, static capacity $c = 1$.

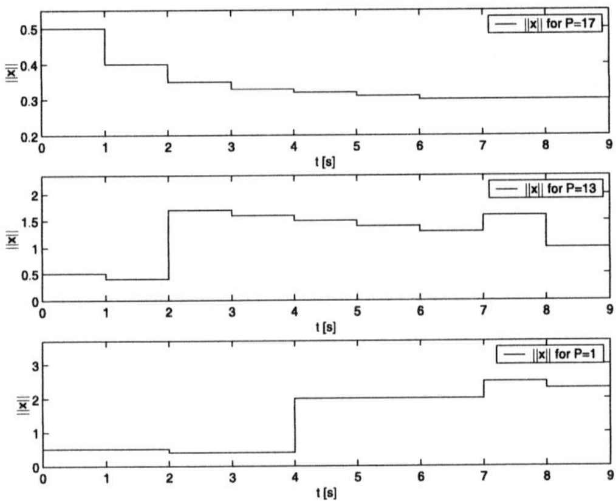

Fig. 8. Norm of the state x for values of capacities $c = 17$, $c = 13$, and $c = 1$.

6. Ephremides A, Hajek B (1998) Information theory and communication networks: an unconsummated union. IEEE Transactions on Information Theory 44(6):2416–2434

7. Hespanha J, Ortega A, Vasudevan L (2002) Towards the control of linear systems with minimum bit-rate. In: Proceedings of the International Symposium on the Mathematical Theory of Networks and Systems, and *http://sipi.usc.edu/~ortega/Papers/HespanhaOrtVas02.pdf.*

8. Hokayem PF (2003) Stability analysis of networked control systems. MS Thesis, Electrical Engineering, The University of New Mexico, Albuquerque, NM

9. Ishii H, Başar T (2002) Remote control of LTI systems over networks with state quantization. 830–835. In: Proceedings of the IEEE Conference on Decision and Control, Las Vegas, NV

10. Kipphan H (ed) (2001) Handbook of print media. Springer-Verlag, Berlin
11. Liberzon D (2002) A note on stabilization of linear systems with limited information. 836–841. In: Proceedings of the IEEE Conference on Decision and Control, Las Vegas, NV
12. Liberzon D (2003) Stabilizing a nonlinear system with limited information feedback. In: IEEE Conference on Decision and Control, Maui, HI
13. Low S, Paganini F, Doyle J (2002) Internet congestion control: an analytical perspective. IEEE Control Systems Magazine 22(1):28–43
14. Mastellone S (2004) Finite-time stability of nonlinear networked-control systems. MS Thesis, University of New Mexico, NM
15. Michel, Wu (1969) FTS for nonlinear discrete-time systems. Int. J. of Control, 9(6):679–693
16. Montestruque LA, Antsaklis PJ (2001) Model-based networked control system-stability. ISIS Technical Report ISIS-2002-001, Notre Dame, IN
17. Montestruque LA, Antsaklis PJ (2003) Stochastic stability for model-based networked control systems. 2:4119–4124. In: Proceedings of the 2003 American Control Conference, Denver, CO
18. Ross DW (1977) Lysosomes and storage diseases. MS Thesis, Columbia University, New York
19. Srikant R (2000) Control of communication networks. 462–488. In: Samad T (ed) Perspectives in Control Engineering: Technologies, Applications, New Directions, IEEE Press, Piscataway, NJ
20. Srikant R (2004) The mathematics of internet congestion control. Birkhäuser, Boston, MA
21. Tatikonda SC (2000) Control under communication constraints. PhD Thesis, Department of Electrical and Computer Science, Massachusetts Institute of Technology, Boston, MA
22. Tatikonda S, Sahai A, Mitter S (1999) Control of LQG systems under communication constraints. 2778–2782. In: Proceedings of the IEEE American Control Conference, San Diego, CA
23. Teel AR, Nešić D (2003) Input-output stability properties of networked control systems. In: IEEE Conference on Decision and Control, Maui, HI
24. Walsh GC, Beldiman O, Bushnell LG (2001) Asymptotic behavior of nonlinear networked control systems. IEEE Trans. Automatic Control 46(7):1093–1097
25. Ying L, Dullerud G, Srikant R (2004) Global stability of internet congestion controllers with heterogeneous delays. In: Proceedings of the American Control Conference, Boston, MA

Index